**Geology of Michigan**

# Geology of Michigan

**John A. Dorr, Jr.**, and **Donald F. Eschman**

Illustrated by Derwin Bell

Ann Arbor   The University of Michigan Press

Copyright © by the University of Michigan 1970
All rights reserved
ISBN 0-472-08280-9
Library of Congress Catalog Card No. 69-17351
Published in the United States of America by
The University of Michigan Press
Manufactured in the United States of America
⊗ Printed on acid-free paper

2008   2007   2006   2005      17   16   15   14

Published with the assistance of grants from the Horace
H. Rackham School of Graduate Studies, the Institute of
Science and Technology, and the Class of 1962 Publishing
Fund (administered by the Institute)

Book design by Quentin Fiore

No part of this publication may be reproduced, stored in a
retrieval system, or transmitted in any form or by any means,
electronic, mechanical, or otherwise, without the written
permission of the publisher.

# Preface

This book has several major objectives. One is to provide identifications and descriptions of the principal geological features of Michigan; another is to explain the origin of those features in terms of the chemical, physical, and biological processes that brought them into being in the past. A third objective is to arrange the events of the past into a history which will portray the geological evolution of the state. Also, certain geological principles will be defined and illustrated. Once understood, these principles can be applied with equal validity to an understanding and appreciation of the geology of other regions as well. The geologist, as a natural scientist, is challenged, excited, and intellectually satisfied by the pursuit and accomplishment of these objectives for their own sake. To help the nonprofessional reader share his feelings, emphasis is placed on scenic attractions of geological significance and on features which the tourist and the student are most likely to be able to observe. Our experience has revealed a need for a book dealing with the geology of Michigan in a broad but elementary fashion. This need has increased in recent years with the growth of interest in Earth Science on the part of students, teachers, and tourists.

State boundaries are artificial; the geologic phenomena of Michigan often extend beyond them, and general geologic principles are applicable anywhere. Whenever the geology of this state is better understood in a wider context we consider features in adjoining regions, elsewhere on the continent, or throughout the world.

The professional geologist may apply his knowledge to the discovery and exploitation of natural resources. Copper, iron, oil, gas, limestone, dolomite, salt, gypsum, coal, clay, building stone, water resources in rivers, streams, lakes, and underground, soils, even the native plants and animals and the natural scenic and recreational attractions of the state, all those resources which contribute so greatly to our economy, owe their origin and distribution, at least in part, to geologic processes and events. We shall examine most of those resources briefly from the geological viewpoint, but that is not a principal objective of this book. Nor shall we consider at great length such subjects as the economics of natural resources or the fascinating history of their exploitation. For those interested in such subjects we recommend, at least as a start, an excellent little book entitled, *Our Rock Riches,* published by the Michigan Geological Survey (1964).

# Geology of Michigan

Little original research is included in this book. We owe a debt of gratitude to all those geologists and other authors whose publications provided source material for this book. We especially wish to thank our colleagues in The University of Michigan Department of Geology and Mineralogy for their valuable advice on chapters as follows: Professors W. C. Kelly and F. S. Turneaure (Ch. IV, Precambrian), Professors L. I. Briggs and E. C. Stumm (Ch. V, Paleozoic), Professors William R. Farrand and Jack L. Hough (Ch. VIII, The Great Lakes in late glacial and post-glacial time), Professor K. K. Landes (Ch. X, Petroleum and Natural Gas), Mr. D. Garske (Ch. XI, Minerals), Professor D. B. Macurda (Ch. XIII, Fossil Invertebrates), Professor C. W. Hibbard (Ch. XIV, Fossil Vertebrates), and Professor C. A. Arnold (Ch. XV, Fossil Plants). Mr. Derwin Bell, department draftsman, prepared the majority of line drawing illustrations. Mr. Robert W. Kelley of the Geological Survey Division, Michigan Department of Conservation, provided helpful advice, particularly in the selection of photographs. We also wish to thank the Publications and Information Office of the Michigan Department of Conservation and the private companies and individuals who provided many of the photographs used. The University of Michigan Institute of Science and Technology provided funds for typing the final manuscript.

For many years Professors George M. Ehlers, Robert V. Kesling, and Erwin C. Stumm, of The University of Michigan Museum of Paleontology and Department of Geology in Ann Arbor, have added to our knowledge of the invertebrate fossils of Michigan. Figures XIII–23 through XIII–33 are made up in large part of photographs of specimens which appeared in their many separate papers. Most of these illustrations were published in The University of Michigan Museum of Paleontology *Contributions* series. Illustrations from papers written under their direction by students also were used. Principal among these individuals were Drs. E. G. Driscoll, W. Humphrey, P. M. Kier, and A. LaRocque. Space limitations make it impossible to cite the source of every photograph, but the authors wish to acknowledge here their indebtedness to the individuals mentioned.

No book such as this could include all of the available geological information or illustrations that might be desired by everyone with an interest in a particular area or subject. Throughout, we have included references to publications at the end of the book. We hope these will assist the reader to expand his knowledge of the geology of Michigan.

# Contents

### I: Introduction     1
Values of Geology, 1; Contents of This Book, 1; General Principles of Historical Interpretation, 2; Major Rock Types and the Rock Cycle, 2; Summary, 12.

### II: Earth History and Geologic Time     13
The Sense of Time, 13; Relative Geologic Time, 13; The Geologic Time Scale, 15; Relative Dating of Igneous Rocks, 16; Absolute Time, 18; Summary, 22.

### III: General Geologic Setting of Michigan     23
The Broad Picture, 23; The Continental Interior in the Paleozoic, 25; The Paleozoic Geosynclines, 25; A Basin beneath the State, 26; The Plastic Earth, 28; The Earth's Interior, 29.

### IV: Precambrian Eras     31
Introduction, 31; Precambrian in Michigan, 38; Iron Ores of Michigan, 61; Keweenaw Copper, 70; White Pine Copper, 77; The Role of Hypotheses in Geology, 79.

### V: The Paleozoic—Era of Inland Seas     81
Introduction, 81; The Paleozoic Rock Record in Michigan, 83; The Cambrian Period—Beginning of the Paleozoic Era, 91; The Ordovician Period, 98; The Silurian Period, 102; The Devonian Period, 113; The Mississippian Period, 123; The Pennsylvanian Period, 126; Paleozoic Era—Conclusion and Summary, 134.

### VI: The Lost Interval     136

### VII: The Pleistocene (Ice Age) Epoch     141
Nature of Glacial Ice, 141; The Work of Ice, 147; General History of the Pleistocene, 158; The Pleistocene in Michigan, 159.

### VIII: The Great Lakes in Late Glacial and Postglacial Time     164
Introduction, 164; The Proglacial Lake Sequence, 168; The Record from Mackinac Island, 176.

### IX: Water and Wind in Michigan     180
Introduction, 180; Water Use, 180; The Hydrologic Cycle, 181; Water Underground, 182; Ex-

traction of Water from the Ground, 184; The Search for Ground Water, 184; Some Ground Water Problems in Michigan, 185; The Work of Ground Water in the State, 185; Surface Waters in General, 187; Rivers and Streams, 187; Inland Lakes and Swamps, 191; Surface Water Conservation Problems, 198; Wind, 198; Shoreline Processes in General, 205; Waves and Shore Currents, 205; Shore Features, 208; The Herring Lakes—A Case History of Dunes and Bars, 213; The Southeastern Lake Michigan Shore, 217.

## X: Petroleum and Natural Gas in Michigan 228

Introduction, 228; Origin and Source, 228; Migration Through the Rocks, 229; Reservoir Rocks, 230; Seals and Traps, 231; Petroleum Exploration, 234; Production of Oil and Gas, 235; Oil and Gas in Michigan, 237; History of Michigan Oil and Gas Production, 237; Geology of Michigan Oil and Gas, 237; Reservoir Rocks in Michigan, 239; Oil and Gas Fields in Michigan, 241; Summary, 243.

## XI: Minerals in Michigan 244

Introduction, 244; Nature of Minerals, 244; Methods of Mineral Study, 245; Physical Properties of Minerals, 247; Mineral Collecting in Michigan, 251; Lower Peninsula Localities and Exhibits, 254; Upper Peninsula Localities and Exhibits, 258; List of Michigan Minerals, 262.

## XII: Rocks 265

Introduction, 265; Main Rock Types, 265; Rock Textures, 265; Rock Identification, 269; Igneous Rocks, 269; Metamorphic Rocks, 269; Sedimentary Rocks, 272.

## XIII: Fossil Invertebrates 285

Introduction, 285; Definition of a Fossil, 285; Types of Fossil Preservation, 285; Naming and Studying Fossils, 287; Classification of Fossil and Living Invertebrate Animals, 288; The Significance of Fossils, 309; Geological Uses of Fossils, 310; Fossils and Ancient Environments, 333; Michigan Fossil-Collecting Localities, 344.

## XIV: Fossil Vertebrates in Michigan 350

Introduction, 350; Relative Abundance of Fossil Vertebrates, 350; Problems of Identification and Restoration, 351; Finding Fossil Vertebrates in Michigan, 352; Special Problems, 353; Documentation of the Michigan Record, 354; The Value of Fossil Vertebrates, 354; General References, 355; General Geologic History and Evolution of Vertebrates, 355; Fossil Vertebrate Faunas of Michigan and Their Ancient Environments, 363; Middle Devonian Faunas, 365; Late Devonian Vertebrate Faunas, 366; Early Mississippian Vertebrate Faunas, 366; Pennsylvanian Vertebrate Faunas of Michigan, 367; Pleistocene and Post-Pleistocene Vertebrate Faunas of Michigan, 371; Early Man in Michigan, 388; Detailed List of Fossil Vertebrates from Michigan, 389.

## XV: Fossil Plants in Michigan 416

Introduction, 416; General Review of Plant History, 416; Precambrian Plant Remains in Michigan, 417; Paleozoic Fucoids, 423; Devonian Plants, 423; Mississippian Plants, 425; Coal Swamp Floras of the Pennsylvanian Period, 426; Mesozoic Floras, 439; Post-Pleistocene Floras of Michigan, 441; Summary of the Fossil Record of Plants in Michigan, 449.

## Bibliography 451
## Index 459

# I
# Introduction

### The Values of Geology

Geology deals with both the nature and history of the earth's crust and interior; it also treats past life on earth as revealed by fossils. Many professional geologists search for economically valuable mineral deposits, oil, and ground water, and their contributions to the economy of the state and the nation are of great material value. It is probably fair to say, however, that most geologists became interested in the field because it seeks to answer questions concerning the earth that have intrigued man for ages. How old is the earth? How and where did the materials that compose it originate? How did the surface landforms come into being? Has the earth remained unchanged through time or has it had an evolutionary history? How and when did life first appear and what has been the course of its development? Man asks such questions because he is inherently curious, a trait that has led to much of his progress. Similar questions may arise from a look at only a relatively small portion of the earth, as we shall demonstrate in this book on Michigan. Such questions do occur to the nonscientist and, to some extent, may be answered by nongeologists. To cover all the geological questions about Michigan that could conceivably be of interest would be impossible because of the vast number of questions and, more importantly perhaps, because not all answers are yet known. You will discover the kinds of questions a person interested in the geology of the state might ask.

Knowledge and understanding, in geology or any other field, have their own values which cannot be expressed in dollars. The sense of pleasure a student of geology derives from an understanding of his natural environment is a never-failing source of intellectual stimulation. Once a person learns how to observe earth features and to ask questions of nature, he may wander at will anywhere in the world and he will seldom be without the thrill of discovery. This book is not written for the professional geologist, but for the person who is curious about the environment in which he lives.

### Contents of This Book

Realizing that geology is concerned with rocks and minerals, many might wonder how an entire book can be devoted to Michigan geology—so much of

# Geology of Michigan

the state is devoid of bedrock outcrops! Of course rock does crop out locally from beneath the glacial material that blankets much of the southern part of the state, and some areas of the Upper Peninsula are free of such glacial cover. Then, too, many quarries in widely scattered localities have exposed bedrock, and oil-well drilling has brought up rock samples from thousands of feet beneath the surface. But do these few glimpses provide enough information for a whole book? Remember that the science of geology goes far beyond rocks, minerals, and fossils, it includes the study of surface landforms such as hills, valleys, dunes, lakes, and beaches; it is concerned as well with the origin of soils and with ground water and oil and even deals with the shape (structure) of rock layers beneath the surface. The table of contents extends this list greatly. For motorists following some of the more popular tourist routes the book also includes descriptions of the geology of some of the outstanding tourist attractions in the state. Geology is most fun when you do it yourself, and those sections will help you get started. The bibliography and text references will enable you to extend your knowledge further if you wish.

## General Principles of Historical Interpretation

Details of the principles used by geologists in their study of the earth, such as the various methods of measuring geologic time, or the means of discovering the nature of ancient geography and the distribution of past environments, are discussed at various places in the chapters that follow, but certain general principles are so frequently utilized that they deserve mention at the beginning.

## Major Rock Types and the Rock Cycle

Geologic history is recorded in the rocks of the earth's crust. The three principal rock types are igneous, sedimentary, and metamorphic, each composed of one or more of the mineral species.

*Igneous rocks* are those which solidify from a molten magma (rock material), either underground (granite for example), or at the surface (as exemplified by the basaltic lava). The magma itself may originate deep within the earth, below the crust, or may form by local melting of preexisting rocks within the crust. If the latter origin is the case, the parent rocks may be igneous, sedimentary, metamorphic, or some combination of them. The original composition of the magma and the conditions under which it solidifies determine the kind of igneous rock that results.

*Sedimentary rocks* originate as sediments deposited at the earth's surface. Subsequent lithification (conversion of unconsolidated sediment) by compaction or cementation turns the sediment to rock. Ordinarily, sedimentary rocks are stratified (layered). The sediments themselves are derived from surface exposures of preexisting rocks of one or more of the three major types. Exposed rocks are attacked by chemical and physical weathering processes. The weathered material, either in the form of fragments of the preexisting rock, as new minerals formed by chemical alteration, or as material in chemical solution, is eroded and transported to a site of sedimentary deposition by such agents as running water (on the surface or underground), wind, waves, glacial ice, or mass movement under the influence of gravity. At the site of deposition, the sediment may simply settle to the floor of deposition as particles or may fall out as a chemical compound precipitated from solution. The nature of the sediments that are deposited depends on the composition of the preexisting source rock, the nature of the weathering and transporting processes that were involved, and the environment at the site of deposition, as well as the subsequent post-depositional history of the deposits. Organic matter in the form of fossils is common in sedimentary rocks. The history of sedimentary rocks is most interesting because those rocks reflect the conditions at or near the earth's surface, conditions with which we are most familiar because they are most readily observed. In contrast, igneous and metamorphic rocks originate under conditions less well known to us by direct observation; except for surface lava flows, these two rock types form at great depths and under relatively high temperatures and pressures.

*Metamorphic rocks* are produced by physical and chemical alteration of any of the three major rock types. Igneous and sedimentary rocks may be changed (metamorphosed) into forms quite different from the parent material. Even metamorphic rocks may be further altered by renewed or later metamorphism. Physical changes in the parent rocks are produced under the influence of heat and pressure. The chemical composition of the original material may remain the same, but new minerals and physical structures are produced. If new chemicals are added as well, then the variety of new minerals formed increases greatly. The character of a metamorphic rock, then, is determined by the composition of the original material, the nature of the chemicals added (or subtracted), the intensity of heat and pressure, the manner or direction of pressure applied, and the length of time over which these various influences act.

Rocks of the earth's crust may undergo change in a cycle, such that they solidify from a molten magma to form igneous rock, are subsequently eroded or dissolved, are transported and deposited as sediment, are changed into rock (lithified), are subsequently metamorphosed, and in some cases even remelted to begin the cycle again. Many variations on this basic pattern of change are possible; for example, an igneous rock may be metamorphosed without going through the sedimentary portion of the cycle. Or sedimentary material may be eroded, transported, and redeposited several times without being metamorphosed. Metamorphic rocks may undergo one or more additional periods of metamorphism before reconversion to molten magma, or they may undergo erosion, transportation, and deposition in another sedimentary cycle.

The type and condition of rocks in any part of the earth's crust at present are the result of the events of geologic history up to the present. Not all rocks have had similar histories. None will remain in its present state indefinitely. This book treats geology in Michigan in terms of the rock materials and fossils found here. Chapters XI and XII deal with minerals and rocks in detail, Chapters XIII, XIV, and XV with their fossil contents. The reader whose geologic background is limited will profit from a preliminary study of those sections before proceeding to the chapters on the history of the state.

*Uniformitarianism.*—This is perhaps the most basic of all the principles of historical geology, that "the present is the key to the past." This principle holds that the cause—effect relationships between the various geologic processes have not changed with time. When the relationships between such processes as stream, wind, wave, or glacial action and their results, such as the various kinds of sedimentary deposits formed by those processes, are understood, then one can interpret ancient rock deposits, which are "effects," to discover the processes or "causes" that formed them. Extended examples of the use of such reasoning appear in many of the later chapters. To illustrate the principle briefly (Fig. 1–1), let us look at sediments resulting from transportation and deposition by modern glacial ice. Sediments laid down directly by a glacier include rock particles ranging from very fine clay-sized particles too small to be seen with the naked eye to boulders several feet in diameter; the surfaces of the larger rock fragments may present flattened and scratched or grooved faces, resulting from abrasion or grinding against the bedrock over which the glacier moved; the ice-deposited sediment is not layered (stratified). That such a glacial deposit differs from wind derived sediment is obvious. Who ever heard of a wind strong enough to blow large boulders? Windblown sediments (e.g., dune sand) are both well sorted and stratified. The geologic agent glacial ice, then, leaves its own unique imprint on its deposits. When an ancient sedimentary rock bearing these same characteristics is found in parts of southern Canada, interbedded with sandstones and conglomerates, the geologist recognizes the rock for what it is—an ancient glacial deposit—through the application of the principle of uniformitarianism.

Although geologists depend heavily on the principle of uniformitarianism, they recognize limits to its applicability. First of all, the present can be the key to the past only if the present is understood, and there are many aspects of the geologic processes that are difficult either to observe or to understand. For example, the surface activity of a volcanic eruption or lava flow can be observed and certain effects thus described. The rock types formed by such activity are available for study. However, little is known directly of what goes on deep within the earth at the time of such an event. Curiously, perhaps, the best information on these deep-seated happenings might be obtained by using the uniformity principle in reverse, that is, by using the past as the key to the present. One might examine the internal anatomy of an ancient volcano, if erosion had bared it to view, in order to get an idea of the nature of the interior of a modern volcano. It would, of course, first be necessary to determine that the deep-seated rocks now exposed by erosion actually had some connection to surface volcanic materials.

# Introduction

Figure 1–1. The principle of uniformitarianism illustrated by a comparison of modern and ancient sedimentary deposits formed by similar agents.

*Left side*—Direct ice deposits called glacial till or "boulder clay." Note lack of stratification and absence of size sorting.

*Top and center*—Moraine at edge of modern Greenland Ice Cap. (Courtesy W. R. Farrand, The University of Michigan Department of Geology and Mineralogy.)

*Bottom*—In Defiance Glacial Moraine of Late Pleistocene age near Ann Arbor (6" ruler for scale).

*Right side*—Water-laid stream deposits showing good size sorting, stratification, and arcuate cross-bedding.

*Top*—Downstream view along Loup River (a tributary of the Platte) at Columbus, Nebraska.

*Bottom*—Pleistocene age glacial meltwater (outwash) deposits in Killins gravel pit dug in a glacial kame on west edge of Ann Arbor (yardstick for scale).

Although geologic processes undoubtedly acted in the same *way* in the past, it does not necessarily follow that they acted at the same rate or that their location and total effect through any given span of time was always the same. Waves beat against the shores of continents in the past just as they do now, and the past and present deposits resulting are similar, but the location and extent of the shorelines and the force and direction of the waves varied widely through time. Today, Michigan is located in the interior of North America, but we shall see that at many times in the past much of North America lay beneath the sea and that ocean waves rolled across the state to pound on sandy shores. Beach sand deposits along such ancient shorelines now form extensive layers of sandstone rock present at many places beneath the state. Naturally, when seas were more widespread wave action and its effects were likewise more extensive, but *waves operated in the same manner then as now.*

Simple as the uniformitarian principle may seem, it has been applied in historical geology studies only during the last century and a half.

*The Law of Superposition and related principles.*—The law of superposition simply states that in a sequence of sedimentary layers, lying one upon another as deposited, the first layer deposited (the oldest layer) is at the bottom and the layers above are successively younger. This principle, establishing the *relative* age of superposed rock layers but not indicating their actual (absolute) age in numbers of years, has been recognized for nearly 300 years, but has been intensively applied during only the latter half of that time. Note that the law applies to rocks *in their original position of deposition*. We shall see later that rocks of the earth's crust can be and have been shifted about, intensely deformed by forces acting from within the earth, and that such changes frequently moved rock layers out of their original positions even to the extent of turning them upside down. Fortunately, however, there are sedimentary rock structures, such as specially shaped strata called crossbeds, ripple marks, mudcracks, old surface soil layers, and many more that indicate which direction *was up* at the time of deposition. Hence the law can often be applied even where the strata have been highly deformed. The relative ages of the sedimentary rock layers beneath the state of Michigan are determined by this basic law. Similarly, sequences of successively younger layered sediments are accumulating from day to day or year to year in modern lakes and ponds, and along the courses of rivers in Michigan. Successively younger strata develop only as conditions favoring deposition prevail. Deposition may be discontinuous, or some of the sediments laid down may be removed if an area suffers erosion, in either case leaving a gap in the sedimentary record. Such a gap in the geologic record of Michigan is described in Chapter VI—"The Lost Interval."

The concept that most sedimentary rock layers originate as nearly *horizontal layers* goes hand in hand with the Law of Superposition. Layers of sediment that accumulate on beaches, along stream or river courses, and on lake and sea floors are so nearly horizontal that deviation (initial dip) from that position usually amounts to, at most, only a few degrees, and more commonly to only a few feet or inches per mile; such a degree of inclination is imperceptible to the human eye where sedimentary rock strata are exposed over a short distance, as in road cuts or quarries. There are exceptions to near horizontality in the case of delta fronts, alluvial fans or talus slopes, but a geologist usually can recognize such exceptional deposits; in any case they are rare. It follows, then, that most sedimentary layers that are no longer horizontal have been deformed by some force that

# Geology of Michigan

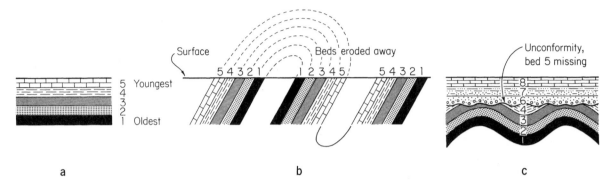

Figure 1–2. Application of Law of Superposition to determination of relative ages of sedimentary rocks.

a—Normal, undeformed, complete stratigraphic sequence.
b—Folded sequence in an anticline (upfold) with right limb overturned and a syncline (downfold) with left limb overturned.
c—Deformation and erosion during time 5, after 4 but before 6.

affected the earth's crust *after* deposition. The amount of deformation and the size of the area involved are indications of the intensity and extent of the deforming forces. Such periods of deformation are historical events. Furthermore, if the geologic ages of the affected layers are known, then the time of deformation must have been *after* the formation of the youngest layer involved and *before* the formation of the oldest unaffected layer (Figs. 1–2, 3, 4).

Two other historical principles are related to those of superposition and horizontality. First, sedimentary rocks form from layers of sediment that originally extended continuously throughout their area of deposition. Such layers end by thinning out to zero thickness at the edge of the area of accumulation or else they lose their identity by merging laterally into some other type of sediment. Picture the floor of Lake Michigan. Water-laid sediments of one type or another cover the lake bottom nearly everywhere. Near the shores, in shallower water, finer grained offshore sediment merges with coarser beach sand and gravel, which in turn thin out up the beach until they either end or merge with materials of non-lake origin such as dune sands. Locally, the lake sediments may grade shoreward into river delta deposits. The area of lake sediment accumulation coincides closely with the area actually occupied by the lake. If the lake were drained, the sediments would define its former limits, just as the dune sands and delta deposits would indicate the former sites of intensive wind and river action.

The second related historical principle deals with disruptions of this continuity. If lithified layers of such lake sediments are found to end abruptly rather than by gradually thinning or by changing character, it is necessary to account historically for the sudden interruption of continuity. One of several explanations might be possible. Figure 1–5a illustrates a situation in which, after several rock layers formed, the region in which they originated was uplifted and streams cut deeply into the earth's surface to expose the underlying layers locally and interrupt their original continuity. A similar effect on a much smaller scale results from quarrying operations and road cuts. Figure 1–5b shows a sequence of events in which compression of the earth's crust produced folds, later eroded to interrupt the continuity of the layers. Figure 1–5c illustrates a break in continuity

Figure 1–3. Undisturbed, horizontal sedimentary strata. (Also see Figs. v–2 and 24.)

*Top*—Rogers City Limestone of Middle Devonian age, in quarry at Rogers City. (Michigan Department of Conservation, photo by R. Harrington, 1955.)

*Bottom*—Rockport Quarry Limestone of the Middle Devonian age Traverse Group in near quarry face. Ferron Point Formation of same group above and behind. Old quarry of Kelly Island Lime and Transport Company, Rockport, Alpena County. (Michigan Geological Survey Division—Berquist File, photo by L. C. Hulbert, 1937.)

# Introduction

7

**Geology of Michigan**

Figure 1-4. Disturbed sedimentary strata. (Michigan Department of Conservation, Geological Survey Division—Berquist File, photos by L. C. Hulbert, 1938.)

*Top*—Erosionally truncated anticline in dolomite of early Middle Devonian age, Detroit River Group. Extension lines show original continuity of bedding. Distant hills consist of glacial moraine overlying bedrock. Kegomic (Mud Lake) Quarry, about one mile east of Bay View, Emmet County.

*Bottom*—Synclinal fold, truncated by erosion, in Petoskey Limestone of Middle Devonian age (in Traverse Group). South wall of old Petoskey Portland Cement Company Quarry on Little Traverse Bay about 2 miles west of Petoskey, Emmet County.

caused by faulting (breaking and slipping) followed by erosion. Figure 1-5d shows an interruption of sedimentary rock continuity resulting from the upwelling (intrusion) of molten rock (magma) from below. In all such instances of discontinuity it is apparent that one or more historical events such as uplift, erosion, folding, faulting or intrusion must have occurred *after* the time of origin of the rock layers affected; in other words, such an event is younger than the youngest interrupted rock unit. Furthermore, the disrupting event must, except in the case of intrusions, be older than the oldest overlying rock unit *un*affected. Thus, if the ages of interrupted or deformed rock

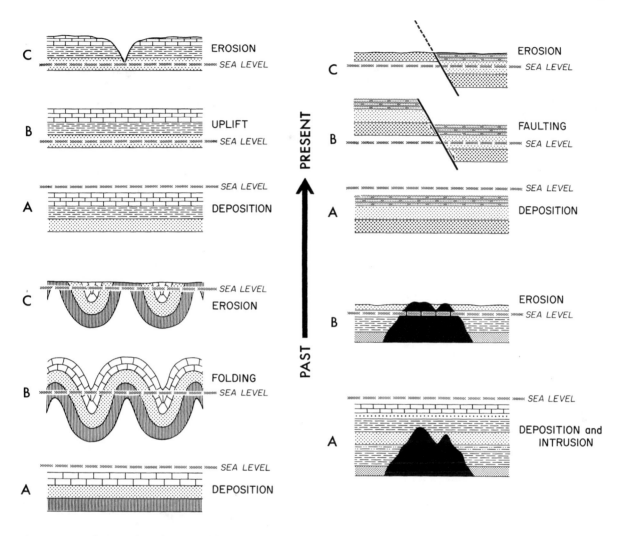

Figure 1-5. Effects of deposition, uplift and erosion coupled with folding, faulting, and igneous intrusion on distribution and age relationships of sedimentary rocks exposed at surface. In each of the 4 cases illustrated, the sequence of events, past to present, goes from A to B to C.

Figure 1-6. Some possible effects of faulting, uplift, and erosion on sedimentary rock strata. Deposition of sediments occurred during times 1—5. During time 6, faulting offset strata (relatively up on left and down on right), and uplift and erosion removed strata 2—5 on left, producing erosional unconformity and break in historical record. Subsidence of the land then allowed deposition of strata 7—9. Finally, uplift and erosion produced present surface features. Thus, in bottom of valley old rocks of age 1 are at same level as younger rocks of age 5. Weakened and broken rocks along fault favored rapid erosion, leading to development of valley there.

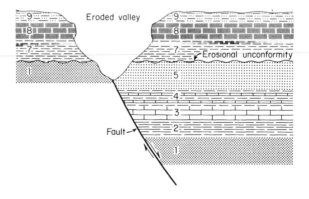

layers are known, the time of occurrence of the events can be determined. Figure 1–6 illustrates this method of relative dating.

*Principles of correlation.*—Rock-forming materials no doubt have accumulated at one place or another, on, or within the earth's crust, throughout all of geologic time. But there is no one place known where rocks of all geologic ages lie one above another with no erosional or depositional gaps in the record. We have also seen that the continuity of rock units is frequently broken. For these reasons the history of the earth must be pieced together by relating, or *correlating*, rocks from one locality to another. At least two different kinds of relationship (physical and time) may be established by correlation. Let us look first at methods of physical correlation. Suppose that a sea invaded an area. Beach sands deposited along the shore would represent the beach environment of deposition even if later buried by other sediment and turned to sandstone. But as the sea moved across the area the position of the beach environment would change and a layer of beach sand representing the many different positions of the shoreline would be spread across an area much broader than the beach occupied at any one time. If, many years later, uplift and erosion locally exposed the sand layer, one might establish by correlation that local sandstone outcrops were all part of one originally continuous layer, remembering that the sandstone layer might be of different geologic age at one place than at another because it took time for the beach environment to move across the area. How is such a correlation established? Figure 1–7 illustrates several methods. One method is *lithologic continuity*, possible in those rare instances where surface exposures of rock layers are so good that individual layers can be traced ("walked out") on the ground or followed on aerial photographs without a break. Another method is through *lithologic similarity*. The lithology of a rock is the sum total of all its characters. If the lithologies of two separated rock exposures are so similar as to leave no doubt that they are part of the same layer a physical correlation would be established. However, if several layers of similar character lie at intervals above one another their several identities might be confused. *Similarity of sequence* might help to eliminate such confusion. The several similar layers might all belong to a sequence whose overall character remained constant from place to place. The order of superposition, or position in the sequence, would help establish the identity of the individual layers. Note that these methods of correlation do not prove that any one layer is the same age everywhere, but merely show that a physical continuity originally existed, as in the case of the ancient beach sand example cited above.

If one seeks to establish that rocks at two localities formed at the same time, in other words are the same geologic age, methods of *time correlation* are needed. A *marker bed* or key horizon is one means of time correlation. We know by applying the principle of uniformitarianism that the materials in certain kinds of rock strata were deposited simultaneously over wide areas, regardless of the surface environment of the time. A volcanic ash from a large eruption, for example, would settle in a very short time over a large area of the earth's surface. Even if several years elapsed while the ash or dust settled this would be almost an instant in the vast scale of geologic time. The sedimentary rock material, regardless of composition, found between any two such ash layers must have accumulated during the same span of geologic time; in this case, the ash layers above and below serve as time planes or boundaries (Fig. 1–8).

Fossils also are used in some cases as indices to *relative* age; in fact they are perhaps the most commonly used means of relative time correlation. More complete discussions of the age significance of fossils are given in Chapters II and XIII. Briefly,

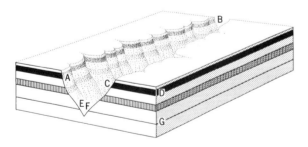

Figure 1–7. Some means of correlation of sedimentary rock strata. By *lithologic continuity*, traced along valley wall, the outcrops at A and B are part of the same (black) layer. By *lithologic similarity*, the rocks at A, C, and D are part of the same layer, like one another, and unlike any others elsewhere in the area or in the remainder of the vertical sequence. By *similarity of sequence*, the layers exposed on valley slope from E up through A are the same as those from F up through C and G up through D.

# Introduction

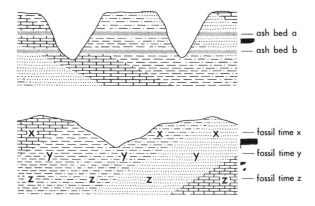

Figure 1–8. Time correlation in sedimentary rocks. Highly generalized. Patterns represent different types of sediments (and rocks derived from them) that accumulated in different types of adjacent but geographically shifting environments.

*Top*—Two ashbeds, resulting from volcanic eruptions, indicate age equivalent parts of several different rock type formations for 2 separate times. Ash from eruptions falls to earth very quickly, hence a single ashbed represents essentially the same geologic time everywhere it is found. Note that any one rock formation, which is a lithologic unit of similar composition everywhere, may "transgress" (cut across) time boundaries because each particular type was deposited over a span of time and the area of accumulation of each type shifted with shifting environments of deposition.

*Bottom*—Same concept as above illustrated by "time parallel" fossils which are types of organisms known to have been widespread but to have lived only during a very limited span of time. Note that during the time represented by Fossil z, for example, different kinds (facies) of sediments were accumulating in different places. Also note that accumulation of the lithologic unit (rock formation) represented by any one pattern, such as the stippled dots, occurred over a span of time (that of Fossils x, y, and z) and at different times in different places.

however, it has been established by the work of thousands of paleontologists over many decades that certain animal types (species, genera, or even larger groups) are restricted to certain parts of geologic time. Once the times of first appearance and extinction and the time span between have been determined their occurrence in rocks can be used as a measure of relative age.

Finally, and most recently, age correlations have become possible through the use of certain radioactive mineral dating methods, described in the succeeding chapter. **Radioactive dating methods** provide an "absolute" measure of time in numbers of years rather than simply in a relative sense.

Thus, there are ways in which it can be shown by correlation that rock materials at various places throughout the world are related to one another either by original physical continuity or by age equivalence, or perhaps both.

A final broad concept useful in the interpretation of earth history is that of *environmental adjustment*. Materials that compose the earth's crust and interior originate under very diverse conditions. For example, beach sands when cemented or lithified become sandstone. Such sands accumulate under normal surface temperatures and pressures in a relatively moist, oxygen-rich environment. Granite, on the other hand, forms from a liquid state deep within the crust under conditions of high temperature and pressure where free oxygen and moisture are rare if not absent. So long as either of these materials, or any other crustal material, remains in its environment of formation it will retain most, if not all, of the characteristics originally imparted to it by that environment. However, if some geologic event, such as burial beneath later sediments, uplift and erosion, or strong deforming stresses, places that material in new environmental conditions, changes may occur. The material, rock or mineral, with certain definite properties will change until it becomes stable in the new environment. Granite, if exposed at the surface by uplift and erosion, will disintegrate physically and decompose (alter chemically) to a new state that is stable in the surface environment. This process of change is called weathering and the product is a soil. In contrast, the beach sand, if deeply buried and thus subjected to extreme pressure and high temperature, will be altered (*metamorphosed*) to form a rock called quartzite in which the original sand grains became intergrown and interlocked in such a manner that the new rock occupies a minimum volume for its mass. This new rock (quartzite) is quite unlike the original sandstone. Examples of the way in which rocks alter to more stable physical and chemical states, developing new characters in the process, could be multiplied at great length. Limestone, formed most commonly as a precipitate from sea water, alters under high temperature and pressure to marble; shale to slate or schist; peat to coal. Directional stress may tilt, fold, or otherwise deform layers of sediment that originally were horizontal.

If some of the original characters of a crustal material remain to reveal its mode of origin and yet the material has developed characters in a new situation, these secondary characters may give clues as to the geologic processes that changed the original rock. The transferral to a new environment and the development of a new set of circumstances that cause a crustal material to alter are both historical events whose effects are evidenced by the adjustments that resulted.

## Summary

Whether geology is studied for some economic reason or simply to acquire an understanding and appreciation of nature and the world around us, many of the questions geology attempts to answer are historical. The special techniques of the historical geologist are many and varied, and the complexity and large scale of the earth may make them difficult to apply in actual practice. This chapter does not tell how to solve all the problems of historical geology, but it does set forth some of the basic principles of historical interpretation. Additional approaches to the problems of earth history are illustrated in later chapters that deal with special aspects of the geology of Michigan. More comprehensive but still elementary accounts of geologic principles and procedures may be found in many modern introductory geology textbooks, such as those by Gilluly, Waters, and Woodford, 1968; Garrels, 1951; Leet and Judson, 1965; Moore, 1958; Clark and Stearn, 1968, all listed in the bibliography.

# II

# Earth History and Geologic Time

## The Sense of Time

Each of us begins in childhood to develop a sense of time; all life, particularly civilized human life, seems to require it. Man's time sense becomes more refined as his need and desire to understand and control his environment increase. We begin measuring time in terms of familiar sequences of events just as the infant learns that bottle and nap follow playtime in that order. Later, day and night become yardsticks of time, and with experience days, seasons, moon phases, or years. But beyond these man's time sense must be supplemented by inventions such as clock or calendar. Science requires special devices for special time problems. The physicist, for example, cannot use an ordinary clock to measure the life span of a part of an atom, a period as small as a millionth of a second, yet such a measurement can be made. The geologist's problem is quite the opposite. He must measure time in thousands, millions, or even billions of years. Geologic processes affecting the earth operate very slowly, and their effects often are imperceptible after a day or year but these effects do become appreciable over longer periods. The waters of the Colorado River system daily erode and transport rock material from their drainage area. The net effect over perhaps a million or more years has been to carve the mile-deep Grand Canyon, but their daily effect is visible only in the constant passage of muddy water past a point along the river. Geologists have devised special methods for dealing with the extremely long time spans of earth history; these fall into two categories, measurements of *relative time* and *absolute time*.

## Relative Geologic Time

Relative dating (time telling) involves placing objects or events in their proper sequence of occurrence. Relative time is not expressed in terms of years. For example, we may say that Lincoln was president after Washington and before Roosevelt without stating (or even without knowing) the actual number of years intervening between their terms of office. Students of historical geology have long been able to reach many important conclusions concerning earth history using relative dating methods, with no more exact idea of time

**Geology of Michigan**

Figure II-1. Late Cambrian age marine sandstone of Munising Formation dipping gently southward in wave-cut Pictured Rocks cliff along south shore of Lake Superior east of Munising. *Original continuity* of strata (as indicated by extension lines) has been interrupted by wave erosion. Formation once extended farther northward an undetermined distance over what now is the Lake Superior Basin. According to *Law of Superposition,* lower layers are older than those above, although difference in age from bottom to top may not, by geologic standards of time, be very great in this case. (Michigan Department of Conservation, photo by Walter Hastings.)

in years than the realization that the earth is very old.

The Law of Superposition discussed in the introductory chapter is one of the important bases for relative dating. Sedimentary rocks consist of layers of sediment that accumulated one after another so that the layers in a sequence are progressively younger toward the top. This relationship automatically establishes the relative age of the layers (Fig. II-1). The introductory chapter stated that no one place on earth is known where rocks representing the whole of geologic time are exposed to view, but various methods of correlation were described by means of which rock layers from many localities and of various ages may be related to one another and the fragmentary rock record of earth history pieced together. The task may be likened to that of reassembling a complete copy of a book on world history if many copies had been torn apart and scattered over a wide area. Separate pages or groups of pages might be put together on the basis of page number. Some pages would find their proper place in the story because the information they contained would logically fill in certain recognizable gaps in what was once a continuous story. Some parts might be duplicates, others never found, and still others too complicated to understand at first. To continue the analogy a little further, suppose it were seen that whenever parts of the story were duplicated certain principal characters always preceded or followed others and all such persons were more or less interrelated, like families of kings or like dynasties. Then those persons or their relatives could come to serve as indices to the proper position of otherwise obscure passages in which mention of them occurred. Fossils are such interrelated characters of the geologic record. The ancestor-descendant continuity and gradual evolutionary change of living things through time have provided means of relating and assembling fragments of the historical rock record. If one repeatedly finds that fossils of types A, B, C, D, and E in one group, and C, D, E, F, G, and H in another group invariably occur in those sequences in superimposed rock layers, then one can construct the longer sequence, A through H. Eventually, one might learn the whole fossil alphabet. A passage from history containing D alone could then be placed between those passages containing C and E. Nearly two hundred years of work by thousands of paleontologists (students of fossils) were required to learn the main sequences and patterns of organic evolution and to establish the validity of the evolution concept, and there still are gaps in the record, but the time relationships of most major animal and plant groups are now well known and documented in thousands of volumes of paleontologic literature in scientific libraries. The time spans of many fossil groups are given in Chapters XIII–XV on fossils.

## The Geologic Time Scale

Modern studies of historical geology began about a century and a half ago in England and Western Europe. Unfortunately, the first studies did not begin at the beginning of the story. Widely scattered exposures of rocks were studied independently by many different people whose observations were later correlated. Sequences of sedimentary rocks representing different parts of geologic time were named after the areas in which they were first studied. Thus, rocks exposed in Devonshire, England, were called Devonian and the region gave its name to the time during which they were deposited. Rocks of similar age elsewhere are said to be of Devonian age. In time, and after much trial and error, major groups of sedimentary rocks representing nearly the whole of geologic time were studied, correlated, and arranged according to their relative age. Their names became the names of the PERIODS in the *Geologic Time Scale*. The relative geologic time scale now in use throughout much of the world is printed inside the front cover of this book where you can find it quickly. You would do well to memorize it if you become seriously interested in earth history. The time scale is like our calendar, in that the whole of earth time is divided first into main subdivisions, called *Eras*, according to the relative primitiveness of the fossils typical of the age. The Eras are divided into Periods, in turn divided into Epochs. Professional geologists use still smaller subdivisions, but these three will serve the purposes of this book; if there is need to be more exact we shall speak of the Early, Middle, or Late parts of those subdivisions. Next to the geologic time scale is a column showing time in millions of years. The methods of determining "absolute" time in actual years are discussed later in this chapter, but for now notice that the eras are not all of the same length, nor are periods or epochs of equal duration. In part, this is because in all history much more is known of the recent past than of earlier times; thus the later part of the time scale is more finely subdivided.

The irregular subdivisions are also the unintentional result of the uncoordinated way in which studies of earth history were begun. In the beginning knowledge was limited to a few, well-studied parts of the world. The early geologists recognized that certain crustal disturbances attended by uplift and erosion had left gaps in the sedimentary rock record, times when no sediments accumulated. They thought the gaps represented worldwide catastrophes and used them as convenient dividing lines for their first time scales. Worldwide exploration later proved that those particular historical gaps were less widespread than expected and that the rock record in other areas often bridged the gaps found in the original study areas. Nevertheless, the names of those old dividing lines

# Geology of Michigan

and times were retained, adjusted for convenience and to reflect increased knowledge, and allowed to stand as the arbitrary but useful boundaries and divisions of the present time scale.

Thus, the geologic time scale grew to its present form. Most of the names are useful throughout the world. However, some of the most ancient rocks and their records cannot yet be correlated from continent to continent, or even throughout all of North America. The names used herein for such ancient times as Keewatin, Huronian, and Keweenawan are applicable only in the Great Lakes region.

No matter what the number of years involved, one can see that the Archeozoic is the oldest era, that an event in the Cenozoic happened after one in the Mesozoic, or that a process that continued from the Proterozoic to the Permian went on during a long span of geologic time. The time scale is a means of expressing relative time and is always written from the oldest period at the bottom to the youngest at the top just as the sedimentary rock record is arranged according to the Law of Superposition.

## Relative Dating of Igneous Rocks

The relative ages of igneous rocks which cooled from a molten state may also be determined and expressed in terms of the time scale, but the methods of age determination are less direct be-

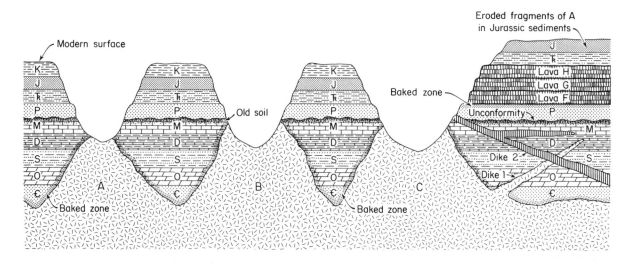

Figure II–2. Age relationships in igneous rocks. Symbols for rock ages are: €–Cambrian, O–Ordovician, S–Silurian, D–Devonian, M–Mississippian, P–Permian, Tr–Triassic, J–Jurassic, K–Cretaceous (Pennsylvanian is missing). A, B, and C are (by lithologic similarity) parts of a single, large igneous intrusion. That intrusion crosscut strata from Cambrian through Mississippian age and the heat baked those layers, hence the intrusion occurred *after* the youngest, Mississippian, rocks affected. A soil developed on the Mississippian rocks as well as on the baked zone, indicating that the region was uplifted, weathered, and eroded (down to the intrusion in some places) before the Permian rocks were deposited, hence the intrusion occurred after Mississippian and before Permian time, that is, in the Pennsylvanian Period for which rocks are missing. Eroded fragments of the intrusion (A-B-C) in the Jurassic rocks suggest that exposures of A-B-C must have persisted somewhere until the Jurassic. Dike 1 crosscut Cambrian through Devonian rocks so is post-Devonian in age (and probably younger). Dike 2 crosscuts Dike 1, the unconformity and the Permian so is post-Permian in age; how young is not determinable because no crosscutting relations are shown between Dike 2 and other rocks younger than Permian. Lava F was extruded upon and baked the Permian rocks, hence is post-Permian. Lavas G and H flowed out one after another upon F, hence, according to the Law of Superposition, the lava series goes F-G-H through time. All three lavas lie beneath the Triassic which was deposited on top of H, hence the lavas formed sometime between the Permian and Triassic periods. The full sequence of events, from oldest to youngest, thus was: Deposition of Cambrian-Mississippian strata; intrusion of A-B-C and Dike 1; uplift and erosion to produce unconformity during Pennsylvanian; subsidence and deposition of Permian with intrusion of Dike 2 and related sill sometime thereafter; extrusion of lavas F, G, and H between Permian and Triassic time; deposition of Triassic, Jurassic, and Cretaceous strata; subsequent uplift and erosion to produce present topography.

**Earth History and Geologic Time**

17

Figure II-3. *Crosscutting principle* for determination of relative geologic ages of igneous rocks, illustrated by specimen outside The University of Michigan Department of Geology and Mineralogy in Ann Arbor. Sequence of events and ages as follows: 1—formation of dark-colored granite (oldest), 2—intrusion of dikes of intermediate age crosscutting granite, 3—intrusion of youngest dike whose margins are continuous and thus crosscut the other 2 dikes.

cause the Law of Superposition does not always apply and because fossils are exceedingly rare in such rocks. The *cross-cutting* principle provides the basic key to relative igneous rock age. Picture a molten mass of rock-forming material (magma) that once moved from deep within the earth into overlying sedimentary rock strata where it cooled to form granite. The molten mass would have "cut across" or *intruded* all the rocks through which it moved (Figs. II-3, II-4). If the magma cut through rocks of Keewatin age (see time scale inside front cover) it must have done so sometime after the Keewatin Period; it would be "post-Keewatin" in age because before then the Keewatin rocks would not have been present. This idea may seem ridiculously simple in theory but in practice it was not

# Geology of Michigan

recognized and applied until less than 200 years ago. True, the magma might have moved upward only for a short distance, stopping before it reached much younger rocks already in existence, but at least it could be *no older than the youngest rock it cross-cut,* even if it were much younger. Thus, its *maximum age* is established. Still further, suppose that the intrusion not only cut certain sedimentary rock layers but later was exposed at the surface during a period of erosion and weathering, that a soil or weathered zone then developed on its exposed parts, and this later covered by more sediments. In this case, the intrusion is not only younger than the rocks it cross-cut but older than the overlying unaffected rocks, so its *minimum age* can be determined. If, in the example above, the intrusion cut rocks of Keewatin age but was eroded at the surface before Huronian rocks were deposited over it, then the intrusion occurred after the Keewatin but before the Huronian period.

If a molten magma reached the surface and was *extruded* as lava or volcanic ash, then the Law of Superposition (see Introduction) applies. The extrusive rock is younger than any sedimentary layers it flowed out upon but older than any later (younger) layers that cover it. Furthermore, if several flows or ashfalls occurred the series would range from oldest below to youngest above.

The situations and conditions in which igneous rocks form are highly varied, but the basic principles of relative age determination are the same. Figure II-2 shows diagrammatically some of the special problems that might be encountered. If this brief account of relative dating methods excites your curiosity, you will find more detailed information and additional references in the books by Shrock (1948), Krumbein and Sloss (1951), and Shaw (1964).

## Absolute Time

Students of earth history have never been completely satisfied to know only the relative ages of geologic objects or events; they have long sought to tell time in terms of solar or calendar years. Geologists call this kind of time *absolute* because it is expressed in terms of numbers of years, but they do not pretend that absolute age determinations are exact to any great degree. Errors as great

Figure II-4. Basic igneous dikes (dark colored) of Precambrian age crosscutting older Precambrian granite along Michigan State Highway M-95 between Humboldt and Republic about 5.5 miles south of junction M-95 and US 41. Dike in lower photograph is offset by a fault.

as several million years may be involved, depending upon the accuracy of the method used and on other things, but on the vast scale of geologic time, involving hundreds of millions or billions of years, such errors may amount to only a small fraction of the total. Absolute dating not only helps establish the position of events in an historical sequence, just as does relative dating, but also makes it possible to determine such things as the actual age of the earth, the rates at which geologic processes act, or the actual time required for events to transpire. Furthermore, proof that

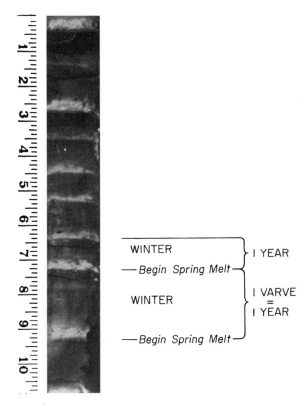

Figure II-5. Varves from small portion of a deep core taken in Lake Superior (also see Fig. VII-14) 35 miles due east of Keweenaw Point in 1049 feet of water and from about 37 feet below lake bottom. Light-colored layer begins to form with first inflow of spring meltwater runoff from land. Dark layer above forms thereafter of mud that settles during winter. One pair of light and dark layers represents one year of deposition. Scale is in inches, hence rate of sediment accumulation here was about 11 inches in 9 years or an average of slightly over 1 inch per year. Younger varves found higher in the sedimentary column indicate that sediment accumulated more slowly later on (Fig. VII-14). Downbending of layer boundaries at core edge due to drag along coring barrel; dark, horizontal cracks due to shrinkage of core with loss of moisture. (Core No. S-62, courtesy W. R. Farrand, The University of Michigan Department of Geology and Mineralogy.)

the earth is very old and geologic time very long lends credence to the claim of Uniformitarianism (see Introduction) that the present can be the key to the past. Uniformitarianism holds that geologic processes, e.g., erosion and mountain building which go on so slowly today that they are scarcely perceptible, are capable of accomplishing vast changes in the surface of the earth. Absolute dating methods demonstrate that the great time spans needed by this doctrine are actually available.

One of the simplest methods of absolute age determination is varve counting. Varves are sedimentary rock layers, or sets of layers, that accumulate during the course of a determinable period of time, usually of one year. A series of superimposed varves can be counted to tell how many years were required for deposition of the whole series. Although some years might be poorly represented, at least the minimum age in years can be determined much as one would count tree rings to tell the number of years involved in a tree's growth. The sediments on the bottom of some of Michigan's lakes and ponds are varved (Fig. II-5). Water accumulated in depressions left as the last Pleistocene ice sheet melted back to the north. Sedimentary deposition could not have begun in these lakes until the ice had cleared the lake site. Thereafter, thick, light, quartz-rich layers accumulated during the period between spring to fall, and thinner, darker, clay-rich layers during the winter months when the water was ice covered and erosion and melting slowed or ceased on the surrounding land. A pair of such layers represents one year. A count of varves in such a lake would provide some measure of absolute time. Varves may consist of more than 2 layers per year, or may result from other conditions, but the principle is the same. For additional information see Knopf (1949).

With the discovery of radioactivity, new methods of absolute geologic dating were made possible. A number of minerals contain radioactive elements, such as Uranium, which undergo a decay process whereby they are transformed into other elements. Uranium decays through a whole series of forms until the end products of Lead and Helium are reached. There are several kinds (isotopes) of Uranium with different atomic weights, each producing its own peculiar isotope of Lead. These Lead isotopes have different atomic weights and can be distinguished not only from one another but also from nonradiogenic Lead. Other naturally occurring isotopes of elements that decay radioactively include: Potassium$^{40}$ which changes to Calcium$^{40}$ plus Argon gas; Rubidium$^{87}$ which changes to Strontium$^{87}$; and Carbon$^{14}$ which changes to Nitrogen$^{14}$. In all of these examples the important point is that the decay rate of each radioactive element is constant, being unaltered by any natural chemical or physical conditions of the environment. Thus, if through some geologic process, a quantity of Uranium is separated from its previously formed decay products and deposited in a new rock or mineral it decays there at a constant rate regardless of environment. The longer it decays the greater the proportion of new decay products formed from it. The slow decay is measurable in the laboratory. Because various elements decay at a different rate there are "clocks" suitable for dating both old and young rocks. One gram of Uranium$^{238}$ produces only 1/7.6 billionth of a gram of Lead$^{206}$ per year, so the decay of Uranium$^{238}$ is a good "clock" for extreme age determinations, but it is a very poor one for recent ages, where the decay time has been too short to produce a measurable amount of Lead. Carbon$^{14}$ decays at a much faster rate, so fast in fact that it can only be used to date deposits up to about 70,000 years old; in older deposits so little Carbon$^{14}$ is left it is too difficult to measure.

Radioactive dating procedures involve collecting samples, isolating any radioactive materials and their decay products, and determining the ratios of decay products to parent elements. The older the sample the greater the proportion of decay products, and the ratios to be expected over different spans of time for any given starting amount can be computed from the known decay rates. The *absolute age* of the sample in years, that is, the time elapsed since the rock and its contained radioactive material were formed, is thus determined. Of course the actual laboratory procedures and the geological problems involved are much more complicated than this brief description suggests. More detailed descriptions occur in any of the elementary geology textbooks listed in the Bibliography. Radioactive age determinations always involve some experimental (statistical) error, but the probable magnitude of error can be calculated and is usually expressed as a plus or minus ($\pm$) figure after the age. Although radioactive dating methods are quite new, many dates have already been established for rocks whose relative ages are known in terms of the geologic time scale. A few of the most significant figures are given in a column next to the time scale inside the front cover. Once the absolute age of any point on the time scale is determined, then rocks of the same relative age everywhere are considered to be the same absolute age as well. Valuable confirmation of the validity of relative dating by use of fossils has come from cases where rocks from different localities were determined to be of the same relative age by their fossil content and subsequently turned out to be of the same absolute age through radioactive dating.

Uranium dating has two main drawbacks. In most cases the method can be applied only to igneous rocks, which then may be related by their cross-cutting relationships to sedimentary rocks. Only then does the absolute date take on significance in terms of the time scale. On the other hand a few sedimentary rocks that can be dated relatively do contain radioactive elements and in those cases the absolute dating is direct. Another drawback is that radioactive decay goes on so slowly in Uranium that the little change occurring over a few millions or tens of millions of years cannot be measured accurately. The slow decay rate, which prevents the use of this radioactive series for dating short periods of time, does make it possible to date very ancient rocks. Some very ancient Archeozoic rocks have yielded ages in excess of 3.5 billion years, and there are older rocks (according to relative dating methods) known that have yet to be dated radioactively.

Fortunately, other radioactive elements are now proving useful in dating sedimentary rocks and relatively young rocks in particular. The radioactive Rubidium/Strontium method can be applied to marine sediments if those elements occur in the mineral glauconite. The Potassium/Argon method can be used to date terrestrial (nonmarine) deposits such as volcanic ash.

The *Carbon$^{14}$ method* of dating, which applies to periods ranging from the present back about 70,000 years into the past, is particularly useful to anthropologists, archeologists, and students of Late Pleistocene (Glacial Age) events. Cosmic rays from space bombard the nuclei of Nitrogen$^{14}$ atoms in the earth's upper atmosphere, transforming the Nitrogen$^{14}$ to radioactive Carbon$^{14}$ (Carbon$^{12}$ is the principal normal, nonradioactive isotope). The Carbon$^{14}$ combines with Oxygen to form Carbon dioxide gas. Although the atmosphere normally contains carbon dioxide most of that gas contains the Carbon$^{12}$ isotope. The radioactive carbon dioxide mixes into the atmosphere and ultimately into the sea in a definite proportion. As plants utilize carbon dioxide in their growth processes, some small amount of Carbon$^{14}$ gets into all plant tissue;

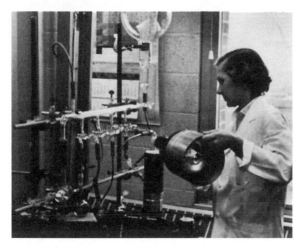
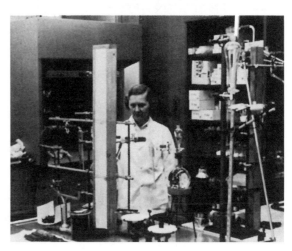

Figure II-6. Radiocarbon (C14) Dating Laboratory, Physics-Astronomy Building, The University of Michigan.

*Upper left*—H. R. Crane, chairman, Physics Department, inserts counting apparatus into lead block compartment which shields radioactive emission counters from nonsignificant background radiation of external origin (cosmic rays, etc.). Level of radioactive emission from fossil sample can then be determined.

*Upper right*—Some fossil materials to be dated, including wood fragment of Late Pleistocene relative geologic age in foreground.

*Lower left and right*—Patricia Dahlstrom, chemist, prepares samples for dating. Carbon in wood or other organic matter must be extracted and converted to gas before introduction within counting apparatus.

because animals eat plants, or at least other animals that have eaten plants, they too accumulate some radioactive carbon. Carbon[14] is constantly decaying radioactively back to Nitrogen[14] but, because the incidence of cosmic rays is nearly constant, more new Carbon[14] is formed as fast as decay goes on. So there is a constant ratio of Carbon[14] to normal Carbon in the atmosphere and also in the tissue of *living* plants and animals. As fast as it decays in living tissue it is replaced by growth of new tissue. However, when a plant or animal dies and growth ceases, the decaying Carbon[14] is no longer replenished. Then, the longer the organism has been dead the lower the ratio of Carbon[14] to normal Carbon[12] becomes. Although the decay goes on rapidly compared with that in Uranium it is slow by human standards, approximately 5600 years being required for one half of any given amount of Carbon¹⁴ to decay. Here then is another built in clock. The ratio of Carbon¹⁴ to normal carbon in ancient organic remains can be compared with that ratio in the tissues of living organisms and, knowing the rate of decay, the length of time since the death of the organism can be determined. Again, the actual technique is quite complicated and a predictable amount of error one way or another from the probable date is involved, but the error can often be held to a few hundred years. In organic remains older than about 70,000 years the radioactive content becomes too low to measure accurately, and dating becomes unreliable, although new techniques constantly improve upon the method. The validity of the dates established from radioactive carbon can be checked by applying the method to materials whose actual age is already known in terms of some human calendar (or by other means). The ages established by the Carbon¹⁴ method are then compared with those of the calendar to see if they conform. This has been done with satisfactory results using the Carbon¹⁴ in wood from old European churches, oak from a Viking ship, funeral ships of Egyptian kings, the variously aged growth rings from old living trees such as the giant redwood and many other datable objects. An extended discussion of Radiocarbon dating will be found in Libby (1955). Figure II–6 shows some of the apparatus in the Michigan Memorial Phoenix Project Radiocarbon Dating Laboratory at The University of Michigan in Ann Arbor.

**Summary**

The geologist has many ways of telling time in earth history. If, in one of the chapters that follow, some rock, mineral, fossil, or historical event is said to be of a certain age either in terms of the relative time scale or in actual years, the age determination was probably based on one or more of the methods described in this chapter. Radiogenic dates based on Lead/Uranium and Potassium/Argon ratios are especially important in Chapter IV on the Precambrian. Carbon¹⁴ dates are used extensively in Chapter VII on The Ice Age, Chapter VIII on The Great Lakes, Chapter XIV on Fossil Vertebrates, and Chapter XV on Fossil Plants.

# III

# General Geologic Setting of Michigan

## The Broad Picture

Some of the details of Michigan geology presented in later chapters are unusual if not unique to this state or to the Great Lakes Region. However, the state shares many aspects of its structure and history with the rest of the North American continent and even with the world as a whole. This book cannot do complete justice to the detailed geology of such large regions—that is a task for general geology textbooks. Nevertheless, the geology of Michigan will be better understood if seen first in broad perspective. It should be remembered that each of the generalizations that follow is a fascinating story in itself and results from the labors of countless geologists. Excellent and comprehensive, if somewhat technical, accounts of North American geology may be found in the recent works of King (1959), Clark and Stearn (1968), and Eardley (1962).

Stable as the earth's surface and its features may seem to most of us in the course of a human lifetime, the geologist knows that through the long ages of the geologic past the earth has been everchanging. Forces acting from within uplift, break, crumple, and otherwise deform the rocks of the crust. Lofty mountain ranges such as the Himalayas or the Rockies rise, and broad lowlands such as the African Rift Valley or the Great Valley in California sink. High standing relief features have formed on the surface time and again, only to be worn away by streams, waves, glacial ice and wind, or to be filled in with sediment carried by those geologic agents. Thus, there is a constant antagonism between constructive forces from within and surface processes tending to level the face of the land. Evidence that this contest has been in progress throughout the geologic past is of many sorts. Many of the highest mountain peaks of the world are composed of sedimentary rocks containing fossil remains of organisms that once lived beneath the sea. Igneous rocks that intruded the crust and cooled at great depths, as well as metamorphic rocks formed deep below the surface under conditions of extreme heat and pressure, are now exposed to view after long ages of uplift and erosion. In contrast, deep canyons, or cores brought up from below by oil well drills, reveal that there are sedimentary rocks, with physical features and fossils formed near or at the surface, that are now buried at great depths in areas where the earth's

**Geology of Michigan**

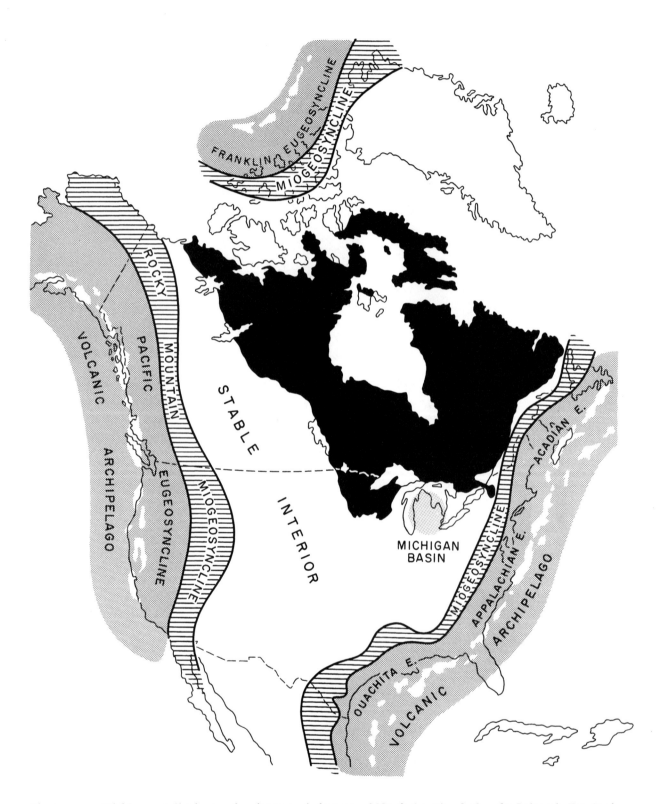

Figure III-1. Highly generalized map of major tectonic features of North America during the Paleozoic Era. Position of Michigan Intracratonic Basin is emphasized. (Compiled, with modifications, from various sources.)

crust has sagged and been filled in. Old beach lines are tilted. Wave cut terraces and other shoreline features now locally stand high above the sea or far below its surface. Streams that once had eroded downward as far as possible toward sea level have been rejuvenated in places by uplift and are again downcutting in steep-walled canyons. Bore holes and geophysical investigations show that beneath the surface of the Mississippi Delta sediment has accumulated in a slowly subsiding trough to a thickness of about 50,000 feet. All these and many other lines of evidence stand in mute but eloquent testimony to the earth's instability. If these are unconvincing, witness such visible evidence of crustal unrest and deformation as earthquakes, land slides, and volcanic eruptions. The face of the earth is truly unstable, and in the long view has shaky foundations.

But not all areas are or have been equally mobile. Certain portions of the continents have tended to remain relatively stable while the stresses within the crust were relieved in others. To understand the role Michigan has played in this great drama let us look backward about 600 million years through time to the beginning of Cambrian time, the dawn of the Paleozoic Era. It may be difficult for some to imagine, but most of the 4.5 or 5 billion years of earth history were already past at that time. The Archeozoic and Proterozoic rocks cropping out in the western portion of the Upper Peninsula of Michigan and buried by later rock elsewhere in the state had long since formed. At the beginning of the Paleozoic Era the North American continent entered a phase in its development which continued for several hundred million years, and, in some cases, even up to Recent time. The map and cross-section in Figure III-1 show the situation diagrammatically and should be referred to during the discussion that follows. First note that the configuration of the continent was quite different from that of today.

## The Continental Interior in the Paleozoic

The interior of North America stood relatively high. At the beginning of the Paleozoic it was already an old land composed of rocks that had passed through a complex history during the 4 to 4.5 billion years of Archeozoic and Proterozoic time. But from the beginning of the Paleozoic this area was destined to remain relatively stable. Although we shall see that portions of the interior frequently sank so that shallow seas could spread inland, these seas were never deep, and the sedimentary layers that accumulated on their floors are relatively thin. This region is known as the *Central Stable Region* or the *craton*. Its northern portion, in northern United States and parts of Canada, where extremely old Precambrian rocks are now exposed at the surface, is called the *Canadian Shield*. The craton was the early Paleozoic nucleus of the continent. Its main deformational events of downwarping, folding, and mountain building had already occurred during the Precambrian eras. Crustal unrest took place thereafter mainly around the margin of the craton. However, certain areas within the craton (*intracratonic* areas) were sites of more than normal activity. The Adirondack Highlands (not to be confused with the modern Adirondack Mountains of the same area), the Wisconsin Highlands, the Ozark Dome, and the Cincinnati and Transcontinental arches were "positive" areas that frequently stood relatively high, and subject to erosion. These interior arches supplied sediment to neighboring lowlands. Elsewhere "negative" areas of the stable interior sagged down, forming local basins in which cratonic sediments of more than average thickness accumulated. Often during the Paleozoic Era these *intracratonic basins* were occupied by marine waters that spread inland across the continent from practically every point of the compass at one time or another. Much of what is now the Great Lakes Region was such an intracratonic basin; it is now called the Michigan Basin. Remember, however, that this great depression and the marine waters that often occupied it in Paleozoic time, bore no relationship to the modern Great Lakes. The Paleozoic history of the Michigan intracratonic basin and the rocks formed there during that era are discussed in detail in Chapter V.

## The Paleozoic Geosynclines

The *continental margins* of Paleozoic North America were, in contrast to the interior, sites of intensive crustal activity. This was to continue in some areas from the early Paleozoic to the present. Often during the Paleozoic Era the present Appalachian Mountain region and much of the now

### Geology of Michigan

mountainous western part of the continent were sites of crustal down-warp where the earth's crust sank deeply to be filled apace by sediments. These long, relatively narrow, deep, marginal troughs are called *geosynclines*—the Appalachian Geosyncline on the east, the Cordilleran (and later the Rocky Mountain) Geosyncline on the west. Parts of southern United States and Mexico also subsided to depths of geosynclinal proportions, particularly during late Paleozoic time. Northern North America may also have done so, but the evidence there is inconclusive. These great troughs were often occupied by seas that spread inland from the Atlantic and Pacific ocean basins and filled with sediment at about the same rate as they subsided, so that although their original floors reached great depths below sea level the water in them rarely was deep. Sediments deposited in these troughs were of relatively shallow marine, or even continental, origin. Some of the sediment was derived from locally high standing portions of the craton, carried outward from the continental interior. However, the major part of the sediment, particularly that in the outer parts of the geosynclines, came from areas of intense volcanic activity. These areas resembled the modern volcanic island arcs or archipelagos which lie in part along the margins of some modern continents and occupy parts of what now are the offshore continental shelves (the Japanese—Philippine archipelago off the eastern coast of Asia). Evidence of the volcanic activity and its marginal location comes from the fact that the sediments which filled the outer parts of the Appalachian and Cordilleran geosynclines were derived from rocks dominantly of volcanic origin. Furthermore, the geosynclinal deposits become both thicker and coarser outward from the continental interior indicating that their sources were marginal to the continent. To continue the modern analogy of the eastern coast of Asia, studies of that area and the sedimentary materials now accumulating in and around it demonstrate its resemblance to geosynclinal areas of the past. The modern geosynclinal area is the Sea of Japan and China Sea area between the archipelago and the continent. The volcanic archipelagos that bordered the Appalachian and Cordilleran geosynclines of North America in the past have long since been beveled by erosion and subsequently lowered beneath the sea and covered by younger sedimentary layers, but the sediment of volcanic origin eroded from those archipelagos and carried continentward into the geosynclines indicates their former presence. Their long buried roots now lie partly beneath the modern continental shelves and partly along the very easternmost and westernmost margins of the modern continent. The lithified sediments of the geosynclines, after a long and complex history the details of which are beyond the scope of this book to relate, were folded, faulted, in some instances intensely metamorphosed, then lifted and eroded to form the modern mountains of eastern and western North America. Many times during their growth and filling the geosynclines were occupied by arms of the sea from the neighboring ocean basins. Often those waters escaped the relatively narrow confines of the geosynclines to flood across parts of the Central Stable Region as shallow inland (epeiric) seas. The epeiric seas often entered subsiding intracratonic basins such as the Michigan Basin and tended to linger in those areas after they had withdrawn from other parts of the interior. Thus, the Michigan and other interior basins received a thicker accumulation of marine sediment than was normal for most areas of the craton. The sediment that filled the Michigan intracratonic basin was derived by erosion from neighboring cratonic highlands such as the Adirondack Highlands to the east and northeast, the Cincinnati, Kankakee, and Findlay arches to the south, the Wisconsin Highlands to the west and northwest, and the Canadian Shield area to the north. It is in that thick sequence of sedimentary rock deposits underlying the state that the long record of Paleozoic geologic history is preserved.

### A Basin beneath the State

The Paleozoic rocks beneath the state are locally about 14,000 feet thick and rest upon a floor, or "basement", of very ancient Precambrian igneous and metamorphic rocks. Were it not for the subsurface shape of the sedimentary layers we would be unable to study most of these rocks except in deep wells or mine shafts as only the uppermost layers would come anywhere near the surface. Fortunately, however, the earth's crust beneath the Great Lakes region sagged downward as the sedimentary layers were deposited. Extensive uplift afterward enabled the agents of erosion to cut

**General Geologic Setting**

27

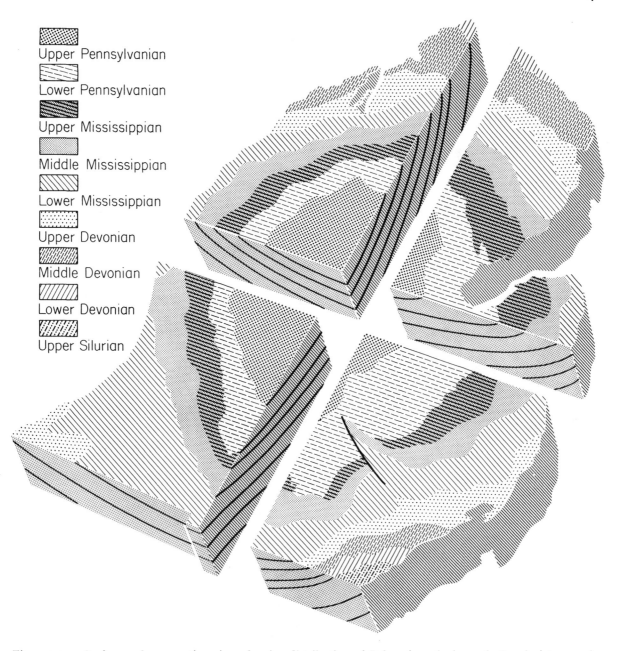

Figure III-2. Surface and cross-section views showing distribution of Paleozoic rocks beneath the glacial cover in the Lower Peninsula portion of the Michigan Basin.

deeply into the crust, slicing across the inclined rock layers so that their upturned and truncated edges now underlie the state as irregular concentric rings. Figure III-2 shows how this structure would appear if we could take a bird's eye view of the state and at the same time look at vertical cross sections cut through the crust. The structure has been likened to a set of nesting bowls or saucers.

Very late in geologic history a continental ice sheet spread a layer of unconsolidated glacial sediment over most of the surface of the state, concealing the bedrock. But bedrock crops out in many places in the Upper Peninsula and locally in the Lower Peninsula where the glacial material is thin or has been stripped away by erosion. Over 26,000 wells drilled for oil, thousands of water wells, and many quarries and mines also cut through the glacial cover into the bedrock below. Thus, the structure and distribution of preglacial sedimentary bedrock layers are well known, although rock is not everywhere visible at the surface.

Remembering the Law of Superposition, notice on Figure III-2 that beneath a point near the center of the Lower Peninsula the oldest rocks are farthest below the surface. Progressively younger layers lie in sequence above. Trace one of the lower layers in the cross sections and see that it rises toward the surface near the edge of the state. Younger, overlying layers approach the surface within the outer rings of older layers. The concentric rings of bedrock just beneath the glacial cover are oldest around the edges of the state and become progressively younger toward the center. Thus, one may encounter bedrock of different ages and types in different parts of the state. The most ancient Archeozoic and Proterozoic rocks of the Precambrian basement approach the surface and crop out locally only in the western part of the Upper Peninsula. The geologic ages of the bedrock layers shown in Figure III-2 may be found on the Geologic Time Scale inside the front cover. Each of these layers is really quite complex and composed of many subdivisions, called formations, consisting of rocks of different type and origin. These are discussed in detail in Chapter V dealing with the Paleozoic history of Michigan. A chart of the Michigan rock formations (Fig. V-2) also appears inside the back cover. A publication of the Michigan Geological Survey entitled "Oil and Gas Fields of Michigan" (Newcombe, 1933) discusses and illustrates the basinlike subsurface structure more thoroughly.

## The Plastic Earth

Much has been said thus far in this chapter concerning radical changes in the earth's crust. Many persons find this difficult to believe. The infinitesimally small part of the earth upon which we stand for the mere span of a human lifetime usually seems quite solid. Surface rocks behave as brittle substances. "Terra firma" describes our normal impression of the earth. Perhaps by now you have wondered how the surface of this same earth can fold, wrinkle, crumple, sag downward thousands of feet, or rise upward locally to form mountains or high plateaus. What forces cause all of this and why does the earth not simply shatter?

The question concerning the deforming forces is the most difficult; in fact geologists do not all agree on the answer. Disagreement is possible because there is still much to learn about the interior of the earth, and the explanations offered so far are based on very little direct evidence. Among the many solutions proposed there are two hypotheses currently in vogue, both of which have their *pros* and *cons*. Some students of the problem suggest that the earth's interior is shrinking, either because of cooling or because of the continual withdrawal of lava from great depths to the surface. They hold that as the relatively thin outer skin or crust of the earth is pulled by the force of gravity against the shrinking interior, stresses of great magnitude are set up buckling the crust, heaving up highlands, depressing lowlands, and causing severe deformation. This buckling is analogous to the wrinkling of the skin of a dried, shriveled apple. Others fail to find convincing evidence for a shrinking globe. Instead, they propose that vast sources of internal heat, presumably from radioactivity, not only forestall cooling and shrinking, but are sufficient to cause the rock material beneath the outer crust to circulate in great convection currents, much like the boils that develop when fluids are heated in a saucepan. Supporters of the Convection Current Hypothesis suggest that when and where such currents rise toward the surface they warp the crust upward, whereas in areas where two circulating currents meet and plunge downward together toward the interior they drag the crust downward, in time compressing it to produce wrinkles, breaks, and other deformational features. Geosynclines would be formed in such downwarped zones. Another school of thought is that the convective flow beneath the crust causes portions of the crust, the continents, to move laterally, like floating rafts on water. No final decision can be made as to which of these, or other possible explanations is correct. Too little is yet known about the nature of heat release and rock composition in the interior. Either of the two basic hypotheses given would be a rather satisfactory explanation of crustal deformation if the basic assumptions on which they are founded could be verified.

The second question—how can the crust respond to the forces acting upon it without shattering—can be answered more easily. The strength and behavior of materials under stress depend upon the size of the mass involved and the conditions under which stress is applied. A small piece of steel, for example, has great strength by normal standards, yet can be *slowly* rolled and drawn into thin wire under proper conditions of confinement, heat, and pressure. Likewise, a small block of ice is a brittle solid but a large thick mass of the same material, confined and under great pressure at the base of a glacier, may deform plastically, flowing either downhill (as in mountain glaciers) or outward from the center of accumulation (as in the case of a continental ice sheet). Under proper conditions, ice may flow over or around obstacles in its path. What has this to do with the earth? Simply this: although the earth is mainly a dense solid, it is also extremely large and is acted upon slowly by forces of great magnitude over long periods of time. The deeper portions of the crust and interior, confined under great pressure at high temperature, can be deformed under stress without excessive breaking. Even so, faults (breaks along which slippage occurs) do occur and minor adjustments develop locally in the form of small joints and fractures found in most surface rocks. It has been calculated, by students of what is called Scale Model Theory, that a model of the earth only six inches in diameter would have to be constructed of some low strength material such as very soft toothpaste or very watery mud in order to simulate in the laboratory the reactions of the larger-scale true earth materials under long-term stress. It can be seen, therefore, that the behavior of the earth as a whole, or some large part thereof, when subjected to great stress over long ages, is quite different from what one would normally expect.

## The Earth's Interior

Geophysicists, students of the physics of the earth, have ways of determining certain of the properties of the earth's interior. One way is through the study of earthquake waves. Forces within the earth are relieved when the crust breaks and rocks on one side of the break (fault) slip past those on the other. Such an event releases energy in the form of shock waves which travel through the earth. These waves follow paths controlled by the form of the wave and the nature of the earth

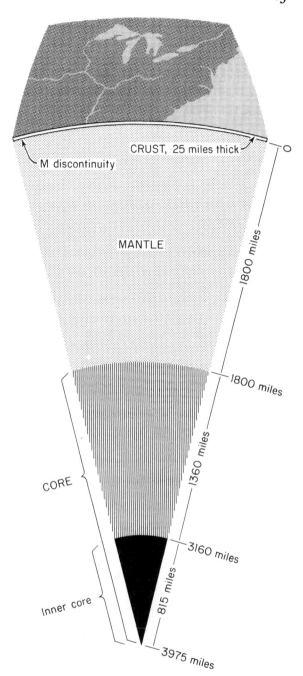

Figure III-3. Zones of the earth's interior beneath Michigan. (Modified from Zumberge, 1963, *Elements of Geology*, 2d Ed., permission of John Wiley and Sons.)

material that transmits them. When they reach the surface, or encounter an abrupt change in material at depth, the shock waves may be reflected or absorbed, or their paths may be altered due to a change in velocity. Seismologists, who are geophysicists specializing in the study of earthquakes, continually record the times and places of occurrence of earthquakes. Studying these records they can determine with considerable precision the density of the earth at various depths, the shape and position of the boundaries between materials of different density, and the strength of those materials.

Studies of the force of gravity, which changes from place to place at the surface, also enable geophysicists to locate and estimate the thickness of portions of the crust that may be composed of especially light or heavy materials.

Through such means the geophysicist can provide a broad general picture of the earth's interior. By setting off explosive charges to create small artificial earthquakes and by very careful gravity surveys they can also determine many details concerning local subsurface structures that are of interest to oil and mining companies. Figure III-3 shows that the crust is only a very thin "skin" compared with the total diameter of the earth. Beneath the continents the crust consists of a relatively light outer layer with a density of about 2.7 grams per cubic centimeter. Beneath this is a heavier layer with an average density of 3.0. These two layers are referred to as the "granitic" and "basaltic" layers, respectively, because their densities approximate those of granite and basalt rock. The crust beneath ocean basins lacks the granitic layer and the basaltic layer may also be somewhat thinner there.

Beneath the crust is the thick *mantle*, which increases in density inward (downward) toward the *core*. The core in turn is even more dense. Its outer part has some properties of a fluid; specifically, it fails to transmit the same type of earthquake shock waves that also fail to pass through fluids at the surface. However, under the great pressures of the interior this part of the core may not be fluid in the conventional sense. The inner part of the core may be solid again.

For our purposes, the most important thing to note in Figure III-3 is that the crust, with whose features we shall deal in the study of Michigan geology, is an exceedingly thin skin. Yet the deepest oil well drilled has failed to pass through this shallow zone, even though it reached a depth of 4.8 miles or just over 25,000 feet. Thus, our "surface" view of geology is really very limited. The very ancient and deep-seated Archeozoic and Proterozoic rocks beneath the bowl-like sedimentary rock layers of the Michigan Basin are well within the relatively thin crust. The National Science Foundation once proposed a project to drill a deep hole (the "Mohole" of popular reports) through the crust. The purpose was to sample the mantle below, but this will have to be done in some oceanic area where the granitic layer is absent and the basaltic layer even thinner than usual.

Another important point to note from Figure III-3 is that the granitic masses of the continents may be thought of as light blocks of the crust that "float" on the denser material below. This may seem difficult to believe, but recall the discussion of earth strength and earth models a few paragraphs back and the picture may become more plausible. Or think of a slab of granite rock floating in a kettle of molten iron and you will have a picture that is closer to the actual situation than if you think of a cork in water. Any good recent geology textbook will develop this point and the evidence for it in greater detail. At any rate the crust is capable of being pulled or pushed downward locally into the underlying more dense layers and of floating back up again when the stress is released. Thus, the forces of crustal deformation, whatever their source may be, can create intracratonic basins such as the Michigan Basin or even geosynclines of much greater depth. In Chapters VII and VIII dealing with the Pleistocene, or Glacial Age, in Michigan and with the history of the Great Lakes, we shall see that the great weight of a continental ice sheet also depressed the crust in the Great Lakes region and that the crust has been springing, or floating, back to its normal level since the ice melted. This particular instance of the gradual deformation of the earth's crust is of great importance in these chapters because it had a profound influence on the geologic history of the Great Lakes.

# IV
# Precambrian Eras

**Introduction**

The term "Precambrian" means all of geologic time before the Cambrian Period. Thus, the Precambrian consists of the Archeozoic (or Archean) and Proterozoic (or Algonkian) eras with their respective subdivisions. There are Precambrian rocks in the Canadian Shield that have produced radiogenic ages of about 3.5 billion years. Yet there are geologically (relatively) older rocks known from that region. Therefore, Precambrian time in the Canadian Shield, of which the western part of the Upper Peninsula of Michigan is a part, must have begun even earlier. According to current astronomical estimates, the earth is between 4 and 5 billion years old. If these estimates are correct, and if the Cambrian Period of the Paleozoic Era began about 600 million years ago, then Precambrian time was exceedingly long and included about seven-eighths of the earth's history. To make this statement a little more meaningful, think of human history in the same proportions. Written human history, including that of the ancient Egyptians, goes back about 6000 years. If we were to discuss a part of that history which, like the Precambrian, included seven-eighths of the whole, then we would deal with all well-recorded human events from the time of earliest Egyptians to the end of the Middle Ages.

Many aspects of Precambrian history are poorly understood, just as in human history there are great gaps in knowledge and very ancient events are often known only in the barest detail. There are many reasons why the Precambrian record is incomplete and difficult to interpret. Many periods of intense deformation of the earth's crust metamorphosed most Precambrian rocks to such an extent that their original characteristics and origins are obscured. Much of the rock record of Precambrian time was removed by erosion following periods of mountain uplift. Folding and faulting make it difficult to relate rocks of one area to those of another, even over short distances, particularly when the deformation was followed by extensive erosion. Varying amounts of crustal uplift and erosion have resulted in rocks of deep-seated origin of a particular age being exposed in one area while age equivalent rocks formed under entirely different conditions crop out elsewhere. Some geologists used to believe that rocks metamorphosed to the same degree were the same age.

# Geology of Michigan

Now we realize that separate and unrelated sets of rocks may have undergone the same amount of deformation at different times, so that degree of metamorphism is no longer regarded as a reliable criterion for age correlation. While fossils commonly are used for age correlation of younger rocks, fossils of Precambrian age are exceedingly rare, and those few that are found are so simple in structure or so badly deformed that they tell us very little about the age of the rock containing them or of the nature of life. Radioactive dating methods provide the best evidence for Precambrian rock ages, and some of the most significant radiogenic ages are shown on the left side of Figure IV-3.

Precambrian rock exposures in Michigan are largely limited to the western part of the Upper Peninsula. In some places there they may be difficult to see because they are covered by dense vegetation, soil, swamps, and glacial deposits. This also makes it difficult to trace rock formations from one place to another across the land surface. They are best known and understood where they are well exposed or have been extensively mined.

Each continent of the world has one or two very large areas whose surface rocks are mainly Precambrian in age, areas called *shields* (Fig. IV-1). The Canadian Shield of North America, centered around Hudson Bay, comprises much of Canada and extends southward into the Great Lakes

Figure IV-1. Precambrian "shields" of the world (in black) with associated stable regions (horizontal ruling). Precambrian rocks are extremely ancient and complex, ranging from 600 million to 4.5 billion years in age, and consisting mainly of igneous and metamorphic rock types. Figure IV-2 shows the Canadian Shield of North America in greater detail. (Modified from Moore, 1958, *Introduction to Historical Geology*, permission of McGraw-Hill Company.)

**Precambrian Eras**

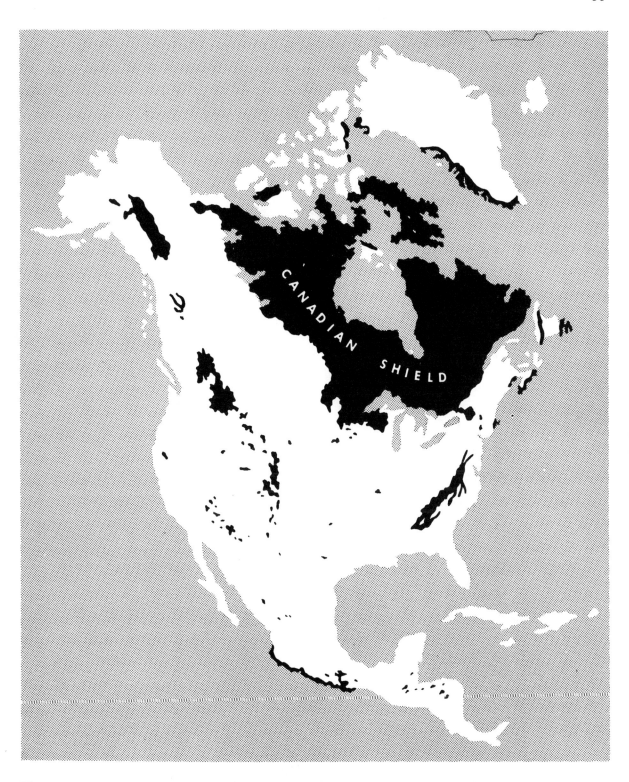

Figure IV-2. Areas of outcrop of Precambrian rocks in North America shown in black. Main area of exposure, centering around Hudson Bay, is the "Canadian Shield" (see Fig. IV-1). Note that the western half of the Upper Peninsula of Michigan falls within the shield and is an area where Precambrian rocks occur at the surface. Figure IV-33 shows this area in greater detail.

**Geology of Michigan**

34

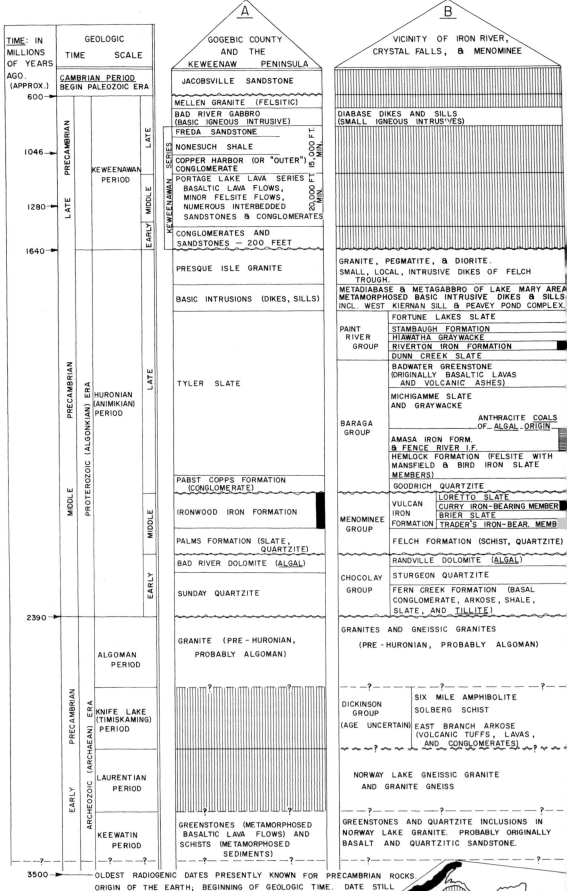

Figure IV-3. Precambrian rocks and geologic history of the Upper Peninsula of Michigan

**Precambrian Eras**

35

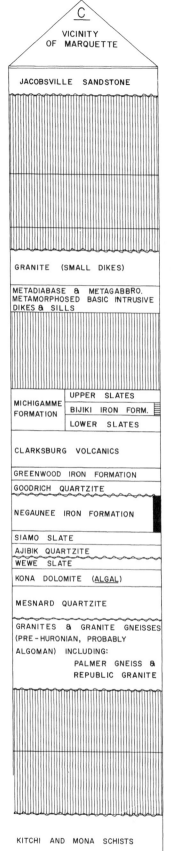

C
VICINITY OF MARQUETTE

JACOBSVILLE SANDSTONE

GRANITE (SMALL DIKES)

METADIABASE & METAGABBRO.
METAMORPHOSED BASIC INTRUSIVE DIKES & SILLS

MICHIGAMME FORMATION
- UPPER SLATES
- BIJIKI IRON FORM.
- LOWER SLATES

CLARKSBURG VOLCANICS
GREENWOOD IRON FORMATION
GOODRICH QUARTZITE
NEGAUNEE IRON FORMATION
SIAMO SLATE
AJIBIK QUARTZITE
WEWE SLATE
KONA DOLOMITE (ALGAL)
MESNARD QUARTZITE

GRANITES & GRANITE GNEISSES
(PRE-HURONIAN, PROBABLY ALGOMAN) INCLUDING:
PALMER GNEISS &
REPUBLIC GRANITE

KITCHI AND MONA SCHISTS

ROCKS OF THESE AGES NOT PRESENT
UNCONFORMITY
IRON FORMATION (I.F.)
MAJOR PRODUCER
MINOR PRODUCER

HISTORICAL EVENTS. INTERPRETED FROM ROCK RECORDS SHOWN TO LEFT. NUMBERS CORRESPOND TO BLOCK DIAGRAMS ILLUSTRATING THESE EVENTS IN FIGURE IV-5.

EROSION OF NORTHERN MICHIGAN (PENOKEAN) HIGHLANDS CONTINUES. PALEOZOIC ERA BEGINS.

15 - MINOR UPLIFT AND CONTINUED EROSION. WEAK CRUSTAL DISTURBANCE. KEWEENAWAN ROCKS TILTED BUT NOT METAMORPHOSED. UNCONFORMITY DEVELOPS.

14 - VOLCANISM. MANY SURFACE LAVA FLOWS AND SUBSURFACE IGNEOUS INTRUSIONS. SOME EROSION, TRANSPORTATION, AND REDEPOSITION OF CONGLOMERATIC MATERIAL BY STREAMS. LAKE SUPERIOR SYNCLINE SAGS TO RECEIVE THESE DEPOSITS. SOME GABBRO & SMALL AMOUNTS OF GRANITE INTRUDED INTO THE PILE. COPPER & OTHER MINERALS EMPLACED BY HYDROTHERMAL HYPOGENE ENRICHMENT.

13 - WIDESPREAD UPLIFT & MOUNTAIN BUILDING, ACCOMPANIED & FOLLOWED BY EROSION. UNCONFORMITY OF WIDESPREAD EXTENT DEVELOPS. NORTHERN MICH. HIGHLANDS FORMED AS PART OF MORE WIDESPREAD PENOKEAN RANGE.

12 - PENOKEAN OROGENY. INTENSE CRUSTAL FOLDING, FAULTING & LOW-GRADE REGIONAL METAMORPHISM. SLATES, GREENSTONES & QUARTZITES PRODUCED BY METAMORPHISM. IRON FORMATIONS STRONGLY FOLDED, FAULTED & LOCALLY METAMORPHOSED. COMPRESSIONAL FORCES DIRECTED TOWARD THE NORTH RESULT IN FOLDS & METAMORPHIC BELTS TRENDING EAST-WEST. SMALL LOCAL GRANITIC INTRUSIONS CROSS-CUT ALL PREVIOUSLY FORMED ROCKS INCLUDING BASIC DIKES & SILLS.
LOCAL BASIC INTRUSIONS IN FORM OF DIKES & SILLS.

PRIMITIVE PLANT LIFE IN FORM OF FLOATING MATS OF ALGAE WERE PRESENT IN SUFFICIENT ABUNDANCE TO PRODUCE LENTICULAR COAL SEAMS IN MICHIGAMME SLATE.

11 - SUBSIDENCE OF LAND. RENEWED SEDIMENTATION. DEPOSITION OF SOME IRON FORMATIONS. SOME SURFACE VOLCANIC ACTIVITY INCLUDING BASALTIC & FELSITIC LAVA FLOWS. GRAYWACKE DEPOSITS FORMED. OTHER SEDIMENTS, CONGLOMERATES, SANDSTONES, & SHALES. ENVIRONMENT OF DEPOSITION TECTONICALLY UNSTABLE. CRUSTAL UNREST INCREASING.

10 - MINOR UPLIFT & LOCAL EROSION OF PREVIOUSLY FORMED ROCKS PRODUCE LOCAL UNCONFORMITY.

9 - SUBSIDENCE OF LAND. SEDIMENTARY DEPOSITION IN SEAS. EXTENSIVE DEPOSITION OF IRON FORMATIONS RICH IN IRON COMPOUNDS & SILICA.

8 - MINOR UPLIFT & LOCAL EROSION OF PREVIOUSLY FORMED ROCKS PRODUCE LOCAL UNCONFORMITY.

7 - SUBSIDENCE OF LAND. SHALLOW SEAS ENTER. DEPOSITION OF BASAL CONGLOMERATES, FOLLOWED BY CROSS-BEDDED SANDSTONES & DOLOMITES IN TECTONICALLY STABLE, SHALLOW SHELF ENVIRONMENT.
SIMPLE ALGAL PLANT LIFE IN EXISTENCE.
(POSSIBLY SOME GLACIATION LOCALLY, SEE BLOCK 7a IN FIGURE IV-5)

6 - WIDESPREAD UPLIFT, MOUNTAIN BUILDING, EROSION, PENEPLANATION. UNCONFORMITY.

5 - ALGOMAN OROGENY. WIDESPREAD & INTENSE CRUSTAL DEFORMATION. HIGH GRADE DYNAMOTHERMAL REGIONAL METAMORPHISM OF ALL PREVIOUSLY FORMED ROCKS, INCLUDING ALTERATION OF ALGOMAN GRANITES TO GRANITE GNEISSES & GRANITIZATION OF ADJACENT ROCKS WHICH GRADE INTO THE GRANITES. FORMATION OF STRONG FOLIATION & LINEATION IN METAMORPHOSED ROCKS. FOLDING & FAULTING. N-S COMPRESSION PRODUCES E-W TRENDING FOLD & METAMORPHIC BELTS. WIDESPREAD, DEEP-SEATED GRANITIC MAGMAS INTRUDED (BATHOLITHIC MAGNITUDE) MAGMAS ENGULF MUCH OF PREVIOUSLY FORMED ROCK.

4 - SUBSIDENCE OF LAND & ENTRY OF SHALLOW SEAS. DEPOSITION OF BASAL CONGLOMERATES, ARKOSIC SANDS IN WAVE AGITATED WATERS. PEBBLES & SAND GRAINS DERIVED FROM GRANITE FORMED EARLIER, NOW BEING ERODED IN ADJACENT LAND. SOME EXPLOSIVE VOLCANIC ACTIVITY PRODUCING BASALTIC LAVAS & ASH (TUFF) LATER METAMORPHOSED TO SCHIST, AMPHIBOLITE, & GREENSTONE. EXACT AGE OF DEPOSITS IN DOUBT.

3 - LOCAL UPLIFT & SUBSEQUENT EROSION PRODUCE STRONG UNCONFORMITY.

2 - LAURENTIAN OROGENY. CRUSTAL DEFORMATION INCL. FOLDING & METAMORPHISM
LOCAL INTRUSION OF GRANITIC MAGMAS PRODUCE NORWAY LAKE GRANITE (LATER METAMORPHOSED TO GRANITE GNEISS) & OTHERS. GRANITE ENGULFS & ALTERS PREVIOUSLY FORMED ROCKS WHICH NOW OCCUR IN THE GRANITE IN THE FORM OF INCLUSIONS OF CHLORITE SCHIST (ONCE BASALTIC GREENSTONE) & QUARTZITE (ONCE SILICEOUS SEDIMENT)

1 - SEDIMENTATION & VOLCANIC ACTIVITY. SILICEOUS SEDIMENTS (LATER METAMORPHOSED TO QUARTZITE) INTERBEDDED WITH BASALTIC LAVAS (LATER METAMORPHOSED TO CHLORITIC SCHISTS & GREENSTONES) SOME OF THESE DEPOSITS LATER ENGULFED IN GRANITE MAGMAS MENTIONED IN NEXT EVENT ABOVE.

SPECIAL NOTE: THE PRESENCE OF SEDIMENTS & LAVAS IN EVENT NO. I IMPLIES THAT AN EVEN OLDER ROCK SURFACE EXISTED ON WHICH DEPOSITION COULD OCCUR. ALSO IMPLIED IS PRESENCE OF ERODING HIGHLANDS SOMEWHERE TO PROVIDE SEDIMENT THAT WAS DEPOSITED. THUS THERE MUST HAVE BEEN A PRIOR EPISODE OR EPISODES OF GEOLOGIC HISTORY OF UNKNOWN LENGTH, BEFORE THE FIRST RECORDED EVENTS. IT IS NOT KNOWN IF THAT EARLIER SURFACE & THOSE STILL EARLIER ROCKS WERE PART OF THE ORIGINAL EARTH'S SURFACE OR NOT. THE OLDEST RADIOGENICALLY DATED ROCKS FOUND SO FAR OUTSIDE MICHIGAN ARE 3.5 BILLION YEARS OLD, BUT EVEN THEY ARE METAMORPHIC TYPES DERIVED FROM SOME STILL OLDER MATERIAL. THUS THE EARTH'S AGE MUST EXCEED 3.5 BILLION YEARS & IS CERTAINLY GREATER THAN THAT OF THE OLDEST ROCKS YET FOUND IN MICHIGAN.

# Geology of Michigan

### Figure IV-4
### COMMON PRECAMBRIAN ROCKS AND ROCK STRUCTURES IN MICHIGAN

| Name | Characteristics | Mode of Origin and Historical Significance |
| --- | --- | --- |
| Algal | Clusters of roughly parallel or concentric bands found in limestone, dolomite, marble. | Formed by simple plants called algae. Evidence of very ancient simple life. |
| Amphibolite | Dark, heavy, hard, sometimes greenish. Parallel arrangement of mineral particles. Long, slim hornblende crystals common. | Metamorphic rock. Strong metamorphism affected either an iron-rich igneous rock like basalt or impure, chemically complex sedimentary rocks of mixed type. |
| Arkose | Sandstone containing 25 percent or more feldspar in addition to quartz. Mica common. Often pink or red. | Sedimentary rock deposited by streams that rapidly eroded a nearby granitic highland. Much vertical uplift and erosion required to expose the granite. |
| Basalt | Very finely crystalline, very dark, dull luster, heavy, often cellular. Rich in iron-bearing minerals such as hornblende, augite, or greenish olivine. | Formed as lava flow or small, near-surface intrusion. Implies volcanic activity. |
| Basic intrusive | An iron-rich, low silica, igneous rock. Usually with medium to coarse, visible crystals. Ex. *Gabbro*. | Subsurface magma cooled and solidified slowly. |
| Conglomerate | Coarse, rounded particles larger than 2 millimeters and often of many sizes. Minerals and rocks composing pebbles often of many types. | Sedimentary rock deposited from swiftly flowing water in streams, rivers, or along margins of beaches. Pebbles moved short distance from source. |
| Diabase | Finely crystalline, medium gray to black or greenish black. Mineral grain size between basalt and gabbro. Iron-rich (ferromagnesian) minerals such as hornblende, augite, olivine. | Molten magma cooled slowly beneath the surface. Sometimes formed at base of thick lava flows. |
| Dike | Small, sheet-like or wall-like igneous rock cutting across sedimentary or other rock structures. | Molten magma moved up along cracks or fissures but cooled beneath the earth's surface. |
| Dolomite | Crystalline grains may be visible or not. Commonly light to medium gray or buff. Soft, easily scratched by knife. Powder fizzes in dilute hydrochloric acid. $CaMg(CO_3)_2$ | Sedimentary. Mainly or entirely formed from limestone ($CaCO_3$) by later addition of magnesium. Most large deposits are marine chemical precipitates. |
| Felsite | Crystal grains below limit of visibility. Hard. Light color. Conchoidal fracture. Dull, stony luster. Contrasts with basalt in being low in dark, iron-bearing (ferromagnesian) minerals. | Surface lava flows or small, rapidly cooled subsurface intrusions. |
| Gabbro | Crystalline grains clearly visible, about equal size. Over $\frac{1}{2}$ dark minerals such as hornblende, augite. Massive. Hard. | Cooled slowly from iron-rich magma at great depth. Exposure at surface implies extensive subsequent uplift and erosion. |
| Gneiss | Coarse mineral grains clearly visible. Elongate grains roughly parallel. Bands of unlike minerals adjacent, often lensing or discontinuous. | Metamorphic. Intense heat and pressure at great depth altered preexisting rock such as granite; followed by extensive uplift and erosion. |
| Granite | Mineral grains clearly visible. Various minerals about equally spaced. Massive, hard. Mainly quartz and feldspar but mica or hornblende common. | Cooled slowly at great depth from molten magma. Exposure at surface implies extensive uplift and erosion to remove formerly overlying rocks. |
| Graywacke | Impure sandstones or conglomerates containing fragments of quartz and feldspar in finer clay matrix. | Sedimentary. Deposited in or near region of intense volcanic activity and rapid surface erosion. |

# Precambrian Eras

Figure IV-4—*Cont.*

COMMON PRECAMBRIAN ROCKS AND ROCK STRUCTURES IN MICHIGAN

| Name | Characteristics | Mode of Origin and Historical Significance |
|---|---|---|
| Greenstone | Finely crystalline, dark, heavy, greenish cast. Massive. Hard. Green due to mineral chlorite. | Metamorphosed diabase or basalt. Originally igneous. First a surface extrusive, then deeply buried, metamorphosed, then exposed again by uplift and erosion. |
| Metadiabase | Composition and characters of diabase, but with elongate crystals oriented roughly parallel to one another and often segregated into bands or folia according to mineral type. | Metamorphic. Intense heat and pressure applied at great depth to diabase. Originally of igneous origin. Events include: intrusion, metamorphism, uplift, and erosion to produce surface exposure. |
| Metagabbro | Composition of gabbro but mineral crystals oriented in one direction. Similar minerals often aggregated into bands. | Metamorphic. Intense heat and pressure applied at great depth to gabbro. Originally of igneous origin. Events include: intrusion, metamorphism, uplift, and erosion to produce surface exposure. |
| Peridotite | Coarse, distinct, mineral grains, evenly spaced. Very heavy, massive. Hard. Black or greenish black. Mainly or entirely ferromagnesian mineral such as hornblende, augite, or olivine. | Deep origin from molten magma. Exposure at surface implies extensive uplift and erosion to remove formerly overlying rocks. |
| Quartzite | Very hard, tough, light to medium colored. Composed mainly of quartz. Granular texture but rock breaks across or through grains, not around them. | Metamorphic. From quartz-rich sedimentary rock. Events include: deposition in water, deep burial and compression, subsequent uplift and erosion of overlying rocks. |
| Sandstone | Angular to rounded fragments of minerals or pieces of preexisting rocks. Particles from lower limit of visibility to pea-size. Stratification often present. Grains commonly quartz but may include other minerals. | Sedimentary. Water or wind deposit. Commonly laid down in streams and rivers, along shorelines, in dunes. Special characters may indicate environment of origin more exactly. |
| Schist | Small, barely visible grains of elongate or platy minerals such as mica, chlorite, talc, or hornblende. Crystals parallel to one another, imparting lustrous, finely laminated, flaky appearance. Tendency to split in one direction. Soft. Quartz and garnet often present. | Metamorphic. Preexisting rocks subjected to heat and pressure under confinement at great depth. Events implied include: formation of original rock, deep burial, metamorphism, uplift, and erosion to form surface exposures. |
| Shale | Dull, soft, invisible grains. Fine layering. Even surface texture. Clayey odor when moistened. Various colors but commonly gray, black, or red. | Sedimentary. Deposited in quiet waters, as in offshore depths of seas and lakes, quiet parts of rivers, floodplains, swamps, tidal flats. |
| Sill | Thin, flat, igneous intrusive between layered structures of preexisting rocks as between sedimentary strata. | Small extensions of magma from larger intrusion. Cooled and solidified at depth. Surface exposure implies uplift and erosion. |
| Slate | Crystalline grains barely or not visible. Homogeneous. Slightly lustrous surface. Splits into very even, smooth, tough slabs that ring when tapped. | Metamorphic. From shale. Surface exposure implies: deposition of shale as surface sediment, deep burial, metamorphism, uplift, and erosion. |
| Tillite | Well-consolidated rock of various sized fragments of various types. Many pieces with smooth, flat faces, surface scratches, or grooves. | Originally a glacial ice deposit. Strongly cemented later. Implies ancient glaciation. |
| Volcanics | Include lava flows, ash, and fragmental pumice, breccia, bombs. | Nearby surface volcanic eruptions. |

# Geology of Michigan

Region where it makes up large parts of Minnesota, Wisconsin, and northwestern Michigan (Fig. IV-2). Geologists may refer to a shield of this sort as a *nucleus,* meaning center of growth. Some believe that the shields went through their periods of most active growth during the Precambrian and since then have tended to remain stable, relatively high-standing, and essentially undeformed.

In contrast, Precambrian rocks also crop out in places outside the shields, but the exposures elsewhere are relatively small. Generally, they are found in regions of great uplift and erosion such as the Appalachian Piedmont and Blue Ridge and the ranges of the Rocky Mountains, or in the depths of such great river valleys as the Grand Canyon of the Colorado, the Royal Gorge of the Arkansas, or the Black Canyon of the Gunnison rivers. In such places extensive local uplift has enabled running water to erode deeply into the earth's crust, removing great thicknesses of overlying rock to expose the deeply buried Precambrian rocks below.

The Precambrian rocks of the Great Lakes Region have been carefully studied for many years and more is known about them than is known about Precambrian rocks in most other parts of the world. Still much of their history is uncertain or unknown and their relationships with the Precambrian rocks elsewhere on the continent and throughout the world are very poorly understood. Therefore, the Precambrian Time Scale in this chapter and the names on it (Fig. IV-3) apply only to Michigan and a very limited surrounding area of the northern Great Lakes Region.

## Precambrian in Michigan

First, let us recall why the most ancient rocks of Michigan crop out mainly in the western part of the Upper Peninsula. The two cross sections in Figure III-2 in the preceding chapter show the subsurface structure of the Lower Peninsula. Note the earth's crust is warped downward beneath the Lower Peninsula so that Precambrian rocks are most deeply buried beneath the center of the state; from there the rock layers are inclined upward and outward. Thus, the older rocks come closer to the surface around the margins of the state. However, it is only in the north that uplift and erosion have been extensive enough to strip away all the younger rocks to bare the Precambrian. This uplift to the north is part of the general rise of the Canadian Shield, of which the western half of the Upper Peninsula forms a very small part. Ancient rocks of the Precambrian form the so-called "basement" beneath the remainder of the state, as well as beneath the rest of the United States south of the Canadian Shield. Very little is known about the buried Precambrian basement in Michigan. Oil wells, the only source of direct information in the form of rock samples, rarely penetrate to the basement because oil is rare in rocks of Precambrian age. Exploration and exploitation of ores and minerals which may occur in the Precambrian basement would not be practical or profitable at present because those rocks are too deeply buried. Geophysical studies suggest that some parts of the Lower Peninsula are underlain by Precambrian rocks of relatively high density similar to the basalts and gabbros of the Keweenaw Peninsula. This was recently borne out by rock cores taken from Precambrian rocks encountered in an oil well drilled in St. Clair County north of Detroit.

Any part of the general geologic time scale shown inside the front cover of this book could be expanded to show greater detail for selected portions of the earth and time. Figure IV-3 does so for the Precambrian of Michigan. The names of the rock units, such as formations, are arranged in conventional fashion from oldest at the bottom to youngest at the top. Sedimentary rocks, and metamorphic rocks once sediments, would occur this way in nature if all were present, undisturbed, lying one above another in a single place. That is to say, they would obey the Law of Superposition (Chapter I). However, mountain uplifts, tilting, folding, faulting, metamorphism, and erosion all have upset the ideal picture so that the age relationships of Precambrian rocks now are very difficult to determine. Rocks that originally were of igneous origin, such as granites, diorites, gabbros, and basalts, also had complicated later histories. The relative ages of igneous rocks that cooled from a molten magma are determined by studies of their cross-cutting relationships with associated rocks (Fig. I-5d). An igneous intrusion, such as a dike, is younger than all other rocks it cut across as it rose through the earth's crust, and it is older than rocks subsequently deposited on any eroded part of it. Such relations are simple in concept but difficult to observe actually; to establish they require careful professional study in the field.

The names of the eras, periods, and other standard Precambrian subdivisions of geologic time used in the Great Lakes Region are shown along the left side of Figure IV-3. The names used in Michigan for the rock formations are shown in separate columns for three general areas of the Upper Peninsula. This breakdown is necessary because the Precambrian histories of the different areas, even in such close proximity as these, differed in details. The environments of deposition were quite varied, and the rocks that are preserved in the different areas after such a long span of geologic time are not always the same. Furthermore, formerly deep-seated rock types have been uplifted and exposed at the surface in some places, but not in others.

Those not familiar with the kinds of rocks shown on Figure IV-3, may find Figure IV-4 helpful. The common Precambrian rock types are listed alphabetically there, with a brief description and a statement concerning the historical significance of each type. The righthand column on Figure IV-3 lists the main historical events that can be inferred from the rock record shown on that chart. Figure IV-5 portrays those same historical events diagrammatically. Let us review the Precambrian history of Michigan by means of those visual aids.

Nearly 3.5 billion years ago, during Keewatin time, streams from bordering highlands deposited sediments and lavas poured out on the surface of the earth in what is now the Upper Peninsula of Michigan and over a broader area as well. Some of those materials, now metamorphosed to schists and greenstones, are represented today by the Kitchi and Mona formations in the Marquette area (Fig. IV-6, 7). Perhaps some of the sediment was deposited in ancient seas, while other sediment was deposited in continental basins. The Mona formation includes *pillow structures* which formed when molten lava flowed out under water. The fact that some of these oldest known rocks were formed as sedimentary deposits implies that even older rocks, of which we have as yet no direct knowledge, must have been in existence to provide the source for the eroded sediment. Lava flows, too, had to have a surface upon which to flow. Perhaps those older and as yet undiscovered rocks were part of the earth's original crust, but this is uncertain; even they may have formed long after the actual beginning of the earth. Keewatin age lava flows, now in the metamorphosed form of greenstones, also are found today in the Gogebic and Dead River Basin areas farther west. Evidence of Keewatin sedimentation and volcanic activity in Dickinson County is slight but important. The Norway Lake Granite Gneiss there has in it some inclusions of chlorite schist (at one time probably basalt) and quartzite (at one time a quartz sandstone). This means that earlier lavas and sediments were engulfed by the magmas that formed the granite. The inclusions are older than the magma that flowed around them (James and others, 1961, p. 12-13).

After the period of Keewatin sedimentation and lava flows, molten magmas began to rise more extensively into the crust in some parts of the Canadian Shield, cutting across and locally obliterating the previously formed Keewatin rocks. Much of the molten magma cooled slowly at great depth to form the ancient granites of Laurentian age. One of these may be that in Sugarloaf Mountain north of Marquette. Few granites of this age have been identified with certainty in Michigan. The Norway Lake Granite Gneiss of Dickinson County, mentioned above, may be Laurentian in age because it contains inclusions of what may be Keewatin rocks, and because the East Branch Arkose found in that same area is a sedimentary deposit of younger age that contains pebbles of that granite (Fig. IV-8). Thus, the Norway Lake Granite Gneiss would be younger than the inclusions caught in it but older than the sedimentary rock formed of pebbles eroded from it. This is a good example of the tenuous line of reasoning that often is involved in an attempt to date and correlate a Precambrian rock formation. You might ask why we have gone to such great pains to discuss a rock formation so poorly dated as the Norway Lake Granite Gneiss. Examine Figure IV-3 again and you will see that if we have been correct in our reasoning and if the data we have used are correct, then that granite is the only rock now known in Michigan that represents the Laurentian Period of the Archeozoic Era. Future discoveries may provide firmer evidence of the geologic events of that time.

Moving on to the next period of the Archeozoic Era, which is the Knife Lake (or Timiskaming) Period we have a similar problem. There is some

12—*Penokean Orogeny.* Widespread deformation, metamorphism, and mountain building.

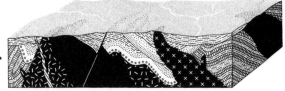

13—Continued uplift and erosion. Northern Michigan Highlands important.

11—*Late Huronian.* Subsidence, sedimentation, some lavas, some iron formation.

10—Minor uplift and erosion.

5—*Algoman Orogeny.* Widespread deformation, metamorphism, intrusions, mountain uplift.

6—Extensive erosion.

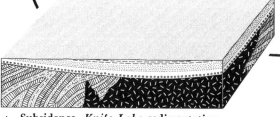

4—Subsidence. *Knife Lake* sedimentation.

3—Erosion.

14—*Keweenawan* lava flows and sediments. Lake Superior Syncline sags.

15—Minor uplift, tilting, and erosion. Precambrian time ends.

9.—*Middle Huronian*. Subsidence, marine sedimentation, origin of major iron formations.

8—Minor uplift and erosion.

7—*Early Huronian*. Subsidence and marine sedimentation.

7a.—Local glaciation.

2—*Laurentian Orogeny*. Local deformation, metamorphism, intrusions, and mountain building.

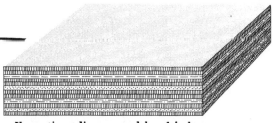

1—*Keewatin* sediments and basaltic lavas.

Figure IV–5. Generalized block diagrams illustrating sequence of events in Precambrian geologic history of Upper Peninsula of Michigan. Numbers on blocks correspond to numbered events described in greater detail in right-hand column of Figure IV–3.

# Geology of Michigan

Figure IV–6. *Top*—Light-colored Mesnard Quartzite (right) downfaulted against Mona Schist (greenstone); both Precambrian in age. Quarry at north edge of Harvey along US 41 about 3.5 miles southeast of Marquette. Geology students at fault show scale. Fault position and movement indicated by dashed line and arrows.

*Bottom*—Clarksburg Volcanics (Late Huronian-Middle Precambrian). Light-colored layers of ash interbedded with thin lava flows, all subsequently metamorphosed to garnet grade. Note offset due to prelithification slumping. North side US 41 just west of Middle Branch of Escanaba River, 3 miles east of junction of US 41 and M–95.

doubt whether any rocks of that period occur in Michigan, although rocks of that age do occur to the northwest across Lake Superior in Minnesota. Examine Column B of Figure IV–3 again and you will find we have placed the Dickinson Group in that period. The evidence for this age assignment is as follows: (a) the East Branch Arkose of that group has pebbles in it that were eroded from the Norway Lake Granite Gneiss, and (b) the Dickinson Group grades into another, younger granite that seems to have crosscut it at a later time and that is thought to be Algoman in age. The relationships of these rock formations to one another are difficult to determine in the field, but if they are correctly interpreted here, then the Dickinson Group, including the East Branch Arkose, is younger than a Laurentian age granite (Norway Lake) and older than an Algoman age Granite (Fig. IV–9). This could mean that the Dickinson Group is Knife Lake in age. However, if the data we have are not correctly interpreted, then we have built an historical "house of cards" which may topple when more reliable evidence is found. Studies of Precambrian rocks often involve such uncertainties.

Regardless of whether or not the sediments of the Dickinson Group accumulated in Knife Lake time, they have been metamorphosed since the time of their deposition. The Six Mile Amphibolite in that group once was a basaltic lava, the Solberg Schist was a fine grained sedimentary rock, and the East Branch Arkose a coarse-grained, sedimentary sand and gravel or conglomerate. The arkose, with its granite pebbles and fragments of feldspar minerals, indicates that somewhere nearby stood a highland that had been uplifted

Figure IV-7. Pillow structure formed in underwater lava flows, now the Mona Schist greenstones of Keewatin (Early Precambrian) age. Roadcuts along US 41 about 1 mile east of junction with M-35 and about 7 miles west of Marquette. Bedrock here, as in most places in Michigan, is covered by at least a local veneer of Pleistocene age glacial deposits, but bedrock surface itself often is smoothed, grooved, and striated as a result of the scouring action of overriding glacial ice. (Photos on right courtesy E. W. Heinrich, The University of Michigan Department of Geology and Mineralogy.)

and deeply eroded so that once deep-seated granitic rock was exposed at the surface. Streams then could pick up the granite fragments, transport them a short distance, and deposit them as an arkosic sediment. If our earlier reasoning concerning the age of these rocks is correct, then the Dickinson Group represents a second cycle of sedimentary deposition, the Keewatin rocks being the first, and the two separated in time by a period of granitic intrusion of Laurentian age. The East Branch Arkose is the basal conglomerate that formed at the beginning of that second cycle, followed in turn by the Solberg sediments and then by the Six Mile lava flows.

The rocks of Keewatin and Knife Lake age locally are thermally metamorphosed and crosscut by granitic rocks younger than the Laurentian granites. These younger granites have been con-

**Geology of Michigan**

Figure IV–8. Cross-bedding in East Branch Arkose of the Dickinson Group (late Early Precambrian), near Norway, Michigan. Matchstick for scale. At time of deposition, cross-beds are concave upward and their open, truncated ends are "up" as illustrated in Figure I–1. This photograph, looking vertically down onto a horizontal outcrop surface, shows that these Precambrian age cross-beds have been steeply tilted and now stand on end. This resulted from strong deformation of the earth's crust, during Precambrian time, but after the beds were deposited. (H. L. James, et al., U.S. Geological Survey Professional Paper 310, Fig. 7.)

sidered to be Algoman in age. They represent a second period of igneous intrusion accompanied and followed by intense crustal disturbance (compression and faulting). All this crustal activity comprised what is called the Algoman Orogeny (Fig. IV–3, righthand column). During that orogeny, dynamothermal metamorphism, resulting from great pressure and heat, altered all preexisting rocks of the region including the Algoman granites themselves. The metamorphism was widespread or "regional" in its effect. Mountains were uplifted in the course of this great crustal disturbance. Subsequent erosion removed much of the rock material formed prior to Algoman time, resulting in a major gap or unconformity in the record between Algoman or older rocks, and younger rocks of Huronian and Keweenawan age. The great Algoman Orogeny occurred about 2.4 billion years ago, as is indicated at the left on Figure IV–3. The Precambrian events discussed thus far are portrayed in block diagrams 1–6 in Figure IV–5.

The Huronian Period, a long time of varied sedimentary deposition, followed the Algoman Orogeny. Some refer to this period as the "Middle Precambrian" or the Animikie Period. A glance at the table of rock formations will give some idea of the complex history of this period. The Huronian began with the deposition of basal conglomerates and sands, probably of marine origin. These are the Mesnard and Sunday quartzites and the Fern Creek Formation. Note the several minor unconformities (indicated by wavy lines on Figure IV–3) resulting from local crustal warping and erosion which occurred at several different times. Formations such as the Clarksburg Volcanics in the

**Precambrian Eras**

Figure IV–9. Coarse-grained and porphyritic pre-Huronian (Early Precambrian) age granite. Pencil for scale in close-up. Roadcut along M-95, 3.5 miles south of junction with US 41. D. F. Eschman in top photo standing on glacially planed bedrock surface. Coarse texture of granite implies slow solidification of original molten magma.

# Geology of Michigan

vicinity of Marquette, and the Badwater Greenstones and Hiawatha Graywacke of Iron and Dickinson counties, represent times of minor volcanic activity; but, by and large, the Huronian was a time of sediment accumulation. The Sturgeon, Sunday, and Mesnard quartzites, which were deposited early in the Huronian, originally were siliceous sediments, later metamorphosed to quartzite. Their exact environment of origin is still in doubt. They may have been either marine sands or river laid deposits, probably the former. In either case, however, they exhibit cross-bedding which resulted from current flow. Pettijohn (1957) studied the direction of original inclination of the crossbeds and concluded that the currents responsible for their formation generally flowed toward the southeast.

The various "Iron Formations" (indicated by special symbols on Figure IV-3) are of particular importance because they are sources of the iron ore which contributes so much to the economy of the state of Michigan (Figs. IV-12, 13). Details concerning iron ore origin, distribution, and production are presented in a later section in this chapter.

The Fern Creek Formation at the base of the Huronian in Iron and Dickinson counties also is

Figure IV-10. Fern Creek Tillite of Early Huronian (early Middle Precambrian) age, nearly 2.4 billion years old, along Fern Creek near Norway, Michigan. Unsorted, unstratified characters of deposit, originally a mixture of granitic cobbles, pebbles and clay now strongly lithified, suggest deposition by glacial ice. Compare with more recent glacial tills or "boulder clays" shown in Figures I-1 and VII-7, 17 and 18. This interpretation is basis for showing a period of glaciation on Figures IV-3 and 5. However, not all geologists accept the glacial origin of this rock unit; mudflows have similar characters. Further proof of glacial origin would include finding faceted and striated cobbles within, and a glacially scoured bedrock surface below, this deposit. (Courtesy F. J. Pettijohn, Department of Geology, The Johns Hopkins University, Baltimore, Maryland.)

# Precambrian Eras

47

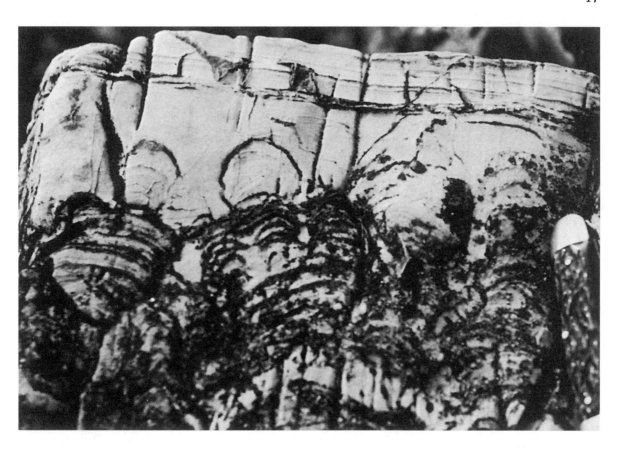

Figure IV-11. Laminated algal deposits in the Randville Dolomite of Early Huronian (early Middle Precambrian) age, in Dickinson County. Pocket knife for scale. See Figures XV-3, 4, and 5 as well. (Courtesy H. L. James, *et al.*, *U.S. Geological Survey Professional Paper* 310, Fig. 32.)

of particular interest because it includes a deposit thought by some geologists to be an ancient *tillite* (Fig. IV-10). Tillite is strongly consolidated (lithified) *till*, an unsorted deposit of many particle sizes, some of which bear smoothed, flattened, and scratched (striated) faces. Till, in turn, is a deposit laid down directly by glacial ice. Therefore, this deposit, and others like it in Ontario, may be the earliest evidence of glaciation in this region. This early glaciation occurred nearly 2.4 billion years before the Pleistocene Ice Age with which most people are familiar.

Although the Huronian rocks now are highly metamorphosed to slates and quartzites, most of them originally were sedimentary rocks, some of which were probably deposited in marine waters. However, the absence of diagnostic marine fossils in these formations makes a marine origin difficult to prove conclusively. The principal evidence for the existence of seas in this area during the Huronian comes from such formations as the Kona, Randville, and Bad River dolomites of Early Huronian age. Dolomite, which is calcium magnesium carbonate, may be deposited directly in seas or may result from the secondary addition of magnesium to limestone. In either case, widespread deposits of this type today originate in a marine environment. The slates (once shales) and iron formations associated with those dolomites also could have had a marine origin.

The *algal structures* found in the Kona, Randville, and Bad River dolomites (Figs. IV-11, XV-3)

Figure IV-12. "Iron Country," vicinity of Ishpeming, in the Marquette Iron District. View from Jasper Hill. Old, abandoned, surface mine pits in foreground. Headframe of underground mine shaft on skyline.

Figure IV-13. Small exposure of Bijiki Iron Formation in the Michigamme Formation of Late Huronian (late Middle Precambrian) age, along side road just north of US 41 near old Champion Mine at Champion. Note slightly folded, alternating iron-rich and siliceous sedimentary rock layers, now metamorphosed.

# Precambrian Eras

Figure IV-14. Siamo Slate of Middle Huronian (Middle Precambrian) age, on south side of US 41 at small park 0.7 miles east of junction of US 41 and M35, about 2.5 miles east of Negaunee. Area of close-up indicated near left edge of upper photo. These rocks, originally sedimentary clay shales, siliceous sandstones and siltstones, were metamorphosed during the Penokean Orogeny and now are slate and quartzite. Directions of original sedimentary stratification and secondary metamorphic foliation (slaty cleavage) are indicated.

Figure IV-15. Anticlinal fold in stratified iron formation of Huronian (Middle Precambrian) age. Taylor Iron Mine, Baraga County. Folding occurred during Penokean Orogeny (see Fig. IV-3).

are especially interesting. Very similar structures are formed today by the activity of lime-secreting algae, simple forms of plant life that precipitate calcium carbonate from water. These structures indicate that plant life, at least in very simple form, existed in Michigan nearly 2 billion years ago. Similar algal structures found in Precambrian rocks in other parts of the world in some cases are even older, those from rocks in South Africa being at least 2.7 billion years old.

Huronian time in Michigan was brought to a close by another episode of igneous intrusion and crustal deformation called the Penokean Orogeny (see event number 12 on Fig. IV-3). Emplacement of small dikes and sills of iron-rich basic igneous intrusives, such as diabase and gabbro was followed by small, local intrusions of granite and of pegmatite dikes (Fig. IV-16). Crustal compression then folded, faulted, and metamorphosed the Huronian rocks (Figs. IV-14, 15). Of course the older Archeozoic rocks that already had suffered earlier periods of metamorphism were again affected so that their structural complexity became even greater and their original characteristics even further altered. Evidence from rocks in Dickinson County (James, 1961), and from Marquette County as well, suggests that compressional forces acted from south to north, forming troughlike downfolds (synclines) and uplifts whose long axes trend east-west. The Penokean Orogeny was widespread throughout the Canadian Shield and produced mountainous highlands in many areas (Fig. IV-

# Precambrian Eras

Figure IV-16. Penokean (late Middle Precambrian) igneous dike of dark granite intruded into and crosscutting Randville Dolomite in Metronite Quarry, about 2.6 miles east of Felch on side road north of M-69. See Mineral Collecting Locality No. 57 in Chapter XI. (Courtesy H. L. James, et al., U.S. Geological Survey Professional Paper 310, Fig. 67.)

17). Radiogenic dating places the time of that orogeny at about 1.64 billion years ago (Fig. IV-3). Although erosion immediately began to reduce the Penokean Highlands, they persisted as sources of sediment into the following Keweenawan Period and even into the Cambrian Period of the Paleozoic Era, when they supplied some of the sediment to form the Cambrian sandstones of the Lake Superior Region (Fig. IV-41 and Chapter V).

The great unconformity separating rocks of the Huronian or Middle Precambrian from those of the following Keweenawan or Late Precambrian time indicates a long period of erosion. The main exposures of rocks of Keweenawan age extend along the northwestern edge of the Upper Peninsula, southeast of Lake Superior, from Wisconsin through the Porcupine Mountains and into the Keweenaw Peninsula. Isle Royale also is composed of rocks of Keweenawan age. Except near the Wisconsin border, the Keweenawan rocks are separated from the older Huronian and Archeozoic formations by a broad outcrop belt of much younger Early Cambrian deposits called the Jacobsville Sandstone (Figs. IV-18, V-5). The Keweenawan Period was a time of extensive surface volcanic activity. A thick series of lava flows resulted when extensive subsurface intrusions periodically reached the surface and flowed out upon it (Fig. IV-23). Most of the lava flows are of the basaltic type, although some felsites, rocks with a chemical and mineral composition closer to that of granite, occur. Basalt is a relatively heavy, dark, finely crystalline extrusive igneous rock commonly derived from lavas the world over. It is rich in ironbearing (ferromagnesian) minerals such as hornblende, augite, and olivine, but these are usually in the form of crystals too small to be identified without a microscope. Basalt is low in free silica (quartz). The finely crystalline texture is the result of rapid cooling. As is the case in modern lava flows, heat loss into the atmosphere was very rapid when thin sheets of molten magma (lava) flowed out upon the earth's surface. In many instances the Keweenawan lavas solidified so quickly that bubbles of enclosed gas were entrapped, thus forming small cavities called *vesicles*. *Vesicular basalt* is the term applied to rocks in

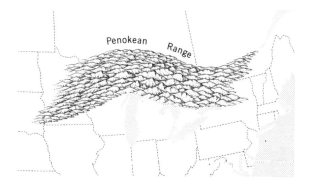

Figure IV-17. Highly generalized paleogeographic restoration of Penokean Range resulting from orogeny and uplift at close of Middle Precambrian time (see Fig. IV-3). Mountains in what now is northern Michigan are referred to in text as the "Northern Michigan Highlands." Vestiges of these ranges locally persisted into Cambrian time at the beginning of the Paleozoic Era. (See Chapter V.)

# Geology of Michigan

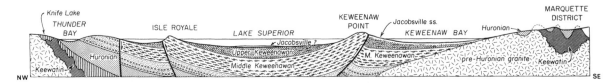

Figure IV–18. Generalized geologic cross section through Lake Superior, Keweenaw Peninsula and Keweenaw Bay, showing Lake Superior Syncline and Keweenaw Fault. (Modified from Leith, Lund, and Leith.)

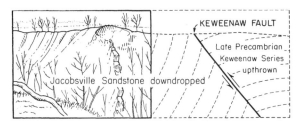

Figure IV–19. Strata of Early Cambrian age Jacobsville Sandstone standing nearly vertically where downdropped and dragged steeply upward along Keweenaw Fault. Fault itself is out of picture to right (northwest), as indicated in sketch. View southwest at what is known locally as the "Natural Wall," along Stable Road about 3.5 miles east of Laurium in the Keweenaw Peninsula. Stairstep appearance of sandstone at right due to joint blocks naturally weathered and broken out.

Figure IV–20. Oblique air view from commercial airliner, looking southwest down all but tip of Keweenaw Peninsula. Geologic features identified as referred to in text. Lower photos, taken along Brockway Mountain Drive at points A and B in lower left corner of air photo, show great ridge (Brockway Mountain) composed of "Middle Conglomerate" of the Late Precambrian age Keweenawan Series. Views, taken looking southwest, show how formations dip northwestward toward Lake Superior and down into the Lake Superior Syncline (see Fig. IV–18). Erosion-truncated ends of formations form southeast facing cliffs.

Figure IV-21. Lake of the Clouds scenic overlook in Porcupine Mountains State Park. People stand on Keweenawan (Late Precambrian) lava flow of amygdaloidal basalt whose eroded end forms cliff face. Extension lines show northwestward dip of beds downward toward Lake Superior Syncline. (Michigan Conservation Department, photo by Robert Harrington.)

Figure IV-22. Close-up of basaltic rock at feet of people in Figure IV-21. Pennies for scale. Dark specks are amygdules (later mineral fillings) in former volcanic gas bubble holes.

**Precambrian Eras**

Figure IV-23. Steeply tilted lava flow in Keweenawan (Late Precambrian) Portage Lake Lava Series. Rock is basalt. Top of flow at left indicated by line. Base of flow is at right of photo. Arrow shows location of close-ups in Figure IV-24. Quarry about 0.5 miles north of center of town of South Range, along west side of M-26.

Figure IV-24. Basalt at top of lava flow in quarry at South Range (see Fig. IV-23). Light, blotchy amygdules are later mineral fillings of former volcanic gas bubble holes. Gas rose toward top of flow while basaltic magma was still fluid. Arrow on Figure IV-23 shows location of close-ups.

which such cavities are very numerous. If these cavities were later filled with mineral matter, such as calcite or copper, the fillings are called *amygdules* and the rock is then termed an *amygdaloidal basalt* (Figs. IV-22, 24). Such basalts are discussed more fully in a later section of this chapter dealing specifically with the copper deposits of Michigan.

As pointed out above, Keweenawan lavas were derived from molten magmas of the "basic" or ferromagnesian type that welled up from far below the earth's surface. Some of the magma failed to reach the surface and instead slowly solidified at great depth. The slow cooling resulted in a coarsely crystalline texture, and the rock thus formed is called *gabbro*. Gabbro and basalt are chemically and mineralogically alike, although they differ in texture because of their different rates of solidification. The Bad River Gabbro of the Keweenaw Peninsula, and possibly some of the dikes and sills found in Iron and Dickinson counties, are examples. A much larger mass of gabbro, several thousand feet thick, crops out in the vicinity of Duluth, Minnesota, where it has been exposed by erosion. This mass, called the *Duluth Gabbro*, also was the result of the extensive Keweenawan igneous activity.

Not all Keweenawan rocks are igneous in origin. Conglomerates and sandstones were deposited in the Lake Superior Region in Early Keweenawan time, the lava flows beginning in the middle part of the period. Between eruptions of lava, local but thick patches of conglomerate and sandstone, generally red or purple in color, were deposited (Figs. IV-25, 26, and 27). The great Outer Conglomerate exposed along the northwest edge of the Keweenaw Peninsula is an example of these conglomeratic stream deposits. Much of the lava seems to have risen through fissures and cracks that opened to the surface in the downwarped area of the Lake Superior Syncline (Fig. IV-18). The interbedded conglomerates were deposited by streams that eroded the lavas and laid down the rounded pebbles and cobbles of volcanic rock in their valleys. The geography of that region was strikingly different then than it is today.

According to W. S. White (1960b), who studied the orientation and direction of elongation of gas bubble cavities in the Keweenawan flows, the lavas were flood basalts which emerged at the surface from fissures and other vents near the center of the Lake Superior Syncline basin and flowed radially *outward* toward the margins of the basin. In contrast, his studies of pebble imbrication, crossbedding, facies changes, and the local abundance of pre-Keweenawan rock derivatives suggest that the Keweenawan conglomerates were deposited by streams that flowed *toward* the basin center across the surface of the basalts. This is confirmed by Hamblin and Horner (1961) who conclude that the principal source for the conglomerates lay relatively close by to the southeast, probably in the general area now occupied by the Huron Mountains. They also conclude that the source for the conglomerates on Isle Royale lay to the northwest in Minnesota and Canada.

The total thickness of Keweenawan rocks in the Lake Superior Region may reach 50,000 feet. As this great mass of material accumulated, layer after layer, the earth's crust beneath sagged still farther, deepening the Lake Superior Syncline. Although Lake Superior itself did not form until over 800 million years later, this structural depression favored the future erosion of a surface basin there. As the crust sagged, the layers of basalt and interbedded sediments were bowed downward. Along the southeast side of the syncline, in the Keweenaw Peninsula, the tilted layers now dip downward toward the northwest at an average angle of about 40 degrees from the horizontal. Their dip direction is reversed on the northwest side of the syncline on Isle Royale and in Minnesota. Much later erosion beveled the edges of those steeply inclined layers. With further erosion, stream valleys were incised into the exposed edges of the softer layers, whereas the harder layers were left standing in relief as long, parallel ridges (Figs. IV-20, 21). These ridges now extend along the length of the Keweenaw Peninsula and continue southwestward through the Porcupine Mountains. Today most of the main roads of the Keweenaw Peninsula follow the long valleys carved by erosion into the relatively soft rock layers. A notable exception is the beautiful Brockway Mountain Drive. This scenic route between Copper Harbor and Eagle Harbor climbs onto a great ridge of erosion-resistant conglomerate in the Keweenaw Series and proceeds for some distance along the crest, reaching a high point at the main scenic pullout station about 735 feet above the present level of Lake Superior (Figs. IV-20a, 20b). Excellent exposures of the Keweenawan basalts and conglomerates

**Precambrian Eras**

57

Figure IV-25. Copper Harbor Conglomerate in upper part of Keweenawan (Late Precambrian) Series, dipping northwestward beneath Lake Superior. At Dan's Point along lake shore drive (M-26) just west of Copper Harbor.

Figure IV-26. Close-ups of Copper Harbor Conglomerate in Keweenawan Series at Dan's Point and Devil's Washtub, just west of Copper Harbor. Cobbles by hammer were eroded, rounded, and deposited by ancient streams flowing over Keweenawan lavas. Raindrop impressions around penny, and ripple marks just to left of coin, formed on exposed muddy sediments deposited in local patches by same streams.

occur along this route. At many places along the way one can look ahead for great distances along the upturned edges of the Keweenawan strata (Fig. IV-28). The steep cliff faces on the southeast are formed by the beveled edges of the strata, whereas the more gentle slopes northwestward toward Lake Superior are formed by the inclined surfaces of the Keweenawan strata that are dipping down into the trough of the Lake Superior Syncline.

Another excellent example of the same sort of control exercised by rock resistance on surface topography is easily seen from the scenic overlook at the Lake of the Clouds in the Porcupine Mountains State Park (Fig. IV-21). There one stands at the top of a great cliff formed by the eroded edge of an amygdaloidal basalt that is part of the Keweenawan Series. Far below lies the lake itself, but directly beneath one's feet are the mineral-filled gas bubble cavities of the ancient lava flow (Fig. IV-22). The fascination of that modern scene is increased when one turns his mind back through the eons of time to the days when that very spot was a heaving lava plain.

Figure IV-27. Ancient stream current ripple marks of the asymmetrical type in a sandy lense within Keweenawan conglomerate. Roadside along Brockway Mountain Drive. W. C. Kelly points in direction of current flow, toward north.

Picturesque harbors, such as Copper Harbor, Agate Harbor, and Eagle Harbor, along the northwestern shore of the Keweenaw Peninsula, occur where waters of modern Lake Superior extend through narrow inlets across the upturned edges of more resistant Keweenawan rock strata, such as the Outer Conglomerate, and then expand into the areas of less resistant and more deeply eroded rocks behind. Lake Fanny Hooe at historic Fort Wilkins State Park rests in a similar depression carved by erosion into the upturned edges of less resistant Keweenawan strata.

A great break in the earth's crust called the Keweenaw Fault extends along the approximate centerline of the Keweenaw Peninsula, crossing Portage River east of the cities of Houghton and Hancock and continuing northeastward past Lake Gratiot and Lake LaBelle to the north edge of Bete Gris Bay near the peninsular tip. There the fault passes beneath the waters of Lake Superior.

Rocks of Keweenawan age, northwest of the fault, were uplifted relative to the younger Cambrian sandstones (Jacobsville Formation) on the southeastern side of the break. The Cambrian sandstones were bent steeply upward by movement along the fault and now stand nearly vertically there (Fig. IV-19). Most of the southeastern half of the Keweenaw Peninsula is floored by the younger Cambrian sandstones (Fig. IV-18). The older Keweenawan age rocks southeast of the fault now lie deeply buried beneath the overlying Cambrian Sandstones. Excellent exposures of the Cambrian-Jacobsville Sandstone occur along U.S. Highway 41 between the towns of L'Anse and Baraga near the south end of Keweenaw Bay (Chapter V).

Recall that the Penokean Orogeny had produced ancient mountains called the Northern Michigan Highlands. Late in Keweenawan time streams continued to erode these highlands. The sediments carried from the mountains by those

Figure IV-28. Keweenawan (Late Precambrian age) basaltic lava flows of the Portage Lake Lava Series, in western part of Upper Peninsula of Michigan.

*Top*—Eroded edge of thick flow forms cliff within view of state highway M-26 northwest of town of Mass.

*Bottom*—Columnar basalt exposed in roadside quarry along north side of M-28 on eastern outskirts of town of Bergland at northeast corner of Lake Gogebic. Vertical columns are defined by shrinkage cracks formed at right angles to the top and bottom surfaces of the cooling lava flow.

# Precambrian Eras

Figure IV-29. Shales and sandstones of the Nonesuch Formation of Late Keweenawan (latest Precambrian) age.

*Top*—Strata dipping northwestward out beneath Lake Superior at mouth of Presque Isle River, west end of Porcupine Mountains State Park, north of Wakefield. (Michigan Department of Conservation, photo by Robert Harrington.)

*Bottom*—Greenwood Falls on Big Iron River north of White Pine.

streams were deposited as a broad sedimentary apron of stream- and lake-laid material along the northern edge of the mountains. Those sedimentary deposits now form the Nonesuch Shale and Freda Sandstone, the last and youngest of the Precambrian sedimentary rocks deposited in the Upper Peninsula. The Nonesuch Shale is particularly important because it served as the host rock for a type of copper deposit to be discussed in detail later on. Later intrusive rocks then cut those sediments to bring the Precambrian Eras of geologic time to a close in Michigan. The Precambrian ended relatively quietly, no major crustal deformation or orogeny disturbing the crust at that time.

As we conclude this highly generalized history of Precambrian times in Michigan, remember that what you have just read is an *interpretation* of the rock record. Recall the many reasons given earlier to account for the incompleteness and complexity of the Precambrian record. Geologists never will be in possession of all the facts needed to write a complete and perfect history of any time in the past. Our interpretation of Precambrian events was read from the rock record as it is presented on Figure IV-3, but not all geologists would agree that the detailed age relations of the various rock formations on that figure are correct as we have shown them. Nevertheless, the broad outlines of Precambrian history are fairly well understood and much as presented here. It was a vast stretch of time which included several periods of sedimentary deposition and intervening times of intense crustal disturbance and igneous activity, a time of accumulation of the thick and important iron formations, a time of very simple life, and it closed with outpourings of lava as great as any later lava flows elsewhere on earth. Thus, the stage was set for the beginning of the Paleozoic Era. That setting included an uplifted and eroding Canadian Shield. Local ranges of mountains stood above the general level of the land and the surface of the earth was underlain by a rock crust that was already exceedingly complex in structure, highly metamorphosed in many places, and rich in certain minerals, especially iron and copper in the Michigan area. Let us take a closer look at the geology of those mineral assets and their contribution to the Michigan economy, before we move on to the Paleozoic Era in the next chapter.

## Iron Ores of Michigan

The great Huronian (Middle Precambrian or Animikie) iron formations of Michigan were briefly mentioned earlier in this chapter. Iron ore was and still is of great importance to the economy of the Upper Peninsula. In 1967 the state's production was valued at $162,610,000. In that year the total production of all minerals in the state was valued at $610,204,000, iron being the largest contributor.

Iron is the fourth most abundant chemical element in the earth's crust, composing about 5 percent of the total mass of the crust. It does not occur as a pure metal in crustal rocks because it is chemically active and combines readily with other elements to form iron-bearing chemical compounds. Iron and oxygen commonly unite to form iron oxides, such as the minerals *hematite, magnetite,* and *goethite* (or *limonite*). If carbon is included with iron and oxygen the compound is an iron carbonate called *siderite*. Iron combined with sulphur commonly forms an iron sulfide ($FeS_2$) called pyrite (fool's gold). If some natural process in the geologic past concentrated such compounds they may now constitute commercially valuable ore deposits. However, iron minerals are also widespread in minor, noncommercial, quantities throughout the rocks of the crust. The oxides of iron impart shades of red, brown, and yellow to rocks and to the soils derived from them. Carbonates of iron produce greenish colors in the rocks. Iron also enters into more complicated chemical compounds to produce vast quantities of such common minerals as

Figure IV-30. White Pine Copper Mine in forest setting near east end of Porcupine Mountains State Park. (Michigan Department of Conservation, photo by Robert Harrington.)

hornblende, augite, and black mica (biotite) which presently are of little or no commercial value.

The iron minerals which, in concentrated form, presently yield significant quantities of iron ore are: *magnetite* ($Fe_3O_4$) and *hematite* ($Fe_2O_3$). *Goethite* (or *"limonite,"* $FeO \cdot OH$) and *siderite* ($FeCO_3$) may occur in association with these but at the moment are unimportant as Michigan ore minerals. Other minor minerals rich in iron, such as the iron silicate, *grunerite,* occur locally. These iron minerals rarely occur in large quantities in pure form. Most often they occur as mixtures in ores and rocks. *Quartz* ($SiO_2$), which is the chemical *silica,* is a common associate.

Although iron minerals are very abundant everywhere, rock formations in some places in the world contain so many layers rich in important iron minerals that they are called *iron formations*. The iron formations of the Great Lakes Region commonly consist of thin, alternating layers of quartz-rich (high silica or siliceous) rock, such as slate or chert (jasper), and layers of almost pure iron minerals, mainly hematite, magnetite, iron silicates, siderite, and goethite. *Taconite* and *jaspillite* are rock names applied to such siliceous iron formations.

Two main types of iron ore now are mined from parts of the iron formations in the Lake Superior Region. These are the "direct shipping" and the "lean" (taconite and jaspillite) varieties. Direct shipping ore can be sent directly from mine to mill. Lean ores must be artificially concentrated before shipment. In the early days of iron mining in the Lake Superior District the percentage of iron in most of the iron formations was too low to be worth mining. Only those parts of the iron formations could be mined in which some natural process had already concentrated the iron to at least 50 percent of the total rock to be removed. Those rich, natural concentrates, the direct shipping ores, were the only "ores" at that time because an ore, by definition, is a natural *mineral substance that can be mined at a profit.* Economic factors, such as closeness to the surface, richness, access to cheap transportation, costs of equipment, value of ore per ton, labor costs, tax costs, competition, all determine that what may be valueless rock in one time or place may become "ore" in another. Recent improvements in mining efficiency and the development of artificial concentrating ("beneficiating") processes now have made it profitable to mine some portions of the "lean" iron formations which contain as little as 25-30 percent iron, although many parts of the iron formations have an even lower percentage of iron than that and cannot yet be mined and beneficiated at a profit. Nor is it yet possible to mine and beneficiate siderite (iron carbonate) at a profit, although some portions of the iron formations are rich in that mineral.

Over the years we have seen a gradual but dramatic change in the types of iron deposits mined in Michigan and in the methods of exploitation. In the early years the miners naturally turned first to rich natural concentrations containing up to 70 percent iron (direct shipping ore), but as these rich local concentrations were gradually depleted, it became necessary to develop methods for profitable mining of the lower grade iron formations (taconites). This was by no means a simple problem because the lower grade ores were

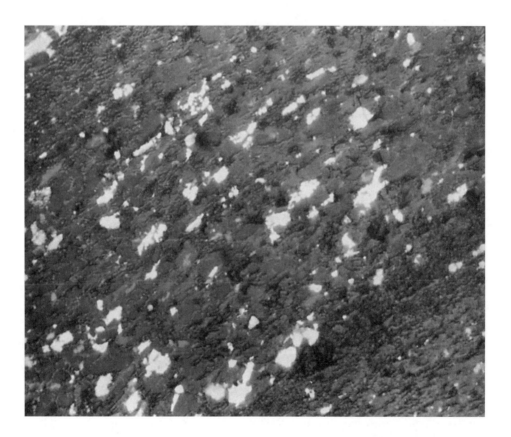

Figure IV-31. Magnified (×250) polished surface of chalcocite copper ore. Irregularly disseminated, light-colored grains of chalcocite occur in dominantly quartz-chlorite-carbonate gangue of darker color. Occurrence typical of cupriferous zone in Nonesuch Shale (Latest Precambrian age) at White Pine Mine. (Courtesy Alex Brown, The University of Michigan Department of Geology and Mineralogy.)

much harder and difficult to drill and, even if this problem could be solved, there was still the necessity of upgrading or "beneficiating" the taconites by artificial methods of concentration.

Beneficiating lean iron formations is a costly process and might never have been considered if it were not for the fact that the rich ores of the Lake Superior District were nearing depletion. It turns out, however, that beneficiated ores have certain advantages, such as ease of transport, controlled quality, and purity, which help to offset the higher cost of their production. Beneficiated ores are formed into pellets of uniform size before being shipped to the furnace. Their coarseness and resistance to crushing make them very desirable in blast furnaces because they permit a more thorough circulation of gases. This leads to a higher and more efficient recovery of pig iron from the ore and this, in turn, to a reduction in production cost. The Michigan iron ore industry in the future may have to rely almost entirely on beneficiation processes to stay in business. In all cases, however, cost is the factor that will determine whether or not beneficiation is feasible.

The iron mineral magnetite is attracted to a magnet, so the process for its beneficiation from associated worthless "gangue" minerals and rock is relatively simple. The magnetite ore is crushed in several stages until most of the particles of magnetite have been broken loose from the other mineral particles. The finely crushed material is passed through strong magnetic separators. In practice, *magnetic separation* is accomplished in several stages. The Empire Mine at Palmer in Marquette County uses magnetic separation for beneficiation.

Unfortunately, many of the Michigan lean ores are composed of nonmagnetic hematite rather than magnetite, so the magnetic separation process is not directly applicable. Beneficiation of nonmagnetic ore may be accomplished by magnetic separation, but the material must first be roasted at high temperature in the absence of oxygen to convert the hematite to magnetite. This adds to the cost of production.

Another concentration method for nonmagnetic ores takes advantage of centrifugal force (angular momentum) and the force of gravity. A mixture of crushed ore and water, with certain other additives, is introduced into the top of a steeply *spiraling trough*. As the sludge whirls swiftly down the spiral, the lighter (less dense) gangue minerals are thrown farther toward the outside of the turns in the spiral, leaving the more dense iron mineral particles to be collected from the inside of the spiral.

In *heavy medium separation* processes, crushed ore is mixed into a liquid that is denser than the gangue minerals but not as dense as the iron minerals. Thus, the lighter, worthless gangue is floated off while the heavy iron minerals sink to be collected.

In the *flotation process,* finely crushed ore is placed in a water bath to which pine oil and other carefully selected reagents are added. The bath is constantly agitated so that bubbles are formed and these rise to form a thick froth at the surface of the tank. The physical and chemical character of these bubbles is carefully controlled so that the iron mineral particles will adhere to their surfaces but the worthless gangue minerals will not. As a result, the iron-rich surface froth of bubbles can be skimmed off as an iron-concentrate. The flotation process is now in use at the Humboldt and Republic mines in the Marquette Range.

Some plants employ several beneficiation processes, the Groveland Mine in Dickinson County, for example. In all cases, however, the resulting concentrate is too finely divided to be usable in blast furnaces; the iron is in the form of a "dust" that would simply blow out the stack of the furnace, or else form irreducible clinker. The iron concentrate must be *"agglomerated"* into larger particles before it is shipped to the furnaces. Several *agglomeration methods* are feasible. *Sintering* involves heating the finely divided particles until they adhere to one another in irregular masses. *Nodulizing* requires that the concentrate be heated to the melting point, where it can be formed into more uniform nodules. *Pelletizing* involves addition of water to the ore, heating, and rolling into pellets of more or less uniform shape and size. Larger *briquets* are formed by pressure molding. In any case the concentrate is dried to remove moisture and to reduce weight, which thus reduces shipping costs.

Regardless of the methods employed, beneficiation is expensive. A ton of beneficiated iron ore produced from taconite or jaspillite, if it does not bring a higher price, may yield too low a margin of profit for the producer to sustain operations.

The several currently used methods of beneficiation result in a product which is composed of between 63-65 percent iron. This has been concentrated from lean iron formation containing only 25-30 percent iron which means that half the rock mined is discarded as waste, yielding only 1 ton of ore from 2 tons of rock. The *direct shipping ores,* in contrast, having an original iron content of slightly over 50 percent, yield a ton of ore for each ton of mined rock. Obviously, the direct shipping or "natural" ores are less expensive to produce and may be rich enough to make underground mining, an expensive process, economically feasible. In contrast, beneficiated ores presently would be too expensive if they were obtained underground, hence are profitable only if obtained from surface pits which are cheaper to operate. Even though direct shipping ores are rapidly being depleted in the United States, very abundant and even richer ores (up to 63 percent iron) now are available from foreign countries. Produced by cheaper labor and shipped in vessels whose foreign registry lowers transportation costs, these foreign ores now can be transported several thousand miles by sea and still cost less at eastern United States furnaces than do the beneficiated domestic ores. Thus, beneficiated ores from the Great Lakes Region will face stiff competition for years to come. Survival of the iron mining industry in Michigan will depend on the production of a high quality ore by constantly improved methods, and the most favorable labor cost and tax conditions. Even so, there is an ultimate limit to the efficiency of the beneficiation processes discussed above. Pure magnetite mineral contains only 72 percent iron, hematite 70 percent, and goethite (limonite) about 60 percent. Thus, even if the beneficiation processes used were 100 percent efficient in recovering all the iron minerals from the gangue rock they still could not produce pure iron, nor, for that matter, a product much richer in iron than some of the best foreign ores available today.

In spite of all these limitations, beneficiation becomes increasingly important in its contribution to the Michigan economy. Production statistics change rapidly from year to year, so the following figures, no doubt, will soon be out of date, but will serve to illustrate the increasing value of beneficiated iron ore. In 1964 the total *value* if iron ore shipped was $144,000,000, 59 percent coming from beneficiated concentrates. In 1967 the total value of iron ore shipped had risen to $162,610,000, the increase being entirely due to an increase in the production of concentrates whose contribution by then had risen to 79 percent; in the same period the total tonnage of direct shipping ore actually decreased each year. Because the concentrated ore is of higher quality, it brings a higher price per ton at the furnace. Thus, it contributes an even higher percentage to the total *value* of ore than is reflected in tonnage figures alone. In 1967 beneficiated ore (11.1 million tons) exceeded direct-shipping ore (3.0 million tons). No doubt the details of the economic picture will change in the future, but beneficiation clearly offers great hope to the Upper Peninsula iron mining industry. The Michigan reserves of low-grade iron formation susceptible to beneficiation are vast, and research may yet discover new methods which, when applied to the even lower-grade iron-bearing rocks, or to those containing siderite, could increase those reserves still further.

Now let us turn to some basic geologic questions about iron in Michigan. How did the iron formations and ore bodies originate? What events in geologic history brought them to their present state?

Most authorities are agreed that the Precambrian iron deposits of the Lake Superior Region originally were deposited as sediments. We have seen that iron is an abundant and widespread element constituting about 5 percent of the earth's crust. Rainwater dissolves and alters (weathers) surface rocks and minerals. Streams, fed by rain, then carry fragments and dissolved materials, including iron, into lakes and ultimately to the sea. More importantly, some rainwater first sinks underground, dissolves much mineral matter there, and then emerges into streams which flow into the sea. Thus, even today, dissolved iron is provided to marine environments where it can accumulate. The rate of accumulation may have been especially high during the Huronian when greater areas of iron-rich igneous rock were exposed at the earth's surface. It is thought that the iron deposits of the Lake Superior Region were deposited as chemical precipitates in shallow waters in vast, semi-isolated bays or arms of the Huronian seas. When iron, dissolved from the land, reached the sea it came in contact with other

# Geology of Michigan

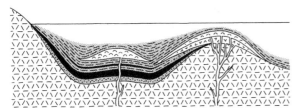

4. Latest Huronian deformation. Thick geosynclinal deposits.

3. Late Huronian. Strong subsidence.

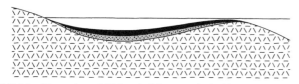

2. Middle Huronian. Mild subsidence of shelf.

1. Early Huronian. Stable shelf adjacent to continent.

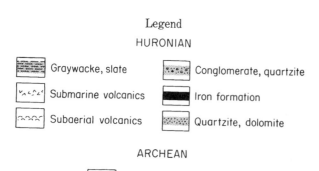

Figure IV–32. Sedimentary origins of Huronian (Middle Precambrian) iron formations. In Stage 1, normal marine quartz sandstones and dolomites accumulate on stable shelf. In Stage 2, mild subsidence of shelf provides basin to receive iron oxides and carbonates. Stage 3 illustrates increasing crustal unrest accompanied by deformation and uplift in adjacent source areas and leading to diminution of iron relative to clastic sediment deposition; volcanism at shelf edge. Stage 4, close of Huronian, is time of strong geosynclinal subsidence, intense crustal orogeny, rapid clastic sedimentation, submarine volcanism, and little iron accumulation. (Modified from H. L. James, 1954.)

concentrated, dissolved elements. As a result many minerals were precipitated, including iron compounds. Some geologists believe that much of the iron reaching the Precambrian seas combined at first with carbon and oxygen to form the mineral siderite (iron carbonate) and that subsequent chemical changes altered some of this to magnetite or hematite. Others believe that the kind of iron compounds that formed depended upon the local composition of the sea water. When large quantities of mud and sand were washed in, layers of siliceous sediment were formed in which the iron content was low. Frequent alternation of such conditions of deposition would have produced the alternating strata of iron-rich, and clayey or silica rich (quartzitic) sediments so common in the iron formations (Fig. IV–13). The series of diagrams in Figure IV–32 illustrates one concept of the conditions under which the iron formations and associated sedimentary strata accumulated. The marine basin shown in the diagrams may have been part of a great downwarped geosyncline (Chapter III). Wave ripple marks and mud cracks found on the surfaces of some of the strata in the iron formations offer testimony to their shallow water environment of deposition. There is no doubt that the iron in the iron formations actually formed in Precambrian time. Eroded blocks and pieces of the iron formation are found in younger sedimentary rocks of Cambrian age that immediately overlie the Precambrian in the same region. This means that the iron formations had to have been in their iron-rich state before the beginning of Cambrian time in order to have been eroded and carried into the Cambrian deposits. A fine example of such a deposit of Cambrian age, containing fragments of Precambrian iron formation, can be seen at the old Quinnesec Mine near the town of the same name (Chapter V).

Earlier in this chapter we saw that after the iron-rich Huronian sediments had been deposited, the Great Lakes Region was involved in a period of intense crustal disturbance called the Penokean Orogeny. Mountain building accompanied that orogeny and erosion continued long thereafter. As a result of compression during that orogeny, the iron deposits were complexly folded and broken by faults so that in places they now extend to great depths below the surface. Figures IV-34-37 show the subsurface structure in the major iron ranges of the Upper Peninsula. Locally, molten magmas invaded the sedimentary iron formations, cooling to form cross-cutting dikes of igneous rock. The present outcrop pattern of such highly deformed strata is very complicated as can be seen on Figure IV-33. The great heat and pressure resulting from deep burial, crustal compression, and igneous intrusions, affected all Precambrian rocks. The muds, turned to shale, were metamorphosed to slates. The siliceous siltstones and quartzitic sandstones were altered to quartzite. The iron minerals in those parts of the iron formations that were strongly metamorphosed were changed to coarser grained magnetite or to the shiny variety of coarse-grained hematite called specularite. These are the so-called "hard ores." Soft ores occur in areas which were less strongly deformed. Long afterward, erosion beveled the earth's surface so as to expose the edges of the complexly deformed iron formations and other rocks at the surface.

We still have not accounted for the origin of the locally restricted but rich *direct shipping ore* bodies. These are natural concentrations of iron minerals which, until the advent of beneficiating processes, were the only deposits of iron that could be mined at a profit. They are far richer in iron than the iron formations in general. These highly concentrated natural ore bodies resulted where locally concentrated flows of underground water leached or dissolved away the silica from a portion of the iron formation, leaving the iron minerals behind as enriched residues. Many subsurface structures of the rocks in the iron ranges favored this process (Figs. IV-34-37). In some places, downward-moving ground water collected and was channeled along the bottoms of downfolded parts of the iron formations, where dense, impermeable slates, quartzites, or igneous dikes prevented the water from sinking further. In the Gogebic Range water was trapped above and against the intersections of faults and igneous rock dikes, where it then flowed in greater volume with a greater leaching effect. Often the intersections of faults, folds, and dikes, in combination, served to channel the waters.

Near-surface concentrations of soft ore were discovered and worked initially. The ore in those relatively small, shallow, open pit mines was soon exhausted, but some of the ore bodies were then followed underground. Mining costs rose as a result. Increased costs had to be offset by improved mining methods and increased efficiency or the companies failed. The downfolded and faulted iron ore bodies of Michigan were followed down in some instances to depths as great as 3000 or 4000 feet. In contrast, in the great Mesabi Range of Minnesota, where the structure is shallow, the ore occurs near the surface in far greater quantity and is still mined by open pit methods there. With beneficiation, the trend now is back to open pit surface mining in the Upper Peninsula because large quantities of leaner iron deposits, left untouched by the early mining methods, are still available there and in fact must be used because the cost of beneficiation is too great to permit underground mining.

Michigan iron ore is or was mined in three main areas called *ranges*: the Marquette Range, the Menominee Range, and the Gogebic Range (Fig. IV-33). Mining operations recently were terminated in the Gogebic Range. The main iron formations and associated ore bodies in Michigan all are of Huronian (Middle Precambrian) age (Fig. IV-3). The Marquette Range extends in a narrow belt for about 30 miles westward from Marquette and includes such towns and cities as Negaunee, Ishpeming, Humboldt, Champion, and Michigamme. A narrow extension runs south to Republic. The Marquette Range structure is a major downfold (syncline) in the crust. The subsurface structure is further complicated by minor

**Geology of Michigan**

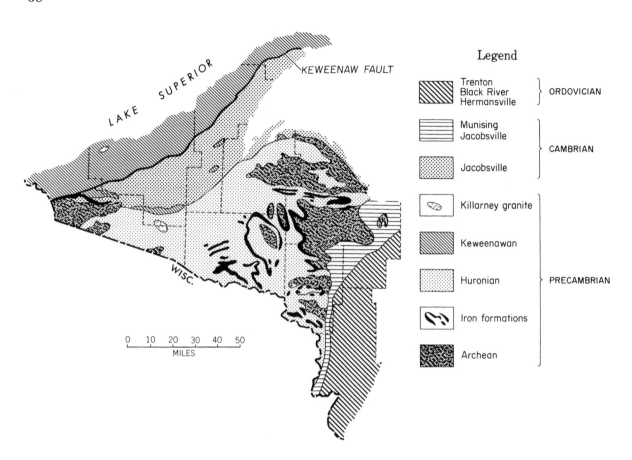

Figure IV-33. Generalized geologic map of western half of Upper Peninsula of Michigan to show iron ranges and distribution of Precambrian and Early Paleozoic rocks. (Modified from Martin, 1957, "Geological Map of Michigan," courtesy Michigan Conservation Department.)

folds, faults, and basic (ferromagnesian) igneous intrusions in the form of dikes and sills (Figs. IV-34, 35).

Ore in the Menominee Range was naturally concentrated by leaching in two iron formations, the Vulcan Iron Formation in the vicinity of Iron Mountain, Quinnesec, Norway, Vulcan, and Loretto and the Riverton Iron Formation in the vicinity of Iron River and Crystal Falls, Michigan, and Florence, Wisconsin. To the north, some ore is mined in the area between Randville and Felch. The concentrated ore bodies occur in downfolds (Fig. IV-36).

The Gogebic Range extends from the western tip of the Upper Peninsula westward into Wisconsin. Ironwood is the principal Michigan city on that range. Ore was produced there from the Ironwood Iron Formation. The Middle Huronian rock layers there are steeply inclined, extend to great depth, are complexly broken and offset by faults, and are cut by igneous intrusions (Fig. IV-37).

The Marquette Range was the first to produce ore in Michigan. Mining began there in 1848. Mines were opened in the southern part of the Menominee Range in the vicinity of Iron Mountain in 1872 and in the vicinity of Crystal Falls and Iron River in 1880. The Gogebic Range was opened in 1884. Mines still are active in the Marquette Range and in the northwestern part of the Menominee Range (Crystal Falls District), but the last operating mine closed in the Gogebic Range in 1965.

There are many millions of tons of low grade iron formation remaining in all the ranges, much of it near the surface. With modern methods of beneficiation some of this may yet be recovered at a profit. Bayley (1958) has estimated that 70

# Precambrian Eras

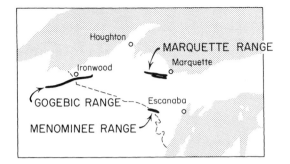

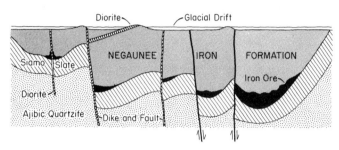

Figure IV-34. North-south cross section across Marquette Iron Range (looking west).

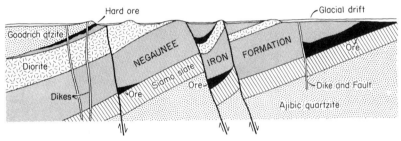

Figure IV-35. East-west cross section through Marquette Iron Range (looking north).

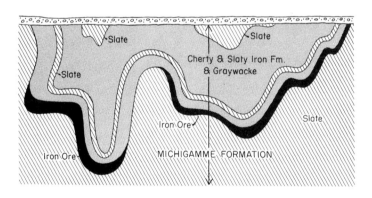

Figure IV-36. Cross section through Menominee Iron Range.

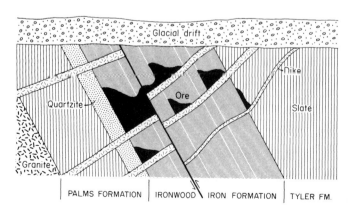

Figure IV-37. Cross section through Gogebic Iron Range.

*(All cross sections generalized from chart with permission of United States Steel Corporation.)*

# Geology of Michigan

million tons of beneficiated ore might be produced from the upper 100 feet of the two iron-bearing members of the Vulcan Iron Formation alone.

In the very early days of iron mining in the Upper Peninsula, local charcoal-burning furnaces smelted some of the ore down to iron. Remnants of the old "Bay Furnace" still stand at what is now the Bay Furnace Camp Ground just west of Munising, as one example. Another, at the town of Fayette (Fig. v–17), on Lake Michigan near the tip of the Garden Peninsula south of Manistique, has been restored as a Michigan State Park. Coal- or coke-burning smelters are more efficient than charcoal smelters, but coal did not occur in sufficient quantity in the northern Great Lakes Region to be used for that purpose. Because the smelting process requires about four times as much coal as iron ore, it is cheaper to ship iron ore to the coal-producing regions of Pennsylvania, Ohio, and Illinois than to ship the coal to the iron. Lack of railroads, and the rapids of the St. Mary's River at the Soo, at first were obstacles to such shipping, but these difficulties were surmounted. Since the opening of the St. Mary's canal in 1855, a steady stream of ships has carried iron ore across the lakes from Minnesota and northern Michigan to the iron and steel mills in the south. Thanks to modern research and development, beneficiation processes now keep that stream flowing.

Those who wish to read more about the geology, mining history, beneficiation processes, and mineral economics of Michigan iron ores should consult the Bibliography, in particular: Bateman (1942), Dutton (1958), Hotchkiss (1933), James (1951 and 1958), Leith, Lund, and Leith (1935), and VanHise and Leith (1911). Those references in turn have their own bibliographies by means of which you might continue your research.

## Keweenaw Copper

Michigan copper deposits occur in two distinctly different forms, which are the metallic or "native" variety and the nonmetallic chalcocite ore. We shall discuss the geology of each variety in turn, beginning with the metallic form which was first to be mined in the state. For 42 years (1845-1887) "Native" Keweenaw Copper was king of the industry in North America. Then production of copper ores at Butte, Montana, and Bisbee, Arizona, surpassed it. Native Copper also occurs in other parts of the world, but nowhere else in the quantity found in the Upper Peninsula of Michigan. Michigan copper of both varieties still contributes much to the economy of the state, and its occurrence continues to arouse the interest of many travelers to the Copper Country.

The copper deposits occur in Late Precambrian rocks of the Keweenawan Series whose geologic history and general nature were discussed earlier in this chapter. The Keweenawan Series crops out in a belt about 100 miles long which extends from the tip of the Keweenaw Peninsula southwestward through the Porcupine Mountains into Wisconsin. The belt of outcrop is relatively narrow, occupying only the northwestern half of the Keweenaw Peninsula (Fig. IV–33). Over 95 percent of the native copper produced there came from a zone about 26 miles long, lying mostly in Houghton County but including parts of Keweenaw and Ontonagon counties. Beneath the surface of the region, the Keweenawan rocks are inclined downward toward the northwest, dipping down beneath Lake Superior at an average angle of about 40 degrees from the horizontal. Locally, these dips may vary, ranging from about 15 degrees in some places to a nearly vertical attitude in others. The trend of the outcrop belt is interrupted around the domal uplift of the Porcupine Mountains where the Keweenawan Series dips outward from the dome in all directions. The dips around the dome are gentle except on the south side, where the rock layers are very steep and in some places even overturned. The series passes beneath Lake Superior, but the dip then reverses itself so that the same rock series reappears at the surface in Minnesota and on Isle Royale, here dipping down to the southeast. Keweenawan rocks can also be traced completely around the southwestern end of Lake Superior where the outcrop belts on the two sides of the lake are joined. Earlier in this chapter we noted that this structure forms what is called the Lake Superior Syncline. Stream erosion in ages past removed the overlying younger rocks that formerly lay above the Keweenawan Series, exposed the series at the surface, truncated the ends of its tilted strata, and formed ridges and valleys in the hard and soft rock layers. Much later, only a few thousand years ago, glacial ice of the Pleistocene

Epoch scoured across the area, but did not completely erase the surface irregularities. When the ice melted off it deposited a blanket of sands, gravels, and clays which still covers much of the region.

The terminology of rocks of Keweenawan age in Michigan is shown in Figure IV–3, and a general cross section of subsurface structure in Figure IV–18. At the place of maximum subsidence in the Lake Superior Syncline, the rocks of the Keweenawan Series are estimated to be between 35,000 and 50,000 feet thick. The point of greatest depth in the Lake Superior Syncline, also the area with the thickest Keweenawan rocks, lies beneath Lake Superior about 20 miles offshore and about due west of Houghton and Calumet. Thus, the center of greatest subsidence is closer to the Michigan side of Lake Superior (White, Fig. 4, in Snelgrove, 1957). The middle portion of the Keweenawan Series includes about 400 distinct lava flows or "traps" with 20 to 30 interbedded conglomerate and sandstone layers. Below these, at the base of the series, lie about 200 feet of conglomerates and sandstone. Above the lava series are the thick Outer Conglomerate, the Nonesuch Shale, and the Freda Sandstone, in that order. Most of the Keweenawan lavas are of the basaltic type, although a few felsites and rhyolites of more granitic composition are included. One can imagine the intensity of the surface volcanic activity that occurred in that region between 1280 million and 880 million years ago. Most of the lava flows poured out in such rapid succession that little weathering or erosion took place between times. However, a few stream-laid conglomerate and sandstone layers did form locally and these later were among the important sites of copper deposition. The source area for the lavas lay within the Lake Superior Syncline, northwest of the Keweenaw Peninsula. Some intrusive igneous rocks, both gabbros and felsites, cross-cut the Keweenawan strata and hence are younger than the latter. The felsites are high in silica and relatively low in iron and magnesium-bearing minerals, whereas the gabbro is of about the same chemical composition as the basaltic surface flows. The Keweenawan strata, including the lava flows, accumulated one layer above another in sequence so that the oldest, that is the first to form, became deeply buried under successively younger overlying layers. Thus, the youngest layers occur in the highest or upper part of the series. This relationship follows the Law of Superposition discussed in Chapter II. The relative ages of the layers can thus be determined from their position in the sequence. During and after their deposition, the layers were downwarped into the Lake Superior Syncline. For this reason they no longer lie one above another in vertical sequence. In the Keweenaw Peninsula they dip steeply down to the northwest. On the opposite side of the syncline and on Isle Royale they dip more gently and in the opposite direction, toward the southeast. It is customary to call those layers deposited first and now cropping out along the southeast side of the exposure belt in the Keweenaw Peninsula the "lowest." The younger strata which crop out along the northwest side of the belt, closer to the Lake Superior shore, are called the "highest" or "upper" beds. The cross-section diagram in Figure IV–38 shows this relationship and some other details of the Keweenawan rock series.

The sedimentary sandstone, shale, and conglomerate layers in the higher part of the section tend to erode more easily than the beds below and thus form a narrow lowland along the Lake Superior shoreline on the northwest side of the Peninsula. The more erosion resistant basalts and conglomerates of the middle part of the series hold up a long, high, central plateau with ridges that stand out in relief along the length of the district.

Deep beneath the earth's surface, near or at the base of the Keweenaw Series, lies a layer (sill) of basic igneous rock of intrusive origin called the *Duluth Gabbro*. This rock formation is over 10 miles thick where it crops out near Duluth, Minnesota. It comes to the surface there because the dip of the strata brings the basal part of the Keweenaw Series upward on the northwest side of the Lake Superior Syncline (Fig. IV–38). The Duluth Gabbro itself does not reach the surface in the Keweenaw Peninsula, but some minor gabbroic intrusives of similar composition, such as the Bad River Gabbro and the intrusive exposed by erosion at Mt. Bohemia, do occur at the surface there. Because gabbro and basalt have the same composition, some geologists believe that molten magma from the same chamber deep below the surface formed both the Duluth Gabbro at depth and the basalts where the magma reached the surface. The magma moved upward

# Geology of Michigan

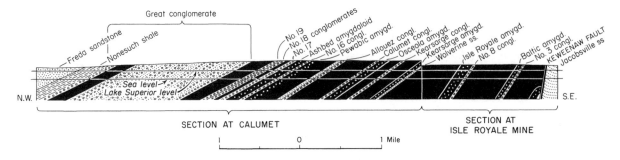

Figure IV-38. Cross section through Michigan Copper Range to show generalized relationships of copper lodes. (Modified from Broderick, 1933.)

through fissures in the crust to form the surface flows.

Underground, where the magma was under great pressure, large quantities of gas and water were dissolved in the molten material. When the magma reached the surface, pressure was removed and the dissolved gas and water separated into bubbles which tended to rise through the lava. The rising and breaking of gas bubbles at the lava flow tops produced a porous or even frothy surface strewn with fragments of partially solidified lava crust. The surface was made even more rough and irregular by movements of still molten lava beneath its partly cooled and solidified surface. As the lava cooled further, some of the bubbles became trapped or "frozen" into the rock to form cavities called *vesicles*. Where mineral matter later filled those cavities, *amygdules* were formed. Thus, *vesicular* and *amygdaloidal* basalt originated (Figs. IV-22, 24). A concentration of vesicles is characteristic of the upper parts of lava flows. In some cases the vesicles joined to form interconnecting spaces, making the rock very permeable. Where the cavities are filled with lighter-colored minerals such rock has a mottled aspect. Copper is an important mineral found in the vesicles, but silver and over 20 other minerals, some very rare, also occur.

The levels rich in native copper are called *lodes*. *Amygdaloidal lodes* contain copper in the gas bubble cavities and fragmental surface material of the basalts. Much native copper was produced from such famous amygdaloidal lodes as the Kearsarge, Baltic, Pewabic, Osceola, Isle Royale, and Atlantic, in that order of importance. *Conglomerate lodes* are another important type. In these, copper and other minerals fill openings around pebbles and grains of sand. The Calumet and Hecla red conglomerate lode, opened in 1864, alone had produced over 40 percent of the native copper from the district by the end of 1929. The amygdaloidal and conglomeratic lodes parallel the inclined surfaces of the strata in the Lake Superior Syncline. They have been mined to great depths. The Calumet and Hecla lode was mined for almost 10,000 feet along the incline, to a depth of nearly 6000 feet below the surface. Mining from such depths is very expensive, however, and increasing costs eventually helped to bring many deep mines to the point where they could not compete economically with the more accessible copper ores of western United States. As one drives through the Copper Country of Michigan he can see many shaft houses with their inclined roofs slanting down toward the northwest. This is the direction and angle of inclination of the rock layers which the shafts follow beneath the earth's surface.

The vein or fissure lode, a third type, contained copper deposited along cracks that cut across the stratification of the Keweenawan rock layers. Production from these never was great, but large individual masses of native copper were especially common in them. One mass weighed over 500 tons. Copper is so soft and malleable that such large masses are difficult or impossible to split up and bring to the surface. One famous mass of copper, a 2-ton chunk called the "Ontonagon Boulder," was torn loose by the Pleistocene glacier as it passed over the Keweenaw Peninsula and was dropped along the Ontonagon River, where it was discovered thousands of years later. After many unsuccessful attempts, this piece finally was moved to Washington, D.C., where it is on dis-

**Precambrian Eras**

73

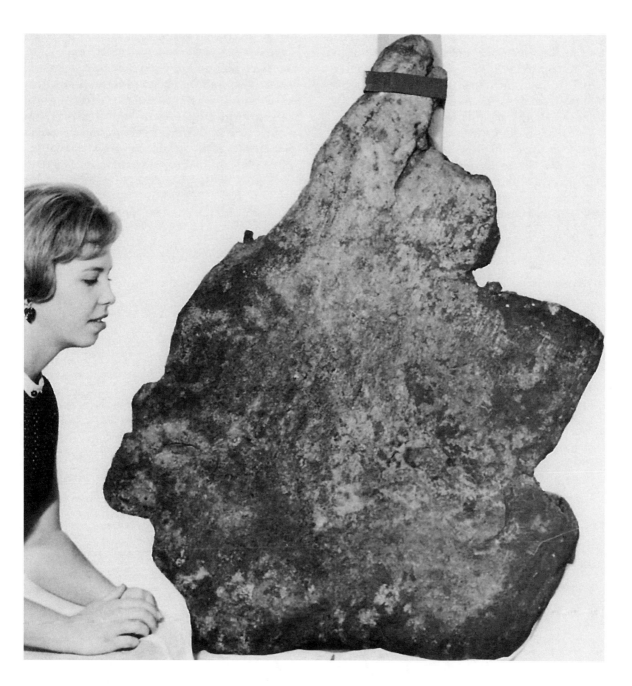

Figure IV-39. Mass of native copper weighing about 483 pounds. Torn loose from bedrock of the Copper Country by Pleistocene glacial ice, it was found, buried in glacial drift, in Houghton County. It apparently had previously been discovered by Indians who had smoothed the edges and added an eye to increase the resemblance to a chief's head. Note glacial scratches (striae) running diagonally from upper right to lower left. (The University of Michigan Department of Geology and Mineralogy.)

# Geology of Michigan

play in the U.S. National Museum. Similar pieces of "float" copper were carried farther south by the glacier. Indians found some of these and revered them as gods or manitous. One such piece, now in the University of Michigan Mineralogy Museum in Ann Arbor (Fig. IV-39), originally was shaped like an Indian head; to enhance this resemblance, the Indians who found it hacked an eye into it and reshaped the profile of the nose. Pieces of native copper and other minerals still can be found on many of the old mine dumps in the Copper Country of Michigan.

There has been considerable argument concerning the mode of origin of the native copper deposits. Some geologists once maintained that the copper had been dissolved from basalts by surface waters which carried it downward through the basalts to precipitate it in the cavities in the lodes. One of the difficulties with this idea is that the lodes occur at the tops of porous layers beneath nonporous zones. This "downward leaching" proposal demands that the copper occur where the water would have concentrated above nonporous zones, rather than below them. The more commonly accepted hypothesis is that of *hypogene* ("carried up")-*hydrothermal* ("hot water") origin. According to this hypothesis, hot waters rich in dissolved mineral matter rose upward along cracks and fissures. The waters carried dissolved copper. Frequently, these rising hydrothermal solutions reached inclined, porous, and permeable zones in the rocks, such as the amygdaloidal tops of lava flows or the conglomerate beds. There the water rose to the tops of the porous zones and then up along them until stopped against faults or against the bottom of nonporous layers in

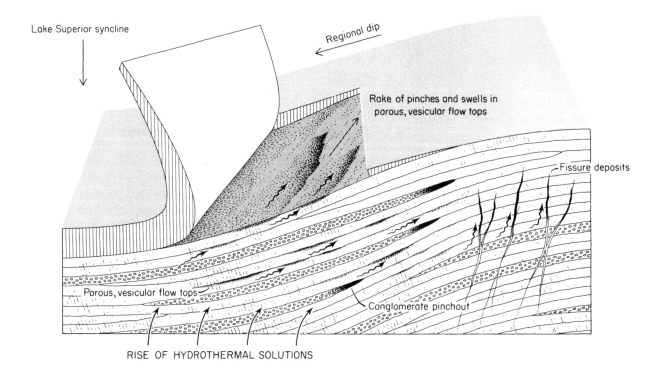

Figure IV-40. Generalized diagram to illustrate some modes of origin of native copper deposits in Upper Peninsula. Rising hydrothermal (hot water) solutions, rich in dissolved mineral matter, travel upward along inclined strata consisting of interstratified conglomerates and lava flows. Darkened areas show potential sites of copper entrapment. *Conglomerate lodes* form at "pinchouts" in conglomerate beds. *Fissure lodes* form at upper ends of cracks in the rock sequence. The tops of many lava flows locally are porous (vesicular) due to accumulation of gas bubble cavities. Zones of such vesicles "pinch and swell" in size and localize movement of copper-bearing solutions. If elongate zones of such porous lava trend diagonally across the regional inclination of the rocks, the solutions follow up the diagonal "rake" of those zones, depositing copper in upper ends of the vesicular zones where the porosity diminishes. The filling of the vesicles produces an *amygdaloidal copper lode*.

"pinch-outs" (Fig. IV–40). If the hypothesis of hydrothermal solutions rising along inclined rock layers is correct, the native copper lodes must have formed during or *after* the tilting of Keweenawan rocks in the Lake Superior Syncline, not before.

From whence did the copper come, how did so much of it get into solution, and what was the origin of the rising waters themselves? It is generally agreed that the original source of the copper was some igneous rock or its parent magma. Gabbroic and basaltic rocks commonly contain minor amounts of copper, about 0.01 percent in the case of those rocks in the Lake Superior Region. If the basaltic lava flows themselves were the source, how did such minor amounts of copper as originally were disseminated through them become concentrated?

One hypothesis suggests that the copper-rich hydrothermal waters were of *deuteric* origin. This means that the water was an original and integral part of the lavas or related, deeper-seated gabbroic magmas themselves. This assumption leads to the conclusion that the copper-rich waters were released from the magma as the latter cooled, the solutions migrating upward into the zones of copper emplacement to become the copper lodes. It is necessary to postulate extensive water migration, rather than deposition restricted to individual lava flows themselves, because, as we have already seen, the copper lodes also occur in conglomerates and in veins and fissures.

W. S. White (in Snelgrove, 1957, p. 11) has suggested another mechanism for incorporating water into the lavas and then driving it upward as a concentrating and transporting agent for copper. White points out that even today the temperature of rocks in the earth's crust near Calumet increases regularly at the rate of about 18 degrees Centigrade for each kilometer of added depth. This fact has been established by measurements taken in mines to a depth of 5488 feet. If this temperature gradient continues down to the base of the pile of Keweenawan lavas and sediments in the Lake Superior Syncline, the rock temperature at 35,000 to 50,000 feet of depth would be on the order of 200°C–285°C. White estimates that during and shortly after Keweenawan time, when the lavas still retained some of their original heat, the rock temperatures at that depth reached at least 300°C (572°F). At those depths the weight of overlying rocks would have exerted a pressure of between 2700 and 4000 times normal atmospheric pressure. At the time of their formation the tops of many of the lava flows were fragmental and vesicular, and the associated conglomerates also were porous. Groundwater would have entered those porous rocks when they lay near the surface and would then have been carried down to great depths as the weight of subsequent flows and sediments continued to depress the crust. White suggests that the tremendous pressures at depth crushed the porous rocks, squeezing the superheated, "buried" groundwater out and driving it upward. In its passage the hot water would have dissolved and concentrated copper and other chemicals from the deep-seated rocks, becoming a hydrothermal solution which then followed fissures and porous passageways toward the surface. There it cooled and deposited the mineral matter, including native copper.

Although it is probable that some copper would be found far below the greatest depths yet mined, it is unprofitable at present to go deeper along most of the older known lodes. Most of the exposed bedrock areas in the Keweenaw Peninsula have been intensely explored by surface prospecting, but exploration is hampered by vegetation and by the surface cover of glacial deposits which often reach 200 feet in thickness and conceal much of the bedrock. Exploratory drilling is expensive. After the discovery of the Baltic Lode in 1882 no major new native copper deposits were found for about 80 years. In 1964 Calumet and Hecla, Incorporated, began construction on the first new native copper mine in the district to be developed in over 30 years; this is the Kingston Mine, a few miles northeast of Calumet between Kearsarge and Mohawk. A second new development is now under consideration by the same company. Both sites were discovered by drilling a large number of shallow, closely spaced exploratory holes. It is possible that still other important deposits remain to be found in the district.

Ancient Indian tribes of the "Old Copper Culture," 3000 B.C.–1000 B.C. (Griffin, 1965, pp. 663–64), long vanished before the earliest European explorers reached the Copper Country, were the original discoverers and miners of the native copper deposits. Copper was traded widely by the Indians throughout northeastern North America.

Cartier heard reports of the copper deposits from the Algonquin Indians that he met, but those Indians held the copper sacred and did not mine it; in fact, none of the modern Indian tribes of that region knew how to mine it. They depended for their supply upon loose surface pieces ("float copper") found along stream beds and in glacial deposits. The Jesuit missionaries who followed Cartier into the Lake Superior Region also heard the legends and found float copper themselves. Reports of the metal reached Paris in 1636 in the "Relacions" of Lagarde. Thereafter, many of the ancient mine pits of the vanished Indian race were found on Isle Royale and along the lodes of the Keweenaw Peninsula. In the pits were found some of the rock hammers, chisels, and wedges left behind by those ancient miners. Piles of charcoal and fire-blackened rocks indicated that the early miners built fires against the rock faces of the pits and then dashed cold water against the hot rock, cracking off pieces that could be pounded apart in the quest for the metal.

Pure or native copper is rare elsewhere in the world. In the Old World, advanced stages of civilization were reached before the discovery of methods for smelting copper metal from natural mineral compounds, such as malachite, could usher in the Age of Metals. In the Great Lakes Region, in contrast, the presence of easily worked native copper made it possible for Indians, otherwise at a Stone Age cultural level, to produce some metal artifacts. One wonders why the later Indians neglected the copper deposits that had been so well known to their predecessors. A partial answer may be that the easily worked surface deposits became exhausted. It also is hard to imagine how that ancient race discovered the copper lodes in the first place, because even in earliest historic times the few rock exposures not covered by glacial drift were for the most part heavily covered with vegetation. It has been suggested that the ancient Indians entered the Keweenaw country so soon after the last glacial ice sheet had melted back that the rocks were still bare of vegetation and the copper veins more easily seen.

The first real interest in the mineral wealth of the Keweenaw developed after 1841. Douglass Houghton, the first state geologist of Michigan, had actually located the copper-bearing rocks there and in that year published a report on the area, urging development of both the copper and iron deposits of northern Michigan. Copper mining began at Copper Harbor in 1844. The first deposits to be worked with at least partial economic success were the relatively unimportant fissure deposits including the Cliff Fissure (1845), the Minesota (1849), and Central (1856). The Isle Royale (1852) and Pewabic (1856) were the first amygdaloidal lodes opened (Fig. IV–42). The Calumet and Hecla conglomerate lode was discovered in 1864, the Osceola and Kearsarge in 1868. The Baltic Lode (1882) was the last important native copper lode discovered until the very recent discoveries of the early 1960's mentioned.

The growth of the copper mining industry in the Keweenaw led to a rapid influx of settlers, but hardships for those pioneers were great and failures were many. Swineford (1876) published a little book reviewing the early history of the copper and iron mining "booms." This would make fascinating reading for those interested in first-hand historical detail.

As the rich native copper lodes became exhausted, production declined. Increasing costs of mining at greater depths also took their toll, as did rising labor and equipment costs. At one time over 100 companies were actively trying to produce copper at a profit from the more accessible parts of the copper lodes, but by 1958 only 12 companies were engaged in copper mining in the state, and today only 3 major companies are active producers. It must be remembered, however, that many of the early mines were very small operations whose combined contributions to the copper production of the state were not great.

Touring the region today, one is impressed with the large number of abandoned workings. However, the Copper Country is far from dead! Eight mines were producing native copper in 1967. Another later development was the recovery of copper by flotation concentration methods from the waste materials discarded during earlier and less efficient mining and smelting operations. Secondary recovery of copper was possible from both the lean waste rock left at the mines and from "stamp sands" formerly discarded into Torch and Portage lakes.

Although Michigan no longer ranks first as a copper producer, "Lake" copper, with its high quality, small but desirable percentage of silver, and low original cost, played a crucial role in the development of our nation at a critical time in history. The native copper of the Keweenaw was available when the electrical industry first began to grow and thus made possible the rapid electrification of our country.

## White Pine Copper

Although native copper production has declined in Michigan, large reserves of copper sulphide in the form of the mineral *chalcocite* now are being mined near White Pine in Ontonagon County, by the White Pine Copper Company (Fig. IV–30). The mining area lies just east of the Porcupine Mountains, about 6 miles south of Lake Superior, 45 or more miles southwest of the major native copper mines of the Keweenaw Peninsula. The chalcocite at White Pine occurs in a finely disseminated state (Fig. IV–31) in the Nonesuch Shale, a deposit of Lake Keweenawan (Latest Precambrian) age (Fig. IV–3). Minor amounts of finely divided native copper are associated with the chalcocite.

The Nonesuch Shale is a stratified, sedimentary rock deposit about 600 feet thick, composed largely of gray siltstones but also including some shale and sandstone layers (Fig. IV–29). Directly beneath this formation lie 2300 to 5500 feet of red sandstones and conglomerates of the Copper Harbor or "Outer" Conglomerate, also of Late Keweenawan age. Below these, in turn, lie the Middle Keweenawan lavas of the Portage Lake Series. The Nonesuch Shale grades upward into the overlying Late Keweenawan Freda Sandstone.

The White Pine mine area is located on the crest of a gently arched anticline which, in turn, is situated along the northeast side of an offset break in the earth's crust called the White Pine Fault. Only the lower 20 to 25 feet of the Nonesuch Shale contain copper in sufficient concentrations to be profitably mined. It is the practice in the mine to divide the copper-rich zone into four members which are, from bottom to top, the "lower sandstone" (actually the upper few feet of the Copper Harbor Conglomerate), the "parting shale," the "upper sandstone," and the "upper shale." The copper ore normally is concentrated in the "parting shale," the "upper sandstone," and the "upper shale." The concentration varies in richness, but locally may be as great as 70 pounds per ton. About 5,610,600 tons of ore were removed from the mine in 1962, averaging 1.15 percent copper per ton. In 1964 this mine produced about 16,000 tons of ore per day. This made the White Pine Mine the second largest underground copper mine in North America at that time (Ensign, 1964).

White and Wright (1954, pp. 676, 680, 688) concluded that the sediments which bear the copper sulphide were deposited in the shallow waters of lagoons, swamps, ponds, and stream channels on an ancient delta. Cross-bedding directions indicate that the source of sediment was an uplifted area to the south. The abundance of quartz, jasper, mica, and specular hematite grains and flakes in all Late Keweenawan rocks including the Nonesuch Shale, indicate that Precambrian rocks rich in iron formations, probably including many formations of Middle Precambrian Huronian age, were exposed to erosion in the source area. The streams flowed generally north or northwest from that highland source area, carrying their sediments down to the margin of the sea that occupied the area of the Lake Superior Syncline. These conclusions receive additional support from the studies of Hamblin (1961, 1965). His interpretation of the paleogeography in Late Keweenawan time is shown in Figure IV–41.

There still is some question concerning the manner in which copper, mainly in the form of copper sulphide, got into the Nonesuch Shale, but research on this problem continues. Movement along the White Pine Fault appears to have cut and offset the ore beds, so the fault, which is a more recent feature, could not have served as an avenue for rising hydrothermal solutions. Most of the copper deposits at White Pine seem to be controlled by, and closely related to, the type, distribution, and shape of the sedimentary layers themselves, not to the fault or the anticline.

Any hypothesis to account for the origin of the chalcocite ore at White Pine must take the close relationship between ore distribution and sediments into account. A satisfactory hypothesis must also explain the fact that much of the copper is in the form of a sulphide compound rather than in the metallic state, and the additional fact that the occurrence of the element does not depend upon close proximity to a lava flow or other igneous rock. It is certain that deposition of the White Pine chalcocite resulted from a quite different set of conditions than those that produced the native copper deposits of the Keweenaw Peninsula. White (1960, p. 402) recognized these limitations when he said, "the distribution of shale ore shows virtually complete independence of the structure and is related in almost all its aspects to the lithology and paleogeography of the host beds." He then summarized 4 hypotheses for the origin of the chalcocite deposits which are: 1—the copper was chemically precipitated into the original mud which later formed the Nonesuch Shale *at the same time* as the mud itself was being deposited; or 2—the copper moved by diffusion from the overlying water into the upper few inches of each layer of mud *after* the mud had been deposited; or 3—the copper was introduced into the deposits *after they had become lithified* (turned to rock) *and deformed* by folding and faulting, but following a pattern closely controlled by the distribution of the original sedimentary layers; or 4—the copper entered the shales from below, after they had accumulated to a great thickness but before they had been deformed. Hypotheses 1 and 2 imply that the copper came from waters that lay directly above the mud and from which the mud itself had come; hypothesis 4 suggests that the copper rose up into the deposits from below; and hypothesis 3 is not specific with regard to direction of movement. Hypothesis 1 implies that the copper was syngenetic, which means it was deposited at the same time as the mud itself. Hypotheses 2, 3, and 4 imply that the copper was *diagenetic,* which means it was emplaced in the sediments of the Nonesuch Shale at some time after the latter were deposited. Hypothesis 2 would require that the copper got into the mud very shortly after deposition of the latter and before many additional layers had accumulated. Hypotheses 3 and 4 would allow the copper to have been emplaced a considerable time after deposition of any particular layer of mud. In both hypotheses 1 and 2, one would expect that the distribution of any particular zone or layer of copper ore would fall within and be limited by the distribution of the layer of mud in which it had been deposited, that is, the boundaries of the individual copper zones should not cut across those of the sedimentary layers. Hypothesis 3 would imply a close but not necessarily absolute relationship between sediment layer and ore zone distribution. Hypothesis 4 does not require a close relationship between sediment boundaries and ore zones and, in fact, one would not expect it in this case. Hypotheses 1 and 2 would require that the copper have been in solution in the surface waters from which it was either precipitated or diffused downward into the sediments at the bottom.

We have listed hypotheses that have been erected after careful initial study of the rocks at White Pine, deducing some of the implications that would stem from each hypothesis. Although a final choice may be premature, recent field and laboratory studies already provide some means to test the hypotheses further against nature. First, geochemical studies have raised serious doubts that copper sulphide in sufficient quantities could have been introduced into the original muds from cool overlying surface waters, even though those waters were carried into the area of copper deposition by streams that had flowed across copper-bearing Keweenawan lavas. This weakens the arguments for both hypotheses 1 and 2. Moreover, recent field studies have shown that, contrary to the implications deduced from hypotheses 1 and 2, the top of the chalcocite rich zone in the Nonesuch Shale *does not conform* exactly to the shape and distribution of the strata in the shale. Instead, the top boundary of that zone cuts across boundaries of the sedimentary strata at a small but appreciable angle. This new data, added to the geochemical problem already mentioned, seems definitely to rule out hypotheses 1 and 2. If copper enrichment did not take place downward from above the shales, then hypotheses 3 and 4, which involve enrichment by waters ascending into the shales from below become more probable, but further studies of the distribution of the copper ore now show that it is not related in its distribution to either the fault or the White Pine anticline. This lack of structural control seems to rule out hypothesis 3. Thus, we are left with hypothesis 4 which suggests that the ore-bearing solutions rose from below, that the solutions entered the strata of the Nonesuch Shale before the latter were deformed, and that the ores were precipitated from those solutions within the shales in a manner partially but not entirely controlled by the shape and distribution of the strata. White himself (1960, p. 402) had come to favor hypothesis 4 at the time he wrote. According to still more recent studies (Brown, 1965) the copper-rich solutions probably rose from depth into the basal sandstones of the Nonesuch Formation, followed these

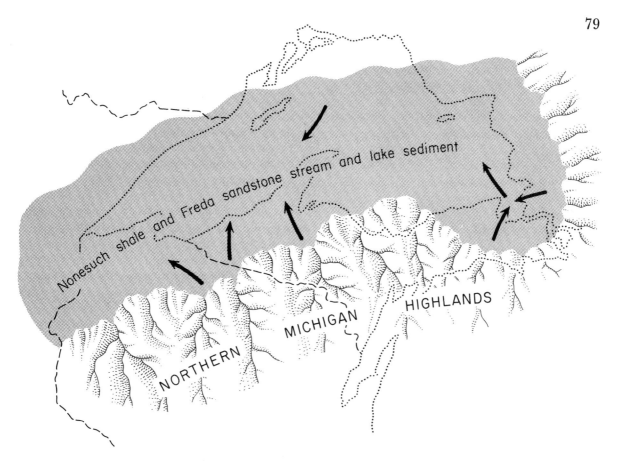

Figure IV-41. Paleogeography of Late Keweenawan (Latest Precambrian) time in Michigan. (Modified from Hamblin, 1961.)

porous avenues further into the formation and then moved by infiltration or diffusion into the adjacent overlying shales. The validity of these conclusions and their predictive use in the location and development of additional ore will be subjected to further field and laboratory testing which may require minor or even radical revisions of the explanation.

### The Role of Hypotheses in Geology

We had a very special reason for going into detail about the hypotheses for origin of copper at White Pine, and the implications of those hypotheses. Throughout this book we attempt to show, through Michigan examples, something about the reasoning processes and working methods employed in the science of geology. The problem of the origin of the White Pine copper sulphides offers an example of the use of a method of reasoning which is very important in geology, more so perhaps than in most other natural sciences. This is the method of *Multiple Working Hypotheses*. The geologist first makes a number of observations of rocks and their contents "in the field" to determine what really is the state of nature. This is mainly a descriptive effort and corresponds to the efforts of a zoologist or botanist who collects specimens, or a chemist or physicist who conducts initial experiments to determine what happens when two substances interact. The initial phases of any geological study involve this data-gathering process and some descriptive studies may stop at this point, particularly when a new field of investigation is first opened. But the important questions still remain to be asked and answered: *how, when,* and *why* did the phenomena that have been observed come to be what they are? At this point the geologist calls upon his knowledge of nature to see if he can recall any modern causes (processes) that are operating today which produce

**Geology of Michigan**

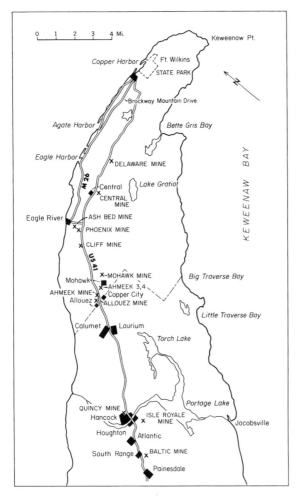

Figure IV–42. Map showing location of some Keweenaw copper mines.

effects (phenomena) similar to those he has observed in the rocks. He may call upon geochemists, geophysicists, or paleontologists to help him if the problem involves chemical, physical, or biological processes with which those specialists would be more familiar than he. However, if he does so he must remember that he is seeking solutions to real historical questions. The geologist then attempts to explain the phenomena. If his knowledge of the real situation is still limited, he may come up with several explanations, each of which has something to recommend it. These are the multiple working hypotheses. The next step involves deductive reasoning. The geologist asks for each hypothesis, "if this were true what other historical implications would be involved; what else must have happened?" The next step is to *test* these implications against further observations. This is the geologist's way of conducting further experiments. He goes back into nature to see if he can find in the rocks additional phenomena which will fit the implications of one or another of his hypotheses. He may go through this process several times, or many geologists and specialists may be involved, each erecting hypotheses, before the most probable answers to the important questions are found. Along the way the multiple hypotheses serve an important secondary function, which is to guide the search for information so that the limited time and energy of the geologist are not wasted on insignificant questions. This process of erecting hypotheses and testing them against nature is not unique to geology but for several reasons is more important there than for laboratory sciences. The enormous size of the earth and its features and the vast length of time required for change make it difficult to conduct meaningful laboratory experiments on any but the simplest and most limited of geological problems. Thus, the geologist conducts his most significant geological experiments by observing what happened to the earth in the course of its history. Testing against nature also is important because geological questions require answers in terms of what *actually did happen* rather than what "might have happened." To make this clear, let us go back to the problem of the origin of the White Pine copper deposits. The chemist might conduct laboratory experiments with copper and sulphur and devise a number of ways to synthesize these into the chalcocite mineral found at White Pine. One of his methods might actually correspond to that by which Nature herself composed the copper sulphide compound. But the chemist would never know which, if any, one of his hypotheses was correct because he would not know the *natural limitations* imposed on his experiments by the conditions *in the real world*. These limitations become clear to the geologist when he determines, by observation, *how* the deposits occur, *what* their relationships to associated rocks and minerals are, and *when* in the complex sequence of historical events that affected the earth the copper deposits were emplaced. Solutions to these questions are important to the economic geologist for their predictive value. In the case of the White Pine copper the correct hypothesis to explain the origin of the chalcocite could guide the further search for ore.

# V

# The Paleozoic—Era of Inland Seas

## Introduction

The Paleozoic Era began about 600 million years ago and ended about 230 million years ago. Vast as a span of 370 million years may seem by human standards, it is short compared with that of the Precambrian; if the Earth is more than 4 billion years old, then the Paleozoic Era represents less than one-tenth of geologic time. The Paleozoic rocks, much younger than those of the Precambrian, have had far less opportunity to be affected by crustal disturbances; consequently they are less altered by metamorphism and less folded and faulted. Furthermore, as less of the Paleozoic rock record has been lost through erosion, a more detailed historical record is available. In addition to these several differences, the Paleozoic rocks of Michigan were mostly deposited in relatively shallow epeiric (epicontinental) seas which frequently invaded the area; as these seas were the home of a great variety of marine life, both plant and animal, the Paleozoic rocks contain abundant fossils. All these characteristics of the Paleozoic rocks together make it easier to interpret a more detailed history of this interval of geologic time.

The maps in Figure v–1 are representations of the geography of the Michigan area during several different periods of the Paleozoic. They are, thus, *paleogeographic* maps. We pointed out in Chapter III that during the Paleozoic the north central part of North America was a relatively stable region where the earth's crust neither sagged deeply to receive great thicknesses of sediment nor rose extensively in lofty uplifts. It was a part of the Central Stable Region of the continent. The northern part of this area was the Canadian Shield, an area of very ancient rocks which had already passed through a complex history of sedimentary deposition, metamorphism, mountain building, and long periods of erosion. Early in the Paleozoic the southern portion of this stable region began to be intermittently submerged beneath seas which spread across the continental platform from the ocean basins and geosynclinal troughs which lay to both the east and the west. Thereafter, throughout the Paleozoic, the configuration of North America, particularly as reflected in the distribution of land and sea, changed frequently. The major changes in land-sea relationships as interpreted by the geologist are shown in the sequence of maps in Figure v–1.

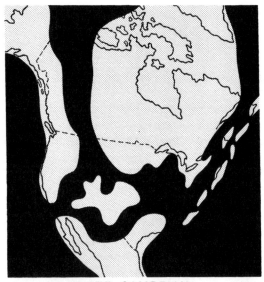
LATE CAMBRIAN

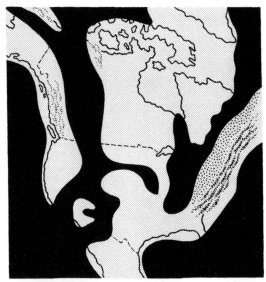
LATE DEVONIAN

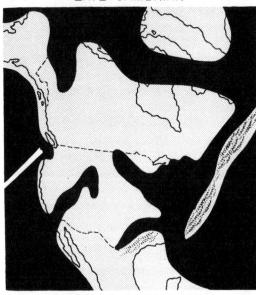
MIDDLE ORDOVICIAN

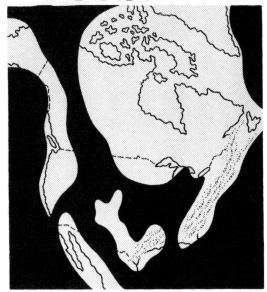
EARLY MISSISSIPPIAN

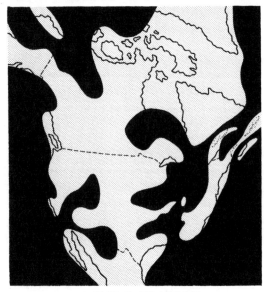
MIDDLE SILURIAN

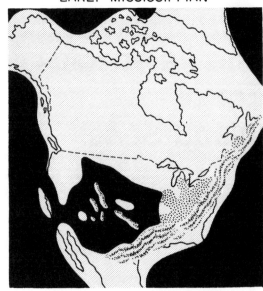
MIDDLE PENNSYLVANIAN

# The Paleozoic Era

Figure v-1. Paleogeography of successive periods of the Paleozoic Era in Michigan. (Modified from C. O. Dunbar, *Historical Geology*, 2d Ed., 1960, John Wiley and Sons, Inc.)

Before explaining in a general way the history of the region as shown in these paleogeographic maps, we should make clear what they represent through explaining how they are made and emphasizing the limits of their accuracy. Although the maps are both interesting and informative, one must remember that they represent a geologic *interpretation*—they are not a simple portrayal of facts. They are an interpretation of the *maximum distribution* of seas during a *selected span* of time. In this one way these maps differ from an ordinary geographic map which represents present-day conditions, at an "instant" in geologic time; boundaries drawn on paleogeographic maps must embody conditions that changed over a period of time, and are thus only approximations or averages. To cite an analogy, a map showing the battle areas of World War II would separate all those positions which saw fighting at one time or another from those positions which saw no fighting, but this map (corresponding to the paleographic map) would not show the position of the battle line at any particular instant (corresponding to the geographic map).

To make a paleogeographic map the geologist must first decide what part of time he wishes to portray. He may elect to show the approximate land-sea boundary and other geographic features for as small a portion of time as the accuracy of dating methods will allow, or he may lump together the conditions that existed over a much longer period. He then plots on a base map the points of known occurrence of all rocks whose ages fall within the selected time interval. Seldom are all rock formations formed during a particular period still preserved; often much of the original rock record has been eroded so the geologist must make an educated guess as to where rocks of that age logically would have occurred before erosion. This "guess" involves the Principle of Original Continuity discussed in Chapter I. Suppose the geologist is attempting to draw the approximate shoreline between land and sea for some time past. It may be clear that the nearshore deposits one would expect are missing. This requires an estimate of how far off and in what direction the shoreline lay. The best of paleogeographic maps are no better than generalized approximations. It can never be said that at any past instant boundaries were exactly as portrayed. If a sea is shown to have occupied Michigan during some past period, this itself will not be in doubt, because rocks of undoubted marine origin are common throughout the state. The uncertainties lie in the exact distribution of that sea and its exact limits in time.

Now observe the place of Michigan on each of the paleogeographic maps in Figure v-1. Events that occurred in Michigan during each period of the Paleozoic are discussed more fully later in this chapter, where more detailed paleogeographic maps are presented at appropriate points. Many "tools" are available to the geologist in his attempt to read the historical record of sedimentary rocks. Some of these have been discussed in Chapters I–III. Additional details are presented in Chapter XII, which deals with rock identification and interpretation. The fossil record of life in the past, as discussed in Chapters XIII, XIV, and XV also provides clues. If your background in geology is slight, you should read those chapters first in order to have a better appreciation of the basis for some of the statements concerning the Paleozoic history of Michigan that follow.

## The Paleozoic Rock Record in Michigan

Figure v-2 summarizes much of the information the geologist has collected concerning the events of the Paleozoic Era in Michigan, information recorded in the sedimentary rocks which underlie most of the state. On this figure are found the names of the Paleozoic rock formations of the Michigan area arranged in order of their *relative* geologic age, according to the Law of Superposition (Chapter II), from oldest at the bottom to youngest at the top. The approximate ages in years (absolute time) determined by radioactive dating methods are shown on the time scale inside the front cover.

How was this information assembled? First, remember that it incorporates the observations of many geologists during a period of over 100 years: men employed by the Michigan State Geological Survey, faculty members of universities and colleges of the state, employees of many independent

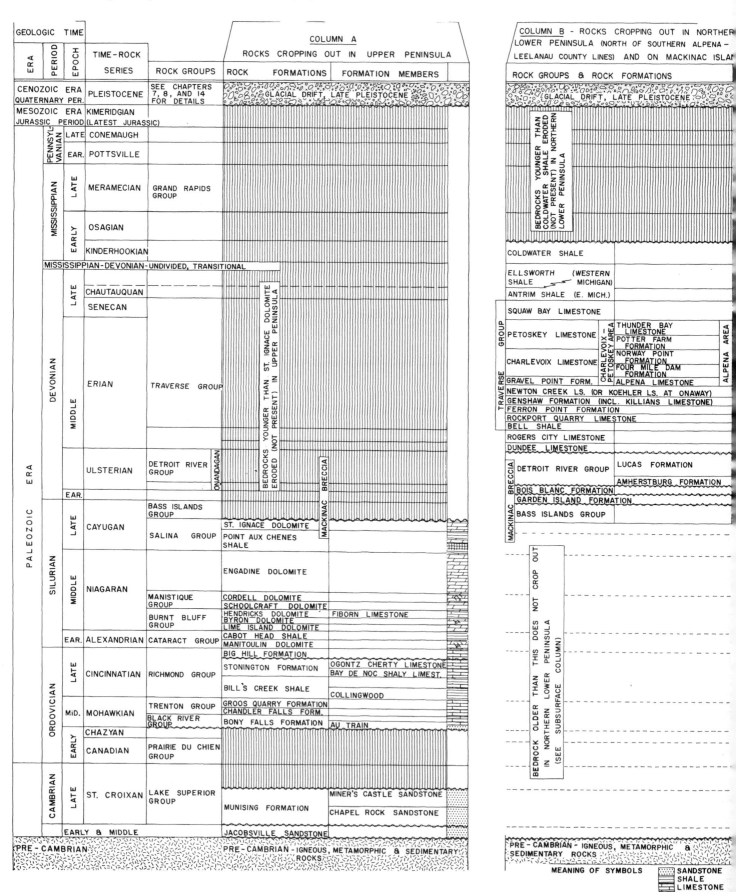

Figure v-2. Stratigraphic chart of Post-Precambrian rocks in Michigan. (Modified, and with additions, from Michigan Department of Conservation—Geological Survey Division Chart No. 1–1964, which was compiled by the Geologic Names Committee, Garland D. Ells—chairman, with the help of H. O. Sorensen, R. W. Kelley, H. J. Hardenberg, L. D. Johnson, and F. W. Terwilliger.)

# COLUMN C - ROCKS CROPPING OUT IN CENTRAL LOWER PENINSULA, OR FOUND IN THE SUBSURFACE BY DRILLING

| ROCK GROUPS | ROCK FORMATIONS | FORMATION MEMBERS | SOME INFORMAL TERMS USED IN SUBSURFACE PETROLEUM WORK | APPROX. MAX. SUBSURFACE THICKNESS IN FEET |
|---|---|---|---|---|
| GLACIAL DRIFT, LATE PLEISTOCENE | | | | |
| | "RED BEDS", UNNAMED, ?LATE JURASSIC? | | | 220 |
| | GRAND RIVER FORMATION | IONIA, EATON, AND WOODVILLE SANDSTONES | | 750 |
| | SAGINAW FORMATION | VERNE LIMESTONE | PARMA SANDSTONE | |
| GRAND RAPIDS | BAYPORT LIMESTONE | | | 160 |
| | MICHIGAN | | "TRIPLE GYP" BROWN LIME STRAY-STRAY SANDSTONE - GAS STRAY DOLOMITE | 600 |
| | MARSHALL FORMATION | NAPOLEON SANDSTONE | STRAY SANDSTONE - GAS, OIL MARSHALL SANDSTONE GAS, OIL | 330 |
| | | LOWER MARSHALL SS. | | |
| | COLDWATER SHALE (WEST) | "MARSHALL" SANDSTONE (EAST) | COLDWATER LIME WEIR SAND - GAS COLDWATER RED - ROCK | 1300 |
| | ELLSWORTH SHALE (W. MICH.) | | "BEREA" (W. MICH.) - OIL, GAS BEREA SAND (E. MICH.) - OIL, GAS | 1320 |
| | ANTRIM SHALE (E. MICH.) | | | |
| TRAVERSE (GENERALLY UNDIVIDED IN SUBSURFACE) | SQUAW BAY LIMESTONE | | SQUAW BAY - OIL, GAS TRAVERSE FORMATION TRAVERSE LIME - OIL, GAS STONEY LAKE ZONE - OIL, GAS (UPPER TRAVERSE IN W. MICH.) | 830 |
| | ALPENA LS. | | | |
| | BELL SHALE | | | |
| | ROGERS CITY LS. | | ROGERS CITY LS. - OIL, GAS | 475 |
| | DUNDEE LIMESTONE | | DUNDEE LIMESTONE - OIL, GAS | |
| DETROIT RIVER | LUCAS FORMATION | | REED CITY ZONE - OIL, GAS SOUR ZONE - OIL, GAS RICHFIELD ZONE - OIL, GAS | 1450 |
| | AMHERSTBURG FM. | | | |
| | BOIS BLANC FORM. | | | 800 |
| | GARDEN ISLAND FM. | | | 100 |
| BASS ISLANDS | | | | 700 |
| SALINA | G UNIT | | | |
| | F EVAPORITES | | | |
| | E UNIT | | E-ZONE (OR KINTIGH ZONE) - OIL | |
| | D EVAPORITE | | | |
| | C UNIT | | | 3150 |
| | B EVAPORITE | | A-2 DOLOMITE - GAS | |
| | A-2 CARBONATE | | A-2 LIME - GAS | |
| | A-2 EVAPORITE | | | |
| | A-1 CARBONATE | | A-1 DOLOMITE - OIL, GAS | |
| | A-1 EVAPORITE | | | |
| NIAGARA | | | BROWN NIAGARAN - OIL, GAS GRAY NIAGARAN - OIL, GAS WHITE NIAGARAN CLINTON SHALE (E. MICH.) | 980 |
| CATARACT | CABOT HEAD SH. | | | 200? |
| | MANITOULIN DOL. | | | |
| RICHMOND | QUEENSTON SHALE | | | 950 |
| EDEN | UTICA SHALE | | | |
| TRENTON - BLACK RIVER (UNDIVIDED) | | COLLINGWOOD SHALE | TRENTON GROUP - OIL, GAS BLACK RIVER FM. - OIL, GAS BLACK RIVER SH. - OIL, GAS VAN WERT ZONE - OIL, GAS | 1100 |
| | | GLENWOOD | | |
| | ST. PETER SS. | | | 260 |
| PRAIRIE DU CHIEN | SHAKOPEE DOL. | | | |
| | NEW RICHMOND SS. | | | 425 |
| | ONEOTA DOLOMITE | | ONEOTA DOLOMITE - OIL | |
| LAKE SUPERIOR | TREMPEALEAU FORMATION | JORDAN SANDSTONE | | 750 |
| | | LODI DOLOMITE | | |
| | | ST. LAWRENCE DOLOMITE | | |
| | FRANCONIA SS. | | | |
| | DRESBACH SS. | | | 1175? |
| | EAU CLAIRE FM. | | | |
| | MOUNT SIMON SS. | | | |
| | JACOBSVILLE SS. | | | 1100+ |
| PRE-CAMBRIAN - IGNEOUS, METAMORPHIC, SEDIMENTARY ROCKS | | | | |

TOTAL 19,125 (NOT ALL BENEATH ANY ONE POINT)

- EVAPORITE (HALITE, GYPSUM, OR ANHYDRITE)
- COAL
- CARBONATE REEFS OR BIOHERMS OF ORGANIC ORIGIN
- MAJOR EROSIONAL GAP IN RECORD. ROCKS OF THESE AGES MISSING HERE; MAY BE PRESENT ELSEWHERE. UNCONFORMITY.
- EROSIONAL GAP IN ROCK RECORD. UNCONFORMITY.

# COLUMN D - ROCKS CROPPING OUT IN SOUTHEASTERN LOWER PENINSULA (WAYNE, WASHTENAW, MONROE, & LENAWEE COUNTIES) AND IN ADJACENT NORTHWESTERN OHIO

| ROCK GROUPS | FORMATIONS | & MEMBERS |
|---|---|---|
| GLACIAL DRIFT, LATE PLEISTOCENE | | |
| BEDROCK YOUNGER THAN THE MARSHALL FORMATION ERODED AWAY (NOT PRESENT) IN SOUTHEAST MICH. | | |
| MARSHALL FORMATION | NAPOLEON SANDSTONE | |
| | MARSHALL SANDSTONE | |
| COLDWATER SHALE | | |
| SUNBURY SHALE | | |
| BEREA SANDSTONE | | |
| BEDFORD SHALE | | |
| ANTRIM SHALE | | |
| TRAVERSE GROUP (ONLY PARTIALLY REPRESENTED, SOME TIME GAPS) | (TIME GAP, NO ROCKS) | |
| | TEN MILE CREEK DOLOMITE | |
| | (TIME GAP, NO ROCKS) | |
| | SILICA FORMATION | |
| | (TIME GAP, ROCKS ABSENT) | |
| | DUNDEE LIMESTONE | |
| DETROIT RIVER GROUP | LUCAS FORMATION | ANDERDON FORM. |
| | | AMHERSTBURG FORM. |
| | | SYLVANIA SS. |
| | BOIS BLANC FORMATION | |
| | TIME GAP, NO ROCKS | |
| BASS ISLANDS GROUP | RAISIN RIVER DOLOMITE | |
| | PUT-IN-BAY DOLOMITE | |
| SALINA GROUP (DOES NOT CROP OUT AT SURFACE BUT IS REACHED BY SALT MINES IN THE DETROIT-WYANDOTTE-WINDSOR, ONTARIO AREA) | | |

BEDROCKS OLDER THAN THIS DO NOT CROP OUT IN SOUTHEASTERN MICHIGAN, BUT MAY BE PRESENT IN SUBSURFACE (COLUMN C)

AREAS REPRESENTED BY VERTICAL STRATIGRAPHIC COLUMNS

(Map of Michigan showing COLUMN A in Upper Peninsula, COLUMN B in northern Lower Peninsula, COLUMN C in central Lower Peninsula, COLUMN D in southeastern Lower Peninsula)

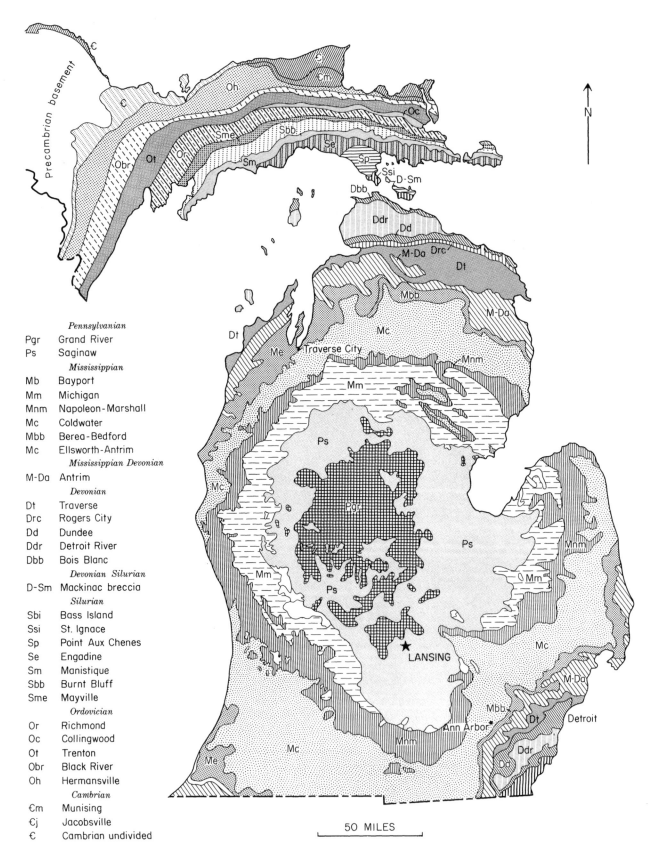

Figure v–3. Bedrock geologic map of Michigan. For western part of Upper Peninsula see Figure IV–33. The distribution of rocks is shown as though the overlying cover of Pleistocene glacial deposits had been stripped off. For a map of the surficial deposits see Figure VII–16. For source of subsurface information see Figure V–4. (Modified from Martin, 1957, "Geological Map of Michigan," courtesy of Michigan Conservation Department.)

oil and mining companies, the United States Geological Survey, and many interested laymen. Some of our knowledge concerning the nature of Paleozoic rocks beneath the state comes from studies of surface outcrops; recall from Chapter III that the sedimentary strata in the downwarped Michigan Basin are like nesting bowls whose eroded edges reach the surface (or are buried beneath a relatively thin surface veneer of glacial deposits) in a series of concentric rings. The younger layers approach or reach the surface near the center of the basin, the older layers around the edges (Fig. v-3). Although bedrock outcrops are rather rare in the Lower Peninsula, occurring mainly in eroded river banks, road cuts, or quarries, extensive collections of rocks, minerals, and fossils made from surface exposures are to be found at many of the universities and colleges of the state. Details on the location of bedrock outcrops are given in a book by Helen Martin (1956). Much additional information regarding the rocks deep beneath the surface comes from cores and cuttings brought up in wells drilled for oil, gas, brine, and water. This subsurface information is collected by several organizations within the state. The Subsurface Laboratory of The University of Michigan Department of Geology and Mineralogy in Ann Arbor is one of several repositories for such records (Fig. v-4); the Michigan Geological Survey is another.

Notice on Figure v-2 that separate *stratigraphic* columns are shown for the Upper Peninsula, the Northern Lower Peninsula and Mackinac Straits Region, the Central Lower Peninsula, and for Southeastern Michigan. The state of Michigan covers such a large area that its several parts had varying histories. Thus, the rocks deposited in one area often differ from those in another. Deposits of a particular age may be nearshore sandstones near Detroit, for example, but these may change either gradually or abruptly to offshore limestones in the center of the state because the environment of deposition changed in that direction. Sometimes erosion erased part of the sedimentary record, or no sediments accumulated in one part of the state during one interval of time, whereas a sedimentary record for that time may be present elsewhere.

Some idea of the amount of sedimentary rock in the Michigan Basin in which evidence of Paleozoic events is preserved, the section summarized diagrammatically in Figure v-2, is given by Cohee (1965). He states that the area occupied by the Michigan Basin is about 122,000 square miles and that the total volume of Paleozoic sedimentary rocks in the basin is about 108,000 cubic miles. About 1 percent of these rocks are of Pennsylvanian age, 5 percent Mississippian, 16 percent Devonian, 30 percent Silurian, 21 percent Ordovician, and 27 percent Cambrian. These percentages neglect, of course, the thousands of cubic miles of rock of each of these ages that have been removed by erosion. The low percentage of Pennsylvanian and Mississippian rocks, for example, is largely due to the fact these younger rocks were deposited on top of the others and hence were the first to be subject to erosion at a later time. Of the rocks that do remain, about 47 percent are carbonates (limestone, dolomite), 23 percent sandstones, 18 percent shale, and 12 percent evaporites (salt, anhydrite, and gypsum).

The column on Figure v-2 which shows the approximate maximum subsurface thickness of all the rock units totals 19,125 feet. Not all of this was deposited in any one place, however; the maximum thickness of one formation may be in a different location than that of another, or some units locally may be partially removed by erosion. A better estimate of the maximum total thickness of all Paleozoic rock units beneath any one point is 14,000 feet. Near the center of the basin, just west of Saginaw Bay, one would have to drill down over two and one half miles to reach the ancient, Precambrian basement! The Paleozoic formations tend to thin toward the basin margins where uplift, resulting in occasional erosion or nondeposition, was more frequent. The rocks dip gently downward into the subsurface toward the basin center (Fig. III-2) at an average inclination of about 60 feet per mile (less than 1 degree). Such a slight angle is imperceptible in any single roadcut or quarry, although minor folding has increased the inclination locally so that some dip may be visible. Over a distance of many miles, however, even the slight regional dip carries individual rock units many thousands of feet into the subsurface.

The Paleozoic sediments overlie the Precambrian "basement." As already pointed out little is known in detail of the nature of these ancient rocks where they are covered by the Paleozoic; few wells penetrate the entire sedimentary column, and those that do fail to extend far into the base-

Figure v-4. Subsurface Laboratory, Department of Geology and Mineralogy, The University of Michigan. Over 26,000 oil and gas wells have been drilled in the state. This laboratory collects, stores, and analyzes records and samples from many of these wells. Information, stored on key-punched cards and magnetic tapes or disks, can be fed into computers, using programs written in the laboratory, to produce numerical or visual analyses of many types.

*Top right*—Graduate geology student consults complete file of drilling records.

*Center right*—Rock chips, brought up in continuous series as a well is drilled, are mounted on cards and stored in special racks. A card is being examined under a binocular microscope.

*At left*—Two views of such chip "logs." Each card, 24 inches long as shown by yardstick, represents about 500 feet of well. At this scale, several cards may be required for a single well. Close-up shows chips, representing black shale underlain by gray shale, at depths between 590 and 726 feet below the surface.

*Bottom right*—Cylindrical rock cores, cut with a special drill bit. From right to left these illustrate sandstone, coralline limestone, lagoonal muds, and coral reef rock. (Yardstick in inches for scale.)

ment. According to Cohee (1965, p. 214-15), the deepest well yet drilled in Michigan, in Ogemaw County just northwest of Saginaw and near the center of the basin, penetrated 13,000 feet of Paleozoic rocks. It "bottomed" in Cambrian sandstones, not reaching the Precambrian. What sketchy evidence there is on the character of the Precambrian surface on which the Paleozoic rocks rest indicates that it is not particularly smooth. Cohee, in the paper cited above, notes that 2 wells drilled only 3 miles apart on Beaver Island, in the northern part of Lake Michigan, reached Precambrian rocks at depths of 4705 and 4000 feet below sea level. This difference in elevation of over 700 feet over such a short horizontal distance suggests that the ancient and now deeply buried erosion surface, the surface on which the first Paleozoic sediments were deposited, has considerable relief. This pre-Paleozoic land surface was marked by hills and valleys, just as is the land surface bordering the modern oceans. Because of the bowl-like shape of the Michigan Basin, Precambrian rocks now are closer to the surface near the basin margin. The basement is only 3626 feet below sea level at the southern end of Lake Michigan and in the area around southeastern Michigan it lies between 1400 and 2400 feet below sea level (Cohee, 1965). In the western part of the Upper Peninsula, where it is exposed at the surface, the Precambrian "basement" is in places well over 1000 feet above sea level. The few samples of Precambrian rocks recovered by deep drilling in the Michigan Basin are of complex rocks of igneous and metamorphic origin, much like those rocks exposed in the Canadian Shield. They probably range in age from about 800 million to about 1.5 billion years (Cohee, 1965, p. 215).

In order to understand and use Figure v-2 to best advantage the reader must understand what is meant by the terms which serve as column headings. You have already been introduced to the terminology of the geologic time scale, with its several orders of subdivisions called "eras," "periods," and "epochs," and with the arrangement of these periods of relative time in accordance with the Law of Superposition, such that the oldest time interval is at the bottom and the youngest is at the top. The next column, going from left to right is headed "Time-Rock Series." The names there, such as Kimmeridgian or St. Croixan, are of areas in North America or Europe where a series of rocks representing this particular part of geologic time is well exposed and serves as the standard reference section of that part of time. The "Rock Group" names in the next column to the right refer to sets of two or more associated and related rock units. Both the number and names of the units within a "group" may change from place to place. For example, the Middle Devonian Traverse Group (Column B) includes different units in the Charlevoix-Petoskey area than in the Alpena area; another example of such variation within the Traverse Group can be seen in comparing Columns C and D; in subsurface, the Group is largely undifferentiated and undivided, although complete, whereas in southeastern Michigan several parts of the Group are missing and those that are present are different than their equivalents in the northern part of the Lower Peninsula (Column B). These differences are in some cases due to nondeposition or subsequent erosion during a part of an interval in one place while there was continuous deposition in another; in other cases the differences are due to the fact the environment of deposition varied from place to place. A rock "group" may also represent a greater length of time in one area than in another

# Geology of Michigan

if the environment in which it accumulated persisted longer there. Consider the Salina Group of Silurian age as an example. In the central part of the Basin (Column C), deposition of the Salina began about halfway through the Middle Silurian and continued to the end of the period; in the Upper Peninsula (Column A) the Engadine Dolomite, representing a different environment of deposition, accumulated during the first part of that same time span and the Salina units (locally called the Pointe aux Chenes Shale and St. Ignace Dolomite) were deposited only during the Late Silurian.

The term *formation* is applied by the geologist to a set of originally continuous rock strata which have a common suite of rock and fossil characteristics. A formation may consist entirely of rock of one type, the Sylvania Sandstone (Column D) for example, in which case the formation name includes the rock (or lithologic) name, sandstone. Often a formation is comprised of two or more lithologies which alternate, and then a lithologic name is not used. Instead, the word "formation" follows a name taken from a particular locality where the formation is well represented, usually the place where it was first studied. The Middle Devonian Silica Formation in southeastern Michigan, for example, is comprised of both limestone and shale; it was first studied in detail at Silica, Ohio, a town a few miles west of Toledo. A formation is usually widespread. Two or more formations which are typically found associated within an area form a "Group," discussed above; the "units" making up the Group are formations. In many cases there are local variations within a single formation, which variation leads to smaller, more restricted subdivisions called *members*. In accordance with the Principle of Original Continuity (Chapter I) a sedimentary formation can be only as widespread as was the ancient environment in which it was deposited; no formation extends indefinitely. Furthermore, a formation is deposited for only that period of time in which the environment of its deposition persisted. If the environment of deposition of a formation shifted its position or disappeared, then deposition of the formation shifted to another place or the formation ceased to accumulate. As part of a given formation might accumulate in one place while part of another formation accumulated elsewhere the two formations, representing different, but age-equivalent, environments, might grade laterally into one another (they might "interfinger" or "intertongue").

In order to further illustrate some of the points just made concerning the intertonguing of different formations (actually, what the geologist refers to as "facies change," the changing from one rock type, or facies, to another), and to give the reader practice in reading and understanding Figure V-2—to practice some four-dimensional geological imagination in time and space—let us go through an example or so. Find the Middle Devonian Sylvania Sandstone in Column D. Note that to the left, in the column representing central Michigan, rocks of the same age are called the Amherstburg Formation. The Sylvania Sandstone is a marine beach sand which accumulated along the margin of a sea in southeastern Michigan at the same time that limestone and dolomite of the Amherstburg Formation were accumulating in the offshore waters of that same sea. The sandstone, limestone, and dolomite are age equivalent rock "facies" which intertongue with one another. Thus, one could describe the Middle Devonian history of southeastern and central Michigan as having been a time of seas when the shore stood near or in southeastern Michigan and offshore waters lay toward the basin center. Also note in Column D that as time passed, the beach and near-shore environment represented by the Sylvania Sandstone gave way to offshore marine waters of the advancing "Amherstburg Sea." The marine origin of those rocks is proven by their marine fossil content.

For another example, locate the St. Peter Sandstone of Early Ordovician age near the bottom of Column C. This is another marine beach and near-shore sandstone. Look horizontally across the chart at the same level (and same time) to Column A representing the Upper Peninsula. There is no St. Peter Sandstone (nor any rocks of equivalent age) shown in Column A. The vertical ruling on the chart signifies a "time gap" or unconformity in Column A. Rocks younger than the St. Peter Sandstone (Bony Falls Formation and Au Train Sandstone) rest directly upon Cambrian Sandstones in the western part of the Upper Peninsula. This unconformity indicates that in the Upper Peninsula, during the time between the Late Cambrian and the Middle Ordovician, either no sediments were deposited or, if they were deposited, were subsequently eroded. In either case this would indicate that while the St. Peter Sandstone was accumulating in marine waters in southeastern Michigan, the Upper Peninsula was emergent and was probably being eroded. An unconformity such as this might be recognized in one or more ways. There might be an irregular, wavy surface of erosion between the rocks above and below it. Possibly an old soil zone, produced by weathering during uplift and exposure, caps the rocks below the unconformity. If the earth's crust was folded or faulted during emergence, the rock layers below the unconformity may have been tilted and beveled by erosion before the later, post-unconformity, strata were deposited across their eroded edges; this type of gap in the record is called an "angular unconformity" because the strata above and below the erosion surface are not parallel. Some unconformities are recognized from fossil evidence; in the case cited above of the unconformity between the Cambrian and Ordovician rocks in the Upper Peninsula, the rocks below the unconformity contain fossils of Late Cambrian age while those above contain Middle Ordovician fossils. Rocks (fossils) of Early Ordovician age are missing, indicating that there is a time gap in the rock record here.

These two examples should serve to demonstrate the manner in which you can use Figure v-2 to interpret the historical events of a part of geologic time in Michigan. We suggest that you try your hand at a similar interpretation; you can check your own interpretations against those found in the remainder of this chapter, which is devoted to a review, geologic period by period, of the historical events of the Paleozoic Era in Michigan. Far more details are known about this period of time than can be presented here, but the most significant and interesting events are included. Some emphasis will be placed on rocks and minerals of economic value that are typical of certain periods (e.g., salt in the Silurian and coal in the Pennsylvanian), but the primary concern in any case will be with the geology of those materials rather than their place in the economy.

## The Cambrian Period— Beginning of the Paleozoic Era

At the beginning of the Cambrian, remnants of mountains initially formed during the Precambrian-Penokean Orogeny (Chapter IV) remained in a belt extending from Ontario, Canada, on the east, across the central and southern part of the Upper Peninsula and into Wisconsin. Hamblin (1961) has called these the "Northern Michigan Highlands." Streams flowing northward from those highlands carried sand and gravel down to lowlands that lay along the present northern edge of the Upper Peninsula. Those lowlands included the area now occupied by the Keweenaw Peninsula and Lake Superior, although neither of those latter features was in existence then. Upon reaching the gentler lowland slopes, the Early Cambrian streams slowed down and dropped their load of sediment in shifting channelways or in local lakes. The coarse sand and gravel deposits and interbedded shales that resulted included fragments of many rocks and minerals eroded from the preexisting Precambrian rocks of the source area in the Northern Michigan Highlands. These deposits form the oldest Cambrian formation exposed in the Upper Peninsula, called the *Jacobsville Sandstone*. The Jacobsville is of Early and Middle Cambrian age (Fig. v-2, Column A). As it accumulated on an irregular Precambrian rock surface that had not been completely smoothed by erosion, its thickness varies from as little as 46 feet in some parts of the Upper Peninsula to over 1100 feet in others, depending on whether it barely lapped over hilltops or filled ancient valleys. Figure v-5, top, shows part of that irregular Precambrian erosion surface and unconformity between Precambrian rocks and the Jacobsville Sandstone on Presque Isle just north of Marquette. Excellent and readily accessible exposures of the Jacobsville also occur in roadcuts, and in the shore bluffs of Lake Superior, along U.S. Highway 41 between the towns of L'Anse and Baraga at the south end of Keweenaw Bay (Fig. v-5, middle). Much surface bedrock of the southeastern half of the Keweenaw Peninsula, and in a broad belt extending from Keweenaw Bay southwestward into the western Upper Peninsula, belongs to this formation (Fig. v-3, map). A combination of characters, including prominent, trough-type cross-bedding, channel-fill lenses, and moderate sorting and rounding of sand grains, indicate that the Jacobsville Sandstone was laid down by terrestrial streams and in lakes. The direction in which the cross-bedding is inclined is an indication of the direction of current flow in

Cambrian, Jacobsville sandstone

Precambrian rocks

Є
pЄ

Figure v–5. Early and Middle Cambrian age Jacobsville Sandstone.

*Top*—Nearly horizontal Jacobsville strata unconformably overlying Precambrian granitic rocks along irregular erosion surface with ancient low hills and valleys indicated by dotted line. Presque Isle just north of Marquette (courtesy Michigan Department of Conservation). Note crisscrossing igneous dikes in Precambrian.

*Middle*—Jacobsville Sandstone along US 41 between L'Anse and Baraga. Keweenaw Bay and L'Anse at left. Inset close-up shows arcuate, trough-type cross-bedding formed by stream currents flowing toward north (right) off an ancient Precambrian (Penokean) uplift, the "Northern Michigan Highlands" (see Fig. IV–3).

*Bottom*—Jacobsville Sandstone deposited on irregular unconformity and lapping up on ancient Precambrian hill of Mesnard Quartzite. Dotted line shows unconformity. Figure on railroad track shows scale. US 41 is beyond tracks. Along Marquette Bay between Marquette and Harvey. Close-up shows wave ripple marks in Mesnard Quartzite, a metamorphosed Precambrian marine beach sand.

those ancient streams (Chapter XII). Hamblin (1958) measured the angles of inclination and direction of many of the crossbeds in this formation and determined that the streams of the area in Jacobsville time were flowing generally northward, moving rather swiftly along fairly straight courses as they left the highlands that lay to the south, but flowing more slowly and with much meandering farther north. The regional slope must then have been rather steep close to the front of the highland sediment source area but diminished northward (Fig. v–6).

The attractive, reddish-brown Jacobsville Sandstone is used locally as a building stone. Its red color is due to the abundance of ferric iron oxide (hematite), much of which was eroded from Precambrian iron formations exposed in the highland source area from which the streams flowed in Jacobsville time. Occasional greenish spots occur where acidic ground water solutions have chemically reduced the iron to the ferrous state.

No Early or Middle Cambrian deposits equivalent in age to the Jacobsville have been recognized in the subsurface in the southern part of the state, so that area may have been high standing then.

After Jacobsville time, and before the Late Cambrian, the region was slightly uplifted and tilted. This is indicated by the fact that the edges of strata in the Jacobsville Sandstone are locally beveled by erosion and lie at an angle beneath the overlying, younger beds of the Munising Formation.

By Middle Cambrian time, seas had begun to invade North America from the west and northwest (Fig. v–1) and from the south. In the Late Cambrian those seas reached Michigan, lapping onto the margins of the Northern Michigan Highlands from both the north and south. At the beginning of the Late Cambrian streams still were flowing both northward and southward from those highlands, eroding the Jacobsville Sandstone as well as Precambrian rocks, and carrying their sediment to the margins of the Late Cambrian sea. Waves, pounding on the shores of that sea, shifted the particles of sand back and forth along the beach, rounding them to nearly perfect spheres, removing the finer particles to offshore waters, and depositing the remaining, dominantly quartz, grains as a near-shore beach sand. That ancient marine sand now forms the *Chapel Rock Sandstone* member of the *Munising Formation*. Those familiar with the scenic attractions of the Upper Peninsula will recognize that these are both names derived from the town of Munising, near which spectacular exposures of the formation occur in the wave-cut Pictured Rocks cliffs of Lake Superior, and from a wave-carved "stack" and cave called Chapel Rock east of Munising (Figs. v–7, 8). The Chapel Rock Sandstone in Michigan is about the same geologic age as the Dresbach Sandstone which was accumulating at the same time in seas along the south side of the Northern Michigan Highlands. The Dresbach Sandstone now is found in the Lower Peninsula of Michigan and in adjoining parts of Wisconsin, Minnesota, and Illinois. The paleogeography of Chapel Rock Sandstone time is shown in Figure v–9. Thereafter, in latest Cambrian time, the Munising Sea spread over most of the Upper Peninsula, drowning the eroded Northern Michigan Highlands (Fig. v–10). The source of sediment had shifted to the east and northeast. Streams, eroding Precambrian rocks and flowing to the sea from highlands then located in

# Geology of Michigan

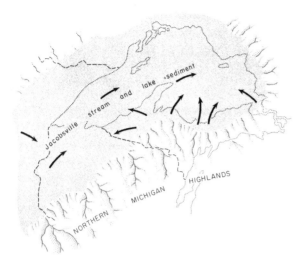

Figure v–6. Paleogeographic map of Early and Middle Cambrian (Jacobsville Sandstone) time in Michigan and adjoining regions. (Modified from Hamblin, 1961.)

Figure v–7. Late Cambrian marine sandstones of the Munising Formation in the Pictured Rocks area east of Munising along the south shore of Lake Superior.

*Top*—Miner's Castle. Two wave-cut stacks related to earlier glacial lake level. Man's figure for scale on wave-cut terrace near base of right hand stack (see arrow). Sea Cave, near water level, is of more recent origin.

*Bottom*—Chapel Rock, a complex sea cave also related to earlier glacial lake level. Collapse of roof would leave supporting columns as stacks (also see Figs. ix–25, 26, 27). The wave-cut shoreline features along modern Lake Superior formed about 500 million years later than the marine sandstones themselves. The ancient Cambrian sea in this area had no relationship to the modern lake features. (Courtesy Michigan Department of Conservation, photos by Robert Harrington.)

Canada, carried sediment of a somewhat different mineral composition, indicating an origin from a source area different than that which provided material for the Chapel Rock Sandstone. Thus, the *Miner's Castle Sandstone* member of the Munising Formation, another near-shore marine beach sand, was deposited above the Castle Rock Sandstone. The Miner's Castle Sandstone is equivalent in age to the Franconia Sandstone of southern Michigan, Wisconsin, and Minnesota. Fossilized marine animals called trilobites (Chapter XIII), found in the Miner's Castle Sandstone near Waucedah in Dickinson County, serve to prove the marine origin of the sandstone. The Munising Formation, with its Chapel Rock and Miner's Castle sandstone members, crops out today as a cliff or escarpment because the original continuity of those strata has been cut by later erosion. As shown on Figure v–11, this Cambrian sandstone cliff follows the south shore of Lake Superior between Au Sable Point and Munising, forming the beautiful *Pictured Rocks*. Both west of Munising and east of Au Sable Point the Cambrian sandstone cliffs swing inland. In places the cliff or "cuesta" is capped and protected from erosion by more resistant, sandy dolomites and dolomitic sands of the Middle Ordovician Au Train Formation. The sandstones are largely covered by glacial sand and gravel of Pleistocene age between Au Sable Point and Tahquamenon Falls. But at the Falls the powerful Tahquamenon River has scoured off the glacial deposits to expose the Cambrian sandstones again in the "Upper" and "Lower" Tahquamenon Falls (Fig. v–12).

Many other beautiful waterfalls along the north edge of the Upper Peninsula are similarly formed where streams plunge over the sandstone cliffs of the Munising and Au Train formations on their journey northward to Lake Superior. Au Sable, Miners, Munising, Laughing Whitefish, and Au Train falls are examples (Fig. v–13).

As the Late Cambrian Munising Sea gradually spread over the eroding Northern Michigan Highlands, the shallow waters of that sea incorporated a coarse gravel, or "basal conglomerate," into its deposits. That conglomerate was composed of weathered and eroded fragments of Precambrian rocks that lay on the surface of those ancient highlands. Many of the rocks exposed on the surface of the highland were Precambrian "iron formations" (Chapter IV). Figure v–14 shows that erosion surface, now an unconformity, and the basal conglomerate of the Late Cambrian Munising Formation resting on the Precambrian Vulcan Iron Formation. The photographs were taken in the old Quinnesec Mine (now called "The Devil's Icebox") just off U.S. Highway 2 near the town of Quinnesec in the southern part of the Upper Peninsula. The folded and beveled Precambrian rocks, strongly tilted and then overlapped by de-

# The Paleozoic Era

Figure v–8. Inside Chapel Rock along south shore of Lake Superior east of Munising (also see Fig. v–7, bottom). Sea Cave carved by wave action in Post-Pleistocene time during higher level of Glacial Great Lakes and further elevated by postglacial rebound of earth's crust. Rock itself is much more ancient marine sandstone, the Chapel Rock Member of the Cambrian age Munising Formation. Note well-developed cross-bedding to left of man's camera and to lower right of his feet. Inset (with millimeter scale) shows well-sorted, well-rounded, partially wind-frosted, marine-beach sand grains composing the rock. For marine fossils (trilobites) from Munising Formation see Figure xiii–31. (Courtesy Michigan Department of Conservation, photo by Robert Harrington.)

posits of the Cambrian sea, provide an excellent example of an *angular unconformity*. Even if one did not know the names of the rock formations nor any of the details of their origin, he could observe the geological features exposed in the old Quinnesec Mine and deduce the following sequence of geological events:

4th—Deposition of the basal conglomerate in a horizontal position on the erosion surface.

3d— Uplift and erosion which truncated the inclined strata below the unconformity.

2d— Strong crustal deformation which folded and partly metamorphosed the Iron Formation.

1st— Deposition of the Iron Formation and associated strata in a horizontal position.

The events are listed in order of occurrence, from first below to last above, so as to illustrate and conform with the Law of Superposition (Chapter I).

No Cambrian rocks are exposed in the Lower Peninsula, but samples taken from deep oil wells indicate that the Cambrian history of the southern part of the state was slightly different from that in the north. No Early or Middle Cambrian rocks have been found there, suggesting that the area may have stood well above sea level then. However, Late Cambrian sandstones of the Mount Simon, Eau Claire, Dresbach, and Franconia formations have been found in Lower Michigan. These, and the sandy dolomite of the Trempealeau Formation, probably, all are marine deposits. Apparently, a Late Cambrian sea moved in from the south and occupied that part of the state at the same time that the Munising Sea lay north of the Northern Michigan Highlands in the Upper Peninsula.

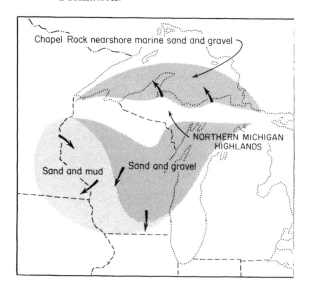

Figure v–9. Paleogeographic map of Late Cambrian time as represented by the Chapel Rock Sandstone Member of the Munising Formation and correlative deposits in Michigan and adjoining regions. (Modified from Hamblin, 1961.)

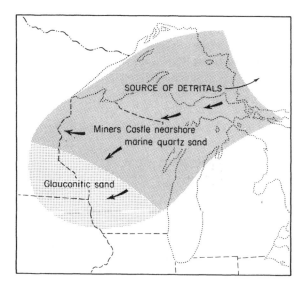

Figure v–10. Paleogeographic map of Late Cambrian time as represented by the Miner's Castle Sandstone Member of the Munising Formation and correlative deposits in Michigan and adjoining regions. (Modified from Hamblin, 1961.)

# Geology of Michigan

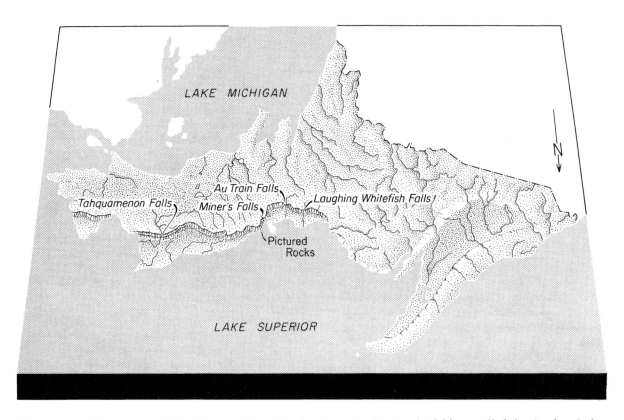

Figure v–11. Escarpment of Cambrian sandstones in the Upper Peninsula of Michigan, called the Cambro-Ordovician Cuesta. (Modified from Hamblin, 1958.) Prominent falls indicated are illustrated in Figures v–12 and 13.

## The Ordovician Period

The Michigan Basin was occupied by marine waters during much of the Ordovician Period (Figs. v–1, 2). The position of the shoreline fluctuated so that Ordovician sediments were at one time deposited in nearshore or beach zones and at another were laid down in offshore waters. Apparently, the sea had remained in southern Michigan from Cambrian on into Ordovician time. In the Upper Peninsula, however, an unconformity separates the Late Cambrian Munising Formation from the Au Train Formation of Middle Ordovician age (Fig. v–2, Column A). This indicates that during Early Ordovician time northern Michigan was an area of uplift and erosion (or at least of nondeposition) at the same time that marine dolomites and sandstones of the Trempealeau Formation and Prairie du Chien Group were accumulating in seas in southern Michigan (Fig. v–2, Column C). In Middle Ordovician time seas again occupied the Upper Peninsula, and a series of marine deposits began to accumulate over much of the state. These included the St. Peter Sandstone, a marine beach and near-shore deposit, and shales of the Black River, Trenton, Eden, and Richmond groups.

Good exposures of Ordovician rocks in the southern part of the Upper Peninsula occur on

Figure v–12. Late Cambrian marine sandstones of the Munising Formation in Tahquamenon Falls.

*Top*—Upper Falls, where rock is the Miner's Castle Sandstone Member.

*Bottom*—Lower Falls, where rock is the Chapel Rock Sandstone Member. Close-up shows ancient current ripple marks found on sandstone surfaces below.

# Geology of Michigan

Figure v–13. Falls of the Cambro-Ordovician Cuesta (see Fig. v–11). Munising Formation sandstones and Au Train Formation.

*Top*—Laughing Whitefish Falls. Late Cambrian age Miner's Castle Sandstone Member of the Munising Formation, capped by thin Middle Ordovician age Au Train Formation.

*Bottom*—Munising Falls. Miner's Castle Sandstone Member of Munising Formation capped by dolomitic sandstone of Middle Ordovician age Au Train Formation.

**The Paleozoic Era**

*Top*—Lower Au Train Falls. Type locality of the Middle Ordovician age Au Train Formation. Lower part of formation here is sandy dolomite, dolomitic and glauconitic sandstone.

*Bottom*—Miner's Falls. (All photos courtesy Michigan Department of Conservation.)

# Geology of Michigan

the Stonington Peninsula and in the valley of the Escanaba River. Marine fossil invertebrates including brachiopods, corals, and trilobites occur there (Chapter XIII).

Ordovician rocks are not exposed in the southern part of Michigan, but an extensive section of rocks representing most of Ordovician time is encountered in the subsurface in oil and gas wells. The variable nature of the rocks there suggests fluctuating marine conditions. The Oneota Dolomite, New Richmond Sandstone, and Shakopee Dolomite, of Early Ordovician age, indicate seas were present in the Lower Peninsula while absent in the Upper Peninsula. An unconformity between the Shakopee and the St. Peter Sandstone indicates a minor withdrawal of the sea, after which the sea readvanced, first depositing the St. Peter Sandstone in a beach and near-shore zone, followed by the offshore marine shales of the Black River, the Trenton dolomites and sands, and the Eden Shale. Another minor withdrawal and resulting unconformity were followed by deposition of limestones and shales of the Late Ordovician Richmond Group over the whole of the state.

The Queenston Shale of the Richmond Group is of particular interest. It thickens eastward from Michigan into Ontario and New York state and is part of a great deposit of deltaic sediment carried into the sea by streams that flowed westward off the Taconic Mountains. Those were the first of several mountain highlands formed in the Appalachian Geosyncline region (Fig. v–1) during the Paleozoic Era.

The Trenton Formation is of special importance because it is one of the major oil producing formations in the state.

The Ordovician, then, was the period during which Paleozoic seas became fully established in Michigan.

## The Silurian Period

Seas persisted in Michigan from Ordovician into Silurian time. They apparently had spread into the state from the northeast, probably along what is today the St. Lawrence Lowland (Fig. v–1). The coasts bordering Silurian seas were low-standing and most of the time lay outside the boundaries of the state, so that the area was occupied by offshore waters nearly the whole of the time. These waters, although in some places shallow, were beyond the reach of all but the finest-grained clastic sediments, so that the Silurian marine deposits of Michigan are mainly chemical precipitates formed in clear seas. Locally, especially around the margins of the Michigan Basin, shallow banks supported reefs, some of which no doubt extended up to or above the surface of the sea, just as do modern reefs. Some Silurian reefs reached thicknesses of over 400 feet. Dolomite is a common type of rock in Silurian deposits. Another and especially interesting type of deposit includes halite (rock salt) and anhydrite or gypsum; these formed under conditions of strong evaporation, in semiisolated marine basins, hence are called "evaporites." The environment of evaporite deposition will be discussed in detail a little later. The Silurian also was a time of accelerated downwarping of the Michigan Basin. Over 3000 feet of Silurian sedimentary rocks are found in the central part of the basin, being thickest in southern Michigan and Ontario.

Rocks of Early and Middle Silurian age lie beneath the surface of southern Michigan (Fig. v–2, Column D), but do not crop out there. They contain reefs of organic origin which in St. Clair and Macomb counties recently have been found to contain significant quantities of natural gas. The Middle and Late Silurian Salina Group, an important commercial source of salt and other chemicals, lies very close to the surface in southeastern

Figure v–14. Old, abandoned Quinnesec Iron Mine, now known locally as the "Devil's Icebox," near town of Quinnesec. Dotted line shows position of angular unconformity and the ancient, irregular erosion surface between steeply dipping strata of older Vulcan Iron Formation (Precambrian-Middle Huronian age) below, with subsequently deposited, younger, and more nearly horizontal strata in basal conglomerate of Late Cambrian age Munising Formation in erosion channel above. Note angular blocks of iron formation in conglomerate in top photo and in close-up (with 6-inch scale).

# The Paleozoic Era

103

# Geology of Michigan

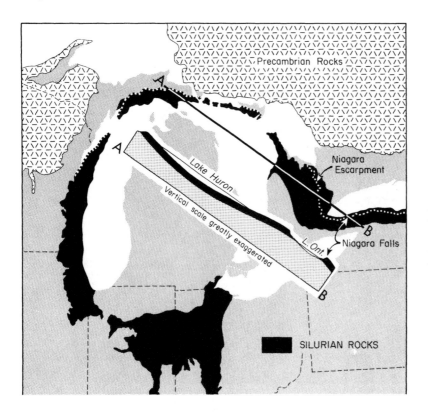

Figure v–15. Cross section and outline map of distribution of rocks of Silurian age in the Great Lakes Region. Position of the Niagaran Escarpment, a cuesta formed by erosion of resistant limestones and dolomites is shown by dotted white line. Note that the same escarpment forms Niagara Falls, Cape Hurd, islands in northern Lake Huron, an upland in the southern part of the Upper Peninsula of Michigan (point A on cross section and map), the Garden Peninsula at Fayette, and the Door Peninsula east of Green Bay, Wisconsin.

Michigan. The latest (youngest) Silurian-Bass Island Group rocks do come to the surface in southeasternmost Michigan in the southeastern corner of Monroe County.

Extensive outcrops of Silurian rocks occur in the southern part of the Upper Peninsula, including the Mackinac Straits Region (Fig. v–2, Columns A and B). Much of the Silurian sedimentary rock section is exposed there. The exposure belt is shown on Figure v–15.

Silurian carbonate rocks, especially the Middle Silurian Engadine Dolomite, are particularly resistant to erosion and tend to form an upland with an escarpment facing outward from the Michigan Basin. Notice that the Silurian outcrop belt shown on Figure v–15 forms the Garden Peninsula southwest of Manistique in the Upper Peninsula, extends southwestward into Wisconsin as an island and peninsular barrier partially separating Green Bay from Lake Michigan, and then continues southwestward into the Wisconsin mainland. This same escarpment also forms Drummond and Manitoulin islands along the north edge of Lake Huron, continues from there into Ontario as the Cape Hurd and Bruce peninsulas separating Georgian Bay from Lake Huron, and thence extends southeastward to the Niagara region between Lakes Erie and Ontario. The Niagara River plunges over the escarpment at Niagara Falls. This long, curving escarpment, extending from near Rochester, New York, to Milwaukee, Wisconsin, around the margin of the modern Great Lakes, is called the "Niagara Escarpment." As a bedrock "high" it is one of the geologic features which helps to determine the outer margin of the modern Great Lakes.

Traveling northward into the Upper Peninsula, the first good exposure of Silurian rocks is encountered in the highway cuts at the north end of the Mackinac Straits Bridge. There the St. Ignace Dolomite and Pointe aux Chenes Shale of the Salina Group are well exposed (Fig. v–16). Following U.S. Highway 2 westward from the bridge, across the southern part of the Upper Peninsula, one sees occasional road cut exposures of Silurian rocks. These continue westward beyond Manistique to Saint Jacques. Thus, U.S. 2 crosses the eroded edges of all the Silurian formations of that region, from the younger ones near the Straits, westward through the oldest which is the Mani-

The Paleozoic Era

105

Figure v-16. Road cut at north end of Mackinac Straits Bridge. A typical collapse filling of Devonian age dolomite blocks forms a mass of "Mackinac Breccia," down-dropped into older, Late Silurian age, Pointe Aux Chenes Shale. Solution of salt and gypsum from shale and collapse of overlying dolomites probably occurred during period of marine withdrawal and emergence of land in Mid-Devonian time. Much more recent (Late Pleistocene) wave erosion around margins of high Glacial Great Lakes in the Mackinac Straits region locally removed the softer shales, leaving the well-cemented and more erosion-resistant breccia masses as free standing inland stacks and arches along those former high shorelines, as shown in Figures ix-27, 28, 29.

toulin Dolomite. Fossils may be found in some of the exposures. They all are of marine types. Corals, brachiopods, and bryozoans are especially common. The Silurian dolomites and limestones of the Upper Peninsula have long been extensively quarried for the manufacture of cement, lime, road aggregate, building stone, paper filler, and agricultural lime. One of their most important uses is as a flux in the manufacture of iron and steel. One of the largest dolomite quarries in the world is located just outside Cedarville, northeast of the Straits of Mackinac, at the eastern end of the Upper Peninsula. There the Michigan Limestone Division of the United States Steel Corporation quarries about 3 million tons of rock per year from the Engadine Dolomite for use as flux stone.

Very picturesque exposures of Silurian rocks occur in the bluffs of Lake Michigan along the northwest side of the Garden Peninsula (southwest of Manistique), where exposures of the Hendricks Dolomite of the Burnt Bluff Group are especially good near the old, abandoned, iron-smelting town of Fayette. This village recently was restored as a Michigan State Park. It is situated on a little bay surrounded by outcrops of Hendricks Dolomite (Fig. v-17).

The Middle Silurian rocks of the Niagaran Series in the Great Lakes Region are world famous geological examples of marine organic reef growth. Lowenstam (1950, 1957) studied the Niagaran reefs and their fossils over a broad area, providing us with the means to examine the Middle Silurian paleogeography of the Great Lakes Region. Reefs are of increasing interest to geologists because they

# The Paleozoic Era

Figure v–17. Fayette State Park on Garden Peninsula southwest of Manistique along north shore of Lake Michigan. Fayette was an early iron smelting town, now restored. Rocks exposed here are part of Niagara Escarpment (see Fig. v–15).

*Top*—View northeast to cliffs of Hendricks Dolomite in the Burnt Bluff Group of Middle Silurian-Niagaran age. Dolomite was used as fluxstone in smelters.

*Bottom*—View southwest from cliffs above to restored village. Smelting furnaces at left. Note uplifted cliff and concave terrace carved on distant headland (Burnt Bluff) by higher waters of one of the immediately post-Algonquin Glacial Great Lakes stages (Arrow a). Glacial Lake Algonquin shoreline, still higher and older, is shown by Arrow b.

often contain gas and oil, a good example being the subsurface reefs or "bioherms" in St. Clair and Macomb counties in southeastern Michigan. Reefs also are fascinating as examples of the enormous biological contribution that lime-secreting marine organisms have made to the rocks of the earth's crust. They represent a very special type of organic and geologic environment. Reef growth in Michigan was not restricted to the Middle Silurian; Figure v–2 shows reefal deposits in the Early Silurian Manitoulin Dolomite and the Middle Devonian Traverse Group as well. Nevertheless, the Niagaran reefs have been thoroughly studied and are, perhaps, best understood.

Modern reefs are built largely by lime-secreting corals and algae, a lesser role being played by associated marine organisms. Abundant fossil remains in Niagaran reefs indicate a similar origin for those deposits. Modern reef-building corals depend on algae and these in turn, being plants, depend on an abundance of light for photosynthesis. Hence, modern reefs flourish in clear, relatively shallow, sunlit waters. Modern reefs are further restricted in their distribution by the fact that reef-building corals grow rapidly only in freely circulating waters where food and oxygen are abundant, and only flourish within the temperature range of 77°–84°F. Thus, extensive modern reefs are restricted to shallow tropical and subtropical seas. Ancient reefs, particularly those of the Silurian, were built mainly by the now extinct stromatoporoids (Chapter XIII) and by certain extinct types of corals. Although the requirements of those extinct animals are not known with certainty, they probably included similarly restricted environmental factors such as shallow water, strong light, and warm temperature. Both modern and ancient reefs typically are rigid structures that extend upward from the sea floor. The "frame builders" of the reef are colonial organisms, such as the Silurian stromatoporoids, corals, and algae, which secrete exoskeletons of calcium carbonate ($CaCO_3$). These form the structural basis of the reef. *Syringopora* and *Favosites* were common Silurian reef-building corals (Chapter XIII). The framework is filled in by the skeletal remains of other organisms. These help to bind the framework together and to fill in the voids. Reef growth begins with the establishment of a colony of "pioneer" frame builders on the sea floor. Growth begins in deep, quiet water below the reach of wave turbulence. Many such colonies, remaining separate from one another, form "patch" reefs. Upon further growth, these may merge to form composite structures. Silurian reefs are most commonly of the "patch" type. The frame builders gradually grow upward into more agitated water, eventually reaching the near-surface zone of strong wave action. There they tend to assume more flattened, mushroom-shaped, wave-resistant forms, or their places are taken by other types that can resist the pounding of the surf. The *core* of the reef usually is an unstratified mass of semiporous carbonate rock formed by the skeletons of the reef builders still largely in their original growth positions. Material torn by wave action from the reef crest accumulates in deeper water along the sides of the reef to form steeply inclined, stratified "flank" beds. These consist of broken reef material intermingled with the skeletal remains of animals that lived and were buried on the reef flanks. Brachiopods, mollusks, crinoids, and a variety of arthropods were and still are common indwellers in the reef and around it. Many of the Silurian reefs appear to have begun their growth on soft, unconsolidated sediment, such as lime mud or fine sand, composed largely of broken shell debris. The reef root often sank into or compressed those sediments as the reef grew upward and increased in weight. The Niagaran reefs studied by Lowenstam ranged from a few tens of feet to several hundred feet in height and cover anywhere from a few square feet to many acres in area. They grade or "interfinger" laterally into stratified and inclined flank beds and those, in turn, extend laterally into horizontally stratified "interreef" deposits. The reefs may consist of either dolomite or limestone. Although dolomite is the more common constituent, there is reason to believe that some of that mineral may have been converted from normal limestone by the addition of the element magnesium, a process called dolomitization and discussed more fully in the chapter on Petroleum and Natural Gas.

The flank beds of the Niagaran reefs commonly are composed of coarser fragments in their upper parts, reflecting the upward growth of the reef into the progressively more turbulent waters near the surface of the sea. The abundance and diversity of indwelling associated animals likewise increases upward in the reef, reflecting the increased complexity of the reef as a natural environment. *Halimeda,* a form of lime-secreting algae which is a common frame-builder in modern reefs, seems to be missing in the Silurian reefs. Its place may have been taken in those times by the now extinct stromatoporoids which seem to occupy structural positions in the ancient reefs similar to those in which *Halimeda* occurs today. Little or no clastic sediment from adjoining lands is found in the Niagaran reefs. Perhaps the lands of that day lay too far off, or perhaps the reef and shoal complex itself served as a screen, catching most of the sediment before it could reach the majority of the reefs. Reefs make their first appearance in Michigan rocks in the Manitoulin Dolomite (Fig. v–2), appear again a little later in the mid-Silurian Byron, Hendricks (Fiborn), and Cordell dolomites, but are best developed in the mid-Silurian Engadine Dolomite. Some also occur in the still younger rocks of the Middle Devonian Traverse Group.

The subtropical or tropical climate in which Silurian reefs flourished, also favored the deposition of evaporites (halite and anhydrite) of the Salina Group. Thus, the Engadine Dolomite and Salina evaporites are age-equivalent facies, at least in part. Examine Figure v–2 closely and you will see that the Engadine Dolomite is correlated in time with the lower part of the Salina Group. A little later we shall see that the growth of reefs on the shallow fringes of the Michigan Basin may have been a partial cause of isolation of the basin, separating its waters from those of the open sea. This, in turn, led to conditions favorable for deposition of the Salina evaporites.

Lowenstam (1950, 1957) plotted the distribution of Niagaran reefs in the Great Lakes Region, studied their relationships to rocks of equivalent age elsewhere, and reached the following conclusions concerning the paleogeography of Middle Silurian (Niagaran) time (Fig. v–15). The southernmost Niagaran reefs are located 4° of latitude farther north than the northernmost reefs of today. Some even occur at 75° North Latitude, which now falls in the Arctic. Some geologists argue that this proves continents have drifted northward toward the pole or that the earth's poles themselves have shifted position through the course of time. On the contrary, Lowenstam concluded that the northerly distribution of Niagaran reefs could be explained as having resulted from southerly winds driving ocean currents farther north from the equator and carrying warmer than normal waters with them. He based his conclusion concerning wind direction in Silurian time on the shape and direction of growth of the Silurian reefs themselves and on the distribution of fine sediment that settled around them. He believed that the reefs grew up in a broad, open, epeiric sea far from land. The vast region in which they flourished included the stable shoal waters around the Wisconsin Upland, the Cincinnati Arch, and the Ozark Uplift (Fig. v–15). Between those shoals and lowlands lay the Bainbridge Basin to the south and the Michigan Basin to the north. Reef growth began on the flanks of the Wisconsin Upland shoals and the north end of the Cincinnati Arch, then spread southward toward the Ozark Island. The fringing reefs, added onto the previously existing shoals, nearly enclosed the Michigan Basin, isolating it from the open sea as an evaporation basin, which we shall discuss next.

The Michigan Basin is one of the greatest areas of *halite* (rock salt, NaCl) accumulation in the world. Cohee (1965, p. 217) estimates that the aggregate thickness of Silurian salt alone is 2000 feet, one single bed of nearly pure halite being about 500 feet thick. The greatest thicknesses originally were deposited near the center of the basin. In some places much of the original salt was dissolved away over the long course of geologic time. For these reasons, the salt beds thin rapidly or even disappear toward the basin margins. A layer 400 feet thick at Detroit has been completely dissolved at Trenton, only 14 miles to the south (Cohee, 1965). Thick deposits of *anhydrite* ($CaSO_4$) also are present. Gypsum ($CaSO_4 \cdot 2H_2O$) forms naturally in near surface rocks by addition of some water of crystallization to anhydrite. Because halite and anhydrite form in evaporating conditions, and also because both are especially abundant in the Silurian Salina Group, their mode of origin will be considered in this chapter although their occurrence is not restricted to Silurian rocks.

The role of Silurian reefs in the isolation of an evaporation basin in Michigan has already been discussed in the preceding paragraphs. The discussion that follows is drawn largely from studies published by Professor Louis I. Briggs (1958) of The University of Michigan, and from the work of Alling and Briggs (1961). We have seen that, during the Paleozoic, Michigan frequently was occupied by seas. Sea water contains many dissolved chemicals, of which sodium chloride (NaCl), calcium carbonate ($CaCO_3$), and calcium sulfate ($CaSO_4$) are most abundant. This dissolved material is not highly concentrated and, under normal conditions, does not precipitate. However, if normal sea water is evaporated, a process that occurs rapidly under warm, dry conditions, the concentration of these dissolved substances increases. If enough water is evaporated, the dissolved materials reach the saturation point and begin to precipitate as mineral solids which accumulate on the sea floor. Experiments show that the three chemicals mentioned as being most abundant in sea water are not equally soluble, so that as a solution becomes more and more concentrated the *least soluble* is the first to precipitate while the most soluble remains dissolved until a higher concentration is reached. Because calcium carbonate is least soluble of the three, it precipitates first to form limestone, or dolomite if magnesium is added. Calcium sulfate is next to precipitate, forming deposits of anhydrite which may later take on water to become gypsum. Sodium chloride is deposited last as halite or rock salt. Thus, in a body of sea water subject to evaporating conditions, the portions or areas in which evaporation is greatest would be the sites of deposition of the most soluble halite, whereas those areas which suffered less evaporation might only receive deposits of anhydrite, limestone, or dolomite. If a single evaporation cycle continued without interruption to completion, all three kinds of deposits would be laid down, one above another, in sequence from least soluble (limestone or dolomite) at the bottom, to anhydrite of intermediate solubility, to the most soluble halite on the top. However, if new water were added at some place in the evaporation basin, dilution would occur. Only the least soluble mineral could precipitate near the inlet where the water had been freshened, and deposition of halite would be restricted to portions of the basin farthest removed from the entry points of freshening waters. Furthermore, periodic influxes of freshwater would interrupt the progression toward increasing salinity and cause it to begin over again. Repeated cycles of deposition of parts or all of the limestone-anhydrite-halite sequence might occur under alternating conditions of evaporation and refreshening. This is not the whole picture, however! The total amount of dissolved solids in a given amount of normal sea water is limited, so that if evaporation were completed without the introduction of more water, only a limited thickness of any evaporite mineral could accumulate. Therefore, if great thicknesses of any of the minerals mentioned are to form, there must be a balance between evaporation of water to the precipitation point within the basin and addition of new water through some inlet from the open ocean.

A modern example of an evaporation basin is the Gulf of Karaboghaz on the east side of the Caspian Sea. That gulf is a shallow, semi-isolated marine basin, in a dry, hot climate near the Desert of Kara-Kum. The gulf is surrounded by land on all sides and joined to the Caspian Sea by only a very narrow, shallow inlet. Sea water of concentration normal to the Caspian Sea can enter the gulf by passing over the shallow sill or threshold of the inlet, but does so only in limited quantity. Fresh-water influx from the desert land is also limited. Under those conditions evaporation goes on very rapidly. As the water in the gulf evaporates, it leaves its mineral matter behind. Additional sea water flows in through the inlet to replace that lost through evaporation because between two interconnected basins (such as the gulf and the Caspian) water seeks to maintain a common level. Thus, the dissolved mineral matter increases in concentration to the saturation point

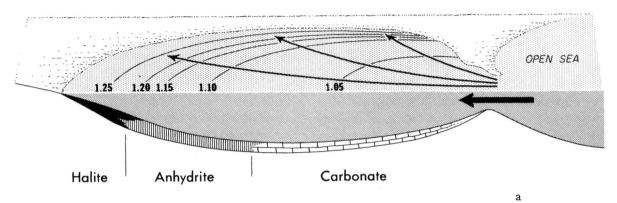

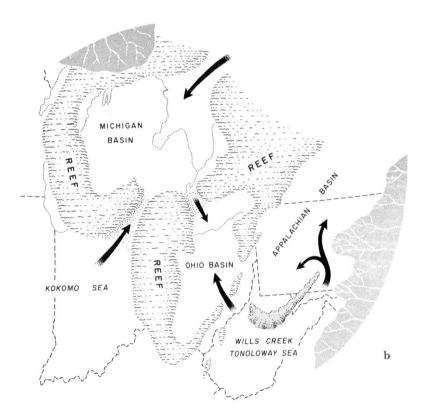

Figure v-18. Deposition of evaporite sediments.

a) Diagrammatic model of the chemical process and physical setting. Water removed from within the basin by evaporation is replenished by saline waters from the open sea which enter the restricted basin through a narrow inlet, over a shallow sill. Dissolved chemicals in the normal sea water become progressively more concentrated by evaporation as the water spreads toward the center and edges of the basin. Least soluble carbonates (limestone and dolomite) precipitate and fall to the bottom near the inlet; more soluble salts are carried farther from the inlet. The chemical precipitates thus form a succession of sedimentary deposits (facies) grading from one to another across the basin sea floor. (Modified from Briggs, 1957.)

b) Paleogeography of Michigan and adjacent regions during the Late Silurian (Cayugan), when extensive evaporites were deposited. Reefs or shallow water carbonate banks isolate the Michigan Evaporite Basin from the open sea, restricting inflow of normal sea water. Narrow inlets through the reefs function as described above. (Modified from Alling and Briggs, 1961.)

and precipitation takes place. Within the gulf, the water near the inlet and around the shores is continually diluted to some extent; hence only the least soluble carbonates can precipitate there, whereas halite precipitates in regions of less dilution and greater concentration farthest from the inlet and shores. Because the rate of water influx varies, the areas of precipitation of the respective minerals shift. Thus, at any given location different evaporite minerals are laid down at different times. This accounts for alternating strata of the various evaporite minerals. However, on the average there is far more halite deposited near the center of the gulf and more carbonate near the margins. Briggs (1958) made a careful study of the chemistry involved; Figure v–18 shows the conditions he concluded would exist in an idealized evaporite-forming basin. He then applied this concept, and the example of the Gulf of Karaboghaz, to the Michigan Basin in order to study the nature, distribution, and thickness of evaporite deposits and other types of sedimentary rocks in the Silurian Salina Formation. His main sources of information were the many well samples and drilling records in The University of Michigan Subsurface Laboratory. Alling and Briggs (1961) later extended this concept, coupled with studies of fossils, to include Late Silurian (Cayugan) rocks over most of northeastern United States and adjoining Ontario, Canada. Figure v–18 shows their interpretation of the paleogeographic conditions in the Great Lakes Region and adjoining states during the time of deposition of the upper part of the Salina Formation, the Pointe aux Chenes Shale, the St. Ignace Dolomite, and other age-equivalent rocks elsewhere. The Michigan Basin is shown as having been occupied by a semi-isolated sea, with two shallow, restricted inlets, each like the sill of Karaboghaz, and one outlet. The outlet is shown leading into another important evaporation basin, the Ohio Basin. The seas to the northeast, southeast, and southwest of those basins presumably opened, at some distance, to the major ocean basins of that time. Thus, there was a limitless source of additional salt water that could flow in the inlets. Extensive "reef banks" are shown as the major features isolating the Michigan Basin from the open sea. We discussed the nature of those reefs and banks earlier in this chapter. The abundance of reef building corals and other organisms indicates that the waters of the basin were warm, a condition which would have accelerated evaporation. Inlets were located where northeastern Indiana and Georgian Bay now lie, but, of course, those modern geographic features were not in existence then. The St. Ignace Dolomite represents part of the enclosing reef "facies." The Pointe aux Chenes Shale is interpreted as a deltaic deposit, spread southward into the basin by a river or rivers flowing from some northern source land. Remembering the limitations of paleogeographic maps, as we discussed them earlier in the chapter, one must not conclude from the map in Figure 18 that conditions in Cayugan time were constantly favorable to evaporation. The Cayugan strata studied represent a considerable span of time, and the map only represents the average conditions of the basin during that time. Some of the strata in the Salina Group, for instance, are normal, fossiliferous, marine limestones, which indicate that the degree of isolation of the basin must have varied.

Similar but less thoroughly studied conditions favoring evaporite deposition occurred after Silurian time in the Michigan Basin area as is evidenced by the salt deposits which occur in commercial quantities in the Middle Devonian Lucas Formation of the Detroit River Group and in the Marshall Formation of Mississippian age.

Gypsum, originally anhydrite, is another common evaporite and is mined commercially from Mississippian Michigan Formation at Whittemore, Tawas City, Alabaster, and Grand Rapids. Salina rock salt is mined at Detroit from a bed 1135 feet below the surface (Fig. v–19). The mine there has miles of passages connecting man-made caverns 22 feet high and 55 feet wide from which 98.3 percent pure rock salt has been removed (Cohee, 1965). There is a similar mine directly across the Detroit River at Ojibway, Ontario. Salt and other chemicals also are recovered from both natural and artificial brines pumped up through wells drilled into the Salina Formation in Manistee, Midland, Muskegon, St. Clair, and Wayne counties. Natural brines are those already present in the rock. Artificial brines are formed by pumping surface water down into an underground salt bed and then back up again. Salt is recovered from artificial brines from the Devonian Detroit River Group at Manistee. Natural brines are pumped from the Mississippian Marshall Formation and from the Devonian Dundee Limestone

Figure v-19. International Salt Company mine, Wyandotte, Michigan.

*Top*—Loading salt. Light patches on roof are blast marks. Note horizontal stratification.

*Center*—"Room and Pillar" mining. About two-thirds of the salt is left in columns to support roof. Along passage, the light bands are pillars, dark areas side rooms.

*Bottom*—Sawing channel for floor before blasting. Light layers are nearly pure salt. Dark layers are salt with very minor amounts of either anhydrite or anhydrite plus dolomite. Light mass at top center right is recrystallized salt in which crystals are very coarse grained, possibly an old channel way formed on an ancient marine tidal flat. (Michigan Department of Conservation, photos by Clyde Allison.)

at St. Louis in Gratiot County. The Salina Formation and the Detroit River Group both provide artificial brines for the Dow Chemical Company plant at Midland.

Salt is a natural resource of great importance and Michigan supplies between 20–25 percent of the nation's output. Some $42,389,000 worth of salt was produced here in 1967. Salt is used for the manufacture of chemicals, meatpacking, ice removal from streets and highways, water softening, and animal feed. Only about 2 percent of it is used as table salt. Associated with the NaCl or halite are other chemical salts, called "natural salines," which occurred in lesser abundance in the original sea water but which are, nevertheless, important sources of bromine, calcium chloride, calcium-magnesium chloride, iodine, magnesium compounds, and potash. Natural salines other than common salt are extracted from brines produced by wells in Gratiot, Lapeer, Mason, Manistee, Midland, and Wayne counties. Wyandotte Chemicals Corporation is a major producer in the Detroit area. A wide variety of chemical products are manufactured from the raw materials in salines.

Gypsum, which forms naturally from anhydrite by the addition of a small amount of water to the crystal structure, is another commercially important evaporite deposit. Although it occurs in many places in the state in rocks of Silurian and Devonian age, it is produced commercially only from the Mississippian Michigan Formation. Gypsum is obtained from surface quarries in that formation at Alabaster and Tawas City in the Saginaw Bay area (Fig. v–20a) and is mined from the same formation underground at Grand Rapids. The Grand Rapids mines extend over about a 300-acre area, 150 feet below the surface (Fig. v–20b). Over 5 million dollars worth of gypsum was produced in Michigan in 1967 making the state one of the top-ranking producers in the nation. The gypsum occurs as several mineral varieties called rock gypsum, "alabaster," satin spar, and selenite. It is used in the manufacture of plaster wallboard, exterior sheathing, lath, and plaster of Paris, and in the raw, uncalcined form for cement retarder.

The surface or near-surface quarries and mines for both Salina salt and Michigan Formation gypsum are located in areas where the "nesting bowl" subsurface geologic structure of the Michigan Basin brings those formations up to or near the surface. Near the center of the Michigan Basin those formations are much thicker and constitute a tremendous natural resource, but they are too deep to be mined except by deep brine wells.

In summary, then, the Silurian Period in Michigan was a time of extensive seas. Bordering lands were low and far removed from the margins of the Michigan Basin. The basin was ringed by shallow shoals and dolomitic reefs which nearly isolated it from the open sea, producing conditions conducive to the deposition of thick evaporite deposits.

**The Devonian Period**

Near the close of the Silurian the seas withdrew from the Michigan Basin and during the Early Devonian the southeastern part of the basin area was emergent. This period of erosion, or at least of nondeposition, is represented on Figure v–2, Column D, by a gap or unconformity signifying the absence of a sedimentary record for that time. To the north and northwest, however, Early Devonian sediments were laid down. These now form the Garden Island Formation. Evidence for the unconformity in southeastern Michigan comes largely from the sequence of fossils found there. Early Devonian types appear to be missing, whereas they are known to occur elsewhere in places where sedimentary deposition was going on at that time.

By the Middle Devonian the Michigan Basin again was fully occupied by seas (Fig. v–1). These

# Geology of Michigan

Figure v-20. Gypsum mining from Late Mississippian age Michigan Formation, a marine evaporite deposit.

*Top*—Surface quarry at National City (Michigan Department of Conservation, photo by Clyde Allison).

*Bottom*—Underground mine at Grand Rapids (photo courtesy Grand Rapids Gypsum Company).

remained in the state throughout the remainder of the Devonian Period. Occasional isolation of the basin led to renewed deposition of evaporite deposits such as had formed so commonly in the preceding Silurian Period. The Devonian evaporites mainly occur in rocks of the Detroit River Group. Devonian fossils in Michigan all are marine types except for some land plant fragments washed into the sea by distant rivers and streams (Fig. v-21). Inspection of Figure v-2 will show that the Middle Devonian shorelines must have shifted frequently and that water depths often changed. The Bois Blanc Formation consists of dolomite and cherty limestone deposited in a marine environment to which abundant dissolved silica was delivered from distant, desert lowlands but in which clastic sediment from the land was minimal. This would indicate an offshore environment. In contrast, the Sylvania Sandstone, found in southeastern Michigan (Fig. v-2, Column D), is an exceptionally pure, well-sorted quartz sandstone. The sand grains are well rounded and have a "frosted" surface. These characters indicate deposition under the influence of wind and waves, thus suggesting a combination of near-shore, dune, and beach environments. Above the Sylvania Sandstone, the limestones, dolomites, and evaporites of the Amherstburg, Lucas, and Anderdon formations indicate that offshore marine conditions had returned again to the area. The Anderdon Limestone is a high-carbonate, reef bank facies of restricted extent within the Lucas Formation. Still higher in the rock section, and later in time, the alternating limestones and shales of the remainder of the Traverse Group represent changes in the environments of deposition from clear to muddy water, due either to changes in water depth or to changing conditions on adjoining lands. The unconformities represent periods of emergence. At the same time, 400 feet of salt accumulated in the Lucas Formation in the northern part of the Lower Peninsula, so that area must have been an evaporite basin.

In Chapter I, in the discussion of the Principle of Original Continuity of sedimentary rock strata, we said that a rock formation might terminate laterally by changing into or "interfingering" with a different type of rock formation of the same age. As an example of this possibility, note on Figure v-2 that rock formations of Middle and Late Devonian age differ from column to column and thus from area to area within the state. These differences in rocks of the same age reflect differences in the environments of deposition, and thus in geography, from place to place during that time.

Column A on Figure v-2 shows that Devonian rocks are missing from the Upper Peninsula. They once occurred there, but were removed later by erosion. The *Mackinac Breccia* offers proof of this statement. The term breccia, meaning "broken," refers to the fact that the formation consists of broken fragments, jumbled together and recemented. The broken fragments include pieces of Devonian rock formations containing fossils that prove their age. The Mackinac Breccia is an ancient collapse breccia or cave breccia found in the vicinity of St. Ignace, in the Straits area, and on Mackinac Island. The Devonian fragments indicate that Devonian rocks must once have occurred in the area so as to contribute to the breccia. The cause of brecciation will be discussed a little later in the chapter in connection with a description of caves which occur in Devonian rocks south of the Straits.

Although all of the Devonian formations of Michigan have been studied extensively, the history of early Middle Devonian time has been very thoroughly reconstructed recently in another study by Briggs (1959). He examined drill cuttings and cores from 111 wells in Michigan and surrounding areas. From those, and with the help of earlier studies which he mentions, he was able to determine the thickness, average percentage composition, and distribution of strata of Bois Blanc, Sylvania, Amherstburg, Lucas, and Anderdon

Figure v–21. *Top*—Rich assemblage of marine organisms on early Middle Devonian (or late Early Devonian) sea floor. This was the time of accumulation of limestones and dolomites of the Bois Blanc Formation and the Detroit River Group (Sylvania, Amherstburg, Lucas, and Anderdon formations) in the Michigan Basin. Letters identify the following: T—trilobites including molted carapaces; N—coiled and straight-shelled nautiloid cephalopods; S—a snail; K—kelp-like seaweed; E—long stemmed crinoid (echinoderm); SC—solitary corals (some polyps with tentacles extended); CC—colonial corals. (Diorama in The University of Michigan Exhibits Museum and Museum of Paleontology.)

*Bottom*—Ocqueoc Falls, the only major natural waterfall in the Lower Peninsula of Michigan. About 15 miles due west of Rogers City. Rock forming falls is Rockport Quarry Limestone of the Middle Devonian age Traverse Group; contains dolomitized fossil corals. (Courtesy Michigan Department of Conservation.)

formations, all of early Middle Devonian age. Figure v–22 shows two examples of the types of maps he first constructed from his raw data. Figure v–23 shows four paleogeographic maps he then drew to *interpret* the meaning of those data. The following excerpts from Briggs (1959, pp. 52–53) summarize some of his conclusions and provide an excellent descriptive interpretation of Michigan's early Middle Devonian history:

The history of the Michigan Basin during the Middle Devonian can be summarized into the following sedimentation framework:

(1) The Bois Blanc (lower Onondagan) sea transgressed across eroded Cayugan rocks in the Michigan Basin from the north or east to about the southwest tip of Michigan. An arid or semiarid climate provided the streams with abundant silica which precipitated as beds in dolomite in the nearshore zones of the basin, and as nodules in limestone offshore.

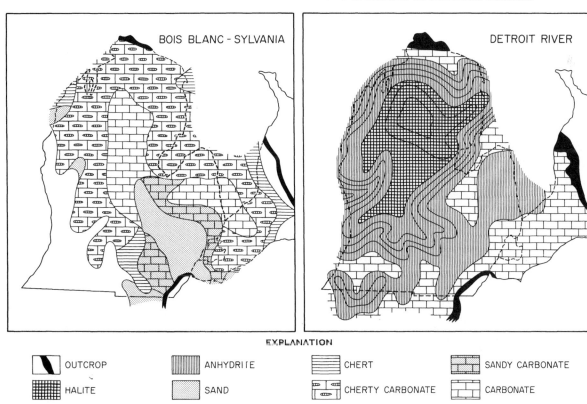

Figure v–22. Sedimentary facies deposited in the Michigan depositional basin during the early part of Middle Devonian time. The Bois Blanc-Sylvania deposits were followed and overlain by those of the Detroit River Group. See Figure v–2 for a stratigraphic chart showing the position of these deposits in time. Note on map at left that the Sylvania Sandstone formed as a facies tongue of nearshore and dune and beach sand interfingering laterally into offshore deposits. (Modified from Briggs, 1959.)

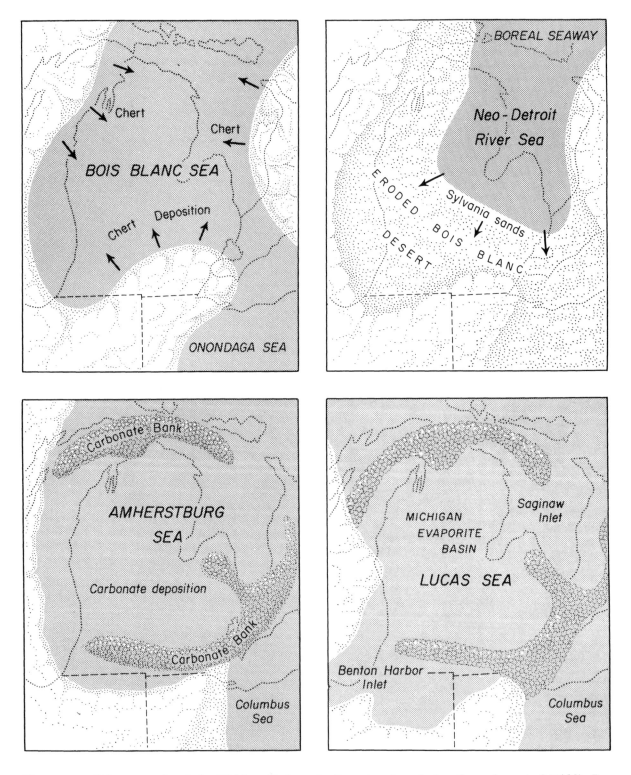

Figure v-23. Paleogeography of the Michigan Basin and adjacent regions during the early part of Middle Devonian time. The Bois Blanc Sea was followed by the Neo-Detroit River Sea, Amherstburg Sea, and Lucas Sea in that order. These interpretations are based on data from maps such as shown in Figure v-22. (Modified from Briggs, 1959.)

(2) Retreat of the Bois Blanc sea to the northeast third of Lower Michigan cut off the connection with the Onondagan sea to the south, and the newly born Detroit River sea apparently was the southern extremity of a boreal seaway. In the beach zone wind-blown sands accumulated while limestones were deposited offshore, and this deposition continued as the sea transgressed to the west and south across the Michigan Basin.

(3) Extensive carbonate banks grew in the Mackinac Straits region and along the southern quarter of Michigan and the Ontario peninsula. The carbonate banks appear to be loci of abundant coral and stromatoporoid biota which characterize the late upper Bois Blanc, and the Amherstburg and Anderdon formations.

(4) Continued transgression, an arid climate, and restricted circulation of the basin brine with the open ocean initiated evaporite deposition which characterizes the Lucas formation. At its maximum extent the Lucas sea covered all of Lower Michigan, part of northern Indiana, the northwest corner of Ohio, and most of the Ontario peninsula. Almost all of the brine in the Lucas sea came into the Michigan Basin through the Saginaw Inlet in east-central Michigan, except for the small Benton Harbor Inlet in southwest Michigan which fed the semi-isolated northern Indiana evaporite basin. A part of the brine from the Benton Harbor Inlet flowed northward into the Michigan evaporite basin. Since the Detroit River sea bordered a landmass near the western edge of the present State of Michigan, the brine in the Benton Harbor Inlet must have originated in the Illinois Basin and spilled northward over the Kankakee Arch into the Michigan Basin.

(5) The youngest Detroit River rocks throughout the depositional basin contain mostly limestone and dolomite that are likely correlative with the Anderdon limestone of the southern carbonate bank.

In the Late Devonian vast quantities of black mud were swept into the Michigan Basin from the east as a result of uplift in the Appalachian Region. These muds formed the Antrim Shale which grades westward and upward into the gray and greenish-gray Ellsworth Shale. Deposition of mud then continued without break into the Mississippian Period. The black muds, rich in organic matter, indicate deposition on a poorly oxygenated sea floor where decay was slow and the waters relatively acidic, thus producing chemically reducing conditions.

Some of the Devonian rocks of Michigan are of special interest. The Antrim Shale just mentioned, for example, is a sedimentary deposit that accumulated during a span of time that included parts of two geologic periods, the Late Devonian and earliest Mississippian. This illustrates the fact that time boundaries on the conventional geologic time scale are human artifacts, defined for convenience' sake to separate what was actually a time continuum in history. There is nothing "sacred" about the boundaries of the geologic periods; certainly there is no reason to accept an older view that deposition of sediments ceased all over the world at the end of each geologic period.

Outcrops of the Dundee Limestone, Rogers City Limestone, and formations of the Traverse Group in the northern part of the Lower Peninsula, are of considerable economic importance. Included are a number of alternating layers of limestone and shale that record fluctuating marine conditions. The Dundee and Rogers City limestones contain oil in many places beneath the surface of the state. The Dundee Limestone in southeastern Michigan, near the town of Dundee, is a major source of limestone for the manufacture of cement (Fig. v–24). At Rogers City in the northeastern part of the Lower Peninsula the Michigan Limestone Division of the United States Steel Corporation operates the world's largest limestone quarry, the Calcite Quarry (Fig. v–24). High grade limestone is produced there for use as a flux in the manufacture of steel and chemicals. The company has established a tourist observation point at Calcite from which an exceptionally fine view of operations is possible and where an interesting exhibit is maintained to show the process of limestone production and its uses. Excellent exposures of the Rogers City Limestone and formations of the Traverse Group occur along U.S. Highway 23 near Rogers City.

At Alpena the Huron Portland Cement Company quarries the Alpena Limestone, also of Middle Devonian age. In the walls of the quarries there, as well as in roadcuts between Alpena and Rogers City, one frequently can see cross sections

Figure v–24. Vast, near-surface supplies of Devonian age marine limestones support large-scale cement, steel flux, and chemical industries in Michigan. Proximity to Great Lakes ports facilitates low-cost bulk shipping. (Also see Fig. 1–3.)

*Top and lower right*—Near Dundee. Flat natural ground surface at quarry top and in distance beyond plant is former glacial lake bed. Rock is Middle Devonian age Dundee Limestone. (Courtesy Dundee Cement Company.)

*Center*—One of many ancient coral reefs (light-colored, lens-shaped body) frequently encountered within horizontal interreef limestone in the Alpena Limestone. Huron Portland Cement Company, Alpena, Michigan.

*Lower left*—Calcite Quarry at Port Calcite near Rogers City on Lake Huron. Rock is Rogers City Limestone. (Courtesy Michigan Limestone Division, U. S. Steel Corporation.)

through ancient Devonian *coral reefs* and smaller *bioherms* (Fig. v–24). These reefs grew in clear, shallow, warm seas during the time of deposition of the limestone. Fossil collecting is good in the Devonian limestones of the Alpena-Rogers City area. Brachiopods, bryozoans, crinoids, corals, and stromatoporoids (Chapter XIII) are common marine fossil types. Colonial coral fragments eroded from these rocks often are found along the beaches of lakes in the Lower Peninsula. When polished by the abrasive action of the waves, these are called "Petoskey stones." These stones derive their name from the fact that they are abundant near the city of Petoskey, where coral-rich, Middle Devonian rocks also crop out.

The use of limestone in the manufacture of cement is especially important to the Michigan economy, with the state ranking fourth in the nation in that industry. In 1967 the cement produced here was valued at nearly $100,000,000. This makes cement manufacture the second largest mineral industry in the state after the iron ore industry. Raw materials required for cement include limestone, clay or shale, some sand and gypsum, and other minor ingredients. All of these occur in the state. Their proximity to one another and to excellent and inexpensive modes of bulk shipping are added factors favoring this industry here. In the northern parts of the state Devonian limestone is supplemented by shale from the Antrim and Ellsworth formations of Mississippian age. In southern Michigan clay is obtained from Pleistocene lake deposits which are geologically much younger.

Limestone and dolomite have many other uses beside those for cement and flux stone. These include concrete aggregate, road surfacing, lime for fertilizers and soil sweetening, paint and rubber filler, asphalt aggregate, railroad ballast, and building stone. Limestone also is used in the manufacture of a host of chemical products. Several factors must combine if limestone or dolomite rock formations are to be of commercial value. These factors include high purity, adequate thickness, closeness to the surface, proximity to cheap bulk transportation facilities, and abundant water. The first three factors depend on the geological history of the rocks, which must have included a long period of deposition of calcium carbonate mud and shell material, clear seas, and later uplift and erosion to remove any overlying rocks. Michigan, of course, is blessed with cheap transportation on the lakes, excellent highways, many railroads, and abundant water. The major limestone- or dolomite-producing centers are at Rogers City, Alpena, Dundee, Petoskey, Charlevoix, Manistique, Cedarville, Drummond Island, and east of Grand Lake. In 40 years, the quarries at Rogers City alone have produced over 300 million tons of limestone.

Limestone dissolves readily compared with the solution rates of other rocks. Regions of near-surface limestone bedrock thus frequently are the sites of *cave* and *sinkhole* formation. Rain water sinking underground, and the movement of groundwater beneath the surface, dissolve the limestone to form subterranean caverns. Sinkholes form when the cavern roofs collapse. Surface stream patterns often are disturbed and streams may even disappear underground in such regions. There are several areas in Michigan where sinkholes are abundant. Perhaps the best of these is in Alpena County about 20 miles west and north of Alpena. The area lies about 7 miles north of Lachine and 6 miles south of Posen, along state road M–65. There the Middle Devonian lime-

# The Paleozoic Era

Figure v–25. Sinkhole in marine limestone of the Middle Devonian age Traverse Group near Posen west of Alpena. Underground solution, followed by collapse (see Fig. IX–5), exposed coralline rock. Inset close-up shows cross section through two colonial coral "heads" (hand for scale), taken near point where man (see arrow) stands in sinkhole.

stones of the Dundee and Rogers City formations and the Traverse Group have undergone extensive underground solution. One can climb down into some of the sinkholes (Fig. v–25) and see coral formations in their original position of growth in the Alpena Limestone in which those particular sinkholes are formed. At nearby "Mystery Valley," about 4 miles south of Posen, a surface stream sinks underground leaving its valley curiously dry. Other sinkholes are located in Presque Isle County to the north. One such area is in Section 36 of Presque Isle Township, about 2.5 miles east of Long Lake, near Ferron Point along the shore of Lake Huron. Another sinkhole lies farther inland to the west, between Rainy Lake and Kelsey Lake, in Section 15 of Case Township. This sinkhole is of particular interest because it proves that Rainy Lake lies in an area of underground solution. This in turn helps to explain the fact that Rainy Lake itself has, on occasion, gone dry when a sinkhole in its floor became temporarily "unplugged," allowing the lake waters to drain off underground.

The rock formation called the Mackinac Breccia (Fig. v–2) is another good example of the effects of underground solution in the Mackinac Straits region (Landes, 1959). The breccia consists of highly broken, and recemented fragments of several Silurian and Devonian rock formations including pieces of the Detroit River Group, Bois Blanc Formation and Bass Island Group. The breccia is found at the surface on both sides of the Straits. Landes concluded that it formed in Devonian time as a result of solution of underlying Silurian salt-bearing strata. Solution occurred during a period of uplift and marine withdrawal after deposition of the Devonian Detroit River Group but before the Dundee Limestone was formed. Subsequent collapse broke up the overlying Silurian and Devonian rocks up through those of the Detroit River Group, but it did not affect the Dundee Limestone or younger rocks because they had not yet been deposited. This indicates that underground solution went on there in the geological past, just as it does today, whenever uplift raised the land above sea level so that groundwater solution could begin. Concern over the bearing strength of the breccia and the possibility of further collapse had to be dispelled by careful geological investigations before construction of the great piers and abutments of the Mackinac Straits Bridge was considered safe. Some of the breccia can be seen in roadcuts at the north end of the bridge and in "St. Anthony's Rock" behind the Homestead Cafe in St. Ignace. Nearby Castle Rock also is composed of Mackinac Breccia (Figs. IX–27–29).

In summary, the Devonian Period, in Michigan, was a time of widespread but fluctuating seas, extensive limestone, dolomite, and shale accumulation, and some evaporite deposition. Near the end of the period black muds of the Antrim Shale were carried into the Michigan Basin from an uplift in the Appalachian Region.

## The Mississippian Period

Mississippian rocks in Michigan are restricted to the Lower Peninsula, extending only as far north as the southern part of Cheboygan County (Fig. v–26). It is probable that at one time rocks of Mississippian age covered areas farther north, but they have since been eroded. In the south, the Mississippian rocks (mostly covered by glacial deposits) extend beyond the Michigan border into northern Ohio and Indiana.

The Antrim Sea had entered Michigan near the close of the Devonian. Initially, the Antrim Shale, a Late Devonian marine deposit, accumulated on the floor of that sea. However, marine waters remained in the state beyond the close of the Devonian into Early Mississippian time (Figs. v–1, 2, 28). Meanwhile, the character of the shorelines and adjacent lands changed and the types of sediments that were deposited reflect those changes. Figure v–2, Columns C and D, shows that in southern Michigan the Antrim Shale is overlain by a time-

# Geology of Michigan

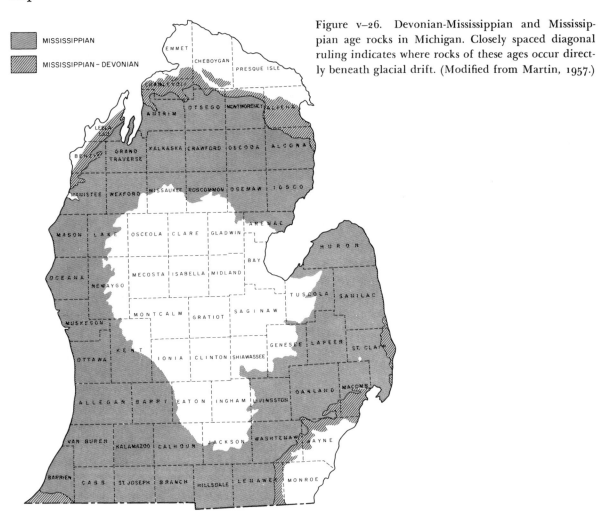

Figure v–26. Devonian-Mississippian and Mississippian age rocks in Michigan. Closely spaced diagonal ruling indicates where rocks of these ages occur directly beneath glacial drift. (Modified from Martin, 1957.)

transitional, Late Devonian to Early Mississippian deposit called the Ellsworth Shale on the west side of the state and by the Bedford Shale, Berea Sandstone, and Sunbury Shale of equivalent age on the east. Expressed in geological terms, the black Antrim Shale on the east and the gray and gray-green Lower Ellsworth on the west "interfinger" near the center of the Michigan Basin. In the transition from Devonian into Mississippian time, the Upper Ellsworth on the west interfingered with the Bedford, Berea, and Sunbury on the east, another good example of a *facies change*. The Ellsworth Shale, partly contemporaneous with the Antrim Shale, reflects the introduction of gray and green muds from a western source area into the Antrim Sea at a time when an eastern source area was providing black muds to that same sea. Silty sands of the Bedford Formation and the Berea Sandstone are interpreted as deltaic deposits carried into the eastern side of the Michigan Basin from the east at a somewhat later time, while the upper part of the Ellsworth Shale was being deposited in offshore waters of the sea in the central and western parts of the basin. The source area for muds of the Ellsworth Shale was the uplifted Wisconsin Highland to the west. The source areas which provided the deltaic sediments of the Bedford and Berea lay to the east and northeast of the basin. Cohee (1965, p. 210) has called that eastern, Early Mississippian delta the "Thumb Delta" because the sediments of that delta are well represented in the subsurface in the region of the Lower Peninsula "Thumb" today. The Bedford and Berea deposits are part of a much larger delta system which occupied large parts of Ohio, and Ontario to the east. A paleogeographic interpretation of those

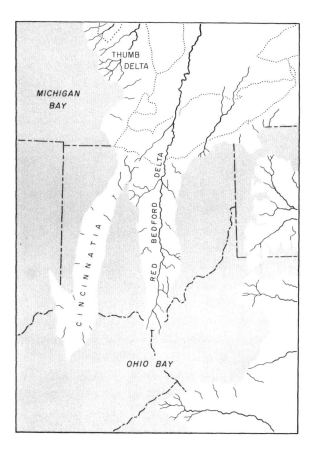

Figure v-27. Paleogeography of Early Mississippian (Bedford-Berea) time in southeastern Michigan and adjoining regions. Inferred area of sea stippled, land unpatterned. "Thumb Delta" shown in area of Saginaw Bay and the "Thumb" region of the Lower Peninsula of Michigan, where deltaic sediments of this age occur. (Modified, with additions by authors, U.S. Geological Survey Professional Paper 259, J. F. Pepper, et al.)

conditions is presented in Figure v-27. Delta growth was retarded thereafter, and most of the southern part of the Lower Peninsula again became entirely an offshore marine environment into which, still in Early Mississippian time, fine grained muds of the Sunbury and Coldwater shales were deposited.

Toward the close of Early Mississippian time a major reduction in the seas caused much of southern Michigan to become a near-shore and beach zone into which the gray, pink, and red sandstones and siltstones of the Marshall Formation were carried (Fig. v-29). The Marshall Formation consists of the Marshall Sandstone and the Napoleon Sandstone.

**The Paleozoic Era**

125

The seas must then have withdrawn entirely from the state because Middle Mississippian rocks are missing here. Above the Marshall Formation lies the Late Mississippian Michigan Formation, a marine deposit of shale, gypsum, dolomite, limestone, and a little sandstone. The Michigan Formation represents a readvance of seas into Michigan in the Late Mississippian. The gypsum of the Michigan Formation is another example of evaporite deposits, such as were described in the section on the Silurian Period. The presence of gypsum indicates that occasionally during the Late Mississippian the Michigan Basin again was a semiisolated marine evaporation basin. Near the close of the Mississippian Period the seas cleared and freshened and the Bayport Limestone was deposited in the basin.

In latest Mississippian time the Michigan Basin was uplifted, the seas withdrew, and parts of the Bayport Limestone and older formations below it were eroded; hence no latest Mississippian age deposits were laid down. This period of uplift and erosion is represented by an unconformity, or gap in the rock record, between the Bayport Limestone and the overlying Pennsylvanian rocks. Cohee (1965, p. 219) believes that this was the most important and extensive time of uplift and folding of the Michigan Basin rocks since the Early Ordovician, that it involved reactivation of ancient Precambrian faults and folds in the crust deep below, and that it produced many of the structures in which gas and oil later accumulated in the Paleozoic rocks of the Michigan Basin (Chapter X).

Mississippian rocks contain a number of natural resources of considerable economic value to the state. Small quantities of oil and considerable natural gas are recovered from the Berea Sandstone. The Napoleon Sandstone is quarried in the vicinity of Napoleon, Michigan, near Jackson (Fig. v-29). The sandstone is used mainly for patios and garden walls.

Mississippian shales, such as the Coldwater Shale in Branch and Calhoun counties, crop out locally at the surface or at least are only thinly covered by glacial deposits. These shales provide clay for brick, tile, and cement manufacture. The Antrim and Ellsworth shales also are important sources of raw material for these uses.

Gypsum for the manufacture of plaster, wallboard, and other products is the most important resource in Mississippian rocks of the state (Fig.

**Geology of Michigan**

Figure v-28. Echinoderms inhabiting sea floor at time (Early Mississippian) of deposition of the Coldwater Shale and Marshall Formation in Michigan. Letters indicate the following: A—bristly looking blastoid; B—a brittle starfish (ophiuroid); C—seaweed; D—inadunate crinoids (arm bases not included in calyx); E—flexible crinoids (tips of arms curved inward); F—camerate crinoids (arm bases integrated into calyx). (Diorama in The University of Michigan Exhibits Museum and Museum of Paleontology.)

v-20). It is quarried at the surface from open pits in the Michigan Formation at Tawas City and Alabaster on the eastern side of the state north of Bay City, and near Whittemore in Iosco County. It is mined from the same formation underground at Grand Rapids on the west side of the state. The Best Wall Gypsum mine at Grand Rapids has been in operation for over 100 years, obtaining gypsum from a 300-acre area 150 feet below the surface. Michigan is the second most important producer of gypsum in the United States. The value of that product from Michigan was $5,090,000 in 1967. Michigan's gypsum, like its rock salt, is a virtually inexhaustible natural resource, one of the few nonrenewable resources for which that can be said.

### The Pennsylvanian Period

Pennsylvanian rocks in Michigan are of particular interest for two reasons. First, as in many other parts of the world, they contain coal. Second, many of the Pennsylvanian rocks of the state were laid down on land as "continental" deposits. We have seen that throughout the preceding periods of the Paleozoic Era the sedimentary rocks of the state were dominantly marine in origin, the only nota-

Figure v-29. Early Mississippian marine sandstones of the Marshall Formation.

*Top*—The Napoleon Sandstone, upper member of the formation, Star Stone Company Quarry, Napoleon, Michigan. Inset shows cross section of a "rippled toroid" in place and right side up. Hammer handle for scale. Toroid is sand filling of whirlpool scoured cavity. Note center cone sucked up by vortex.

*Center left*—Inverted artificial rippled toroid produced by pouring plaster into a cavity formed by a whirlpool in laboratory sand table. Ridges are ripple marks. Foot ruler for scale.

*Mid-center*—Inverted natural rippled toroid from quarry above.

*Center-right*—Marine clams and a nautiloid cephalopod from the Lower Marshall Sandstone (5 centimeter scale).

*Bottom*—Views of cross-bedded Lower Marshall Sandstone Member along I-94 near Battle Creek.

### Geology of Michigan

ble exception being the streamlaid, Early and Middle Cambrian Jacobsville Sandstone found in the Upper Peninsula. There may have been other continental sediments deposited here during the Paleozoic because the periods of uplift and erosion represented by unconformities on Figure v–2 were times when streams wore away parts of the previously deposited formations. Certainly, those streams must have deposited sediments somewhere, at least temporarily, just as do the streams of today. However, such deposits may never have been widespread or thick, and if they ever existed were subsequently eroded. During the Pennsylvanian Period, in contrast, the seas that occupied portions of the state were shallow and of short duration. Their shorelines frequently fluctuated back and forth across the state, and during times of marine retreat streams meandered sluggishly across the emergent lowlands, depositing sediments locally in their broad channels. It was in the Pennsylvanian Period, then, that the long dominance of the Paleozoic seas came to an end in Michigan.

As already noted, the land had risen at the close of the Mississippian, so the Pennsylvanian Period actually began with Michigan in the emergent condition inherited from that prior period. Thus, when deposition of sediments finally began again, later in the Early Pennsylvanian, the basal deposits were laid down upon an erosion surface cut across the folded and slightly inclined layers of preceding periods. A slight angular unconformity resulted. Later in the Early Pennsylvanian, in Late Pottsville time, the first of many minor marine advances carried the sea across the state from the north and west and some marine Pennsylvanian strata accumulated. Deltas then grew from east to west into the Michigan Basin (Cohee, 1965, p. 219). A relict of the sea persisted along the west edge of the state in spite of this delta growth, and it received red shales while elsewhere the basal sands of the Saginaw Formation were accumulating (Fig. v–2, Column C). The margins of this shallow sea then began to fluctuate, new seaways extended into the state from the south and southwest, and the central and western parts of the Michigan Basin were alternately marine areas, then swamplands, then emergent coastal plains traversed by streams. A series of sedimentary deposits, including the marine Verne Limestone, accumulated under those varying conditions to form the *Saginaw Formation*. That formation now consists of alternating and intertonguing stream and river channel sands, river floodplain silts and clays, shallow water marine or tidal swamp shales and limestones, and swamp-laid coals. The occasional marine limestones and shales contain the fossil remains of salt or brackish-water animals, thus proving that those deposits originated in a marine or brackish-water environment. The stream channel sands and river floodplain clays, and swamp-laid coals are rich in terrestrial plant fossils. Many of these are illustrated and described in Chapter XV. Some of the shales contain both marine fossils and broken plant remains, suggesting deposition either on marine tidal flats or deltas. Kelly (1936) recognized several "cycles" of deposition represented by alternating continental and marine or brackish-water sediments. Several minor unconformities in the sedimentary record represent brief times of emergence of the land. The Pennsylvanian deposits of the Saginaw Formation with the included unconformities thus represent a series of marine transgressions and regressions often referred to as "overlap" and "offlap" deposits (Chapter XII). Typical examples of cyclic deposition in the Saginaw Formation can be seen today in some of the shale quarries in the vicinity of Grand Ledge near Lansing. Figure v–30 shows one such quarry with the rock types, common fossils, and other features of historical significance identified and interpreted. The sandstones of the Saginaw Formation often are discontinuous lenses of highly variable thickness. They include abundant fossilized leaves, tree trunks, and roots of land plants and represent stream channel deposits. The medium to light gray shales and siltstones, also rich in land plant fossils, and with poorly developed and irregular stratification, are river floodplain and swamp deposits. The upper parts of some of the shales often are bleached to a light gray, the stratification destroyed, and ironstone concretions common. These zones are called "underclays" and are thought to be old soil zones perhaps formed under cover of swamp waters. Burrowing soil dwellers, such as worms, and root growth are thought to have been the causes of destruction of the stratification. Darker gray to black siltstones and shales with thin, regularly bedded layers that split easily and evenly, and containing both land plant fragments and marine or brackish-water invertebrate animals, such as the mud-loving brachiopod *Ling-*

Figure v–30. Pennsylvanian age Saginaw Formation, Grand Ledge Clay Product Company clay quarry, Grand Ledge, Michigan. Numbering of rock units corresponds to that on Figure XIII–40 where an interpretation of the history of deposition of these beds is provided. Level *A* indicated at several points in Unit 3 contains the marine brachiopod *Lingula carbonaria* (See Fig. XIII–28 (19)). Arrow at *A* in lower right photo shows location of same *Lingula* zone at far (north) end of quarry. *B* on upper photo shows location of close-up in lower left where vertical streaks in sandstone are "stigmaria" (roots of *Lepidodendron* and *Sigillaria*) in original position of growth (6-inch scale).

*ula*, were deposited on muddy marine tidal flats. Coal seams are common. These often are lensing or discontinuous, no one layer continuing for any great distance before thinning out and ending. The coal formed from accumulated generations of dead plant remains in lushly vegetated, swampy lowlands. The plant remains usually are too fragmentary to be identified exactly. Some of the swampy depressions were oxbow lakes and sloughs left behind when ancient rivers abandoned one meander bend for another. Other depressions were tidal swamps formed in lowlying areas near sea level.

The Verne Limestone near the middle of the Saginaw Group is widespread in the western and central parts of the basin and represents a period of extensive marine invasion; fossil marine invertebrates are common in it. An unconformity occurs at the top of the Saginaw Formation. Above it lie the Woodville, Ionia, and Eaton sandstones of the Grand River Group. Studies of the fossil plants of the Saginaw Formation and Grand River Group (Arnold, 1949) indicate that the latter group is much younger than the Saginaw Formation. Rocks of Middle Pennsylvanian age seem to be missing, perhaps because of uplift during that time. The sandstones of the Grand River Group thicken and thin rapidly, even disappearing locally. Their discontinuous, lensing character, the presence of current ripplemarks, and the abundance of waterworn land plant trunks, twigs, and leaves, together suggest that these sandstones of the Late Pennsylvanian were deposited in meandering rivers and streams that flowed sluggishly across the state near the close of Pennsylvanian time. Evidently, the land had risen still further, the seas had withdrawn, and even the swampy vestiges of those seas had largely disappeared. The long Paleozoic history of marine occupation and sedimentation in the Michigan Basin finally was drawing to a close.

The Pennsylvanian coal deposits tell an especially interesting story. Compressed and altered plant remains occur as coal deposits in rocks of many different geologic ages throughout the world, but Pennsylvanian coals are particularly abundant and widespread. Studies of modern conditions in coal-forming regions, such as the Dismal Swamp of Virginia and North Carolina along the Atlantic Coastal Plain, indicate that partially decayed vegetal materials accumulate most readily and extensively in low-lying, swampy regions of high rainfall, poor drainage, and lush plant growth. Although warm temperatures favor plant growth they are not absolutely essential, as can be seen from the fact that peat, the first stage in coal formation, is forming in swampy areas of Michigan and even in the Arctic today. Paleobotanists have determined, from studies of the structures of Pennsylvanian fossil plants that those plants were not cold climate types; this is indicated by the fact that they lack annual growth rings which, if present, would reflect summer growth and winter dormancy. On the other hand, tropical climates are generally unfavorable to the accumulation of peat because bacterial decay there proceeds at such a rapid rate that dead plant remains are destroyed soon after they fall to the forest floor. Plant material does not ordinarily accumulate, year after year, on the floor of a tropical forest or swamp. Thus, we already have a partial picture of climatic conditions in Pennsylvanian coal swamps. The presence of shallow, stagnant water is necessary because if plant remains fall on dry land they soon decay, regardless of climate, and the carbon that would constitute the most important ingredient of coal is consumed by slow oxidation in the atmosphere. Stagnant water prevents decay by excluding air and limiting the extent of bacterial action. Decay bacteria produce waste products (humic acids) that are poisonous to the bacteria themselves. Such toxic wastes accumulate in stagnant waters so that in the lower levels of peat deposits bacterial action ceases. On the other hand, if the water is too deep, the growth of plant life is inhibited and conditions become unfavorable for extensive peat accumulation. The formation of peat depends upon a delicate balance of climatic, water, and biological conditions. Nevertheless, peat, a porous, spongy, partially decayed mass of plant fragments, is common in swampy areas today. Apparently, similar conditions favoring peat formation were typical of vast lowland areas in the interior of North America during the Pennsylvanian Period. The paleogeographic map for the Pennsylvanian (Fig. v–1) indicates the location of some of the seas and bordering swamplands of that period.

Many varieties of ancient and for the most part primitive plants contributed their remains to the Pennsylvanian coals. More easily identified fossil remains of the same plants are scattered in lesser

The Paleozoic Era

131

Figure v-31. Pennsylvanian age coal swamp forest. Letters indicate the following: Am—small amphibian; As—*Asterophyllites*, the leaves of the large sphenopsid called *Calamites*; C—*Calamites*, a large sphenopsid; D—giant dragonfly; L—*Lepidodendron*, a lycopsid; N—*Neuropteris*, a pteridosperm ("seed-fern"); Si—*Sigillaria*, another genus of lycopsid; Sp—*Sphenophyllum*, a small sphenopsid. See Chapter xv for descriptions of Pennsylvanian fossil plants. (Diorama in The University of Michigan Exhibits Museum and the Museum of Paleontology.)

quantities throughout associated types of sedimentary rocks. The plants ranged from small, low, sprawling types to trees several feet in diameter and 50 to 100 feet high. Figure v–31 shows a diorama in The University of Michigan Museums which depicts a scene in a Pennsylvanian coal swamp. Descriptions and illustrations of the plants themselves will be found in Chapter XV.

The peat formed in the Pennsylvanian swamps was altered to coal by heat and pressure. These agents drove off excess water and volatile gases, compressed the remaining vegetal matter, and thus increased the percentage of carbon in the deposit. As the Pennsylvanian peats became more and more deeply buried beneath younger sediments, the weight of overlying deposits supplied the pressure required; the necessary heat resulted from the fact temperature increases with depth beneath the earth's surface. It can be seen in modern coal-forming regions, and observed in laboratory experiments, that moderate heat and pressure alter peat to lignite (brown coal). Further alteration produces bituminous or "soft" coal, the type found in Michigan. Anthracite or "hard" coal forms under conditions of extreme heat and pressure such as occur only in regions of strong crustal compression or igneous intrusion. Anthracite occurs in the Appalachian Mountains to the east, but coals of the same age in the relatively undeformed Appalachian Plateau and in Michigan remain in the lower, bituminous *rank*. Coals of high *grade* are those low in residual ash after burning. Michigan coals are generally of low and quite variable grade. The grade of a coal was determined by the original environment of deposition. If little or no sediment was carried into a coal swamp, the vegetal material accumulated free of ash-forming substances, whereas the ash content increased in those coals which became mixed with mud, silt, and sand carried by streams flowing into the swamps. Actually, coals may range in grade from pure coal with practically no ash-forming sediment, to sediments in which only scattered, coalified plant fragments occur.

Knowing the environmental conditions that favor the formation of coal, one can view the deposits in the Saginaw Formation of Michigan in a much more interesting light. Picture in your mind's eye, the heavily vegetated tidal swamps and the sluggish streams meandering across broad floodplains, close to an inland sea and cloaked in the moisture-laden atmosphere of subtropical or warm-temperate climate.

Pennsylvanian strata underlie about 11,500 square miles in the central part of the Lower Peninsula (Fig. v–33). The basin-like structure of the strata carries the Pennsylvanian rocks down to their deepest level, 960 feet below the surface, in northcentral Midland County. There is where they also are thickest, in some places reaching 750 feet. Outward from there, erosion has reduced their original thickness progressively to zero toward the margins of the Michigan Basin. The basal 15 to 150 feet of sandstone is noncoalbearing. The few natural exposures of Pennsylvanian rocks occur in stream valleys—that of the Grand River near Grand Ledge for example. Elsewhere, they are covered by younger deposits, mostly of glacial origin.

There are no commercial coal mines presently operating in Michigan, but in the past coal was produced in Bay, Saginaw, Tuscola, Shiawassee, Genesee, Ingham, Eaton, Calhoun, and Jackson counties. Coal-containing beds are numerous, 14 having been reported in Bay County alone, but not all known seams were mined. Three seams, called the "Saginaw," "lower Verne," and "upper Verne" were mined locally; the Saginaw seam, locally 3 feet thick, was the principal producing horizon. Few seams were thicker, although one in the Wenona Mine north of Bay City was 4.5 feet thick. The thinness of the individual seams, their limited lateral extent, the excessive amount of overburden, the low quality of the coal itself, and the weak structure of the coal which caused it to break down to fine powder in transit were all undesirable factors which made Michigan coal uneconomical to produce. Cohee (1950) estimated the unused coal reserves of the state at about 220 million tons, but only about half of that could actually be recovered by normal mining methods. Coal was first discovered in Michigan in 1835 in Jackson County, and some mining continued there until 1910. Mining near Grand Ledge, not far from Lansing, began in 1839. In the southeastern part of the Michigan Coal Basin, in parts of Ingham, Eaton, and Shiawassee counties, the coal was close enough to the surface, and the glacial cover thin enough, to permit surface strip mining in open pits. Some such suitable areas probably still exist in those areas. Underground shaft mining was carried on mainly in Bay and

Saginaw counties and about 59 mines operated there at one time or another. Maximum annual production, 2 million tons, was reached in 1907. Production declined thereafter and the last mine, the Swan Creek Mine northeast of St. Charles in the Saginaw area, closed down in 1952. The future of coal mining in the state at present is unfavorable in spite of the reserves that remain. As Arnold (1965) has suggested, it is probable that only vastly increased local demand brought about by depression conditions could revitalize this industry.

In view of this dim picture for the coal industry of the state, it is interesting that Michigan has recently become the nation's foremost producer of peat, the material from which coal is produced by natural processes. The Michigan peat being produced today is, of course, geologically much younger than the original peat from which the Pennsylvanian coals formed. The Michigan peats on the market today are of Post-Pleistocene age and were deposited in swampy depressions left behind by the last Pleistocene glacial ice sheet. Michigan produced about $2,300,000 worth of peat in

Figure v-32. Grand Ledge Clay Product Company tile plant at Grand Ledge, Michigan. Gas-fired kilns in left background. Nearby quarries (see Fig. v-30) in Early Pennsylvanian age Saginaw Formation provide clay shale raw material. (Michigan Department of Conservation, photo by Clyde Allison.)

# Geology of Michigan

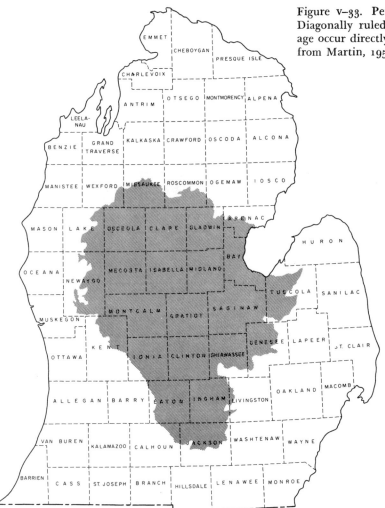

Figure v-33. Pennsylvanian age rocks in Michigan. Diagonally ruled area indicates where rocks of this age occur directly beneath the glacial drift. (Modified from Martin, 1957.)

1967, which constituted about 38 percent of the national output.

*Clay*, often associated with coal in Michigan, is a much more important contributor to the present economy of the state. Some of the clay comes from lake deposits of Post-Pleistocene to near Recent age, like the peat mentioned above. The remainder comes from Paleozoic deposits such as the Antrim, Ellsworth, and Coldwater shales of Devonian and Mississippian age, and the clays and shales in the Saginaw Formation of Pennsylvanian age. Some clay is produced in nearly every county in the state. About 75 percent of it goes into the manufacture of cement, 15 percent into clay products such as tile and brick, and 10 percent into lightweight "expanded" aggregate used for construction purposes. Clay also is used for oil well drilling mud, molding sand bonder, and as a soil conditioner. Several tile manufacturing plants are in operation in the Grand Ledge area near Lansing. These produce a wide variety of clay products, some of which are illustrated in Figure v-32.

## Paleozoic Era—Conclusion and Summary

The Paleozoic chapter of Michigan geologic history ended with the deposition of the last sediments of Pennsylvanian age, the stream-laid sandstones of the Grand River Group (Fig. v-2). Elsewhere in North America deposits of the Permian Period constitute the last pages of that chapter, but by then Michigan was an upland, receiving no Permian sediments that have been preserved, and probably undergoing erosion. Throughout most of the Paleozoic Era the Michigan Basin was occupied by shallow epeiric seas. The uplift that brought the Paleozoic Era to a close in Michigan probably was associated with the widespread rise of the whole of eastern North America as a result of the Appalachian Orogeny, which at the same time intensely deformed the sedimentary rocks of the Appalachian Miogeosyncline to form the ancestral Appalachian Mountains.

Thus, we come to the next chapter in the geologic history of Michigan. It really is *not* a chapter at all, but rather is a great gap in the record. Therefore, we shall call it the "Lost Interval."

# VI
# The Lost Interval

The geologic record in Michigan is almost completely blank from the end of the Pennsylvanian Period until the last stage of the Pleistocene. Throughout most of the state, glacial drift of latest Pleistocene (Wisconsin) age lies directly upon bed rock ranging in age from Precambrian through the Late Pennsylvanian. Rocks of the following ages are missing in between: latest Paleozoic (Permian); nearly all of the Mesozoic Era (except the very Late Jurassic) including the Triassic, Early and Middle Jurassic, Cretaceous; and all of the Cenozoic Era except the latest Pleistocene. We shall call this great gap in the historical record the "Lost Interval." About 280 million years of geologic time are involved.

This vast time span is not a complete gap, however. Sandwiched in between the Paleozoic rocks and the much younger Pleistocene glacial drift are some sedimentary strata which for many years have been referred to simply as "red beds." Sometimes they have been called "Permocarboniferous red beds." The stratigraphic position of these rocks in Michigan is shown near the top of Column C in Figure v–2 in the preceding chapter. These strata occupy only a small area within the central part of the Michigan Basin, as is shown in Figure vi–1. Nowhere are they actually exposed; their presence beneath the glacial drift is demonstrated only by well samples and records. The "red beds" consist mainly of sandstone, shale, and clay, with minor beds of limestone and gypsum. The age of the "red beds" long was uncertain. No datable fossils of any kind had been found in them, and they were referred to as "Permocarboniferous" mainly because they lie above Late Pennsylvanian strata, hence occupy a stratigraphic position that could have been Late Carboniferous (Pennsylvanian) and Permian. Being red, they resembled Permian red beds known to occur elsewhere in the United States. Recently, Professor Aureal T. Cross and his students at Michigan State University have found fossilized microscopic plant spores in well samples from the red beds. Some of these spores are illustrated in Chapter XV. They have identified these as being latest Jurassic (Kimmeridgian) in age. Thus, it appears that the red beds really are mid-Mesozoic in age, and form a tantalizing fragment of history from the Age of Reptiles. The Jurassic red beds vary in thickness, normally being about 100 feet thick, but in places reaching thicknesses of 300 to

# The Lost Interval

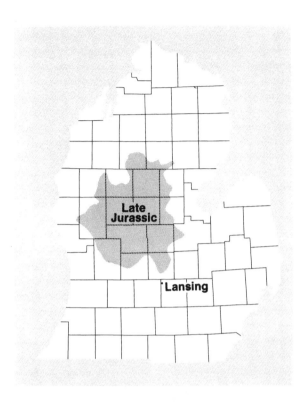

Figure VI-1. Late Jurassic rocks in Michigan. Diagonally ruled area indicates where rocks of this age occur directly beneath the glacial drift. These are the rocks referred to on earlier maps and charts as the "red beds." (Modified from Cohee, 1965.)

400 feet (Cohee, 1965). They occur in lense-shaped patches and in many places disappear completely. The red beds overlap the boundaries of the underlying, older rocks. Thus, they appear to be sedimentary fillings laid down in depressions, possibly in ancient lakes, ponds, and stream valleys, on the older Paleozoic rock surface. The presence of gypsum indicates that the climate in that area, at least locally, during latest Jurassic was semiarid or arid.

Actually, the absence of rocks in the Lost Interval is a form of negative historical evidence from which certain conclusions can be reached regarding conditions in the state. Apparently, this part of North America has been an emergent area, no doubt subject to erosion, since some time in the Late Paleozoic. Elevation of the land led to erosion rather than to sedimentary deposition. Although subsidence of the land and subsequent marine invasion occurred elsewhere, and epicontinental seas are known to have been widespread across other parts of the continent, it is apparent from the absence of marine deposits of that age that Michigan was not the site of extensive marine invasion then. The era of the Paleozoic seas had ended and the earth's crust beneath the Michigan Basin, for so long subject to downwarping, had become more stable. The place of Michigan in the paleogeography of North America during the Lost Interval is shown in Figure VI-2.

Although there is no record of marine deposition in the state during this interval, it is difficult to believe that *no* sediments of *any* kind were ever laid down here then. Surely, at certain times and in certain areas of Michigan, some stream deposits must have formed, just as stream-laid sediments are accumulating on portions of the continent that are above sea level today. Conceivably, small patches of such deposits could lie concealed somewhere beneath the Pleistocene glacial deposits that now blanket so much of the state, and the lenticular Late Jurassic red beds may represent just such deposits, although too little is known about them at present to reach reliable conclusions concerning their nature and history. It is well established that, in general, over the long course of geologic time deposits laid down on land by rivers and streams, or in lakes, are not permanent, but are sediments halted only briefly in their long but inevitable journey to the sea. If any quantities of land-laid, continental sediments were deposited in Michigan during the Lost Interval, it is likely that most of them would have been swept away again by erosion in such an area of general uplift.

Rocks belonging to the ages missing from Michigan during the Lost Interval *are* found in many other areas of the world outside Michigan. The remainder of this chapter will deal very briefly with some of the biologic changes and geologic events that can be demonstrated to have occurred by evidence found elsewhere. Some of the animals and plants that evolved and were living then must have inhabited this state even though there are no rocks here in which their fossil record is preserved. Figure VI-3 illustrates the events discussed next.

As an uplifted area of dry land, the state certainly must have been covered by many of the ancient forms of *land* vegetation that we know lived during the Lost Interval. These would have included some of the primitive types of plants that

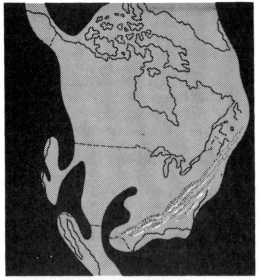
EARLY PERMIAN

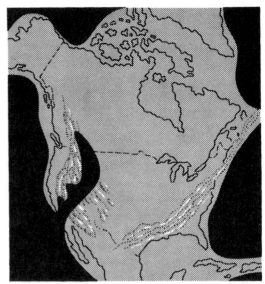
EARLY TRIASSIC

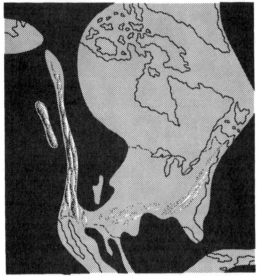
MIDDLE LATE JURASSIC

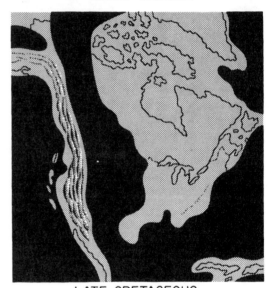
LATE CRETACEOUS

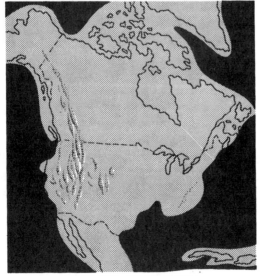
EOCENE (CENOZOIC)

Figure VI–2. Paleogeographic maps for Permian to Pleistocene time. This is the so-called "Lost Interval" in Michigan geologic history. The only rocks in Michigan known to represent this vast, 280-million-year-long span of time prior to the Pleistocene Ice Age are the recently recognized Late Jurassic deposits (Fig. VI–1) formerly called simply "red beds." The latter do not crop out anywhere from beneath the glacial deposits in the state, occupy a limited area beneath the glacial drift, and are so far known only from well records. The maps show that the reason for this great historical gap in the state lies in the fact that the Great Lakes region was not occupied by extensive lowlands or seas during that time. (Modified from C. O. Dunbar, *Historical Geology,* 2d Ed., 1960, John Wiley and Sons, Inc.)

# The Lost Interval

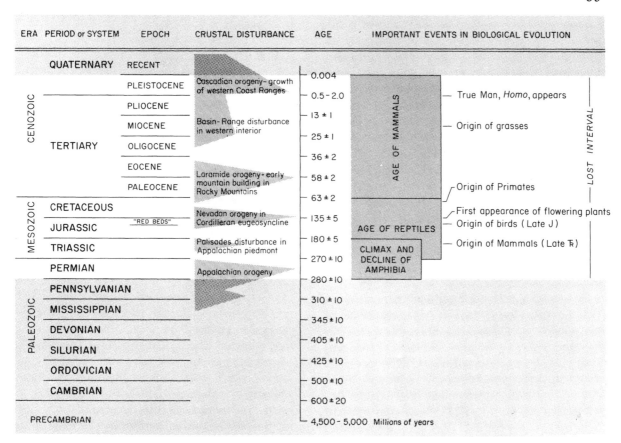

Figure vi-3. Time chart showing events *not* recorded in the historical gap called the "Lost Interval" in Michigan. Note that during Permian to Pleistocene time many important crustal disturbances and events in biological evolution occurred and are recorded elsewhere in rocks of many ages not found in Michigan. See Figure vi-2 for maps that explain reason for gap.

had lived here previously in the Pennsylvanian coal swamps (Chapter V), and that persisted for some time thereafter. Also present might have been the first conifers which also appeared in the Pennsylvanian Period and which have persisted, in one form or another, to the present day. The more advanced Flowering Plants, or *angiosperms*, which first appeared during (or just before) the Early Cretaceous, and the important *grasses* which originated much later, in the Miocene, must have occurred here, too. One can imagine that each of these new evolutionary developments in the plant kingdom would have resulted in striking changes in the vegetative cover of this state, as we know they did in areas of the world where fossil records of such plants happen to be preserved.

The same can be said for changes in the animal kingdom. During the Pennsylvanian and Permian periods, amphibians, some of giant size, were the dominant form of vertebrate life on land. Some of those ancient amphibians persisted on into the Triassic Period. Reptiles first appear in Middle Pennsylvanian and some of their oldest known fossil remains are found in rocks of that age in Illinois and Indiana. By the Triassic Period reptiles had become the dominant forms of land-dwelling vertebrates, replacing amphibians in that role, and holding their position of dominance throughout the remainder of the Mesozoic Era until the close of the Cretaceous Period some 63 million years ago. The Mesozoic Era often is called the Age of Reptiles. Noteworthy amongst the ruling reptiles of that time were the many types of dinosaurs, but other types including marine swimmers and mammal-like reptiles, also flying reptiles, were abundant. The oldest known mammals occur in rocks of Late Triassic age, and the oldest known birds are found in rocks of the

# Geology of Michigan

Late Jurassic. Mammals and birds "lingered in the wings," so to speak, during the Mesozoic Age of Reptiles, but with the passing of the dinosaurs those two more advanced groups assumed positions of dominance amongst the land vertebrates. The Age of Mammals began about 63 million years ago, at the beginning of the Paleocene Epoch of the Cenozoic Era. Among the mammals, lower primates first appear in mid-Paleocene rocks. In the Old World the first higher primates of the anthropoid group appeared in the Oligocene; the earliest so-called "ape men" or prehumans in the Miocene and Pliocene. True man, *Homo sapiens,* finally appeared early in the Pleistocene Epoch. Man probably first reached North America during the last (Wisconsin) stage of ice advance in the Pleistocene.

Certainly, one must conclude that many of these evolutionary changes in the constitution of the world fauna and flora during the Lost Interval had some effect on the nature of life in Michigan, even though no actual records of those events are found here, because where fossil records are available we know that, given enough time, most major animal groups have sent at least some migrant representatives of their type into the farthest corners of the earth.

Many important changes in marine life, both vertebrate and invertebrate, also occurred during the Lost Interval in areas far removed from the state (Chapter XIII).

Vast changes of a geologic nature occurred in the earth's crust during the Lost Interval (Fig. vi–3). In the Pennsylvanian and Permian periods in North America, the Appalachian Geosyncline Region (Chapters I and V) was complexly folded, faulted, and uplifted. Although those early mountains were eroded away, many of the basic structures formed at depth still are preserved in the Appalachian Mountain Region today; at the same time, crustal deformation and mountain building occurred in Oklahoma and Texas, and some early precursors of the Rockies, the so-called "Ancestral Rockies," rose in New Mexico and Colorado. But all of those uplifts were worn down again by erosion. During parts of the Jurassic and Cretaceous Periods, extensive crustal deformation, volcanic activity, igneous intrusions, and mountain building again occurred in what is now the western third of the continent. Beginning in the Late Cretaceous, and continuing at intervals at one place or another right up to the present, the entire western half of North America was the site of extensive crustal deformation, mountain building, volcanic activity, and erosion. The Rocky Mountains, Sierra Nevada, and Pacific Coast ranges all took form during this time; all had very complicated geologic histories. The Appalachian Mountain Region, too, was reelevated several times during the Cenozoic Era. In fact, during the Late Cretaceous and Cenozoic, most of the major mountain regions, lowlands, and other distinctive geographic features of the modern world initially took shape.

These events, biologic and geologic, and many others as well, are recorded in far greater detail in any good general historical geology textbook. We cannot do all those far-flung events justice in this brief chapter. Furthermore, few of those important happenings affected Michigan. But it is important to realize, from what little has been said, that what we have called the "Lost Interval" in Michigan's geologic past was a vast stretch of time during which there was opportunity for many geologic events of great historical importance to occur. Yet there is no direct record of these here in the state; the unconformity between Late Paleozoic and Late Pleistocene times in Michigan is thus a large historical gap, broken only briefly by the poorly known Late Jurassic red beds. Throughout this time Michigan, as part of the stable interior of the continent, was neither greatly uplifted and deformed nor downwarped to any great extent. Toward the very end of this long period, within about the last 500,000 to 2,000,000 years, the Pleistocene Epoch began. At least four times, major continental ice sheets moved southward from the Canadian Shield Region into northern United States, only to melt back again. Michigan probably was covered by ice each time, but, as discussed in the next chapter, only those deposits of the last, or Wisconsin, stage of the Pleistocene have been definitely recognized here.

It is a great temptation to continue speculating about what "may have been" during the Lost Interval in Michigan, or to expand upon details of earth history from other regions, but this book is mainly about Michigan, so we must pass on to the last chapter of geologic history actually represented in the record here, that is the Pleistocene Epoch.

# VII
# The Pleistocene (Ice Age) Epoch

## Nature of Glacial Ice

Of the various geologic agents important in shaping the topography of the earth, glacial ice is particularly well suited to serve as an illustration of the use of the Principle of Uniformitarianism. The apparent haphazard character of glacial deposits, their general lack of topographic regularity, and their "freshness" clearly indicate to the geologist that they are the result of a unique set of circumstances acting in fairly recent time. A brief history of the growth of our understanding of glaciation will, perhaps, demonstrate for the reader the use that the geologist makes of the Principle of Uniformitarianism, even though he may do so in an unconscious way.

As late as the 1830's the scientific world first realized that the recent geologic past had been a period of extensive glaciation. Interestingly enough, many of the untrained inhabitants of the Swiss Alps had long assumed that the boulders and soil, as well as the Alpine topography, of their homeland were the result of extensive glaciation—they could see relatively small masses of ice still lingering in the upper reaches of the valleys. Between 1821 and 1835 two European engineers, Venetz and Charpentier, separately arrived at the conclusion that the boulders scattered across the mountainous area must have been carried there by extensive alpine glaciers. At first their ideas were generally rejected. In 1836 Charpentier talked one of the disbelievers, a Swiss naturalist by the name of Louis Agassiz, into taking a field trip through a part of the Rhone Valley to look at some of the evidence of glaciation in its lower regions. Although he came to scoff and dissuade, Agassiz soon realized that, if anything, Charpentier had underestimated the effects brought about by glaciation on the area. Scurrying back to Paris, Agassiz published soon thereafter a treatise putting forth the idea of extensive glaciation in the recent geologic past. While by no means the first man to have the idea, Agassiz deserves much of the credit for spreading the gospel of glaciation throughout the scientific world.

As he visited areas well removed from the Swiss Alps, areas such as Great Britain, the northern plains of Germany, and, ultimately, North America, Agassiz became convinced that not only had there been a period of extensive glaciation by glaciers of continental size but that this glaciation

# Geology of Michigan

actually involved several complex oscillations of the ice masses—the idea of "multiple glaciation." The evidence in support of this concept includes: (1) the superposition of glacial deposits, a glacial till of one particular lithology resting on top of glacial till of different lithology, with the differing lithologies suggesting that the two ice sheets advanced along different routes picking up different rock material enroute; (2) the presence of interglacial or nonglacial deposits, or of a well-developed weathering zone, between two glacial drifts; (3) a distinct difference in the apparent ages of topography of two areas, youthful or relatively recent ones marked by many knobs and kettles, hills and closed depressions, and older, more weathered and eroded ones characterized by relatively flat, well-drained topography, even though both areas are underlain by glacial till; (4) a statistical difference in the depth of leaching, or the depth of weathering, on till found in two different areas. These and other less important things indicate to the trained geologist that an area has been involved in more than one glacial advance, each frequently separated from the next by a relatively long period of time.

But what is the mode of formation of the ice which goes to make up the large glaciers that affect large portions of this world of ours? There are two prerequisites for the formation of glacial ice: (a) adequate precipitation in the form of snow, and (b) sufficiently cold temperatures so that the snow which falls in a winter does not all melt in the succeeding summer. That both of these factors affect the amount of snow that accumulates, and thus the amount of glacial ice that may form is indicated by Figure VII–1, a graph showing the generalized distribution of the elevation of the regional snowline relative to latitude. In general, average annual air temperature increases as one goes from polar regions toward the equator; it also, of course, increases with a decrease in altitude. If temperature, as related to latitude, were the only control in the elevation of the perennial snowline, the snowline would rise regularly to higher altitudes from the poles to the equator. As the figure shows, however, the highest elevation of the perennial snowline is in the vicinity of latitude 20°, and its elevation is lower in areas of still lower latitude. To explain this fact one must consider the other factors that might affect snowline elevation. The latitudinal area near 20° is marked by a very dry climate all around the world—this is why the major deserts of the world, for example, those of Australia and Africa, lie within this zone. In the still lower latitudes, while the temperature is somewhat higher, precipitation is so much greater that snow which falls in the high mountains is able to persist from one winter to another at a lower elevation—thus the perennial snowline drops down again as the equator is approached. Figure VII–2 illustrates in yet another way that both adequate snow and low temperatures are essential for the formation of glacial ice. Note that in this figure showing the distribution of ice in the northern hemisphere during the Pleistocene large areas of Siberia and Alaska were unglaciated, even at times when the continental ice reached the Ohio River Valley. Both Siberia and Alaska were sufficiently cold to support glacial ice—in fact, many parts of these areas are even now underlain by perennially frozen ground (permafrost) at a relatively shallow depth. However, these two areas were largely unglaciated, primarily because they are both relatively arid regions in which there is so little precipitation that whatever snow falls in the winter disappears in the succeeding summer even though the summer is quite cool.

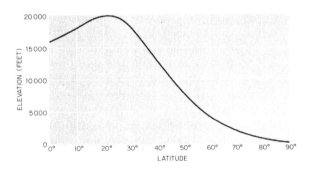

Figure VII–1. Variation in elevation of permanent snow line with latitude, greatly generalized. In polar regions (90° latitude) snow line extends to sea level, because of cold climate. Southward the snow line rises to progressively higher elevations because only there are mean annual temperatures low enough for annual snowfall to exceed annual melt. Note that highest elevation of permanent snow line is at about 23° latitude (in both northern and southern hemispheres). From there to the equator (0° latitude) the snow line falls again in elevation because, although temperature continues to rise, precipitation, including snow, increases even more rapidly and offsets increased melting rate.

Given snowfall sufficient that some which falls in the winter is not melted in the succeeding sum-

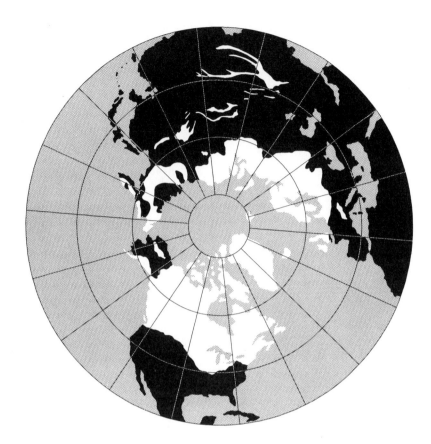

Figure VII-2. North polar projection showing approximate maximum extent of Late Pleistocene (Wisconsinan) glacial ice (in white). (Modified from Flint, *et al.*, 1945.)

mer, snow will begin to accumulate on the ground. As this snow is exposed to the sun's rays, it undergoes some changes—it is metamorphosed. The flakes of the new-fallen snow become somewhat compacted and recrystallized; in a matter of days, or at most weeks, small ice granules are formed, called "old snow" by the glaciologist and "corn snow" by the skier. This change is accompanied by an increase in density from 0.1 to 0.3 or more (the density of pure water being 1.0). With time and the accumulation of additional snow, additional compaction of the granules occurs, and the density of the mass increases still further. When the density of the old snow reaches 0.55, both the rate and the mechanism for density increase changes; apparently the density of 0.55 is as great as can be accomplished by a simple rearrangement of the ice granules. Granular ice with a density of greater than 0.55 is called *firn* by the glaciologist. As the firn accumulates further changes take place through deformation of the grains, local melting, refreezing and recrystallization, and the mass becomes still more dense and compact, with a corresponding decrease in permeability. As a result of this metamorphosis the lower part of the snow accumulation eventually becomes totally recrystallized to form glacial ice, with a density of between 0.8 and 0.9. Glacial ice differs in several ways from ice formed in the refrigerator or by the freezing of water in a pond or lake. Typically, it contains dust particles and other solid material that became mixed with the snow while at the surface. Also, glacial ice usually contains much air that became entrapped between the grains and flakes of snow and ice in the accumulative process.

There are two basic types of glaciers found on the surface of the earth. The glaciers studied by Agassiz and Charpentier in the Swiss Alps are what are called *alpine, mountain,* or *valley* glaciers; they are long, stream-like masses of ice confined to rather steep-walled mountain canyons. Such masses of ice flow downhill, from the accumulation areas high on the flanks of the mountains to the snouts of the glaciers at lower elevations. On the other hand, the extensive, roughly circular masses of ice which presently

# Geology of Michigan

cover the Antarctic continent and the island of Greenland, or the smaller patches covering portions of Iceland, are quite different sorts of glaciers. They typically flow across the surface of the earth as though it had no relief, even crossing mountain ranges, which they engulf and override as though these were only minor obstacles in their path. The larger of these roughly circular masses of ice, called *continental* glaciers, are thousands of feet thick, with their bottoms frequently well below sea level.

The simpler to understand of the two types of glaciers is the valley glacier. The upper portion of such an ice tongue, situated above the perennial snowline (firn line) lies within the zone of accumulation. More snow falls there in the winter than melts in the summer, and new glacial ice is continually forming. The lower portion of the ice tongue, below the firn line, is within the zone of wastage, where melting exceeds precipitation of new snow. Actually, wastage is a broad term, including much more than ordinary melting; the term *ablation* is used by glaciologists in referring to all the forms of wastage, including melting, evaporation, and sublimation. Some ice tongues, particularly those in Alaska and along the Labrador and Greenland coasts, terminate in the sea where they form high ice cliffs that extend a hundred or more feet above the water's surface and a thousand or so feet below the surface of the water. Such an ice front wastes by *calving,* the term applied to the undercutting and breaking off of masses of ice through melting and wave action which act on the snout of a glacier extending into water. Calved blocks of glacier ice floating in water are called *icebergs.*

In the case of a valley glacier, as long as more snow falls in the winter in the zone of accumulation than melts in the summer, there will be a net addition to the amount of ice in the glacier. This ice will flow toward the glacier's snout. As long as the advance of new ice downglacier is balanced by wastage of ice at the snout, the terminus of the glacier will remain fixed, even though the ice continues to move down valley. If the rate of ice formation is greater than the rate of wastage at the lower end, the ice front will advance down valley. On the other hand, if the amount of new snow at the upper end of the glacier were to decrease while the rate of wastage at the snout remained constant the snout of the glacier would retreat up valley,

Figure VII-3. *Top left*—Glacial grooves on dolomite bedrock of Late Silurian age Bass Islands Group, west side of South Bass Island in Lake Erie near Put-In-Bay. (Michigan Geological Survey—Berquist File, photo by L. C. Hulbert.)

*Center left*—Glacial grooves on Middle Silurian age Engadine Dolomite in quarry along M-123 near Ozark, about 30 miles north of St. Ignace in the Upper Peninsula. Arrow points to man for scale. Ice moved in direction of grooves toward distance.

*Bottom left*—Glacially scoured and smoothed igneous bedrock along M-95 north of Republic in the Upper Peninsula.

*Top and center right*—Glacial erratics of Precambrian age metamorphic and igneous rocks carried south into Michigan by glacier from Canadian Shield to north. Light-colored igneous dikes, more resistant to weathering than the host rock, often are etched in relief. Found in cemetery on west edge of Ann Arbor, these are typical of many similar erratics commonly used as headstones, driveway markers, or yard ornaments in regions of glacial moraine. Pen for scale is 6.75 inches long.

*Bottom right*—Arcuate "chattermarks" formed when boulders held in the grip of moving glacial ice bounced or chattered over bedrock. Arrow shows direction of ice movement. Rock is Late Mississippian age Bayport Limestone in quarry south of Bayport, Huron County. (Michigan Geological Survey Division—Berquist File, photo by L. C. Hulbert.)

# Geology of Michigan

and the ice tongue would be shortened, even though the direction in which the ice was moving (downhill) remained the same.

Just how glacial ice flows is not clearly known; to say the least, the mechanism of flow is very complex. Much glacial motion is the result of the slipping of the basal portion of the ice across the surface of the ground; such movement is evidenced by the ice-scoured bedrock surfaces across which glaciers have moved (Figs. VII-3, 4). Some glacial flow, however, is the result of internal movement. There is evidence that many of the ice crystals glide along sheets of atoms parallel to the base of the crystal; this form of intragranular movement is accompanied by recrystallization so the crystal outline remains essentially undistorted. Other crystals "flow" past one another by granulating and shearing, a form of intergranular movement. Phase changes (solid to liquid to solid, for example) within the ice mass, resulting in recrystallization, are important in glacial flow; such recrystallization is shown by the fact that most ice crystals near the snout of a glacier are many times larger than those near the snowfields which nourish the glacier. The shearing of one band of ice past another is another means by which internal flow of ice occurs. For further details on ice movement see Sharp (1960).

The fact that glaciers flow indicates ice is a very weak rock, so weak that it flows when only a few

Figure VII-4. Glacial grooves on limestone bedrock exposed on Kelly's Island in Lake Erie.

tens of feet of ice have accumulated beneath a relatively thick cover of firn and snow. Glacial flow is too slow to be seen directly, but its occurrence can be easily proven by observations spaced over successive years, such as by the dislocation of lines of stakes placed across an ice mass and subsequently resurveyed. The speed with which glacial ice flows varies considerably, from a maximum of more than 150 feet per day to speeds measured in terms of a hundred or so feet per year. An average rate for glacial flow probably is something of the order of a few thousand feet per year.

Flow occurs only at considerable depth within the ice, and the overlying ice is under slight enough pressure to remain brittle. This upper, brittle zone, riding on top of the more plastic, active ice is frequently marked by deep cracks, called *crevasses*, extending down into the ice a distance of 100 to 150 feet.

## The Work of Ice

Ice accomplishes a great deal of work. It erodes the country across which it moves, transporting the soil and rock debris it has picked up, and eventually depositing that load either directly (by melting) or indirectly (by glacial meltwater) on the countryside.

Erosion by glacial ice is accomplished by plucking and by abrasion. The first process occurs in the following way. As a glacier moves across the ground, ice near its base melts locally and the resulting water sinks into cracks in the bedrock beneath. As this water subsequently freezes it expands, exerting a considerable prying force on the rock. Blocks loosened by the expansive force of the freezing water become frozen into the base of the glacier to be plucked or quarried as the ice moves away from the bedrock knob. The second way in which ice erodes is through the abrasive action of rock material frozen into the bottom of the ice. This material acts as a giant rasp on the topography, and the abrasive particles striate, scratch, and groove the bedrock surface across which the ice moves. Frequently, glacial striations may be seen to mark the surface of the bedrock beneath glacial drift (Fig. VII–3). Likewise, the particles (cobbles and boulders) carried in the ice become striated and faceted. (Perhaps the best-known glacially grooved surface in the Great Lakes Region is that

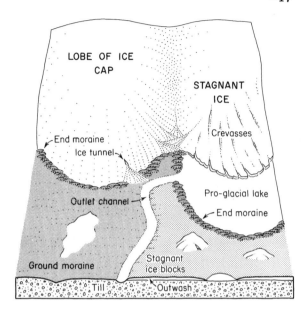

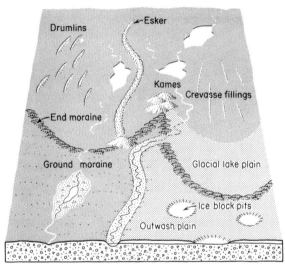

Figure VII–5. Highly generalized "before" and "after" diagrams to show the relationships between glacial ice and the deposits and land forms found in glacial drift. Lower diagram portrays conditions after ice has withdrawn. (Modified from Zumberge, *Elements of Geology*, 2d Ed., 1963, John Wiley & Sons, Inc.)

found on Kelly's Island, off Marblehead Peninsula in Lake Erie Fig. VII–4).

Nearly all the debris transported by the ice is carried in suspension, permanently held in the ice as a result of the rigidity of this flowing mass. Most of the material is carried near the base of the ice because this is where it is picked up. In a valley glacier much material is also carried along the

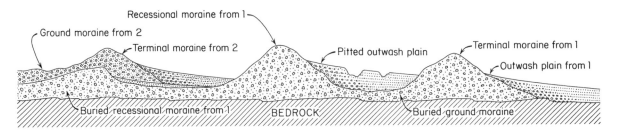

Figure VII-6. Diagrammatic cross section showing relationships of deposits formed by 2 stages of glacial advance and retreat. Ice advanced from left to right. Deposits were laid down during successive stages of retreat as ice front melted backward from right to left. At time of maximum advance during Stage 1 deposition at ice front produced Terminal Moraine 1 and Outwash Plain 1. Retreating from that point the ice laid down ground moraine (later buried). Balance between rate of forward ice movement and backward melting of ice front led to temporary stillstand of the ice front at the position where Recessional Moraine 1 and the Pitted Outwash Plain were deposited. Renewed backmelting and another stillstand of the ice front led to deposition of more ground moraine and then a third recessional moraine, later buried. Ice eventually withdrew entirely from the area, but then readvanced during Stage 2 to a position where Terminal Moraine 2 and its related outwash were deposited. Retreat of the ice front from Terminal Moraine 2 led to deposition of Ground Moraine 2. These later deposits (Terminal Moraine 2 and Ground Moraine 2), related to the second stage of ice advance and retreat, were laid down upon ground moraine and a recessional moraine formed during the first stage of glaciation. In Michigan, a good example of a partial readvance of ice and deposition of later stage glacial deposits over those of an earlier stage would be that of Valders ice deposits laid down over earlier and older Wisconsinan deposits. (See Figs. VII-16, 18.)

edges of the ice, as lateral moraine, or on the surface of the ice, having cascaded down onto the surface of the ice mass from the bordering, steep mountain slopes. There is practically no limit to the size of the material which can be carried by ice. Many glacial boulders, frequently known as glacial "erratics" (lithologic "strangers," unlike the bedrock of the area in which they are found), are of rather large size. Boulders ten feet or more in diameter are not unusual (Fig. VII-3).

The material carried by the ice is deposited wherever the ice melts. These glacial deposits are of two general types, *till* and *outwash*. Figures VII-5, 6 show the relationships of these glacial deposits in diagrammatic form. Glacial till is material deposited directly by the ice and, therefore, unsorted and unstratified—a heterogeneous mixture of all shapes and sizes of particles, without layering or strata (Fig. VII-7). Till typically forms one of several types of moraine, including *end moraine* or *lateral moraine*, ridges built along the end or side of the ice respectively, and *ground moraine* which is left as a general cover over broad areas.

The term moraine is used both for rock material as it is transported within and on ice (lateral, medial, and ablation moraine) and for deposits, usually of till, left by the ice after its retreat (Fig. VII-8). An end moraine is a mass of glacial debris (drift) laid down at the end of a glacier while the front remained in approximately the same position over a period of years. Occasionally, geologists refer to *terminal moraines*, which mark the outermost point of a particular glacial advance, and *recessional moraines*, which are end moraines marking lines of temporary halt in the general retreat of the ice front. Ground moraine is left as a thin drift cover when the ice margin moves across the countryside at a fairly rapid rate. Most of the higher hills south of the Straits of Mackinac in Michigan are parts of end moraines. In a few places, areas of ground moraine are the sites of extensive groups of elliptically shaped *drumlins* primarily composed of till. These streamlined hills, up to a mile or so long and as much as 100 feet high, are molded by overriding ice; their long axes are aligned in the direction of ice flow. Many well-developed drumlins are found on both sides of Grand Traverse Bay, particularly in Leelanau and Charlevoix counties (Figs. VII-9, 10). U.S. Route 31 between Torch Lake and Charlevoix runs along the length of first one and then another of these elongated hills. Another area of drumlins may be seen along U.S. Route 2 between Harris and Waucedah in the Upper Peninsula.

Frequently, at the snout of a glacier the material dropped by the melting of the ice is reworked,

**The Pleistocene Epoch**

149

Figure VII–7. Comparison of typical glacial deposits in the Fort Wayne and Defiance moraines of Pleistocene age near Ann Arbor.

*On left*—(yardstick for scale) are stratified and well-sorted sands and gravels in outwash deposits laid down by streams of glacial meltwater. Note cross-bedding, channeling, and lensing of bedding.

*On right*—(6-inch scale in lower right) are unstratified and unsorted tills (boulder clays) deposited directly from ice. Many of the cobbles are faceted and striated by glacial scouring (see also Fig. XII–8).

Figure VII–8. Typical glacial moraines in Michigan.

*Top*—Northeast of Battle Creek. Note second morainic ridge on distant skyline.

*Next*—West of Battle Creek, along M-89, about 2 miles east of Gull Lake.

*Next*—View east from "Old US 27," near the Cheboygan-Otsego County line between Wolverine and Vanderbilt. Morainic ridge is in far distance, glacial outwash valley ("valley train") in middle distance.

*Bottom*—Knob and kettle (swell and swale) topography in moraine near Chelsea. (Courtesy Michigan Geological Survey Division—Berquist File, photos by L. C. Hulbert.)

washed, sorted, and transported by the glacial meltwaters draining from the ice. *Outwash* is the general term applied to the sorted and stratified material laid down by glacial meltwaters (Fig. VII–7). It comprises the second major group of glacial deposits. Outwash deposits tend to be coarse grained and may take many forms including those of *outwash plains* and *valley trains,* and the various ice-contact deposits such as *pitted outwash plains, kames* (Fig. VII–12), *kame terraces, eskers* (Fig. VII–13), and *ice-channel fillings.* Finer-grained lacustrine (lake) and marine outwash deposits are laid down quite far from the ice margin. Outwash deposits frequently are deposited in and around stagnant blocks of ice which became separated from the general ice margin by wastage; the former sites of the ice blocks are represented by depressions called *kettle holes* (Fig. VII–11), now occupied by swamps and lakes. Most of the smaller inland lakes of Michigan occupy kettle-holes. The steep edges of the outwash deposits around such depressions are the slumped ice-contact slopes. Nearly all the sand and gravel deposits of Michigan are the result of deposition by glacial meltwaters derived from the melting of the last (Wisconsinan) ice sheet. Occasionally, areas mapped as moraines are underlain by outwash, particularly in the southcentral part of the Lower Peninsula and in the eastern half of the Upper Peninsula.

*Proglacial outwash* is deposited in front of the ice snout as an apron-like *outwash plain* sloping more and more gently in a downstream direction; as in other stream deposits, these outwash sediments become finer and finer grained as the distance from the ice increases. Excellent examples of outwash plains are found in front of many of the moraines in Michigan. *Valley train* is the term applied to a similar deposit confined within valley walls; many of the sand- and gravel-filled valleys between high moraines (e.g., the valley extending from Mancelona to Kalkaska and beyond) are good examples of valley trains.

As the term implies, ice-contact deposits are formed directly against the ice margin; they are usually formed around the margins of ice that has stagnated as the result of thinning. The surface across which the meltwater deposits are laid down lies in part on isolated blocks of ice and in part on stream deposits resulting from aggradation or fill of the areas between the ice blocks. With subse-

**Geology of Michigan**

152

Figure VII–9. Part of drumlin field in Charlevoix and Antrim counties just north of Torch Lake. Town of Norwood in northwest corner of photo. Drumlins are ice-molded ridges consisting mainly of ground morainic till. They trend south-southeastward in direction of glacial ice movement (see arrow at left) and in general appear as long, cigar-shaped or pencil-shaped dark, vegetated ridges. Ploughed fields with same trend occupy lowlands between the drumlins. Highway US 31, just north and south of the county line, follows the crest of a drumlin. Lakes with the same trend occupy areas between drumlins. Note low drumlin forming an island or underwater bar in Lime Lake. Also note pattern of offshore sand transport in Lake Michigan. See Figure VI–10 for ground photos of same area. (U.S. Department of Agriculture aerial photograph.)

# The Pleistocene Epoch

153

Figure VII–10. *Top*—Cross section of a typical drumlin east of US 31 between north end of Torch Lake and City of Charlevoix. Note that drumlin interior is an unsorted, unstratified clay and boulder till, indicating direct ice deposition.

*Bottom*—View to north shows a similar cigar-shaped drumlin ridge diagonally crossing a county road in the same general area. Ice flow was from left to right (toward south-southeast) along ridge. See Figure VII–9 for air photo of same area showing many such ridges.

quent melting of the buried or partly buried ice, what was a continuous surface representing a stream valley or valley floor is disrupted. The collapse of the deposits around the melting ice blocks produces relatively steep ice-contact slopes bordering undrained depressions (Fig. VII–11). Because the isolated blocks of ice are of a variety of shapes and sizes, with some steep and some gentle sides, the resultant ice-contact deposits show great variety of form. They include a continuous series ranging from the *pitted outwash plain* on the one extreme to an area of *kames,* isolated hills of outwash sand and gravel surrounded by ice-contact slopes, on the other; from an area of much flat-topped outwash with a few kettles to an area that is mostly kettles separated by a few isolated patches of outwash. The pitted outwash between Whitmore Lake and Brighton in Livingston County and that traversed by Michigan Route 33 south of Onaway are but two of the many excel-

Figure VII-11. *Top*—Typical kettle hole (ice block pit). Lack of exterior drainage allows water to collect, in this case supporting lush growth of cattails. Steep sides formed in contact with the stranded ice block which lay at a moraine front and was surrounded by outwash gravel and sand. Cherry Hill Road southeast of Dixboro, just east of Ann Arbor.

*Bottom*—Rich agricultural land on flat, former bed of Glacial Lake Saginaw. View near Reese, east of Saginaw and about 10 miles south of Saginaw Bay. Detroit and its suburbs in southeastern Michigan are built on a similar, flat lake plain formed during early glacial lake stages in the Erie Basin. (Michigan Geological Survey Division—Berquist File, photo by L. C. Hulbert.)

lent examples of pitted outwash in Michigan. Excellent examples of groups of kames are found in Stony Creek Park near Rochester, the gravel hills near Oxford, and the Irish Hills near Walter J. Hayes State Park southeast of Jackson. Many isolated kames like that near Lachine west of Alpena on Michigan Route 32 (Fig. VII–12) are found throughout the state.

An elongate ridge of outwash bounded on both sides by ice-contact slopes is called an *ice-channel deposit*. It may be the result of deposition between ice blocks, of collapse of outwash of variable thickness, or it may be a deposit laid down in a tunnel in or under the ice (the classical "esker"). Excellent examples of such ridges of outwash (Fig. VII–13) are the Blue Ridge, along U.S. 127 six miles

Figure VII-12. Well-sorted, stratified and cross-bedded glacial outwash gravels and sands are an important natural resource in Michigan (also see Fig. VII-7).

*Top*—Hydraulic dredging in Greenoaks Pit about 4 miles southeast of Brighton. (Michigan Department of Conservation photo.)

*Center*—Killins Gravel Pit, in a glacial kame (see Fig. VII–5) at west edge of Ann Arbor.

*Bottom*—Old Flanders Pit, in a conical glacial kame west of Alpena near town of Lachine. (Michigan Geological Survey Division—Berquist File, photo by L. C. Hulbert.)

**Geology of Michigan**

Figure VII–13. Glacial eskers in Michigan (see Figure VII–5).

*Top*—Roadcut through a small esker crossing U.S. Highway I-94 just east of a rest stop, 2.5 miles east of the Race Road (No. 147) exit, between Chelsea and Jackson.

*Center*—Blue Ridge Esker, one of many Michigan sources of high grade sand and gravel. Note stratification in these meltwater deposits. Piles in foreground are processed gravel ready for commercial use. View from Blue Ridge Road east of US 127 about 6.5 miles south southeast of Jackson.

*Bottom*—Same Blue Ridge Esker cut by US 127 south of Jackson. Close-up shows glacial meltwater washed gravel. Many of the cobbles are the fossiliferous, brown Marshall Sandstone of Mississippian age, a local bedrock formation scoured by the glacial ice in its southward passage. Shovel for scale.

**The Pleistocene Epoch**

157

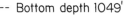 Modern Lake Superior surface

---------- Bottom depth 1049'

1059' below water surface, 10' below sed.-water interface
LAST STAGES OF ANNUAL VARVE FORMATION

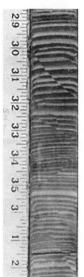

1080' below water surface, 31' below sed.-water interface
WELL DEVELOPED SPRING MELT (Light) AND
   WINTER (Dark) LAYERS

1086' below water surface, 37' below sed.-water interface
EARLY STAGES OF ANNUAL VARVE FORMATION

Figure VII–14. Annual varves (see Fig. II–5) in selected sections of a deep water core from bottom sediments in Lake Superior. Core taken about 35 miles due east of tip of Keweenaw Peninsula. Latest Pleistocene or early postglacial time is represented. Light-colored, sandy layers formed during spring meltwater runoff; darker mud layers during winter. One pair of layers, light plus dark, represents one year's worth of sediment deposition. In the complete core, varves between the 1058–85 foot levels totaled 27 feet in thickness and counted up to 1338 years. Ten feet of sediment at top lacked varves. Minor offsets, down-bowing along edges, and dark cracks are due to core barrel disturbance. (Courtesy W. R. Farrand, The University of Michigan Department of Geology and Mineralogy, Core No. s–62.)

south of Jackson, and the Mason esker, which lies just east of U.S. 127 and extends from Mason to DeWitt, in Ingham County.

The quiet water deposits resulting from deposition of glacial debris in a lake or in an ocean are in general similar to ordinary lake or sea deposits. *Glacial varves* (Fig. VII–14), however, are an exception to this general rule. Lakes closely associated with glaciers and fed by meltwater streams are characterized by cyclic deposition. During the summer months much clastic sediment is brought into the lake from the melting glacier. The coarser fraction, down to very fine sand and silt size, will settle to the bottom soon after entering the lake, but the stirring of waves resulting from wind action and from stream inflow keeps the finer-grained materials in suspension. It is only during the winter months, when the lake is frozen over and free from wave action, and when there is little influx of meltwater, that the water is quiet enough for the fine-grained clays and silts to settle. The result is a pair of layers, one relatively thick, made of sand and coarse silt, and the other, a thin one, composed of clay and fine silt; such a pair of layers represents the deposits of one year and is called a glacial varve. Glacial varves representing annual deposits have been used as a basis for an absolute postglacial chronology (Flint, 1957, Chapter 17). Varves are not formed in the marine environment, primarily because the salts in solution cause the fine-grained particles to collect (flocculate) as larger aggregates which quickly settle to the bottom with the coarse clastic particles. However, a distinctive feature of both lacustrine and marine glacial deposits is the occasional lense of interbedded coarse material (pebbles, cobbles, and boulders) resulting from the melting of floating icebergs which contained coarse debris.

## General History of the Pleistocene

The Pleistocene Epoch, between 500,000 and 2,000,000 years long, was different from much of the rest of geologic time in that it was a time of rigorous climate during which glaciers covered more than one-third of the land surface of the earth. The cold climate was so atypical, so unlike that of the greater part of geologic time, that the Pleistocene is considered by many to have begun when the temperature first dropped. Since such a climatic "deterioration" may have begun long before the advent of the first major continental ice sheet, geologists differ as to when the Pleistocene "officially" began. This is why two figures are given above for the age range of the Pleistocene. Actually, the Pleistocene Epoch was marked by several major climatic variations that were worldwide and synchronous—in all there were at least four major periods of cold ("glacials") separated by at least three periods of climate considerably warmer than that of the present (the "interglacials"). The cooling off that produced the periods of glaciation in the higher latitudes also resulted in times of greater humidity ("pluvials") in the lower latitudes, outside the limits of ice cover. This effect was produced by the decrease in rate of evaporation which resulted in an increase in "effective precipitation"; there is some evidence that the pluvials were periods of greater actual precipitation as well.

The subdivisions of the Pleistocene as recognized in North America are shown in Figure VII–15. The subdivisions of the Wisconsinan shown on the right are those used in the literature on the Michigan glacial deposits. The time terminology of the Wisconsinan glacial is currently undergoing considerable change (Frye and Willman, 1960), but to introduce the new terms into a nonprofessional book of this sort would confuse rather than

Figure VII–15. Subdivisions of the Pleistocene in North America.

| Glacials | Interglacials | Substage |
|---|---|---|
| Wisconsinan | | Valders |
| | | Two Creeks |
| | | Port Huron (Mankato) |
| | | Cary |
| | | Tazewell |
| | | Iowan |
| | | Pre-classical Wisconsin |
| | Sangamonian | |
| Illinoian | | |
| | Yarmouthian | |
| Kansan | | |
| | Aftonian | |
| Nebraskan | | |

# The Pleistocene Epoch

clarify; instead, we will use that terminology employed in the maps and reports available on the Wisconsinan deposits of Michigan.

Note that Figure VII–15 is another example of a relative time scale (Chapter II). Since the development of Carbon$^{14}$ radioactive dating methods in the mid-1940's (Chapter II), the Wisconsinan portion of this time scale is becoming accurately dated. C$^{14}$ dates are possible on organic material 50,000, and perhaps as much as 70,000, years old. Dates obtained on material from "preclassical Wisconsinan" deposits are greater than 30,000 years old. All organic material from Illinoian deposits is inert because the age limit for C$^{14}$ dating is exceeded in them.

Any explanation of the climatic changes that characterized the Pleistocene must take the following into consideration: (a) the Pleistocene contrasts with nearly all the rest of geologic time because similar glaciation occurred only on a very few other occasions dated as Precambrian and Paleozoic and all separated in time, (b) the Pleistocene itself was a time of complex climatic variations, with at least four main glacials (each in itself apparently multiple or complex) separated by interglacials, and (c) all Pleistocene climatic changes affected the entire world at approximately the same time. Many a hypothesis has foundered on one or another of these points. Some explain fairly well the complexities of the Pleistocene but call on an alternation of the climatic changes between the southern and northern hemisphere. Others explain the worldwide nature of the glacial climate but do not explain the complexities of the glacials. None of the proposed hypotheses adequately explains what started the Pleistocene in the first place. At last count over 35 different hypotheses had been offered to explain the climatic peculiarities of the Pleistocene; in the eyes of most glacial and Pleistocene geologists no one of these proposals seems adequate.

The majority of the hypotheses proposed fall into one or more of the following categories. (a) those calling on some change in the atmospheric material causing interference with the receipt of solar energy by the earth (increase in volcanic or interstellar dust to cool the climate, increase in carbon dioxide gas to insulate and warm the earth, etc.); (b) those calling on some variation, either cyclic or intermittent, in the output of energy by the sun (sunspot cycles, solar flare activity, etc.);

(c) those calling on perturbations of the earth in its movements within the solar system such as precession of the equinoxes, variation in the eccentricity of the orbit, and variation in the angle between the earth's axis of rotation and the plane of the earth's orbit about the sun); (d) hypotheses calling on various geographic changes on the earth itself, which would change both the circulation of oceanic water and precipitation on the continents, as well as many other factors. It seems probable to most workers that the initiation of the glacial epoch might best be explained by one hypothesis, whereas the climatic variations within the Pleistocene may have been due to a totally different mechanism. One fact that seems of utmost importance is that the major ice ages of the geologic past (those of the Precambrian, Late Paleozoic, and Pleistocene), if they are at all correctly dated, coincide with periods of very extensive continental uplift and relatively restricted seas. This seems to be more than merely coincidental. Future studies of upper atmospheric circulation, which strongly affects the climate of the world, may provide some of the answers to the general question of the causes of Pleistocene climatic fluctuation. Flint (1957), a basic reference on glacial geology and the Pleistocene Epoch, is a good source of additional information on the causes of glaciation.

## The Pleistocene in Michigan

There is positive evidence in Michigan, in the form of well-dated deposits, for only the last of the four Pleistocene glacial advances. However, Illinoian and Kansan (?) glacial deposits are found to the south of Michigan in Ohio and Indiana. Therefore, it is reasonable to assume the state was overriden by at least these two earlier advances as well. Leverett and Taylor (1915) spoke of indurated (cemented and hardened) drift underlying Wisconsinan deposits in Michigan and separated from the Wisconsinan deposits by a strong zone of weathering; they suggested the underlying drift is Illinoian, the weathered zone representing the subsequent Sangamon interglacial. As local cementation of Wisconsinan deposits, particularly of outwash sands and gravels, is a common occurrence in many parts of the state, some, or even most, of what they referred to as Illinoian drift

## Geology of Michigan

Figure VII-16. Glacial map of Michigan. (Modified from the *Glacial Map of the United States East of the Rocky Mountains*, Geological Society of America, 1959.)

may be Wisconsinan in age. The buried peat and soil reported by Leverett and Taylor to overlie this hard till could not be found by the authors, although a diligent search was made. Separate maps of the surface glacial deposits of the Upper and Lower peninsulas of Michigan, in color, can be purchased from the Michigan Geological Survey, Department of Conservation, Lansing. These two maps are essentially Leverett and Taylor's work plotted on a new base.

Figure VII–16 shows the control that the major, preexisting topographic features of the Great Lakes Region exerted on the shape of the ice front. Note that the end moraines are looped around each of the several major lowlands now occupied by Lake Superior, Lake Michigan, lakes Huron and Erie, and Saginaw Bay. The ice flowed more readily into those depressions, scouring them to increased depth and forming thickened ice "lobes" in them. From a broad viewpoint, the ice moved generally from north to south across the Great Lakes region, but the lobate character of the ice mass resulted in some anomalous local directions of ice movement. For example, when the ice front of the Saginaw lobe stood at the Port Huron moraine, extending through the city of Saginaw, ice expanded in a fan-shaped manner out of the lowland, so that in Arenac and Gladwin counties ice locally advanced from the east, although the general direction of advance was from the northeast. The series of moraines (Defiance and Fort Wayne, particularly) trending northeast-southwest in southeastern Michigan similarly were deposited by ice moving locally to the west and the northwest out of the Erie basin. Between the areas of well-developed moraines representing the three north-south trending ice lobes are two *interlobate areas,* one along the axis of the Thumb and the other extending north-south through the center of the state. These interlobate areas contain much sand and gravel in the form of both high kames and extensive outwash plains called the "Sand Barrens." These features resulted from meltwater deposition.

During much of the Wisconsinan glacial the ice margin was well south of the state, and all of Michigan was covered by several thousand feet of ice. It was not until the very latest part of the Tazewell stadial that the Wisconsinan ice front melted back to northern Indiana. All glacial deposits of southern Michigan south of (beyond) the Port Huron moraine are of Cary age. This interval, extending approximately from 16,000 years to about 13,500 years before the present, was a time of general ice retreat marked by periods of still-stand during which recessional moraines were laid down. The general retreat was interrupted by some relatively minor readvances as shown by areas where thin till is found overlying outwash gravel, as in the Ann Arbor (Fig. VII–17) and Northville areas, and by oscillations in the elevation of Glacial Lake Maumee (Chapter VIII).

The Port Huron moraine is the terminal moraine of the Port Huron (Mankato) ice advance which occurred about 13,000 years ago. This readvance of the Cary ice front also is reflected by changes in the various proglacial lake levels in southeastern Michigan; this particular readvance resulted in the formation of Glacial Lake Whittlesey (Chapter VIII).

Following the Port Huron (Mankato) ice advance (Port Huron moraine), there was another period of general, though somewhat intermittent, retreat which resulted in the withdrawal of ice from the whole of the Lower Peninsula, uncovering the Straits of Mackinac, and probably uncovering a large portion of the Upper Peninsula as well. This period of withdrawal is called the Two Creeks interstadial. During Two Creeks time, the climate of east central Wisconsin, in the vicinity of the Door Peninsula separating Green Bay from Lake Michigan on the southeast, was mild enough that a forest of spruce, pine, and birch trees developed on weakly weathered Cary till. The Two Creeks interstadial, which lasted for only some 1000 years and ended about 11,850 years ago (Broecker and Farrand, 1963), was brought to a close by a final readvance of ice. This ice advanced from the north down the Lake Michigan basin to approximately the latitude of Holland, Michigan, overriding the above-mentioned forest in Wisconsin and burying it beneath red glacial drift (Fig. VII–18). That overridden forest area now serves as the type locality of the Two Creeks interstadial. This last Wisconsinan ice advance is called the *Valders stadial.* It covered the northern part of the Lower Peninsula of Michigan. The red color of the Valders drift results from the fact that the ice had picked up and incorporated much red silt and clay from the bottom sediments of Lake Superior and from the Precambrian iron formations of the western part of the Upper

Figure VII–17. Thin cover of glacial till over outwash sand and gravel in Whittaker-Gooding Pit off Cherry Hill Road just south of Dixboro, east of Ann Arbor. The till, an unstratified, unsorted ice deposit, consisting here of clay and faceted and striated cobbles, shows typical character of a weathered and eroded till. Note numerous deep, narrow, closely spaced, nearly parallel rivulets. This gravel pit is on the flank of the Defiance Glacial Moraine. The presence of till over outwash may record a minor, local readvance of the ice front, or may simply be a mud flow of till from the moraine front out over the flanking outwash deposits (see Fig. VII–6).

# The Pleistocene Epoch

163

Figure VII-18. Two Late Pleistocene (Wisconsinan) age glacial tills near Gaylord. Unstratified material in lower quarter of photo is a Pre-Valders till. Dark, cross-bedded outwash sand and gravel in center formed when ice had temporarily retreated. Above the outwash is a second till, red colored in outcrop, called the Valders Till. The latter indicates a readvance of ice. Thin, dark band at very top of exposure is a weathered soil zone developed on the Valders Till. (See Figs. VII-6, 15.)

Peninsula. This, in turn, is evidence that the Lake Superior basin had previously been cleared of ice and at least partially filled with water.

The Valders ice deposits entirely lie north of the northern (inner) side of the Port Huron moraine, that is, north of a line from Muskegon to a point within 12 miles west of Thunder Bay on Lake Huron. From Thunder Bay the drift border turns toward the north, reaching the lake near Rogers City. Some of the moraines in the eastern part of the Upper Peninsula are recessionals of the Valders ice. The Valders advance is the one most responsible for the formation of the north-northwest—south-southeast trending drumlins in Leelanau and Charlevoix counties (Figs. VII-9, 10).

From the time of most extensive Valders ice, about 11,850 years before the present, the general pattern was one of ice retreat. Thus far, it has been impossible to state with confidence the date ice last covered any part of the state, but Broecker and Farrand (1963, p. 800) make the general statement that all of Michigan was free of ice by 10,000 years ago.

Those interested in further details concerning the glaciation of the Michigan area should consult Wayne and Zumberge (1965).

The events of the late-glacial and postglacial history are best learned from a study of the proglacial and postglacial lakes which formed in the three upper Great Lakes basins, which is the subject of the next chapter.

# VIII

## The Great Lakes in Late Glacial and Postglacial Time

### Introduction

The basins containing the five large lakes we call collectively the Great Lakes are primarily the result of glaciation during the Pleistocene, discussed in the preceding chapter. Like all lakes, the Great Lakes will be, geologically, rather short-lived. They will be filled with sediment and vegetation, or their outlets will be incised and the basins drained in a matter of only a few hundreds of thousands of years—a short period of time when contrasted with the several tens of millions of years represented by each Paleozoic period.

Despite the fact the basins were covered with glacial ice until little more than 10,000 years ago the histories of the lakes that occupied the several basins since they were first uncovered are very complex. But before turning our attention to the bodies of water, or lakes, that have occupied the five basins, we should explain how these depressions came into being.

Prior to the Pleistocene and its several glacial advances the sites of the present lake basins were stream valleys related to the general drainage of the midcontinent region. The stream system was well adjusted to the structure of the underlying rocks—that is, the axis of each valley system was along a belt of weaker rock and the divides, the areas between the major valleys, were developed on more resistant rock (Figs. IV-18, and V-15). The geologist uses the term *subsequent* to define such a valley developed on weak rocks. Note in particular that the basins now occupied by Lakes Michigan, Huron, and Erie are located along a belt of relatively weak Devonian rocks, in large part shales and limestones which overlie the resistant sequence of Silurian rocks that holds up the Niagaran escarpment. This escarpment is traceable from the Niagara Falls region in a broadly curving arc along the Bruce Peninsula and the islands which separate Georgian Bay from Lake Huron, through the eastern part of Michigan's Upper Peninsula, and south along the western edge of Lake Michigan, where it forms the peninsulas which separate Green Bay from Lake Michigan. The Lake Ontario basin, Georgian Bay, and Green Bay are located along a belt of relatively weak Ordovician and Silurian sediments which underlie these same stronger Niagaran rocks. The position of the Lake Superior basin is, at first glance, more difficult to explain—until one remembers that strong and weak are relative terms. Even within the Canadian Shield there are sedimentary basins and structural basins which are the site of "weaker rocks." The Lake Superior basin is largely developed along the length of a structural sag, a syncline, with the resistant Keweenawan rocks of volcanic origin forming in part the sides of the basin (in Keweenaw Point and Isle Royale). Recent work on the floor of the central portion of this lake basin has demonstrated that it is underlain by sandstones of Cambrian age, which rocks are considerably softer than the Keweenawan rocks on either side.

Thus, through stream erosion over a long period of time a valley system (or systems, since the exact relationship of the several valleys is not clearly known) was developed on the zones of weaker rock. It is most probable that the flow was to the north and northeast, as is the present drainage of the Great Lakes, but even the direction of stream flow is not clearly known. At any rate, stream erosion had developed a valley system which was at least in part aligned in a direction parallel to the flow of the Pleistocene ice sheets as they advanced from the Canadian region.

Glacial ice flow is strongly affected by variations in ice thickness—in general, the thicker the ice at a given point across the ice front the faster the flow will be. With more rapid flow there is greater erosion. It can be readily seen, then, that any original topographic irregularity, such as a stream valley which parallels ice flow, will be accentuated through continued glaciation. The preglacial stream valleys in the Great Lakes region were, thus, the loci of considerable glacial erosion. The elevation of the floor of all the lakes but Lake Erie is well below present sea level; the deepest point in Lake Superior is 700 feet below sea level, in Lake Huron, 170 feet below sea level, in Lake Michigan, 343 feet below sea level, and in Lake Ontario, 532 feet below sea level, while that of Lake Erie is 362 feet *above* sea level. Lake Erie is in several other aspects quite different from the other Great Lakes, for it has an average depth of less than 60 feet and contains only a little more than one-thirtieth the amount of water found in Lake Superior (Hough, 1958, p. 7). As all the lake basins are bounded by high land, and relatively high bedrock, on all sides it is evident that the erosion by glacial ice was great.

When the ice front began to retreat, uncovering the most southerly portions of the basins, lakes began to form between the ice front and higher ground to the south. These relatively small depressions south of the glacial ice dam filled with glacial meltwater and normal runoff from the surrounding areas until the level of the lowest sag in the surrounding rim was reached, which sag then became the point of discharge for the basin. The geologist refers to such a lake held in by an ice dam blocking the valley as a proglacial lake (one out in front of a glacier). With continued retreat of the ice front, uncovering more of the land area bordering the lake basin, lower outlets to the basin may be uncovered; the level of the proglacial lake will then suddenly drop to that of the newly uncovered outlet. One would expect, therefore, that the sequence of lake levels would be from higher to lower with ice retreat, and this is what generally happened. However, the details are much more complex. As discussed in the last chapter, following retreat of Port Huron (Mankato) ice there was a resurgence of the ice front during Valders time; some of the Great Lakes region was reinvaded by ice at this time. Such a readvance caused a return to high lake levels in several of the basins after a general progression to quite low levels. Other, more minor readvances of the ice front, even before Port Huron time, covering up once more outlets that had been exposed by earlier retreat, made the proglacial lake sequence in the several basins still more complex.

A second important factor that complicated the lake history in postglacial time is somewhat more difficult to explain. Simply stated, the earth's crust in the Great Lakes region has risen vertically since glaciation, with the greatest amount of uplift affecting the more northern part of the area. At times of extensive glaciation, when nearly all the continent north of the Missouri and Ohio river systems was covered with ice to a maximum depth of as much as 10,000 feet, the earth's surface was depressed under the great weight of the ice. Rather simple calculations, ignoring many factors which would tend to reduce the amount of depression, would suggest that the amount of crustal depression was about one foot for every three feet of ice cover, and that beneath the thickest ice, near the center of the continental glacier, located somewhere in the vicinity of Hudson Bay, the total depression may have exceeded 3000 feet. As the weight causing this crustal depression was removed, through wastage of the ice cover, the earth's crust began to rebound, much as a rubber balloon returns to its original shape when we stop squeezing it with our hands. The earth, being a mass with considerable strength, composed of solid material for the most part, did not respond immediately to the addition and subsequent removal of weight—instead, there was a lag in both the depression and rebound of the crust, and much of the uplift resulting from wastage of the continental glaciers at the end of the Pleistocene took place over a period of several thousands of years. There is even considerable evidence that some of the rebound has yet to be accomplished in the Hudson Bay region.

While such depression and rebound of the earth's crust is of great interest to the geophysicist (the geologist primarily interested in the "physics" of the earth) in that it sheds intriguing light on both the physical properties and the structure of the earth, we must turn our attention to the subject under discussion here, the effect such movement had on the lake basins of the Great Lakes region in postglacial times. When ice first disappeared from the region, and before much crustal rebound had occurred, the northern parts of the basins were at a much lower elevation than they now are. Early in deglaciation, therefore, many of the basins drained toward the north through outlets now abandoned because they since have been raised high above other outlets farther south. Postglacial uplift in the southern part of the region was much less, or it did not even occur there, because the thin ice cover typical of the marginal portion of the glacier did not depress the crust. Shifts in the outlets resulting from such changes in altitude due to differential uplift added greatly to the complexities of the postglacial history of all of the Great Lakes region.

One other complication resulting from such uplift which should be pointed out at this time is the effect the increase in the amount of uplift, as one goes north in the Great Lakes region, had on the elevation of the shore features developed around earlier lakes—these features, beaches, wave-cut cliffs, and the like, were also differentially uplifted. To cite an example, a relatively continuous and traceable strandline, developed along the shore of one of the largest of the proglacial lakes to occupy the Lake Michigan basin (Algonquin) stands at an elevation of 619 feet at Traverse City and at 813 feet above sea level on Mackinac Island (some 80 miles to the northeast). What was once a horizontal feature representing approximately the surface of the lake is now tilted so that it slopes an average of 2.4 feet per mile over a distance of 80 miles. The hinge line (zero isobase), the line south of which no uplift occurred and south of which the lake features are still horizontal, for this particular proglacial lake crosses the Lake Michigan basin, extending from a point a little south of Kewaunee, Wisconsin, through Manistee, Michigan. Because uplift due to rebound took place over a long time the older lake features are in general tilted by a greater amount than are the features related to younger lakes. Figure VIII–1 shows the hingelines for several of the important lake stands in the Michigan area.

Figure VIII–1. "Hinge lines" (zero isobases) for some representative glacial lake stages in the Great Lakes region. North of its hinge line the shoreline of each lake stage rises progressively above the horizontal due to crustal rebound resulting from the removal of the weight of the glacial ice through melting. South of its hinge line each shoreline is horizontal, because there the earth's surface was not depressed at the time the lake was in existence. The several lake stages whose hinge lines are shown are of progressively younger age toward the north. For further details on the subjects of hinge lines and crustal rebound due to ice melting consult the text. (Modified from Wayne and Zumberge, 1965, in *The Quaternary of the United States*, edited by Wright and Frye, permission of Princeton University Press. Copyright 1965 by Princeton University.)

Still another factor which resulted in complexities in the lake sequence was the lowering of some of the proglacial lake outlets through erosion by discharge streams. In many cases the lowest outlet available to a particular proglacial lake was across a sag in the crest of a glacial moraine. As overflow first began at a particular point there was a relatively stable lake surface graded to the elevation of the outlet, and normal shore processes around the margin of the lake formed wave-cut cliffs, beaches, and the like, at that elevation. With continued overflow, however, the level of the outlet was cut down or incised until the underlying bedrock was uncovered, at which time a second stable lake level formed with related shore and wave features developed along its margin. Some of the workers on the problems of the proglacial lake history have also accounted for a cessation of such outlet incision, as indicated by well-developed strandlines, for example, as being due to the development of an armor of boulders too large for the outlet stream to remove; with increase in discharge of the stream, probably due, these workers say, to the addition of water draining from another of the lake basins, this boulder lag was subsequently removed and incision to a still lower level occurred.

All these several factors—the retreat and readvance of the ice front uncovering and covering once more outlets for the several basins, the rebound of the earth's crust resulting from deglaciation which tilted the lake basins, and less importantly, the incision of the various outlets by lake discharge—served to make the late glacial

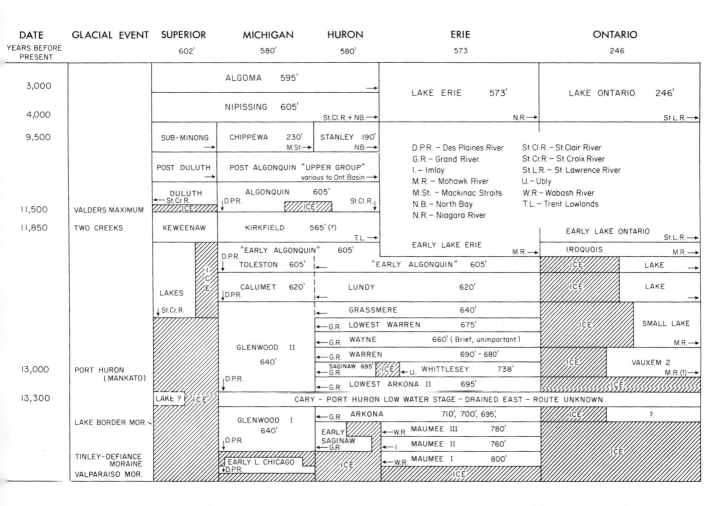

Figure VIII-2. Geographic and time relationships of some of the major stages of the Glacial Great Lakes. Note the shifting location of the outlets as indicated by arrows.

## Geology of Michigan

and postglacial history of the Great Lakes region a very complex one.

The reader should remember that the historical discussion which follows is based on interpretation of the geologic record of many lakes, left in the form of beaches, wave-cut cliffs, stacks, bottom sediments, and sand dunes, all features which are discussed in Chapter IX. This lake history is actually the result of the interpretation of past events through the analysis of the geologic record—analysis based on the principle of uniformitarianism, where the present effects of the several geologic agents are used to interpret the happenings of the past.

It should be noted by the reader that all elevations cited in the following discussion, unless otherwise qualified, are "pre-uplift" elevations, with the warped portions of the lake features restored to horizontality.

### The Proglacial Lake Sequence

*Introduction.*—As will soon be seen, the many lake stages in each of the several Great Lakes basins make a single, straightforward story nearly impossible; since each stage in a particular basin is dependent on the position of the ice front and also on the elevation of the outlet for the particular basin and, often, that of the outlets to several other basins, no readily coherent story can be told. Figure VIII-2, of itself highly simplified and generalized, gives the reader some idea of the complexity of the problem. In the following brief treatment the more important of the lake stages will be discussed in chronologic order, beginning with the oldest. For further information the reader is referred to several articles and books dealing with the subject and to the still more complete bibliographies they contain. Of particular importance are Leverett and Taylor (1915), Hough (1958), and Hough (1963).

In order to follow the discussion better the reader should frequently refer to Figure VIII-3, which shows the present lowlands in the Great Lakes area; many of the stream valleys shown served as outlets during late glacial and postglacial times.

*Pre-Two Creeks Time.*—With general retreat from the terminal moraines of Cary age, located

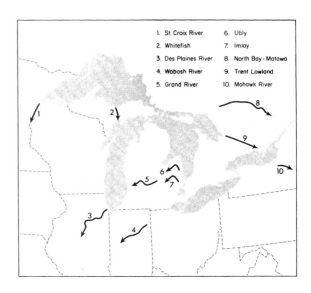

Figure VIII-3. Relationship of modern drainage features and former glacial lake outlets in the Great Lakes Region.

south of the Great Lakes area, the first basins to be uncovered were those of Lake Michigan and Lake Erie. In these basins two high-level proglacial lakes formed, Chicago and Maumee, respectively (Fig. VIII-4). Chicago, which lasted for a long time, actually consisted of several different lakes at various elevations (Glenwood, 640 feet; Calumet, 620 feet; Toleston, 605 feet), all of which drained through the Chicago outlet, along the Des Plaines and Illinois rivers and into the Mississippi River. The early Chicago (Glenwood) had an elevation of 640 feet controlled by outlet stabilization due to the development of a boulder lag resulting from initial rapid downcutting, according to Bretz (1951b). Maumee I also drained into the Mississippi River system, by way of a gap in the moraine near Fort Wayne, Indiana, and into the Wabash River; this lake had an elevation of about 800 feet. With further retreat of the late Cary ice front the level of Maumee dropped to 760 feet as water drained across the Thumb area near Imlay City, Michigan, and the Fort Wayne outlet was abandoned (Fig. VIII-5). This westward drainage continued across the central part of Michigan by way of the valley now occupied by Grand River and emptied into the Glenwood stage of Chicago, still at 640 feet elevation. Subsequent readvance closed off the lowest westward drainage from the

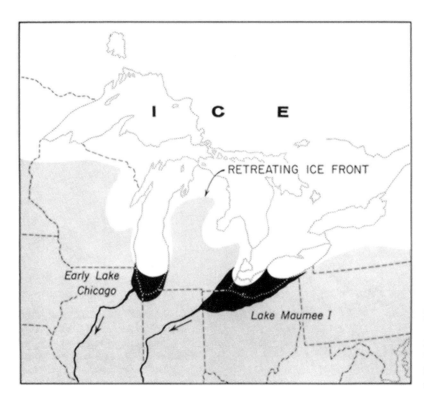

Figure VIII–4. Glacial lakes Chicago and Maumee I. (Modified from Hough, 1963, "The Prehistoric Great Lakes of North America," permission of *American Scientist*.)

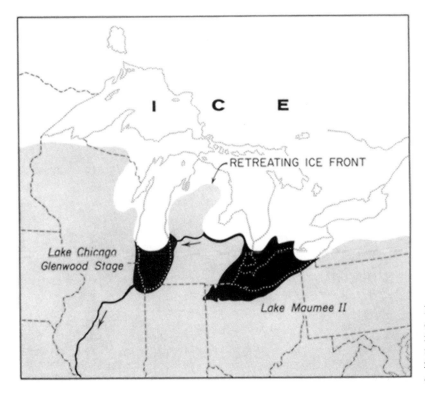

Figure VIII–5. Glacial lakes Chicago and Maumee II. (Modified from Hough, 1963, "The Prehistoric Great Lakes of North America," permission of *American Scientist*.)

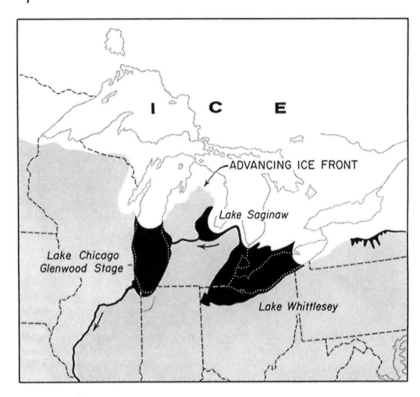

Figure VIII–6. Glacial lakes Chicago, Saginaw, and Whittlesey (Modified from Hough, 1963, "The Prehistoric Great Lakes of North America," permission of *American Scientist*.)

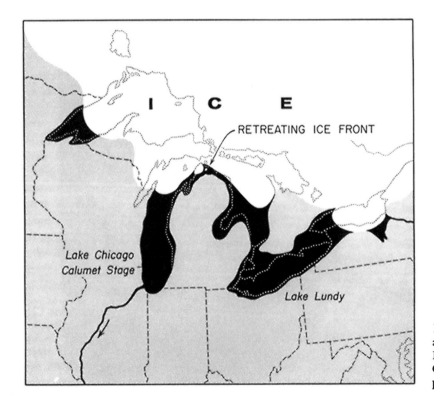

Figure VIII–7. Glacial lakes Lundy and Calumet. (Modified from Hough, 1963, "The Prehistoric Great Lakes of North America," permission of *American Scientist*.)

Erie Basin and caused a rise in the level of Maumee to 780 feet; Maumee drainage apparently was once more largely through the gap at Fort Wayne and down the Wabash Valley with the outlet lowered some 20 feet by stream action. While it is not yet clear whether part of the Ontario Basin was free of ice during any of the time represented by the three different levels of Maumee, a lake certainly began to form in the Saginaw Bay lowland as it was uncovered.

Subsequent retreat of the ice front brought to an end Maumee, as still lower outlets across and around the Thumb area were uncovered, and the next fairly stable lake surface in the Erie-Huron-Saginaw Bay Basin, Arkona, came into being. This lake, as did Maumee, drained west down the Grand Valley and into Lake Chicago, according to Hough (1963), still with an elevation of about 640 feet. The several levels of Arkona are apparently related to periods of downcutting of the Grand River outlet. During this interval, and at other times when the Grand River drainageway into Chicago was operative, a large delta was built at the then existing mouth of the Grand River in the vicinity of Allendale, west of Grand Rapids. This delta remnant, with an area of about 100 square miles now heads about 13 miles east of the present shore of Lake Michigan.

Still further retreat of the ice front is indicated by the presence of shallow-water sediments, well below the lowest Arkona level, covered by deeper-water deposits near Cleveland, Ohio, and by evidence of a low-water stage in the Michigan Basin separating two different higher lakes representing the Glenwood stage. The exact limits of the lakes occupying the several basins at this time are not known. Drainage is presumed to have been to the east, perhaps by way of the Trent Lowland and eventually along the Mohawk Valley into the Hudson. It is likely that there were lakes in at least a part of each of the five basins during this interval, but evidence has been found for such a lake in only the Michigan and Erie basins.

At this point in our story we must digress briefly to explain that there are two differing ideas concerning the precise pattern of lake shores and connections through the next several lake stages. Bretz (1951a and b, 1953, 1964) believes that the increased discharge through the Chicago outlet resulting from the addition of extensive drainage from lakes in the Huron-Erie basin through the Grand River channel caused incision of the outlet to Lake Chicago such that it was lowered to the Calumet level (620 feet). Hough (1963, 1966), on the other hand, claims that the lowering of the Chicago outlet to 620 feet did not occur until much later. We recognize that controversy exists at this point, but in order to continue this simplified historical treatment of the sequence in all five Great Lakes basins in a manner hopefully most understandable to the reader we shall follow the sequence used by Hough (1963). Those who wish to become fully aware of both sides of the controversy should refer to the publications cited above.

The interval of relatively low lake level mentioned above came to a close as the ice advanced once more across the Great Lakes region to deposit the Port Huron moraine (approximately 13,000 years ago). The readvance caused a return to the Glenwood level in Lake Chicago (640 feet elevation) and eventually recreated two separate lakes to the east, one called Saginaw, (695 feet elevation) in Saginaw Bay, and the other Whittlesey (elevation 738 feet) in the Erie basin (Fig. VIII–6). Whittlesey drained northwest and west across the Thumb through the Ubly channel and the water from both these lakes eventually flowed into Glenwood by way of the Grand River. Whittlesey is perhaps one of the best represented of all the proglacial lakes which occupied the Erie basin. Its beach is a gravel ridge, in many places standing 10 to 15 feet above the flat lake plain to the north and east, which can be readily traced for a distance of more than 400 miles from the Ubly outlet southward and eastward to a point east of Buffalo, New York. The beach ridge was the site of Indian trails and of the early roads because it was dry during all parts of the year, particularly in comparison with the clayey lowlands on either side; this route determined the pattern of towns and even of major highways until fairly recent times.

The resurgence of the ice to form the Port Huron moraine and the return to the relatively high lake level represented by Whittlesey was rather short lived, and with wastage the lake levels began to drop once more as lower outlets around the Thumb were uncovered. The first post-Whittlesey lake level of any consequence was that of Warren, which was a single lake occupying both the Saginaw and the Erie lowland, with its surface at an elevation of 682–690 feet. Subsequent crustal rebound has lifted the Warren beach to an elevation as high as 760 feet at Bad Axe, Michigan. Warren drained west across the state by way of the Grand River as did several slightly lower lakes that existed during the retreat of the Port Huron ice.

With still greater retreat freeing much of the Huron basin of ice the lake level dropped intermittently, eventually reaching 640 feet; then for the first time a single lake (Grassmere) occupied both the Michigan and Huron basins, connected by a narrow, ice-bound strait somewhere in the vicinity of the present site of Mullet and Burt lakes and the Indian River. With the resultant great discharge through the Chicago outlet this channel was soon cut down to bedrock and a lower stabilized level, Lundy in Huron Basin and the Calumet stage of Lake Chicago in the Michigan Basin were formed with a surface elevation of 620 feet above sea level (Fig. VIII-7).

Sometime during this same interval, in fact probably both before and after the readvance to Port Huron moraine, a lake (Keweenaw) formed in the southwestern part of the Superior Basin as it became ice free. The early waters in this basin drained south past Duluth and down the St. Croix River to the Mississippi. Likewise, as ice withdrew from the Ontario Basin water was impounded between the ice front and the Niagaran escarpment to form various proglacial lakes which drained eastward along the ice margin and the Mohawk Valley and eventually reached the sea by way of the Hudson or Susquehanna rivers. The best developed of these several lakes held in by Port Huron ice, formed quite late in this retreat; this lake, Iroquois, has recently been dated as a little over 12,000 years old. It should be noted here that Iroquois was considered to be much younger in age until about 1960, as it was assumed that the last ice dam which blocked the St. Lawrence Valley, an ice dam necessary to the formation of Iroquois, was Valders. Several workers have recently shown that the Valders ice nowhere crossed the St. Lawrence, but instead terminated somewhat north of the lowland.

This is a good point at which to emphasize that a relative chronology, built in large part on interpretation and deduction, is quite subject to rearrangement as positive evidence is obtained through the gathering of additional data. In this case additional field work, coupled with $C^{14}$ dates, showed conclusively that Port Huron (Mankato) ice was the last ice to dam the St. Lawrence lowland. It was the ice that held in the water of Iroquois, and with the Port Huron (Mankato) ice being the last obstacle to the general eastward drainage for all the Great Lakes, all of the lakes Warren through Lundy, which drained through the Chicago outlet rather than to the east, must be post-Port Huron (Mankato) but not post-Valders.

Contemporary with Iroquois in the Ontario Basin a single large lake developed in the Michigan–Huron basin, with a surface elevation of 605 feet. This lake, Early Algonquin, drained both south through the Chicago outlet and down the St. Clair River into the Erie Basin and eventually over the Niagaran escarpment into Lake Iroquois. A very strong system of shore features—some beaches, wave-cut cliffs, and stacks, and even some sand dunes—was developed along the margin of this lake, particularly in the northern part of the Michigan and Huron basins. At this time the Upper and Lower peninsulas of Michigan were separated by a broad strait (covering much of Emmet, Cheboygan, and Presque Isle counties) with several large islands interspersed; each of these former islands bears rather distinct notches cut by wave action during Algonquin time.

With further ice retreat, resulting in the freeing of the St. Lawrence Lowland and the uncovering of successively lower drainageways from Georgian Bay eastward to the Ontario Basin, the lake levels dropped to form the Kirkfield stage in the Huron Basin (elevation 565?) (Fig. VIII-8). This shrinkage occurred in the Two Creeks interval, at which time a forest of spruce, pine, and birch flourished at Two Creeks, Wisconsin. At this same time of relatively low water the Allendale delta, built by the Grand River into the higher stage of Lake Chicago west of Grand Rapids, was trenched by streams flowing down to the lowered lake. Concurrently, as ice withdrew from the St. Lawrence lowland, marine waters flooded the still-depressed valley to produce the "St. Lawrence" Sea, represented by well-dated brackish- and marine-water deposits; and the Superior Basin was still occupied by a lake (Keweenaw) in which was deposited a great thickness of red silts and clays derived from the iron-rich rocks of the surrounding area.

*Post-Two Creeks Time.*—The Two Creeks interval was brought to a close by a readvance of the ice during Valders time. This readvance

**The Great Lakes**

173

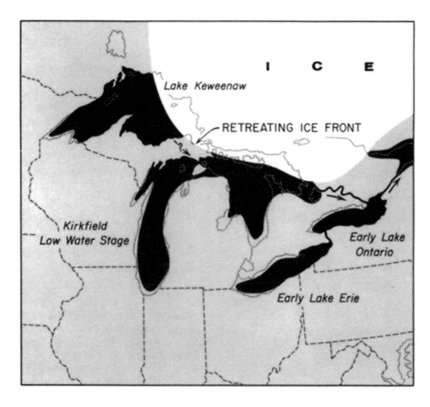

Figure VIII-8. Kirkfield low water stage of Glacial Great Lakes. (Modified from Hough, 1963, "The Prehistoric Great Lakes of North America," permission of *American Scientist*.)

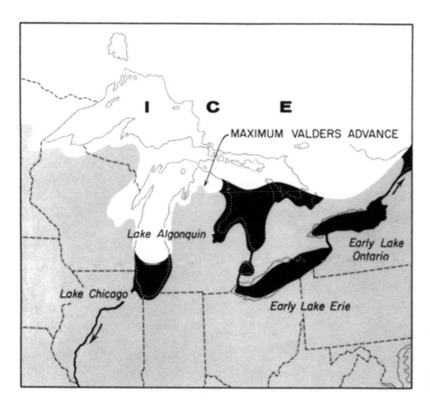

Figure VIII-9. Glacial substage called the Valders Maximum. (Modified from Hough, 1963, "The Prehistoric Great Lakes of North America," permission of *American Scientist*.)

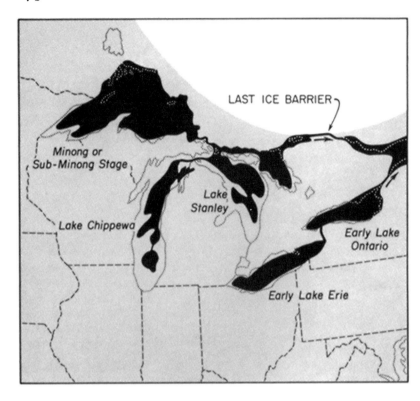

Figure VIII–10. Post-Valders lakes of the Chippewa-Stanley low-water stage. (Modified from Hough, 1963, "The Prehistoric Great Lakes of North America," permission of *American Scientist*.)

Figure VIII–11. Nipissing postglacial lake stage. (Modified from Hough, 1963, "The Prehistoric Great Lakes of North America," permission of *American Scientist*.)

# The Great Lakes

effected only the upper lakes (as pointed out the Valders ice did not cross the St. Lawrence lowland and thus did not strongly affect either of the Erie or Ontario basins), but here great changes occurred as the ice advanced down the Michigan Basin to approximately the latitude of Milwaukee and Grand Rapids and deposited a red till rich in the lake sediments it had picked up as it crossed the Superior basin. This readvance caused a return to higher lake levels in the Michigan and Huron-Erie basins, and separate lakes were formed in the two basins at an elevation of 605 feet, draining south down the two outlets used by the earlier single lake Algonquin (Fig. VIII–9). The return to a higher lake level in the Michigan basin resulted in the deposition of fine-grained lacustrine sediments in the valleys cut in the Allendale delta during the low-water stage representing Two Creeks time.

Once more relatively continuous retreat of the ice front resulted in successively lower lake stages. The several levels of the lake named Duluth which again formed in the Superior basin, were related to downcutting of the St. Croix outlet and, later, to the uncovering of lower outlets across the Upper Peninsula of Michigan (for example, along the valley of Whitefish River, extending from near Marquette and Au Train bays to Little Bay DeNoc). Apparently, ice filled the eastern part of the Superior basin for a long time, for there is no record of Algonquin, or of a body of water at the elevation of that lake, although such a lake level was high enough to have completely covered the St. Mary's River area at the Soo.

This glacial retreat soon uncovered the Trent Lowland, and the lake in the Michigan-Huron basin dropped from the Algonquin level. There were many different lake levels below the Algonquin, presumably related to the opening of various outlets across Ontario between the St. Clair River and the North Bay-Mattawa outlet, which was still ice-covered. Various workers have recognized six such levels, ranging from 540 feet to less than 400 feet in elevation, and locally, as on Mackinac Island, evidence of more than twice that number can be seen.

With the uncovering of the low outlet through the site of North Bay, the lowest of all the lake levels to form in the Great Lakes basins came into being. This series of lakes (Fig. VIII–10) at first was merely inferred from the fact the levels of the immediately post-Algonquin lakes, represented by beaches in the Georgian Bay region, passed well above the level of the North Bay outlet when projected northward; it was later verified by studies of cores taken from the floor of some of the modern lakes. The shorelines shown in the figure are based on the elevation of layers of shallow water sediment found beneath lacustrine sediments of deeper origin related to later, higher lake levels. Drainage between Chippewa (Michigan basin) and Stanley (Huron basin) was by a deep valley cut through the Straits of Mackinac region, which valley is now represented by a submerged channel. Concurrent with these low lake levels in the Michigan and Huron basins there was a low lake level in the Superior Basin which drained across a sill at Sault Ste Marie and into Stanley. Note that the lakes occupying the Erie and Ontario basins were nearly the same shape and size as are the modern lakes in these basins; the major variance in the geography of the eastern part of the Great Lakes area that is readily noticeable is that the St. Lawrence Lowland was still below sea level at this time. It should be noted that from this time (about 9500 years ago) the entire Great Lakes area has been free from the direct influence of glacial ice—from this time on we deal with "postglacial time."

All major changes in the lakes which subsequently occupied the several Great Lakes basins have resulted from changes in outlet levels brought about by uplift of the area due to isostatic rebound of the earth's crust. The early results of such uplift were the withdrawal of marine waters from the St. Lawrence Lowland and a slow but continual rise in the level of the various lakes (particularly those in the Michigan and Huron basins) resulting from the rise in the elevation of their outlets. The first higher lake level of which there is any real record is that of Nipissing Great Lake which came into being, with its surface at an elevation of 605 feet, when the North Bay outlet had been raised 415 feet to cause drainage from the Michigan-Huron basin through the two old southern outlets—the Chicago outlet and the St. Clair River (Fig. VIII–11). Nipissing actually had three outlets, as for much of its history the North Bay outlet was still operative. During Nipissing time wave action once more strongly shaped the shoreline region, and extensive wave-cut flats, backed by strong cliffs, were developed. In many

**Geology of Michigan**

Figure VIII–12. Algoma postglacial lake stage just prior to modern Great Lakes. (Modified from Hough, 1963, "The Prehistoric Great Lakes of North America," permission of *American Scientist*.)

places, particularly along the eastern and northern shore of Lake Michigan, the Nipissing strandline is marked by very large sand dunes. Tower Hill, in Warren Dunes State Park south of Benton Harbor, and Mt. McSauba, just east of Charlevoix, are examples of such large sand dunes related to wind action during Nipissing time. The dune-capped beach which separates Round Lake from Little Traverse Bay, and is traversed by Michigan route 131 between Bayview and Harbor Springs, is an excellent example of the work of both types of shore processes during Nipissing time. In similar fashion arms of Grand Traverse Bay were cut off during this same time to form Torch and Elk lakes. Sleeping Bear Dune, north of Frankfort, at least in part glacial moraine and outwash veneered with windblown sand, was shaped to nearly its present form in this same interval.

With continued uplift the North Bay outlet became inoperative and the greater discharge through the two southern outlets caused downcutting. As the St. Clair outlet was floored with unconsolidated glacial deposits, while the Chicago outlet was developed on bedrock by this time, the St. Clair outlet quickly became the dominant one;

relatively rapid downcutting occurred until bedrock was reached, and Lake Algoma came into being with surface elevation of 595 feet (Fig. VIII–12). This lake occupied all 3 of the upper lake basins, as its surface was so high that the still slightly depressed Soo area was submerged. About 3200 years ago the St. Clair outlet was lowered still further, after a lateral shift of the river from its bedrock sill to an area underlain by till, to give rise to the present Michigan-Huron lake surface with an elevation of about 580 feet. With this lowering (and some further rebound) the Superior sill emerged and Lake Superior became a separate lake at an elevation of approximately 602 feet.

For at least the last 2500 years the Great Lakes have scarcely changed at all—a most remarkable and long-lived stability when the complex history of the several lakes is considered.

### The Record from Mackinac Island

One of the more famous tourist attractions in Michigan, this island situated in the Straits of Mackinac a short distance east of the Straits

# The Great Lakes

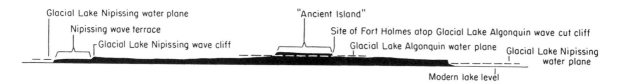

Figure VIII–13. Mackinac Island, looking eastward from St. Ignace (also see Fig. IX–29).

Bridge is an excellent spot in which to see well-developed shore features related to the lake stages of Algonquin and later age. In fact, Mackinac Island is in many respects a perfect place for one to try his hand at geologic interpretation—at interpreting the past from what he can observe going on at the present. In this brief section we summarize the lake history as shown in the features of the island (after Stanley, 1945) and explain the more important points of interest.

A view of the island as seen from the approach to the Straits Bridge (Fig. VIII–13) shows two rather pronounced notches cut into the bedrock of which the island is made. The upper notch, with only a small hill ("Ancient Island") rising above it not far from the southern end of the island, is the result of wave action in Lake Algonquin; the lower notch, only about 50 feet above present lake level, for the most part is the wave-cut cliff related to Nipissing. On the northeastern side and along part of the west side of the island much of the Nipissing cliff has been destroyed by wave action in post-Nipissing time; for example, the cliff extending between Carver Pond and Robinson's Folly (Fig. VIII–14), the promontory on the southeastern tip of the island, is an active one.

A closer look at the island quickly proves that the lake sequence is much more complex than is evident from afar. Along the Short Rifle Range, on the hill slope between Fort Mackinac and Fort Holmes are found 11 separate beach ridges; as they are between 759 and 799 feet in elevation, these ridges are all intermediate between Algonquin and Nipissing in height. Along Custer Road a short distance to the west 14 beaches from 764 feet to 806 feet in elevation can be seen. Another place in which many beach ridges can be found, within the same general elevation range, is along British Landing Road south of its intersection

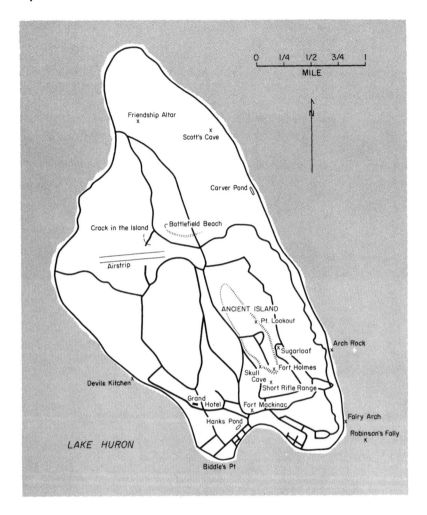

Figure VIII–14. Map of Mackinac Island showing features of geological interest related to the Glacial Great Lakes stages. (Modified from Stanley, 1945.)

with Annex Road. The best-developed single beach intermediate in elevation between the Algonquin and Nipissing level is Battlefield Beach (elevation 716-718 feet), which crosses the golf course and British Landing Road about 500 feet north of Leslie Avenue (Fig. IX–32). That the post-Nipissing lake sequence is complex is indicated by the numerous beach ridges found between the Nipissing level and the present lake nearly everywhere on the northern one-third of the Island; 28 different ones are visible near Pointe Aux Pins.

Many of the points of particular interest to the Island visitor are the direct result of wave action, either the result of wave erosion or the result of transportation and subsequent deposition of the debris eroded by waves from the shore. For a discussion of the origin of these several types of features the reader is referred to Chapter IX; in the discussion which follows stress is laid on the relationship of wave-formed features on Mackinac Island to the proglacial Great Lakes.

The two prominent cliffs mentioned earlier, one related to Algonquin and the other to Nipissing, are the best developed wave-cut cliffs on the Island. In several places low cliffs are found which are related to one of the intermediate lake levels; east of British Landing Road, Battlefield Beach can be traced to a wave-cut cliff related to the same water plane.

Associated with the two major wave cut cliffs are a variety of landforms, arches, stacks, caves, etc., resulting from differential wave erosion. On Mackinac Island most such features are developed in Mackinac Breccia, a breccia developed long ago through the collapse of the roofs of limestone caves and subsequent recementing of the broken fragments with limestone (Chapter V). In general, the breccia is more resistant to weathering than is the surrounding, unbrecciated rock, and through erosion the breccia is made to stand out in relief. As some of the large limestone blocks within the breccia are weaker than the rest, pits and small caves sometimes are formed in these otherwise resistant masses.

The best known wave-cut arch on the Island is Arch Rock (Fig. IX–29), formed by wave erosion during Nipissing time, which undermined the cliff and left a mass of breccia bridging a chasm. A smaller arch, Sanilac Arch, is found near the eastern base of Arch Rock. Fairy Arch is similarly related to differential wave erosion along the "sea" cliff.

Sugar Loaf, rising about 75 feet above the wave eroded flat some 300 feet east of the Algonquin wave-cut cliff at Point Lookout, is a prominent stack (Fig. IX–29). A second stack resulting from wave erosion in Lake Algonquin is the one pierced by Skull Cave, the hiding place of Alexander Henry following the massacre at Fort Michilimackinac in 1763 (Fig. IX–29). This stack, located about 50 feet from the cliff surrounding "Ancient Island," rises nearly 30 feet above the surrounding wave-cut flat. Skull Cave is cut into the western side of the stack. Two other stacks which may be seen on the Island are related to Nipissing erosion. One of these, Friendship Altar, is located 600 feet north of British Landing Road near British Landing; about 8 feet wide and 13 feet high, it is 10 feet in front of the Nipissing cliff. The second Nipissing stack is located about 500 feet northwest of Scotts Cave on the northeast side of the Island.

Other masses of breccia form the many promontories and pillars, not true stacks because they are not completely separated from the wave-cut cliff, visible from Lake Shore Boulevard. Among these features are Lover's Leap, Sunset Rock, Pontiac's Lookout, Chimney Rock, and Robinson's Folly. Another can be seen along the walk from the ferry landing to Ft. Mackinac. Most of these date from Nipissing time, although some have been modified by later wave erosion.

The two best-known caves resulting from wave erosion are Skull Cave, mentioned above as resulting from erosion in Lake Algonquin, and Scott's Cave. Scott's Cave, 15 feet long, 8 to 10 feet wide, and about 9 feet high is developed in the Nipissing wave-cut cliff that is the most prominent feature in the northeastern part of the Island. Devil's Kitchen is a group of small caves, or niches, and wave-rounded rock surfaces related to erosion in lower, post-Nipissing lakes; it is located on the west side of the island at the base of the post-Nipissing wave-cut cliff.

Some of the areas in which abandoned beaches can be seen have already been pointed out, but one or two other areas are significant enough to merit mention. The pointed lowland terminating in Biddles Point, not far from the Grand Hotel, is a compound spit formed in Nipissing time. The most prominent beach ridge in this part of the island is the one behind the old Astor House; it cuts off Hanks Pond on the southeast. Perhaps the single most prominent Nipissing beach, however, is the gravel bar which parallels State Road and lies a short distance to the east. This beach formed as a baymouth bar closing off a bay that lay to the east when the lowering lake approached the Nipissing level. The cut-off bay area is now the site of a swamp that can be seen on both sides of British Landing Road.

The last type of geologic and geomorphologic feature to be discussed is the result not of wave action but of ground water solution along major joints in the carbonate rock of which the island is composed. The best known of these solution-widened joints (fissures) is "Crack-in-the Island" located a short distance north of the airport. There are other, smaller solution fissures at a number of spots around the island. Several small sinks may also be seen on the island, particularly along the east side of State Road in the edge of the woods lining the golf course and a short distance south of Pt. Lookout on "Ancient Island," that portion of the island within the Algonquin cliff. Some of this ground water solution is post-proglacial lakes in age, as judged from the relationships of the solution features to proglacial beach ridges.

# IX

# Water and Wind in Michigan

## Introduction

This chapter deals with the work of water and wind and with the geologic effects that these agents accomplish. What is said, while it applies to the action of these agents everywhere, is necessary for an understanding of their role in Michigan geologic history; on the other hand, the treatment herein is by no means complete and the several general geology texts cited in the Bibliography should be consulted if more information is desired.

Geologists are interested in water and wind mainly because they are capable of doing work; they move earth materials. The movements of water and wind are dynamic processes, producing results which are ordered in time and thus have histories. In fact, many of the geologic effects of moving water and wind are predictable under certain circumstances.

This chapter deals first with water underground, where it is encountered in springs and wells and is important in the formation of caves. Surface waters in streams, lakes, and swamps, and their action in shaping valleys and shorelines are then considered. Also included is a discussion of the action of wind, the effects of which are most important, in Michigan, along lake shores. Surface and ground water are related to one another as parts of a single interconnected system, as will be seen through a discussion of the hydrologic cycle.

## Water Use

One of the reasons for acquiring an intelligent understanding of the geology of water is that one can then better understand the problems created by its use. It is obvious that both surface and underground water "pay no heed" to arbitrary, man-made boundaries, such as lot lines and political boundaries. Unfortunately for the easement of "water problems," most of the world has always followed a doctrine of water "rights" based on a principle of first come, first served. In the more arid parts of the world, the doctrine of "first use," which gives water rights to the first landholder in a region, is the doctrine usually followed. In a water-rich region, such as the state of Michigan, the typical doctrine applied is that of "riparian rights," which simply means that one has the right to unlimited use of all water which lies on or under, or passes through, property which he controls.

Until very recently, Michigan had no water problem—there was enough water for all. However, with the growth of population and the great increase in water demands per capita, the situation is rapidly changing, and a good and adequate water supply is becoming less and less common. Remember, the problem is not that there is less water to be used but, instead, that the amount used in a particular area has increased to the point where the available supply is no longer adequate. With change in the balance between supply and demand, the people of Michigan will soon have to realign their thinking on the subject of water-usage priorities. New legislation on water and much action in the several courts of law will be required as the "water shortage" becomes more and more acute. Until that time each controversy concerning the use of water in this state must be resolved independently on its own merits, a difficult and time-consuming practice.

## The Hydrologic Cycle

Water flowing across the earth's surface possesses energy to do geologic work. That energy comes from the sun and from the force of gravity. Heat from the sun evaporates water, raising it into the air, while the force of gravity tends to pull it back down, returning it to the land and causing it to flow downslope toward the sea. The effects of these two opposing forces are often described in terms of the hydrologic cycle (Fig. IX–1). Because about 70 percent of the earth's surface is covered by seas, the major part of evaporated water comes from the ocean surface, but lesser amounts also come from lakes, streams, vegetation (by transpiration), and even from the soil.

Water in the atmosphere often forms clouds. Such water in the air has potential energy because of the pull of gravity. It may condense as rain and begin to fall, thus releasing some of this potential energy. When rain strikes the land, it tends to flow toward the low spots, expending more of its potential energy and accomplishing geologic work in the process. Some rainwater runs directly off the land surface into streams and rivers and thence back to the sea. Some of it sinks underground, remaining there for varying lengths of time before it ultimately reaches the sea again. Some of it is also retained by vegetation, but plants "transpire" much water back into the atmosphere during life

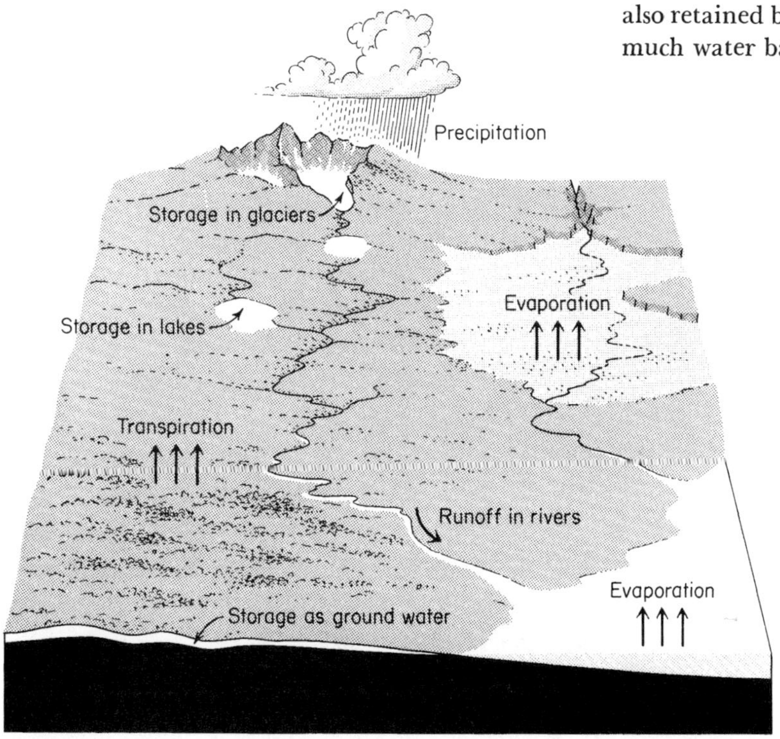

Figure IX–1. The hydrologic cycle. Moisture, evaporated from water surfaces (mainly the seas) enters atmosphere, precipitates as rain or snow, flows off the land surface or underground, and eventually returns to the sea. Enroute, it may be temporarily cycled through plants and transpired into the atmosphere again or stored for long periods in snow, glacial ice, or underground. The driving forces are solar energy which causes evaporation, and gravity which pulls moisture down again.

# Geology of Michigan

and the remaining water is released upon death. Some water may also be trapped temporarily in glacial ice or in the form of snow during the wintertime. Nevertheless, through the long geologic ages, the process is essentially a cycle—water rising, falling on the land, returning to the sea, and rising again; the sun's energy, which is more or less constant, is stored in water, released through the pull of gravity, and ultimately put back in again.

## Water Underground

We who live on "dry land" seldom realize that almost everywhere beneath us there is water, although such features as springs and wells, certainly prove its presence. The water underground (ground water) is almost entirely that portion of precipitation (rain and snow) which sinks into the ground rather than immediately running off the surface into streams and lakes.

Rain water soaks into the ground and, under the force of gravity, slowly moves downward until it reaches a level below which all the pore spaces are filled. This level, the top of the saturated zone, is referred to as the *water table*. Figure IX-2 shows the water table in a humid region in relationship to the zone of moving water above and the saturated zone below; note that the water table is a subdued replica of the earth's surface, higher beneath the hills and lower beneath the valleys. A well will produce sustained quantities of water only if it is drilled below the level of the water table into the zone of saturation and even then only if the material penetrated is permeable enough to yield the contained water. Any such porous and permeable material containing water is called an *aquifer*.

The water table is irregular because of the slow rate of flow due to friction between the water and the pore walls—the water surface cannot flatten quickly enough to offset the continued addition of water from above. Underground water tends to flow from the high spots on the water table toward the lows. The rate of movement may vary from a few inches or feet per year to many feet per hour, depending upon the water supply, the gradient (angle of slope) of the water table, and the permeability of the underground material, but the "usual" rate of flow is of the order of a very few thousand feet per year. In some materials such as clay or shale, pore spaces, while comprising a large part of the total volume, are so small that water encounters a great deal of frictional resistance to flow and it barely moves at all. Such materials are said to be "tight"; they do not *produce* large quantities of water in wells even

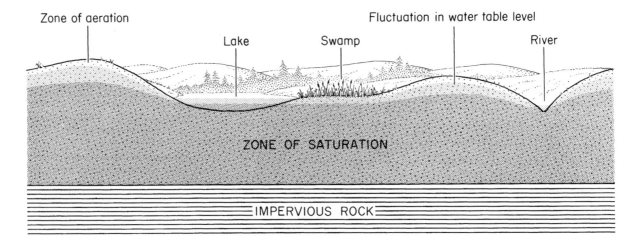

Figure IX-2. The top of the zone of underground water saturation is called the water table. In humid regions such as Michigan the shape of the water table surface is a subdued replica of the ground surface above. The water table surface fluctuates up and down with seasons and climatic changes; locally at lakes, swamps, and permanent streams, it intersects the ground surface to feed permanent standing surface waters.

though they may *contain* much water. This contrasts with materials such as sand and gravel through which water may flow rather rapidly and freely, providing water in abundance to the bottom of a well. Water-saturated sand and gravel thus are good aquifers, and a good aquifer is the goal of all well-drillers.

If the surface of the earth is sufficiently low, it may intersect or pass below the ground water table level (Fig. IX–2). In such cases the water table lies exposed at the surface; thus streams, swamps, and lakes are formed. Most lakes are "spring-fed," that is, supplied throughout the year by ground water flow. Those streams and rivers which do not dry up during the dry seasons are also fed by ground water; in them the surface of the water generally represents the water table. On the other hand, streams which lie above the water table flow only during times of rapid surface run-off; they are *intermittent*.

The shape of the water table fluctuates with the seasons and with climatic changes. During dry summer months or when surface waters are frozen in winter, preventing sink-in, the water table may drop and "flatten out" because the escape of water from the water table into surface waters may be more rapid than the addition from above. In contrast, during times of spring thaw or heavy rains, the addition of water from above may exceed the subtraction and the water table may rise. Such fluctuations in the level are indicated by the dashed lines in Figure IX–2. Obviously, a well drilled into the "zone of fluctuation" will go dry periodically; wells should be deep enough to reach the lowest possible level of the water table.

If ground water is free to move in or out of the zone of saturation, as in the above cases, it is said to be "unconfined." However, if a layer of impermeable material, such as clay or shale, lies above an aquifer and the water is not free to rise, it is said to be "confined." If the level at which water flows into the open end of a confined aquifer is higher than the level from which a well taps the aquifer at another point, the "hydrostatic head" may cause water to flow at least part way to the surface, or even to pour out on the surface. Such a well is called an *artesian* well. Note that it is not necessary that water flow to the surface but only that it rise in the well to qualify as artesian. Theoretically, if unhindered, the water would rise in an artesian well to the same level as the open upper end of the aquifer system. However, the re-

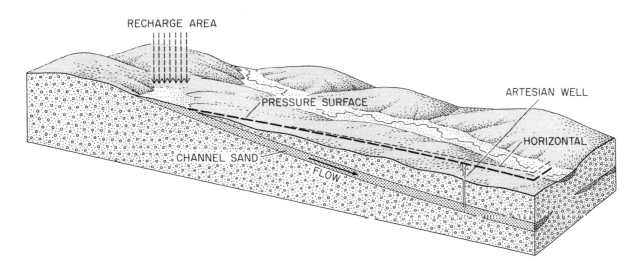

Figure IX–3. A typical artesian ground water system in glacial deposits of Michigan. A Pleistocene age glacial meltwater channel sand, being porous, is the water-conducting aquifer. Water is confined therein by the impermeable, clayey glacial till surrounding the channel. Pressure down the aquifer results from hydrostatic head due to water inflow in the higher recharge area. Water will rise in a well to the pressure surface (heavy dashed line) which falls progressively below the horizontal level (light dashed line) of the recharge area because of frictional energy loss in aquifer.

tardation of water flow due to friction in the porous aquifer expends some of that hydrostatic energy, resulting in an artesian rise which is somewhat less than would occur in a frictionless system. Figure IX-3 shows a typical condition in Michigan which would lead to artesian water. Numerous filled stream beds and valleys lie within and below the relatively impermeable glacial till of the state, producing artesian systems.

There are many small artesian systems related to gravels buried beneath glacial drift in Michigan. To cite only a few as examples, the municipalities of Ann Arbor, Northville, Alma, St. Louis, and Cadillac all get some of their water from such systems. Artesian systems within the state that are developed in bedrock are exemplified by the large flowing well on Grosse Ile in the Detroit River, by nonflowing wells tapping the Marshall Sandstone (Figs. v-2, 29) to supply water for Battle Creek, and by flowing bedrock wells found in some areas of the northern part of the Lower Peninsula of the state.

## Extraction of Water from the Ground

While some wells may be of the flowing artesian type, the majority must be pumped in order to produce water; that is, to lift water from the zone of saturation to the surface. Withdrawal of ground water through a well produces a "cone of depression" around the well (Fig. IX-4), the radius and depth of the cone depending on the rate of water withdrawal. This cone increases the slope of the water table, causing the water to flow more rapidly into the well. However, if the rate of withdrawal is too rapid, the cone of depression may drop below the bottom of the well, and if such rapid withdrawal is continued, the well may go dry. Notice that if the cone of depression spreads sufficiently far, it will lower the water table below the bottom of neighboring shallow wells, cutting off their supply of water. A large number of wells in a small area, as, for example, in a town, produce a composite cone of depression which tends to lower the water table beneath the whole area of withdrawal. If and when wells go dry in such an area, it does not mean there is no more ground water, but simply that the rate of withdrawal has exceeded the rate of inflow from adjoining areas. A decrease in withdrawal usually allows the water table to rise again, decreasing the depth of the cone of depression and permitting ground water to reenter the wells of the system.

A cone of depression fluctuates most during times of excessively heavy water use, especially if such times of heavy use—summer lawn-sprinkling, for instance—coincide with times of drought when the addition of surface water to the saturation zone is small. At such times rationing of the available ground water may be necessary. Another way in which a water crisis may be prevented is through artificial recharge of the zone of saturation by diverting "used water" back into the ground.

## The Search for Ground Water

Michigan is fortunate in that adequate ground water occurs almost everywhere beneath the surface of the state. It is thus relatively rare that an individual is unable to obtain water from a well on his property, and it is ordinarily unnecessary to engage the services of a ground water specialist, although a well-driller will likely assist in the actual placement and drilling of the well. However, as the demand for water increases in Michigan, the demand for the services of professionally trained ground water geologists who apply geologic principles plus an understanding of hydraulic engineering in their search for ground water will increase.

Now a word about "water witching" or "dowsing"—the search for ground water by using forked sticks, divining rods, or other gadgets. That such methods seem successful in locating water underground, both in Michigan and else-

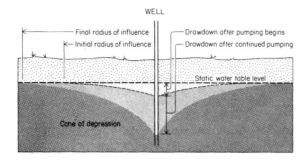

Figure IX-4. Cone of depression on water table surface, resulting from pumping of a water well. Size of cone and effect on adjacent wells depends on rate of water withdrawal as balanced by rate of inflow.

where, stems from the fact there *is* water underground almost everywhere. Any person with an elementary understanding of the principles of ground water occurrence, such as an experienced well-driller or almost any observant person, usually will be successful in discovering water supplies in Michigan.

There is absolutely no scientific evidence that water exerts any kind of a force on a divining rod or any other kind of instrument such as dowsers are accustomed to use. In fact, elaborate and extensive statistical studies by the United States Geological Survey and other organizations have conclusively shown that the chances of a water-witcher discovering water with a divining rod are no greater than the chances of anyone who simply drills a well by pure chance. Nevertheless, one continually hears of cases where a person is unable to get water—and where water was found only after engaging the services of a dowser. Unfortunately, dowsers rarely keep statistics on their proportion of failures; one hears only of their successes.

It is true that in some cases an individual property owner may be unable to obtain suitable water on his particular property. Here in Michigan this may be due to peculiar local circumstances stemming from the glacial history of the state; it may also be that the individual has simply not drilled enough wells or gone deep enough in his search. Perhaps one more well might find water. This is the sort of chance the dowser takes—if he recommends the one additional well which happens to be successful, his success is advertised; if the well he recommends fails, he moves on. A recent group psychology study indicates that the degree of reliance on dowsers is almost directly proportional to the need for water. In water-poor areas, so much depends upon the discovery of water that people are willing to take almost any steps that might help, and after all other moves fail, they may engage a dowser simply because their financial need for water is far greater than his small fee—"I can't lose and I might gain." Even in such cases there is no evidence that the dowser's success is statistically greater than pure chance.

### Some Ground Water Problems in Michigan

In Michigan the majority of water wells are shallow and are pumped by ordinary means. Occasionally, however, the water table is so low or the depth to the aquifers so great that deep wells are required. In such cases special pumping equipment is needed.

There are also areas which do not contain a ground water supply adequate to meet the needs of the area. The limited nature of the water supply often is the result of the glacial history of the region, since most of the state's ground water is produced from surficial glacial sands and gravels. As described in Chapter VII, one result of the Ice Age in Michigan was the deposition of a very complex series of deposits of sand, gravel, clay, boulders, and mixtures thereof which coats the bedrock of the state to varying thicknesses. In areas underlain by well-sorted, porous, and permeable glacial outwash sands and gravels, there is likely to be an abundant subsurface supply of water which may be readily produced. However, in areas of "boulder clay" (glacial till), the materials may be much less porous and permeable so that, though there is water in the subsurface, it cannot be produced in sufficient quantity. In some areas, glacial lake clays or other fine-grained sedimentary materials may be so "tight" (impermeable) that it is nearly impossible to get water from wells.

An additional problem with the ground water of Michigan stems from the impurities, material in solution, contained gases, and so forth, it may hold. Nearly all the ground water extracted from the glacial drift of the Lower Peninsula is "hard" due to the lime it contains in solution, whereas water taken from the bedrock of the state frequently contains sulfur, iron, and various salts in solution. Various chemical treatments are used to remove these dissolved materials.

### The Work of Ground Water in the State

Ground water accomplishes work in many ways. In this discussion we will treat only those features resulting from ground water action which are well displayed in Michigan.

Nearly everyone is familiar with the ability of ground water to dissolve such soluble rock material as limestone to produce caves. While Michigan is not ordinarily thought of as a region of caves, there are caverns at some places in the state. During the Paleozoic Era, as described in detail in

**Geology of Michigan**

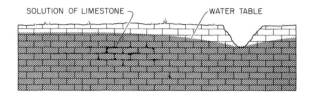

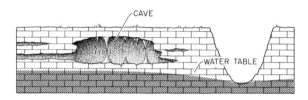

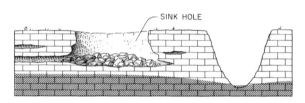

Figure IX–5. Cavern and sinkhole development. Solution of rock, usually soluble limestone, by ground water begins along natural fissures and bedding surfaces, either above or below the water table. Enlarged spaces coalesce to form caverns. Cavern roof collapses to expose open sinkhole. See Figures V–25 and IX–7.

tured by minor movements of the earth's crust, providing routes for the entrance of ground water. The subsurface structure of the bedrock layers of Michigan, as described in Chapter III, is such that in some places these soluble limestones now lie directly below the glacial deposits. Several times in the geologic past these soluble rocks were subject to ground water solution during times of uplift and withdrawal of the seas. Then, just as at present, the bedrock of Michigan was elevated above sea level, facilitating the movement of ground water which dissolved parts of the soluble rocks. The ground water percolated downward from the surface, following along cracks and fissures in the limestone, dissolving the rock along those openings and enlarging them, in some instances to cave size. In cave-forming regions, it is common for surface waters to move down widened joint cracks into the subsurface and to flow in subterranean channelways. Figure IX–5 shows how a cave may develop in such a region and later collapse to form a sinkhole at the surface. It should be noted that there is not yet complete agreement on whether most solution to form caves occurs above, below, or at the water table.

Chapter V, the state was covered by seas in which great thicknesses of potentially water-soluble marine limestone, dolomite, and rock salt were laid down. These now form bedrock layers beneath the state. In most cases these layers have been frac-

In Michigan there are several very interesting areas of sinkhole development. One is in the vicinity of the towns of Leer and Posen, west of Alpena, particularly within a few miles radius of Sunken Lake and Fletcher County Park. Figures IX–6 and IX–7 are a map and a photograph of some of the sinkholes found there. Figure V–25 in Chap-

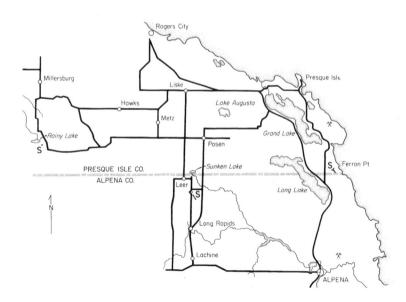

Figure IX–6. Location of some sinkholes (Black dots and S-symbol) in Lower Peninsula of Michigan. Crossed pick and hammer symbolize a major quarry.

ter V shows details of the wall of one of the sinkholes near Posen. Many other sinkholes, some occupied by lakes, are found in Presque Isle County. One of these areas is in Bismarck and Case townships a few miles south of the town of Millersburg. Another is just east of Long Lake, near Ferron Point, in southeastern Presque Isle County. Still another area of sinkhole development is north and west of Monroe in the southeastern corner of the Lower Peninsula; here the sinkholes are extremely broad and shallow, having been partially filled with sediments related to former high lake levels in the Lake Erie basin.

Rainy Lake, about ten miles southeast of Onaway, has had an especially interesting history. It apparently occupies a sinkhole depression (Fig. IX–7) the bottom drainage exit of which is usually plugged with sediment. On occasion, however, the "plug" is naturally released, allowing the lake waters to drain away underground.

Near the Straits of Mackinac there is evidence of *ancient* ground water solution. Millions of years ago, during the Devonian period, ground water solution in this region produced caves which subsequently collapsed. Blocks of the rock which formed the roofs filled the caves; on cementation a formation known as the *Mackinac Breccia* was produced (Chapter V and Figure V–2). St. Anthony's rock in St. Ignace is composed of this breccia, as are Castle Rock, several miles further north, and Arch Rock on Mackinac Island (Figs. IX–27, 28, 29). The presence of this ancient collapse breccia (cave breccia) in the Straits region caused concern when the construction of piers and footings for the Mackinac Bridge was contemplated, and a careful geologic study was necessary to prove there was no danger of further collapse in the region.

## Surface Waters in General

The preceding discussion demonstrates that surface water in streams, lakes, and swamps is related to ground water beneath the land, and that all such waters are part of a single, interconnected system to which the water table is related. Compared with many other areas of the United States and the world, the land surface of Michigan has been shaped but slightly by running water because the state was covered by a Pleistocene ice sheet until some 10,000 to 15,000 years ago. The surface features of the state came into existence only as that ice sheet melted back in a generally northward direction. The distribution of Michigan surface waters, as shown by the abundance of lakes, is related to this youthful, as yet unorganized condition; geologic processes and events which are normal in the developmental history of land drainage have not had time to proceed very far here.

As the glacial ice receded, a great deal of meltwater flowed across parts of the state; hence, many streams and lakes which came into existence in early postglacial times no longer exist. This chapter is mainly concerned with existing features, their origin, history, and possible future. Chapters VII and VIII include a discussion of some of the more important ancient glacial drainage ways and lakes.

## Rivers and Streams

The relatively recent exposure of the state from beneath glacial ice partly explains why there are so few through-flowing streams and so many areas of ponded water. The postglacial topography of the state was exceedingly irregular, with many undrained depressions occupied by lakes and swamps. As these filled with water and overflowed, they became interconnected into a very haphazard drainage network. The low areas through which streams flowed from one lake to the next are not part of a regularly developed stream valley system, but are simply accidental and randomly located low spots within the irregularly piled glacial deposits. When one considers that this was the way in which most Michigan stream systems developed, it is easy to understand why few Michigan streams are major ones, and why so few flow for long distances across the state into the basin of one of the Great Lakes. Since the recent retreat of the glaciers, stream erosion has had too little time to produce new valleys such as are necessary for a well-integrated drainage system.

Because stream action has not been a dominant process in the shaping of Michigan's landscape and is poorly illustrated here, the reader should refer to a general textbook in physical geology if he wishes a complete treatment of stream processes. Presented here are only those aspects of

Figure IX-7. *Top*—Two men (arrow) peer into end of one of many sinkholes near Posen. Cavern roof has collapsed exposing effects of underground solution in Middle Devonian age Alpena Limestone (also see Fig. V-25).

*Bottom*—Rainy Lake during one of several times it drained dry. A natural "plug" in a sinkhole within the lake basin failed allowing water to disappear underground. Devonian age limestones in this region are subject to extensive groundwater solution just as in the region around Posen to the east. Copied from an old picture postcard (no. 14) by William B. Gregg of Onaway. Note people on lake bed in distance. Erosion channels lead toward sink-hole and underground outlet.

stream action which are essential to a general understanding of the formation of the topography of Michigan.

Streams are the result of the force of gravity pulling water downhill; as the water flows across the land, it expends some of its energy eroding and transporting sediment, ultimately depositing it elsewhere. A stream is a very complex system in equilibrium; in nearly every stream there is a balance between ability to do work and work done. The energy available in a stream depends on the amount of water (discharge) and the speed with which the water flows. This available energy is balanced exactly by an expenditure of energy equal to that necessary to allow flow (to overcome friction within the water and between the water and the ground across which it flows) plus that energy necessary to "do work." The fact that only some 2 to 4 percent of the total energy is utilized in transporting a load in a typical stream certainly gives evidence that a stream is a highly inefficient transportive "machine." If the energy of the stream increases due to increase in discharge, velocity, or both, the amount of work the stream does increases. A change in the amount of work a stream can do will affect not only the total load or carrying *capacity* of the stream but also the maximum size of particles the stream can move. *Com-petence* is the term used for the maximum-sized particle a stream can move under a given set of conditions. A stream system is actually much more complex than indicated so far, because if any change in the channel shape or roughness is brought about by erosion, or if deposition of sediment decreases the work done, these changes alter greatly the friction within the system, which in turn affects the amount of energy available to do further work. Because each stream changes almost continually in all these ways—due to changes in rainfall, the crumbling of a bank, and so forth—it is best to consider stream equilibrium as constantly changing, or "shifting."

All of us are familiar with the fact that in times of flood, streams actually overflow their banks, covering at least a part of the "bottomland" or *floodplain* which borders the channel; yet we know that much of the time the stream surface is well below the top of the channel banks. Less familiar is the fact that there is a great increase in both the velocity of the water and the caliber (particle size) of the load during times of flood. A stream which flows at 2 miles per hour at low water stage may have a velocity of 10 miles per hour at flood. The caliber of the particles carried may increase from 1 or 2 inches in diameter to 3 feet or more. A rule of thumb frequently cited is that competence (maximum caliber) varies as the square of the velocity.

A much longer term variation inherent in any stream system is the change in nature of erosive activity, which occurs as the valley system is lowered by erosion toward *baselevel*, the lower limit of stream erosion. Baselevel normally is determined by the level of the sea, lake, or master stream into which the given stream flows. All streams normally tend to lower their valleys by downward erosion. At the same time they tend to widen their valleys through lateral migration of the channel due to the irregularities (curves) in the channel. As water flows around a curve, its force is exerted more strongly on the outside and downstream sides of the bend, and greater erosion at those points, coupled with deposition in the shallower, slack water on the inner and upstream side of the bend, results in both lateral and downstream movement of the curve. In time such channel migration widens the valley. The regular but everchanging curves of a stream are called *meanders*. The migration of meanders across the val-

Figure IX-8. St. Clair River delta, with Lake St. Clair in foreground. Retouched snapshot from airliner (courtesy John Galley). View toward north. Symbols are: A—Algonac; SR—St. Clair River; WI—Walpole Island; SC—South Channel; DI—Dickinson's Island; NC—North Channel; dotted line—U.S.-Canada International Boundary. Delta growth is one important process filling lake basins. Islands between major distributaries grow where lowered velocity of water currents allows vegetation to trap sediment.

ley floor is what gives rise to the flood plain typical of all stream valleys except those youthful canyons in which the stream channel, itself, occupies all of the valley bottom. Lateral migration is most prevalent along low-gradient streams, those that are working at or near baselevel in their lower reaches. Broad, meander-filled valleys are typical of nearly all major stream systems near the sea, such as the Mississippi, Nile, and Amazon rivers; to a lesser degree the same features are found along the lower reaches of the Kalamazoo, Grand, Clinton, and Huron rivers where each approaches its baselevel, the particular Great Lake into which it empties.

As baselevel—the ocean, lake, and so forth—is approached, the velocity of the flowing water is checked, causing a decrease in the ability of the stream to carry a load, and the material in transport is deposited. Stream load deposited in the ocean, lake, or a relatively quiet body of water forms a delta, which grows gradually outward from the shore. A classic example of a modern delta in the Michigan region is that of the St. Clair River where it empties into Lake St. Clair (Fig. IX-8); the many branching stream channels separating low lying islands on the delta are typical of such a feature. Delta growth at the lake margins and deposition of the finer sediment load offshore may eventually fill the entire lake basin; the growth of vegetation is also important in the filling of many smaller lake basins (Figs. IX-12, 13).

The top of the delta and the stream valley immediately adjacent to the lake are controlled by and hence "graded" to the lake surface elevation. If subsequent to the development of a broad val-

**Water and Wind**

191

Figure IX-9. Ypsilanti delta formed by ancestral Huron River where latter flowed into Glacial Lakes Maumee and Whittlesey (see Figs. VIII-4, 5, 6) when waters in the Erie Basin stood higher than at present. The delta surface (indicated by dashed line) forms a long, gentle slope just above modern, artificial Ford Lake and below the skyline. Houses line delta edge. Younger river terrace related to a lower glacial lake stage is indicated by arrow. Modern Ford Lake is formed by dam on Huron River. Cars in foreground head east on I-94. Ford Plant at edge of Ypsilanti is just out of sight to left. Terraces related to the same high glacial lake stages as the delta occur well upstream along the Huron River at Ann Arbor (Fig. IX-10). Later erosion by Huron River (when glacial lakes lowered) dissected delta which formerly covered the area now occupied by Ford Lake.

ley floor or delta graded to the lake, the water in the lake basin drops to a lower level the stream is "rejuvenated," begins downward erosion again, and incises itself into its former floodplain and deltaic deposit. Once this incision has occurred, the surface of the delta and the former floodplain are represented by terraced remnants left above the water level along the newly incised stream valley. Such remnants serve as evidence for the earlier, higher lake level. Good examples of such incised deltas and former broad floodplains are found in many places in Michigan, to mention only a few, along the Clinton River between Rochester and Utica, the Au Sable River west of Oscoda, along the Huron River near Ypsilanti (Fig. IX-9) and the Grand River in the vicinity of Allendale, west of Grand Rapids. Incised terraces are common along the lower reaches of rivers such as the Huron and Rouge and its branches found in southeastern Michigan. Those living in the Detroit metropolitan area can see these in Rouge Park (Fig. IX-10). Incision of the deltas and valley floors and their abandonment at inland locations around the margins of the Great Lakes basins resulted from a drop in baselevel from the high proglacial lakes to the lower levels of the modern Great Lakes (see Chapter VIII).

**Inland Lakes and Swamps**

The history of the Great Lakes is discussed in detail in Chapter VIII. Therefore, we are here concerned only with smaller inland lakes and swamps.

The sources of water for these are three. First, and relatively unimportant, is the rain which actually falls on the surfaces of lakes or swamps. Second is runoff, the water which falls as rain or snow on land and instead of sinking into the ground runs directly into surface streams and thence into depressions. This is a major source of supply during spring thaw when frozen ground prevents downward percolation of water, or when heavy rains provide surface water at a rate faster than it can sink into the ground. However, if a lake or

Figure IX–10. Old river terraces and meander erosion bluffs related to higher water levels of Pleistocene Glacial Great Lakes in the Erie Basin.

*Top*—Island Park along Huron River in Ann Arbor. Modern river flood plain in foreground. Higher flood plain terrace seen through trees in distance at right. Field stones in foreground wall, gathered locally, are glacial erratics, mainly Precambrian igneous and metamorphic rocks, brought down from north by Pleistocene glacier.

*Center*—High terrace (foreground) and related bluff (distance right) along the Rouge River in Rouge Park north of Plymouth Road in Detroit.

*Bottom*—Same terrace and bluff in Rouge Park at toboggan slides. Detroiters enjoy winter sports in their "backyard," on features resulting from postglacial events.

swamp is to receive water throughout the year, there must be a third very important source of water, which is groundwater flowing into the surface part of the system. Earlier in this chapter it was pointed out that the water table, in humid regions like Michigan, is a subdued replica of surface topography. In most places the water table lies below the land surface, but where the water table intersects and passes above the earth's surface, as it does in the case of certain depressions and stream valleys, then the underground water serves as an important source of water for the surface part of the water system (Fig. IX–2). Actually, all bodies of permanent surface water are fed to some extent by underground water, either by minor seeps or actual springs. It is the flow of ground water which keeps such permanent water bodies in existence during the dry periods of the year when there is no surface runoff. If the bottom of a natural depression is below the water table level only during the wet season, the depression will be dry during much of the year.

There are many ways in which depressions may come into existence. Here we are interested only in those which are important in Michigan. Streams may form small depressions along their flood plains. For example, a meandering stream may cut off a meander loop, leaving an "oxbow" depression behind. Solution by ground water, as previously discussed, may form sinkholes, which frequently hold lakes. Inland lake depressions formed in another way along the shores of the Great Lakes where lakes such as Kalamazoo Lake, Manistee Lake, and Crystal Lake, all found along the Lake Michigan's shore, occupy closed depressions formed by the growth of sandbars and dunes across the mouths of bays (Fig. IX–11). An example of this process is given at the end of this chapter in the history of the Herring Lakes Embayment.

The most widespread and important cause of inland lake and swamp depressions in Michigan was glaciation. Topography resulting from glaciation is typically irregular. The haphazard dumping of material by melting ice leaves many undrained depressions, some of which subsequently hold surface waters, either as lakes or swamps. Moreover, as the glacial front recedes, blocks of ice are left stranded (Chapter VII). These ice blocks may subsequently be partially covered with glacial deposits; but when they finally melt away, they leave behind ice block pits or "kettle holes." This is the most common sort of depression in Michigan (Chapter VII, Fig. VII–11). The bottoms of these depressions are at different levels with respect to the ground water table, hence some of the depressions may be occupied by surface waters of an open lake, others by a swamp, and still others may be dry much or all of the year. Although kettle hole lakes are found in many places throughout Michigan, one of the greatest concentrations occurs in a broad belt which extends from just north of Pontiac southwestward to Jackson. This broad belt includes the recreation areas so valuable and easily accessible to the Detroit Metropolitan Area and adjoining communities. Included in this area are such large state projects such as the Waterloo and Pinckney Recreation areas. Lake Orion, Orchard Lake, Whitmore Lake, and many other lakes occupy kettles within this belt. This belt of lakes is in a morainic and interlobate area which developed between the margins of two lobes of the wasting Wisconsin ice sheet, a glacier which stagnated as it wasted (Chapter VII).

Another, somewhat different, origin may be

Figure IX-11. Shore processes along Lake Michigan. Light areas are those of active sand movement in offshore bars, beaches, modern foredune ridges, and finger-like "blowouts" within largely stabilized high Nipissing dunes.

*Left*—Crystal Lake north of Frankfort occupies a lowland between two high, tree covered, east-west trending glacial moraines. The embayment between the moraines, formerly open to Lake Michigan on the west, was closed by growth of baymouth bar. Dunes related to Glacial Lake Nipissing grew atop the bar to increase and complete the closure. Note tendency toward closure of Betsie Lake, another embayment, fed by Betsie River. Offshore breakwaters deflect sand and inner parallel walls accelerate water current through channel to flush out sand.

*Right*—Hamlin Lake, on Big Sable Point north of Ludington, is fed by 4 streams. Big Sable River enters an embayment which in turn drains through narrows at *A* into Hamlin Lake. Other much smaller streams enter embayments and thence flow into Hamlin Lake at *B, C,* and *D*. Another embayment, not joined to Hamlin Lake, is seen at *E*. Formerly, the Lake Michigan shoreline extended along the line A-B-C-D-E. Then hooked spits grew out from the shore above *A* and below *E*, probably fed by sand carried into the main lake by major rivers at those two points. Spit growth produced two "arms" which enclosed Lake Hamlin, first as a bay and then as a separate lake. High dunes related to Glacial Lake Nipissing grew atop the spits, completing the closure. Note the three prominent, light-colored underwater bars ("balls"), and the intervening deeper water and darker colored "lows," offshore in the upper and lower parts of the photo. These lows and balls are typical and important shore-line features as shown in Figure IX-45. (U.S. Department of Agriculture aerial photos.)

attributed to certain lake basins, especially those of Burt Lake, Mullet Lake, and Black Lake in the northern part of the Lower Peninsula. These lakes occupy depressions scoured out by glacial erosion. During higher water stages just after the ice had retreated the entire area was covered by waters of some ancient, proglacial Great Lakes. With a drop in water level and a rise in the level of the land (Chapter VIII), the shallower portions of those deeply scoured basins became dry land, but the deeper portions remained submerged as inland lakes isolated from the present Great Lakes basins. If the level of Lake Michigan were to drop still farther, dry land would form in similar fashion across the northern end of Grand Traverse Bay isolating the deeper portions of the east and west arms of that bay to form two separate large inland lakes similar to Burt, Mullet, and Black lakes.

Lakes and swamps are geologically short-lived features. The shallower depressions are soon filled with sediment and by vegetation which grows, dies, falls to the bottom, and produces peat. For a time the growing vegetation and the porous peat may act as a sponge, holding the water table somewhat higher than it would be otherwise, but eventually the depression becomes filled and surface water no longer stands there. Often man drains such depressions before they become filled by natural processes in order to provide rich agricultural land for the raising of crops, particularly for truck farming. Chapter XIV discusses cases where fossil remains of extinct Pleistocene mammoths, mastodons, musk oxen and other animals, trapped in these bogs, have been uncovered by man.

Such artificial draining, of course, lowers the ground water table in adjoining areas, which ultimately may decrease the value of adjoining uplands.

Most of the deeper depressions begin their history as open-water lakes; these, too, eventually fill in, but the time required is longer than for swamps. The actual length of time required for filling depends upon the size and depth of the lake, the ability of the bottom to support plant growth, and the rate at which sediment is carried in by streams and slope wash. There is a sequence of vegetation in such lake filling, beginning with aquatic plants such as cattails and rushes and surface-floating types such as lily pads and duck-

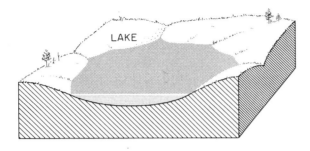

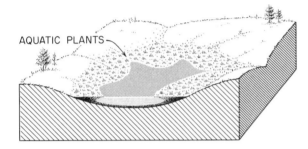

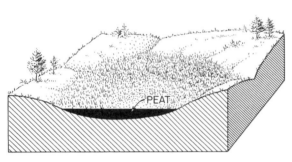

Figure IX–12. Typical stages in the filling of a small glacial lake in Michigan. The depression may have originated as a glacial ice block pit (kettle hole) or simply as an undrained depression amongst haphazardly dumped glacial deposits. The open-water stage (*Top*) is followed by the encroachment of plants, some of the vegetation being aquatic or semi-aquatic and even floating around the water's edge to form a quaking bog. Finally, open water is eliminated and a peat bog (*Bottom*) occupies the entire depression.

weed. The dead remains of these plants fall to the bottom and decay to form peat which begins to fill in the lake, particularly around the margins. Other plants then take root in the soggy new ground; in turn these are followed by water-loving shrubs or trees such as leatherleaf or tamarack. Eventually, some of the larger trees such as maples may invade the boggy ground. In this manner roughly concentric rings of different vegetation types establish themselves around the lake depression with the aquatic forms near the open water and the forms preferring drier conditions farther inland. The succession of vegetation develops as illustrated in Figures IX–12 and 13. The aquatic vegetation next to open water forms a floating mat which one may walk upon but which shakes or "quakes" when disturbed. In time the rings of vegetation extend themselves toward the center of the depression and the lake becomes a swamp. Artificial water bodies such as mill ponds and reservoirs go through a similar evolution by sediment and vegetation filling.

Figure IX–13. Michigan inland glacial lakes in various stages of filling, vegetation contributing to the process. Smallest and shallowest depressions fill fastest. All photos in Pinckney State Recreation Area; scale about 1 inch to 4500 feet, taken September, 1963, U.S. Department of Agriculture. North is toward bottom of page. Note tendency toward rounding as lake in original depression becomes reduced to pond. Many depressions are glacial kettle holes or ice block pits, with steep, ice-contact edges.

*Top*—Two ponds near Hell, Michigan. Original depressions outlined. Dark is open water. Narrow band at waters edge is quaking bog of partially floating vegetation (mosses, cattails, arrowhead, pitcher plants, duckweed). Rough-textured zone of small, light-colored trees and shrubs includes tamarack, poison sumac. Large trees at outer edge of depression are maples and yellow birches. Large, dark trees outside original depression on upland are hardwoods (here mainly oak, hickory, and walnut). Highland Lake in lower right, artificially dammed at Hell, has had less than a century to fill and hence still retains irregular outline of original depression.

*Bottom left*—Just northeast of junction of North Territorial and Hankerd roads. Depression half filled with dark-colored vegetation. Light-toned aquatic vegetation half covers remaining water.

*Center right*—Between North Lake and east end of Halfmoon Lake. Note floating clusters of lily pads out from water's edge.

*Bottom right*—Depression on west side of Hankerd Road, completely filled with cattails.

**Water and Wind**

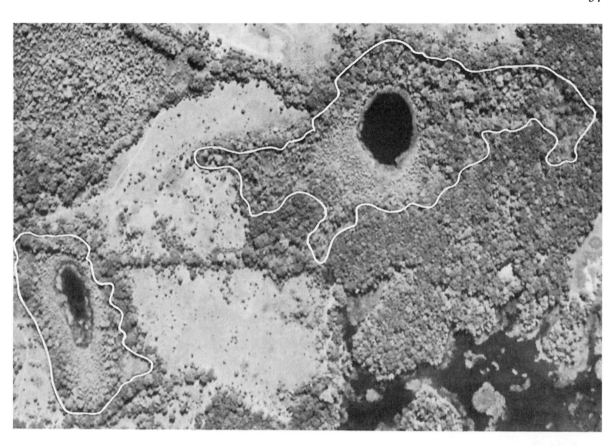

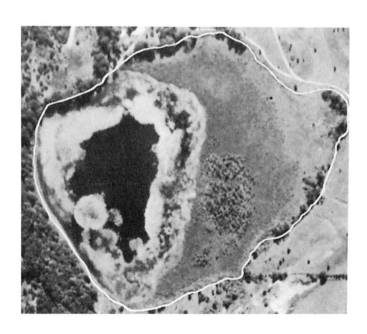

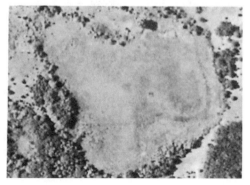

## Surface Water Conservation Problems

The increase in population in Michigan has raised a number of problems related to surface water because human occupancy changes the characteristics of natural streams and lakes. The clearing of the land in some cases caused accelerated erosion; it also facilitated floods and resulted in deposition of sediment downstream. The construction of surfaced roads, driveways, rooftops, and city streets has resulted in an increased rate and quantity of runoff; less water remains to sink into the ground, to enter the ground water reservoirs, and thus to serve as a year-around source of water for streams. Consequently, there is a greater variation in stream flow now than there was a century and a half ago. At the same time, man has drained swamps, and both raised and lowered the water levels in lakes to achieve some particular end—all of which have an important effect on the water table in the surrounding area.

Water pollution is another problem related to man's presence. As more and more cities dump waste, both organic and industrial, into the existing surface water, the pollution levels tend to rise. If, at the same time, stream flow is decreased because of water withdrawal, the ability of the streams to transport the waste products is seriously impaired. In many areas control of drainage systems by joint community or area effort is fast becoming necessary. The problem of water pollution is one to concern all of us, for we are rapidly ruining one of our most precious and essential natural resources through carelessness.

## Wind

Winds are masses of air in motion. They result from pressure differences due to differential heating of the atmosphere. Whenever pressure differences develop there is movement of air from the areas of high pressure toward those of low pressure. Because air is not very dense it is only competent to move sediment particles of relatively small size; sand is the particle size wind moves most readily. Winds must blow about 8 miles per hour to move fine sand, over 25 mph to move coarse sand over 1 millimeter in diameter, and over 75 mph to move 2–3-inch pebbles. The inability to move coarse material, coupled with the fact that material finer than sand size is carried in suspension for long distances along with the wind, makes wind a very effective sorting agent—wind-blown deposits characteristically are well sorted (Fig. IX–14).

Wind erodes by both *deflation* and *abrasion*. Deflation means the picking up of loose particles. Once solid particles are set in motion by the wind they abrade the area across which they are moved, causing still further erosion. The bouncing (saltation) of these grains, which move within a zone only a foot or two above the ground surface, abrades not only the ground but the grains themselves, the latter becoming rounded and frosted in the process. One can readily observe the narrow zone of moving sand on any beach or dune on a windy day. Large pebbles left behind because they were too large to be moved by wind may form a *lag gravel*. Boulders and pebbles in a lag gravel

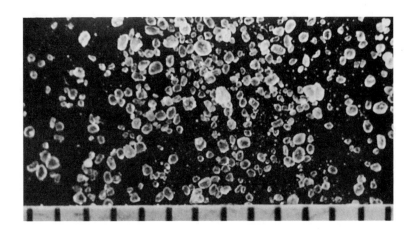

Figure IX–14. Well-sorted (all nearly 1 size), well-rounded, partially wind-frosted quartz grains from a Michigan dune sand. The few more angular grains were more recently added from beach to dune, hence have had less time to become rounded. Millimeter scale at bottom shows degree of enlargement (about ×8).

Figure IX–15. "Lag gravel" of pebbles left behind after wind has blown fine-grained material out of a glacial deposit near Lake Michigan. Perched dune sand in background rests atop the glacial material. (Photo by W. H. Sharp, from I. D. Scott File of Michigan Geological Survey Division. Originally published by R. W. Kelley, geologist, in Michigan Conservation Department booklet, *Michigan's Sand Dunes, a Geologic Sketch*, 1962.)

may be polished and faceted by sand blasting to form *ventifacts* (Figs. IX–15, 16).

Material transported by the wind is deposited whenever the wind ceases to move or where its velocity is checked. Thus, sand-sized material collects around an obstruction, such as a bush, tree, or grass clump, in the lee of a wall or building, or around some windbreak, such as a snow fence, and a sand dune begins to form. The classical sand dune, one developed without the influence of vegetation, is asymmetrical, with the windward side quite gentle and the lee side having a slope

Figure IX–16. *Left*—Wind polished limestone glacial erratic, on the wind-deflated morainic surface upon which Sleeping Bear Dune is perched. A survey bench mark is set in top. (Courtesy R. W. Kelley, geologist, from Michigan Geological Survey Division—Berquist File. Photo by L. C. Hulbert.)

*Center and right*—Wind-faceted pebbles (dreikanters) from a lag gravel. (Courtesy R. W. Kelley, geologist, Michigan Conservation Department booklet, *Michigan's Sand Dunes, a Geologic Sketch*, 1962.)

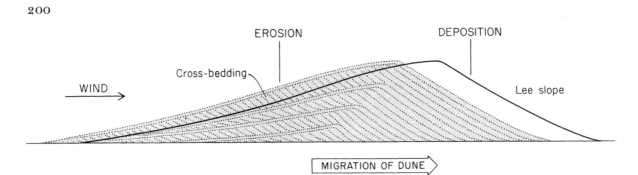

Figure IX-17. Cross section to show anatomy, erosion, and deposition in a sand dune.

of about 30°, equal to the angle of repose of dry, sand-sized particles (Fig. IX-17). The internal structure of such a dune, also seen in the figure cited, is the result of the fact the dune migrates in a downwind direction, due to the rolling and bouncing of sand up the windward side and its subsequent fall over the crest to come temporarily to rest on the steep lee side. The geologist refers to bedding such as that seen in a deposit of wind-blown sand as "crossbedding"; the attitude of the crossbeds is useful to the geologist in distinguishing the flow direction of the current which formed them.

If a dune becomes covered with vegetation, it

Figure IX-18. *Left*—Looking down a blowout on Mt. Baldhead, a stabilized high dune, at Saugatuck. Note lower, straight-crested, poorly stabilized foredune ridge in distance. High dunes in this area are older and related to higher water level of the Nipissing stage of the Glacial Great Lakes. Foredune is younger and forming along modern Lake Michigan shoreline. This old photograph, taken in 1938, illustrates sand stabilization just beginning in the blowout as a result of evergreen plantings. (Michigan Geological Survey Division—Berquist File, photo by L. C. Hulbert.)

*Right*—Small blowout, with incipient vegetation stabilization in foreground. Nine miles west of Gros Cap along US 2 in Upper Peninsula.

Figure IX-19. "Blowouts" due to wind action on shoreward side of old "high dunes" related to higher water level of Glacial Lake Nipissing at Warren Dunes State Park (also see Fig. IX-24). Arrow on lower aerial photograph indicates location and direction of upper photograph. The older, high dunes for the most part are stabilized by vegetation, but blowouts form locally where vegetative cover is destroyed by fire, disease, or drought, or where wave or stream erosion at base of dune causes sliding. (Aerial photo from U.S. Department of Agriculture.)

may be *stabilized*. Such a dune may be reactivated locally by destruction of vegetation, in which event *deflation hollows* or *blowouts* form, most often on the windward side. As these depressions expand and deepen, the external shape of the dunes is greatly changed; often the blowout dune is steeper on the windward side (Fig. IX-18). The steep windward face on Tower Hill, the large dune just east of the parking lot at Warren Dunes State Park about 15 miles south of Benton Harbor is a large blowout resulting from reactivation of an ancient sand dune (Fig. IX-19).

The average layman thinks of sand dunes whenever he thinks of a desert. It is true that the desert is a good place for wind action to be effective, because such an area is relatively free of vegetation, is frequently marked by fairly strong winds, and is also a place where running water is not very frequent. Nevertheless, most desert regions are relatively free of windblown sands. On the other hand, one of the most common sites of sand dunes is along the shore of a large lake or the ocean. There the three things so necessary for the formation of sand dunes are present, namely (a) a good, fairly stable wind, resulting from the free, unimpeded sweep across a large body of open water, (b) a source of sand provided by the beaches along the body of water, and (c) a place to put the sand downwind which is the land to leeward of the beach in the case of onshore winds so common to coastal areas.

Dunes are in an almost constant state of change.

# Geology of Michigan

Moreover, they are affected by processes other than wind action, such as wave erosion, lake level, groundwater changes, and climatic changes. Thus, their history may be very complex. The final sections of this chapter discuss examples of dune growth and change in the light of all these factors.

The beach sand from which the wind builds dunes comes primarily from two sources, the sand in glacial deposits along shore which waves loosen and carry into the lake, and sand eroded from inland glacial deposits and carried into the lake by rivers and streams. About 90 percent of the mineral grains in Michigan dune and beach sands are composed of the mineral quartz. The remainder are grains of epidote, garnet, magnetite (black sand), hornblende, calcite, ilmenite, orthoclase, tourmaline, and zircon. Quartz is dominant for several reasons. It is an abundant mineral in the upper part of the continental crust, is chemically stable, hence, little affected by weathering, and is very hard, thus resisting wind abrasion.

The distribution of sand dunes in Michigan is shown on Figure IX-20. Note that some dunes lie far inland from the present shores of the Great Lakes. These *inland dunes* generally are the oldest dunes of the state. They formed around the margins of glacial lakes and glacial outwash plains (Chapter VIII) when ancestral Great Lakes levels were higher, and the lands themselves lowered due to ice depression, shortly after the Late Pleistocene (Wisconsinan) continental ice sheet had retreated from the state. Nevertheless, even these inland dunes are no older than about 13,000 years. The inland dune areas no longer are the sites of extensive dune growth. In most cases they were stabilized by vegetation long ago, although it is unfortunate that man's poor farming practices

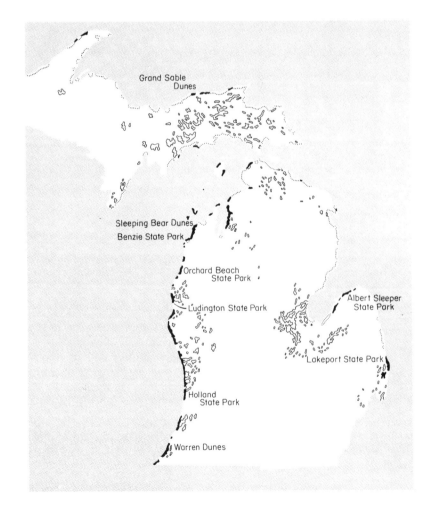

Figure IX-20. Blown sand and dune areas in Michigan. Dots along shore lines of present Great Lakes indicate areas of extensive modern sand movement and foredune growth. Black areas are older, high dunes related mainly to Glacial Lake Nipissing higher water levels. Irregular areas of sand, including dunes, located well inland from the present lakes originated at still earlier times when proglacial predecessors of the Great Lakes stood at much higher levels and even farther inland (see Figs. VIII-4–12). In particular, note the dune field southwest of Saginaw Bay in the area formerly occupied by various levels of Glacial Lake Saginaw. (Modified from map by R. W. Kelley, geologist, Geological Survey Division, Michigan Department of Conservation.)

Figure IX–21. The "Sleeping Bear," a dune related to higher water level of Glacial Lake Nipissing (see Fig. IX–20) and now partially stabilized by vegetation. Surface of modern Lake Michigan, at right, lies about 450 feet below dune crest, but dune itself is "perched" well above lake level with its base (just below tree line) resting on a platform of glacial moraine. Cobbles and pebbles too large to have been removed by wind and originally contained within the moraine, can be seen forming a "lag gravel" in foreground and on steep slope in right foreground. (Michigan Department of Conservation, photo by Walter Hastings.)

locally destroy vegetative cover and reduce soil moisture so as to make such sand areas again susceptible to wind erosion.

An especially interesting group of inland dunes is that situated just southwest of Saginaw Bay. Sands of this group of low dunes and associated glacial lake beaches are clearly visible along U.S. Highway 10, between Midland and Clare, especially as one approaches the latter city. These dunes lie on an ancient, proglacial lake plain and along what was the shore line of Glacial Lake Saginaw and related stages (Chapter VIII). Many inland dunes located well inshore from Lake Michigan, on the western side of the state, are related to older and higher stages of Glacial Lake Chicago which occupied the Lake Michigan Basin.

The *coastal dunes* are younger than the inland dunes. They formed along the shores of later Glacial Great Lakes stages (particularly Nipissing) and at the modern Great Lakes shore line, hence are generally less than 4500 years old. Coastal dunes fall into 2 main categories. These are the *foredune* ridges which are low dunes (30 to 50 feet) close to the present water's edge, and the *high dunes* (over 100 feet). Some high dunes may lie at the water's edge, but more often they are located slightly inland, behind the foredunes. Some of the old, high dunes consist of sand that was deposited on the tops of glacial moraines and other high glacial features that lay near the water's edge at a time when Great Lakes levels were slightly higher relative to the land elevation. These are called *perched dunes* (Fig. IX–21). The top of Sleeping Bear Dune stands 450 feet above Lake Michigan, and Grand Sable Dunes in the Upper Peninsula at 380 feet above Lake Superior, but these are perched dunes. The dunes themselves are not that thick.

The foredune ridges, which may be one or more in number, are the youngest of the coastal dune group; often they still retain their straight, relatively unmodified, original crestlines. They are related to relatively low water levels and shore lines of the modern Great Lakes. The high coastal dunes are older than the foredune ridges, but younger than the inland dunes. In most cases the high coastal dunes were stabilized long ago by vegetation, including a heavy forest growth, but accidents of nature such as wave erosion, destruction of plants by fire, drought, or disease, and the works of man, in many instances have allowed wind erosion to begin again locally on the high dunes. As a result, extensive U-shaped blowouts and blowout clusters have formed. These, in turn, have greatly modified the original crest lines of the high dunes into highly irregular, often sinuous shapes (Fig. IX–19).

A blowout deepens, enlarges, and extends farther inland as wind carries sand from the front or windward side of the blowout, up and over the dune crest. This sand then falls down the lee or backslope of the dune, where wind currents are too weak to transport or erode. The sand is deposited in a series of inclined layers which dip steeply downwind at the natural *angle of repose* of loose sand. The deposit behind the blowout grows inland, layer after layer, in the wind direction, often engulfing trees that had established themselves there earlier when the dune was stabilized (Fig. IX–22). The rate of inland movement varies greatly but rarely exceeds 5 feet per year. This inland movement of sand behind blowouts is a common and easily observed phenomenon which no doubt gave rise to the *mistaken* notion that the coastal dunes as a whole are moving inland, and will continue to do so indefinitely, engulfing everything in their path. Those who once believed this, pointed to the inland dunes as proof of their concept, but we have already seen that the inland dunes lie where they do because they formed along the shore lines of formerly higher and more extensive ancestors of the modern Great Lakes. Thus, the inland dunes still are located close to their original position of deposition. They are not migrating now and did not do so to any

Figure IX–22. Trees being buried by sand on back (inland) side of modern dune along east shore of Lake Michigan, Michigan City, Indiana.

great extent earlier. Some supporters of the concept of dune migration point to the well-established fact that dunes in desert regions *do* migrate downwind. However, they neglect to recognize that vegetation is sparse in dry deserts, but is abundant in the humid Great Lakes region, where it readily stabilizes moving sand. The coastal dunes of Michigan are not migrating inland as a whole, but are simply being modified locally in blowout areas. The overall effect of blowouts is to change the original, straight-crested dune form to a more sinuous and irregular shape and thus to broaden the dune belt. There are, however, natural limitations to the extent of dune movement. The farther inland a blowout grows, the farther its windward face gets from the source of fresh, loose beach sand. Moreover, the strong winds that build up their potential erosive energy across the wide, relatively unimpeded *fetch* of the open lake, lose energy rapidly when they strike the rough surface of the land. The farther any portion of a dune moves inland from the shore, the greater the energy winds must lose before reaching the dune front. Beyond a certain point then, other factors remaining constant, the wind moves too slowly to pick up and transport sand. Then stabilizing vegetation quickly takes over, increasing the roughness of the land surface and further diminishing the force of the wind. This means that wind velocity, vegetation, dunes, and erosion are in an uneasy state of equilibrium which restricts sand movement and dune modification within a relatively narrow, irregular belt rarely over one-half mile wide, just inland from shore. Minor changes may occur locally, but major shifts in the position of the coastal dune belt can occur only if there is a major change in lake water level or climate. Within historic time the lake levels of the Great Lakes have fluctuated as much as 5 to 6 feet. Changes of this order of magnitude are sufficient to account for considerable local modification of existing coastal dunes, but fall far short of the major, long-term changes in lake levels and shore lines which have occurred over the last 13,000 years due to lowering of glacial lake outlets and postglacial rebound of the land. The inland dunes, for example, are related to glacial lake levels over 200 feet higher than those of the present Great Lakes (Hough, 1963).

## Shoreline Processes in General

Along the shores of larger lakes, such as the Great Lakes, the constant work of waves, currents, and winds is particularly effective in shaping the land. Shorelines "evolve"; they have predictable futures. An understanding of shoreline processes is important, not only for an appreciation of modern shorelines but also for an understanding of the history of the proglacial Great Lakes (Chapter VIII). While this chapter stresses the effects of those agents in Michigan, many of the features mentioned are also found along shorelines throughout the world.

## Waves and Shore Currents

Water waves derive their energy from moving air (wind), which in turn derives its energy from dif-

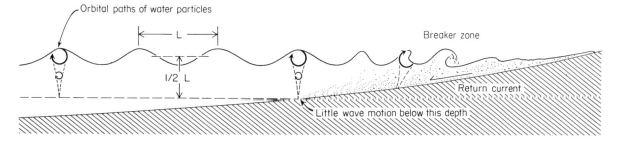

Figure IX-23. Water waves. L– Wavelength, crest to crest (same as trough to trough). 1/2L indicates the water depth, equal to one half a wavelength, below which the velocity of water particles, around their orbits with the passage of each wave, becomes too slow to stir fine sand. In depths less than 1/2L, waves begin to markedly "feel bottom," shorten, heighten, and eventually break, expending the major part of their energy on the shore line in that narrow shallow water band and on the beach. Finer-grained sediment, put into suspension by wave action in the shallower waters, settles farther out in water depths below reach of the waves. (Also see Fig. XII-13.)

ferential heating by the sun. Waves expend their energy against any obstacle they strike, and particularly against the shore.

Perhaps you have noticed that a floating object bobs up and down with the passage of each wave but does not move along with the wave; when still-fishing from a small boat one's bobber stays nearby even as the waves roll onward. This illustrates that while energy is transmitted through water in the horizontally progressing form of waves the water itself moves very little in that direction. Instead, each water particle revolves in an orbital path with the passage of a wave (Fig. IX–23). The diameter of the orbit decreases with depth. During the interval of time required for the passage of a wave the water particles travel once around their orbits. Those traveling the larger orbits near the surface must, of course, move faster than those traveling the smaller orbits at depth. Below a certain depth, depending on the size of the waves, there is very little water movement. We can readily appreciate, therefore, that the energy in waves decreases rapidly with depth. Note that there is a general relationship between wave amplitude (height) and wavelength—large, high waves have a greater wavelength than do smaller ones. Studies have shown that the orbiting water particles move fast enough at a depth of about one-half wavelength to stir fine sand. At lesser depths water particles in the wave move fast enough to disturb coarser sand, gravel, or even boulders. Wavelengths vary, and, of course, the longer the wavelength the greater the depth of "wave action," that is, the greater the depth of water in which the wave does any work.

Waves approaching a shallowing shore begin to "feel bottom" when they enter water depths which are about one-half their wavelengths; inside this depth the orbital path is distorted. Because of this interference by the bottom the waves begin to slow down. The wavelength decreases and the waves become higher and more asymmetrical until they eventually fall forward or collapse as breakers. The distance from shore at which breakers form depends upon the size of the waves and the shape of the bottom. The zone inside the breaking point is called the zone of breakers. It is in this zone that most wave energy is expended on the bottom, and here much sediment is moved along the bottom.

The limit to which waves rush up the beach marks the point where energy ceases to be expended against the shore. As the water flows back down the beach it expends some of its potential energy in the opposite direction, by moving particles away from shore. In this manner wave action is limited in its effect to a rather narrow band along the shore, within which it accomplishes a great deal of work. It may drag particles backward and forward, up and down the beach, or even along it when waves strike the shore obliquely. Particles are abraded, rounded, and eventually worn away. The abrasive effect of waves is shown by the rounded brick, concrete, and glass fragments so common along some beaches. The larger particles such as boulders, cobbles, pebbles, and coarse sand remain near shore or even on the beach, where normal wave action is not strong enough to move them. The finer particles, such as fine sand, silt, and clay, may be shifted away from shore into deeper water. In this manner sedimentary particles tend to become sorted according to particle size along a shore, from relatively coarse materials near shore to finer and finer particles offshore.

The actual range in particle size on any particular beach is, in part, related to the nature of the source material. For example, along the shores of Lake Michigan, where waves cut into cliffs of glacial till, the boulders and cobbles form a coarse-grained beach. Elsewhere along the same shore waves cut into cliffs developed in dune sand. There the coarsest material available is medium to coarse sand. Materials are also carried along the beach from distant source areas and these contribute to the nature of the material found at any point. During a storm very coarse materials may be thrown high on the beach to form a *storm beach*. During storms some sections of a beach may be torn out by the waves only to be replaced by the deposition of material during times of calmer water. The same thing may occur seasonally, with sand beaches being destroyed during the winter only to be replaced during the summer. Although the waters actually dissolve some material on the shoreline itself, they exert their strongest effect mechanically as they dash against cliff bases. Their hydraulic force pries at the cliff, and particles held in their grasp grind away at the coast. The power of a large storm wave is tremendous, and those developed along the shores of the Great Lakes are capable of moving masses of solid rock many feet in diameter.

**Water and Wind**

207

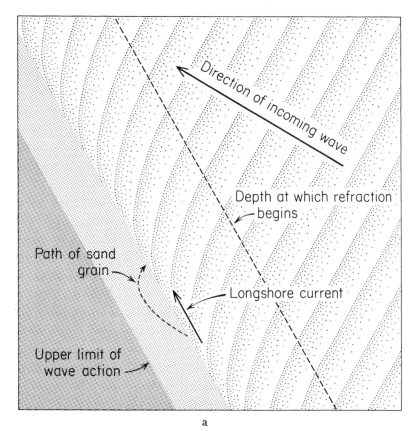

Figure IX-24. Wave refraction along Lake Michigan at Warren Dunes State Park. (Also see Fig. IX-19). Note bending of wave fronts as they approach shore. Diagram illustrates refraction and movement of particles on beach. Wave fronts which approach shore at an angle slow down at their inner ends, where they first enter shallow water and encounter frictional drag on the bottom. The outer portions of each wave continue to travel faster. Thus, the wave fronts tend at the last moment to swing nearly parallel with shoreline and to direct their energy more nearly (but not quite) at right angles to the shore. (Photo from Michigan Department of Conservation, 1940.)

# Geology of Michigan

Waves approaching the beach at an angle tend to swing parallel to the beach; the end in shallower water tends to "feel bottom" and to slow down sooner than does the outer end and the wave is bent or "refracted," as illustrated in Figure IX–24a, b. This can be seen whenever one views waves from above, from a high cliff or an airplane. Wave refraction also results in a concentration of wave energy against protruding headlands and a diminution of energy in bays.

Despite this tendency for refraction, waves are seldom bent entirely parallel to the beach, and therefore some wave energy is usually expended in an oblique direction, setting up currents which move along the shoreline. These are called *longshore currents*. Longshore currents may also be produced by wind literally blowing masses of water along the shore. Longshore currents transport much sediment along a shoreline. This movement is assisted by *beach drift,* resulting from particles washing up and down the beach under the oblique attack of waves (Fig. IX–24a).

## Shore Features

After this brief summary of wave and current action, let us now examine some of the results they produce. The force of waves approaching and breaking against the shore tends to develop a "wave-cut cliff" (Fig. IX–25). The eroded material may be transported toward deep water where it is deposited; this deposition and the cliff retreat due to wave action produce a nearly flat surface, in

Figure IX–26. Sea caves, cliffs, and stacks cut by wave erosion along shores of Great Lakes.

*Top*—Miner's Castle along Lake Superior in the Pictured Rocks area east of Munising (also see Fig. V–7). Sandstones here are the Late Cambrian age–Munising Formation, an ancient marine deposit. Columnar stacks at top of castle were carved by waves of an earlier and higher glacial lake whose shoreline was subsequently uplifted by crustal rebound after removal of the depressing weight of glacial ice sheet (also see Fig. IX–33).

*Bottom*—Pointe Aux Barques at the tip of the "Thumb" of Michigan on Lake Huron. Sandstone is of Early Mississippian age. (Photos courtesy Michigan Conservation Department.)

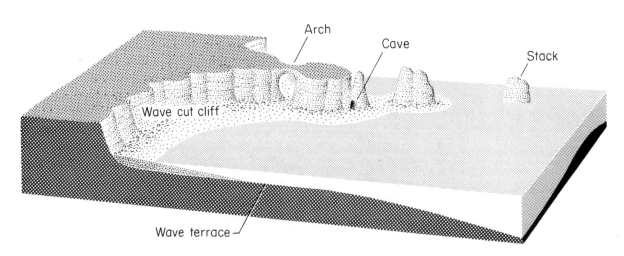

Figure IX–25. Wave erosion features along a Great Lakes shore line. As cliff is eroded inland a wave terrace is left behind. Note concave-upward profile of cliff and terrace combined. A sea cave may erode in softer parts of the cliff. A cave may enlarge, lengthen, and cut through to form an arch. Collapse of an arch leaves a free-standing stack. See the following figures for photographs of sea cliffs, sea caves, arches, stacks and terraces, both modern and ancient: Frontispiece; II–1; V–5, 7, 8, 17; VIII–13; IX–26–33.

# Geology of Michigan

Figure IX-27. Castle Rock, a wave-cut stack related to the Glacial Lake Nipissing shoreline. Along I-75 just north of St. Ignace. Rock is the Mackinac Breccia (see Fig. V-16, and IX-25, 28, 29, as well).

*Top*—Bluff just behind castle, and at right, is sea cliff along former Lake Nipissing shoreline.

*Bottom*—Note the angular blocks of Devonian age dolomite composing the well-cemented breccia. Superior resistance by breccia to wave erosion accounts for preservation of this stack. The breccia itself formed millions of years earlier when solution of rocks below caused local collapse of overlying dolomites. Surrounding softer rock was eroded away more rapidly, receding to position of cliff behind stack.

# Water and Wind

Figure IX–28. Isolated sea stacks of resistant Mackinac Breccia along wave-cut cliff of ancient Glacial Lake Nipissing shoreline. Note angular blocks of well-cemented Devonian age dolomite. (See Mackinac Breccia in text.)

*Top*—View south from roadside park along US 2 at Epoufette, 6 miles west of St. Ignace. Lower Peninsula of Michigan on skyline across Mackinac Straits.

*Bottom*—St. Anthony's Rock, in alley behind Homestead Cafe in St. Ignace.

Figure IX-29. Arch Rock, Skull Cave, and Sugarloaf Stack, wave erosion features along uplifted glacial lake shore lines on Mackinac Island (also see Figs. VIII-13 and IX-32). Modern Lake Huron to east, also visible below through the arch. Arch Rock is related to Glacial Lake Nipissing; Skull Cave and Sugarloaf were formed along the shore of Glacial Lake Algonquin.

part depositional and in part erosional, immediately offshore from the beach. This *wave terrace* develops at the base of the wave-cut cliff. If cracks, fissures, or zones of soft material accelerate wave erosion at certain points along a cliff, then caves may form; such caves may join to form a tunnel or *arch*. Often a column, called a *stack*, of resistant rock may be left standing offshore from the cliff. The fact that cliffs, beaches, wave-developed flats, caves, and stacks are all normal and common shoreline features is important for an understanding of the topography along the coasts of the Great Lakes. In Chapter VIII dealing with the glacial history of the Great Lakes, attention is directed to such shoreline features, both erosional and depositional, now abandoned as a result of changes of lake level. Miner's Castle near Munising in the Upper Peninsula is a good example; so, too, are Castle Rock, between Saint Ignace and the Soo, and the many features found on Mackinac Island. All these and many other shoreline features (Figs. IX-26-33) were left standing high and dry as a result of the lowering of Great Lakes levels and an accompanying rise in the level of land after the glaciers had retreated.

Where wave action is intense, strong longshore currents pick up loose material and transport it along the shore. Wherever the longshore currents slow down, as may happen in a bay or where a river enters the lake, or even around a breakwater, the material in transport is dropped to form a bar, spit, or other form of beach. A current moving off a point of land into deeper water may build out a *spit*; often the action of onshore waves deflects the current so that the spit becomes hooked. The sediment which forms the spit is deposited because the moving current which carries the sediment loses its energy (momentum) when it encounters the still-standing mass of deeper water. Longshore currents frequently build spits from one or both sides of the mouth of a bay. Such spits may grow to form a baymouth bar, closing off the mouth of the bay entirely to isolate an inland lake. Mention was made earlier in this chapter of several such lakes. These processes are illustrated diagrammatically in Figure IX-34 and by photographs in Figures IX-35 and 36.

Waves and longshore currents can be very damaging to shoreline property. Common methods of controlling these effects include the construction of *breakwaters* parallel to the shore to interfere with the expenditure of wave energy on the shore, and *groins* built perpendicular to the shore to cause deposition of beach sand. Figure IX-37 shows groins built near the stream flowing from Lake Hamlin into Lake Michigan at Ludington State Park. Here the groins prevent sand from clogging the stream, build sand beaches, and protect the beach road from wave and current erosion. The placement of such structures as breakwaters and groins must be studied with care, for while they protect the shore in one place they may cause accelerated erosion elsewhere.

As pointed out earlier in this chapter, sand dunes are commonly found near the shores of large bodies of water, because conditions there are perfect for the formation of such deposits. There are few places in the world where dunes are more common, or more spectacularly developed, than along the shores of the Great Lakes.

## The Herring Lakes—
## A Case History of Dunes and Bars

In 1937 I. D. Scott and K. W. Dow described the development of dunes and baymouth bars in the region of the Herring Lakes, Benzie County, Michigan, about 6 miles south of Frankfort. Some of the details of this study are summarized here to serve as an example of how the complicated effects of Glacial Great Lakes history, wave and longshore current action, and wind action may be combined in the historical development of a particular area. The Herring Lake embayment includes Upper Herring Lake and Lower Herring Lake. At present these are connected by a narrow stream, Herring Creek, but are separated along most of their common border by low, sand

Figure IX-30. *Top*—Recessional beach ridges formed along shore line by progressively dropping glacial lake water levels in Lake Huron Basin. Waters along west shore of modern lake appear dark gray. North is to left. (U.S. Department of Agriculture photo taken north of Port Huron.)

*Center*—Burt Lake, looking northwest from bluff (glacial lake wave cliff) one half mile southwest of Indian River. Same elevated, wave-cut cliff (see arrow), with terrace below, is seen in profile on spur across bay. Cliff, part of Glacial Lake Algonquin shoreline, formed at former high water stage when Burt Lake depression, now an inland lake, was connected with the major Great Lakes. (Michigan Geological Survey—Berquist File, photo by L. C. Hulbert.)

*Bottom*—View northwest across Little Traverse Bay from near Petoskey toward Harbor Springs. Arrow indicates uplifted Glacial Lake Nipissing wave-cut cliff with broad wave terrace below and to left.

**Water and Wind**

215

Figure IX–31. Wave-cut cliffs with concave terraces below, along glacial lake shore lines that have been uplifted by crustal rebound after retreat of Wisconsinan ice sheet. North shore of modern Lake Michigan.

*Top*—View west from Cut River Bridge along US 2, 26 miles west of St. Ignace.

*Bottom*—Burnt Bluff at tip of Garden Peninsula near Fayette State Park.

## Water and Wind

Figure IX-32. Mackinac Island. *Top*—Fort Mackinac at edge of uplifted Glacial Lake Nipissing wave-cut cliff. Note stack of Mackinac Breccia standing free of cliff.

*Center*—Town of Mackinac Island. Fort Mackinac and upper level of homes stand at edge of Nipissing wave cliff (stack indicated by arrow). Lower town is on Nipissing wave terrace at foot of cliff. Ft. Holmes (not visible) stands on highest point of island behind Fort Mackinac on a small, high area called "Ancient Island." (See Fig. VIII-13.) Ancient Island alone stood above the high-water level of Glacial Lake Algonquin.

*Bottom*—Man standing on "Battlefield Beach," a minor glacial lake shore-line feature between the Glacial Lake Algonquin and Nipissing levels. View taken on golf course.

ridges. Upper Herring Lake, on the east, stands about 13 feet above Lake Michigan. Lower Herring Lake stands at the level of Lake Michigan and is connected to it by a short outlet. The Herring Lake area has not been affected by postglacial uplift because it lies south of the hinge line of postglacial uplift (Chapter VIII). All changes are, therefore, only the result of fluctuating lake levels and the growth of shoreline features. The Herring Lake embayment within which the 2 lakes lie is a U-shaped depression enclosed on all but the west side by the Manistee Moraine of glacial origin. Figure IX-38 is an aerial photograph of the area; Figure IX-39 is a diagrammatic representation of the historical development of this region. Scott and Dow concluded as follows (p. 450):

> The Herring Lake embayment lies within a U-shaped moraine open to the west.
> 
> The embayment was first occupied by water in late Lake Algonquin time, during which stage Upper Herring Lake was isolated by a series of mid-bay bars. Dune formation during this stage is represented by a single ridge on the north side of the embayment.
> 
> The embayment, except for Upper Herring Lake, was drained during an early stage of Lake Nipissing.
> 
> At the time of late Lake Nipissing the embayment was again flooded and remained open to Lake Michigan. During this stage not only the series of low dune ridges along the north shore of Lower Herring Lake, but also the largest parabolic dunes of the area were formed.
> 
> The embayment was finally closed by the present baymouth bar in post-Nipissing time. Upon this bar and adjoining shores lower dunes of recent origin have been formed.

Another detailed discussion of the growth of dunes in the Manistique area is presented by Stevenson (1931). The changes which go on along shorelines have many interesting legal implications for water and land use. An interesting and detailed discussion of the processes of growth and destruction along shorelines as they may affect man is presented in a short paper by Scott and Stevenson (1931) listed in the bibliography.

### The Southeastern Lake Michigan Shore

J. S. Olson (1958a, b, c) recently studied the factors controlling the development of shoreline features and the history of those features at the southeastern end of Lake Michigan. His studies, conducted mainly in the Indiana Dunes State Park in Indiana and in Warren Dunes State Park in Michigan, provide a detailed example of the general principles that govern the work of water, wind, and waves throughout the Great Lakes Region.

Wind velocity studies in that region (Olson, 1958a, b) proved that the flow of air over the beach dunes and inland is as shown in Figure IX-40. The wind force and direction are represented in that figure by arrows proportional in length to wind velocity. Winds from the lake are deflected upward and made turbulent by the relatively young, low, foredune ridge or ridges, and by the older high dunes behind. Turbulent back eddies form behind both types of ridges, in the lee of trees and grass, or any other form of *surface roughness*. High and low velocity stream lines, normally

## Geology of Michigan

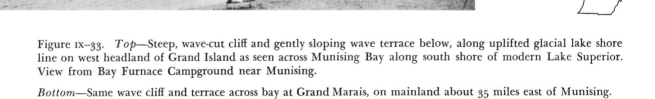

Figure IX-33. *Top*—Steep, wave-cut cliff and gently sloping wave terrace below, along uplifted glacial lake shore line on west headland of Grand Island as seen across Munising Bay along south shore of modern Lake Superior. View from Bay Furnace Campground near Munising.

*Bottom*—Same wave cliff and terrace across bay at Grand Marais, on mainland about 35 miles east of Munising.

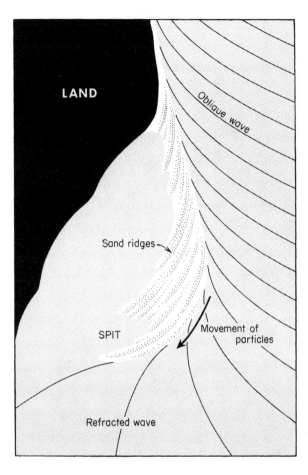

Figure IX-34. Progressive growth of a hooked spit by successive addition of sand ridges. Compare with Tawas Point illustrated in Figure IX-35.

more widely spaced, are compressed or crowded together where forced up along the windward face of the dunes. This brings higher velocity winds closer to the ground and increases their erosive effect where they pass over the dune tops. Back eddies of turbulent, low velocity air are especially characteristic behind the high dune *blowouts* where much sand deposition occurs. The sand is carried from blowouts on the windward face, up over the dune crest, and then is dropped in the lee, where wind velocities are low or even reversed. The high velocity winds flow off the dune tops into a region of the air well above ground and hence too high to cause erosion behind the dune. Irregularities in the "perfect" dune profile have a variety of causes including concentration of wind in gusts, local storm effects, the earlier shapes of the dunes themselves, growth of vegetation rang-

ing from low grasses to trees, and even the constructions of man.

It has long been known that friction between moving currents (wind or water) and the surface they pass over slows those currents down. The frictional drag increases progressively downward until the velocity reaches zero before the ground itself is reached. Thus, there is a thin zone of motionless air near the ground. That zone may range from a mere film up to several inches in thickness. Its thickness is controlled by the size of the irregularities of the ground surface, including those due to vegetative cover. If surface dust or sand grains are smaller in diameter than the thickness of the stationary air zone, the moving air currents above cannot strike them, and they lie below the reach of wind erosion. It is for this reason that dust particles are hard for wind to pick up; unless a local gust forces moving air down closer to the ground, the dust particles lie within the motionless air zone. Once up in moving air, however, dust will remain suspended indefinitely to be carried completely out of the area. Large diameter particles (pebbles, cobbles) may project up into the moving air zone but be too heavy for wind to move. Sand grains whose diameters are intermediate, are large enough to protrude through into the moving air above the motionless zone, yet are small enough to be pushed, rolled, bounced, or even carried by wind. Olson found that artificial plantings of marram grass (*Ammophila*) so greatly increased surface roughness on sand areas as to increase the thickness of the still air film by 30 times, thus retarding erosion and effectively trapping sand blown into its vicinity. Needless to say, the effect of higher growing vegetation on wind velocity would be even greater. Thus, vegetation (or any other obstacle) increases surface roughness, retards wind erosion, and initiates sand deposition. Olson (1958b) showed that in those areas where they grow or have been planted, the following are effective sand stabilizers: marram grass, wheat grass (*Agropyron*), dune willows (*Salix*), cottonwood or balsam poplar (*Populus*), wild rye (*Elymus*), sand cherry (*Prunus*), sand reed grass (*Calamovilfa*), little bluestem bunchgrass (*Andropogon*), grape (*Vitis*), red ozier dogwood (*Cornus*), choke cherry (*Prunus*), basswood (*Tilia*), and others. All are able to withstand and even thrive on sand burial. White pine and jack pine locally are effective sand stabilizers, too. Olson found that

**Water and Wind**

221

Figure IX-35. Tawas Point near Tawas City. A complex hooked spit with long history of growth. First examine present sites of growth where sandbar at 4 is enclosing a lake at D. In future, a bar may enclose a lake at E. In past, Lake A became enclosed by Sand Ridge 1, Lake B by Ridge 2, and Lake C by Ridge 3, in that order. Now examine Point 5, which is beginning to enclose Bay H. Then note the similarity to Points 9 and 8, Point 7 and Bay F, and Point 6 and Bay G, which formed in that order in the past. Point 8 and Ridge 2 clearly are parts of one former "hooked" sand ridge on which a lighthouse now stands. Hooks and bays at 6-G and 7-F are especially similar. Arrows show general direction of current flow and sand transport. Currents at 5 are deflected into Tawas Bay by force of waves and wind coming in off open lake, thus "hooking" the spit. (Michigan Conservation Department, photo by R. Harrington.)

Figure IX-36. Sturgeon Point on Lake Huron just north of Harrisville. Sand spit in foreground. Just inland from the shore-line road note succession of closely spaced tree lines paralleling the modern shore line. Trees reflect a series of former beach lines formed by recession of high waters of glacial lake stages down to modern water level. See Figure IX-30 for airphoto of similar beaches farther south. (Michigan Conservation Department, photo by R. Harrington.)

Figure IX-37. Groins along the Lake Michigan shore line near Ludington, Michigan, trap sand and retard beach erosion. (Photo courtesy H. N. Pollack, The University of Michigan.)

# Geology of Michigan

Figure IX–38. Herring Lakes area, Benzie County. City of Frankfort lies on opposite side of Betsie Lake, north of town of Elberta in extreme northwest corner. Geologic history of Herring Lakes shown in Figure IX–39. Dark, tree-covered areas north and south of the lakes are higher glacial moraines enclosing the Herring Lakes Embayment, which originally opened to Lake Michigan. A third depression east of Upper Herring Lake occupies part of the same original embayment but has been filled by vegetation and sediment carried by streams whose radial pattern can be seen feeding toward the center of that area. Betsie River embayment likewise is sediment filled in its upper reaches above bridge, forcing river to meander broadly. Light-colored "fingers" extending inland from Lake Michigan are sand blowouts in the high dunes related to Glacial Lake Nipissing and now largely stabilized by vegetation. Note sand transport along Lake Michigan shore. Outer breakwater at Betsie Lake deflects sand offshore. Parallel retaining walls accelerate current out of Betsie Lake thus flushing channel free of sand. (U.S. Department of Agriculture photo.)

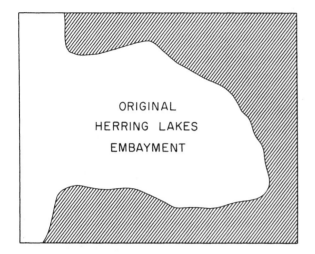
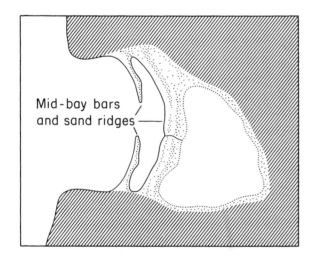
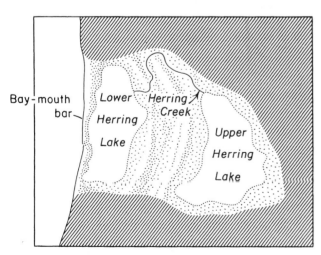

Figure IX–39. Diagrammatic sequence to show evolution of features in Herring Lakes area, Benzie County. Compare with Figure IX–38. (Modified from Scott and Dow, 1937.)

the different modes of growth of these various kinds of plants caused different types of sand deposits to form. Wheat grass, for example, is very low growing but sends off long and abundant vegetative extensions laterally, hence favors growth of low but broad dunes. Marram grass grows upward more strongly, often in clumps, but also sends lateral rhizomes off widely, hence traps sand to form broad but somewhat higher dunes. Such a grass may become nearly or completely buried in winter, yet with renewed growth in spring sends up its stems to the surface to form new grass clumps which trap another layer of sand (Figs. IX–41–44). Willows grow still higher and also send off propagative side shoots, hence form broad dunes which can grow to much greater heights before the trees are overwhelmed by the sand. Cottonwoods stabilize dunes which are very high because those trees grow upward rapidly, but such dunes are narrow and steep because the cottonwoods do not send off lateral growth. Destruction of any stabilizing vegetation exposes the dune sand to renewed wind erosion. Vegetative growth and related dune growth or destruction may reflect climatic changes, rise or fall of ground water table with rise or fall of lake levels, and a variety of vegetation-destroying accidents. A rise in lake level, or a period of strong storms, may initiate wave erosion at the foot of a dune. This in turn destroys vegetation, leads to slumping, and exposes sand to destructive erosion as in the case of blowouts. A drop in water level exposes a new sand supply on the broadened beach along the dune front, thus initiating a period of dune growth. If erosion in blowouts locally lowers the sand surface down close to the water table, the rates of plant colonization are increased by the more readily accessible water supply and the increased vegetation in turn stabilizes the sand more rapidly there, halting further erosion. Many of these changes are cyclical in nature, dune growth and subsequent destruction occurring over and over again. Olson (1958c, p. 473) summarizes the nature of these changes as follows (Fig. IX–45): "During periods of several successive years when lake levels are several feet below the mean level, sand is blown from the wide beach into a new foredune ridge. But much of it may be eroded by waves during ensuing years of unusually high mean lake level, aggravated by storms. When the level again falls and the beach widens, the begin-

# Geology of Michigan

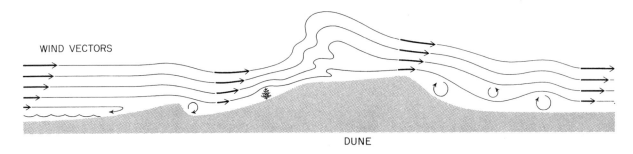

Figure IX–40. Flow of air over Michigan beaches and shoreline dunes. Note anticurrent eddies behind dunes. Wind vector arrows indicate relative wind velocity, longest arrows representing highest velocity. A dune tends to cause a "pile up" on its windward face forcing higher velocity air closer to ground and increasing erosive effect of wind. In lee of obstacles, such as dunes, fences, and trees and other vegetation, high velocity air is far above ground; coupled with the effect of reverse eddies, this reduces wind velocity at ground surface and sand deposition rather than erosion results. (Modified from Olsen, 1958, in *Journal of Geology*, permission of University of Chicago Press.)

ning of a new ridge may cut off the sand supply from the ridge behind it. This 'shore-dune cycle' may be repeated several times in a few decades. Protracted lowering of the lake may also help underwater sand bars emerge as barriers; these may be permanently colonized by vegetation if net long shore deposition of sand is sufficient to prevent their destruction during later episodes of high lake level. The connections between beach and dune forms and lake level and between lake level and precipitation and temperature account for many details of shore-line history around Lake Michigan for the past century and possibly for earlier episodes of the late post-glacial climatic history."

Olson (1958c) has shown that a number of factors govern lake level and shore-line locations. Normal tides account for changes of 1 to 3 inches in lake level. Wind friction may drive or drag water from one side of a lake to the other. In October 1929 a strong storm and associated winds caused a rise of 1.95 feet in slightly over a day at one end of the Lake Michigan Basin. High atmospheric pressure at one end of a lake may depress the water and cause it to rise at the other end where air pressure is lower. If the high pressure cold front passes quickly, the pressure is released and the water tends to move suddenly backward into hydrostatic equilibrium. Thus, it may overshoot its mark and then oscillate back and forth several times before reaching equilibrium level. Such oscillatory waves are called *seiches*. Seiches on Lake Michigan are reported to have caused local changes in water level as great as 5 feet, and as great as 9 feet on Lake Huron. The rise and fall of water level due to such oscillations may take several minutes or many minutes to occur. High evaporation during an exceptionally dry summer or high precipitation during an exceptionally rainy period may cause changes in water level, but the effects of these factors are not yet completely understood. Heavy rainfall in one year may not flow immediately into the lake basin in that year or even the next, but may be retained if the runoff in a particular part of the basin is retarded by exceptionally heavy vegetative growth or other factors. Lake levels tend to drop in the winter when much precipitation is frozen in the form of ice and snow on land and then rise in the spring with the inflow of meltwater. Summer droughts tend to lead to lowered lake levels. Plants transpire large quantities of water into the atmosphere. Many people have argued that diversion of water through the Chicago Sanitary Canal into the Mississippi River causes excessive lowering of Great Lakes levels. Records of the U.S. Lake Survey, however, indicate that such lowering amounts to only about 3 inches compared with the natural variations of 5 to 6 feet. Thus, it is clear that lake levels are controlled by a wide variety of complexly interrelated factors not all of which are well understood. Olson has clearly shown, however, that whatever their cause, lake level changes do lead to modifications in shore-line features.

Figure IX–45, taken from Olson (1958c, Fig. 2) shows that the rotational movement of water waves offshore tends to produce a series of lov

Figure IX-41. Blowouts in Rosy Mound Dune, an old, stabilized, Glacial Lake Nipissing shoreline feature. View east from modern foredune ridge along shore of Lake Michigan about 3 miles south of Grand Haven. High Nipissing dune crest is 230-40 feet above modern lake level. Stabilizing vegetation is gaining foothold in foreground. (Michigan Geological Survey—Berquist File, photo by L. C. Hulbert, 1939.)

Figure IX-42. Same location as in preceding figure. View north along modern foredune ridge just inland from wave-washed strand line. Older and higher, forest-stabilized, Glacial Lake Nipissing dunes in background. Sparse stabilizing vegetation is seen on the actively growing foredune. (Michigan Geological Survey—Berquist File, photo by L. C. Hulbert, 1939.)

Figure IX-43. Ludington State Park along Lake Michigan. View south. Low, straight-crested, foredune ridge in middle distance is related to level of modern lake. High, forested, stabilized dunes, related to level of Glacial Lake Nipissing, seen in distance.

## Geology of Michigan

Figure IX-44. Sand stabilization along shore lines of Great Lakes.

*Top*—Heavy inland forest growth arrests sand movement at Great Sand Bay on Lake Superior shore of Keweenaw Peninsula, 5 miles northeast of Eagle Harbor.

*Center*—Grasses and low willows, with extensive root systems, arrest sand at Warren Dunes State Park along southeastern Lake Michigan shore. Pocket knife at arrow tip for scale.

*Bottom*—Snow fence and grasses trap sand at Warren Dunes State Park. Large blowout in distance is in high dune related to Glacial Lake Nipissing. (See also Figs. IX-19, 24.)

ridges called "balls" and intervening depressions called "lows." There may be 3 ridges, and they often can be traced for many miles alongshore. Olson believes that if lake levels remain stable the positions of these "lows" and "balls" tend to remain stable but that they may shift if lake levels drop. Specifically, if water levels fall, the innermost ball may be subject to wave and wind action which causes it to migrate inland, eventually onshore, and finally up against preceding onshore dunes to form a new foredune ridge. A rise in water level with associated wave erosion might then cause the destruction of part or all of such a dune. Figure IX-45, from Olson (1958c, Fig. 3), shows his historical inferences concerning the effects of lake level changes as they can be documented from about 1876 through 1949 and as he has inferred them for earlier periods in the twelfth and fourteenth centuries. If his conclusions are correct, then it is clear that over the course of many centuries there have been repeated wet and dry periods and associated high and low lake levels. We have already seen that the high dunes are related to far older geologic times when glacial lakes Algonquin and Nipissing formed after the retreat of glacial ice.

In summary then, we have shown in this chapter that water and wind are involved in a variety of geologic processes which tend to modify the surface of the land, the ground beneath us, our water supply, and the shore lines and other features of our many lakes.

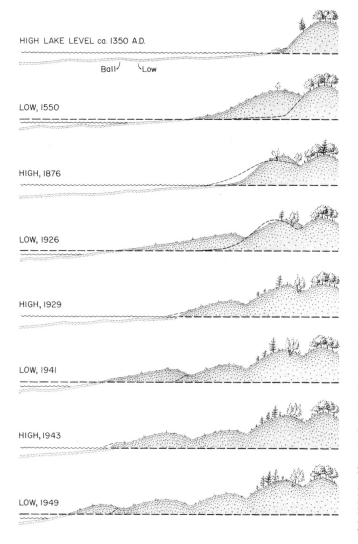

Figure IX-45. Sequence of dune development and sand accretion postulated to explain southeastern Lake Michigan shore-line features. Offshore "balls" and "lows" (also see Fig. IX-11, right) are important in process. During high-water stages, waves attack base of shoreward dune; windward face is modified by slumping and increased wind action; vegetation is destroyed and blowouts form; sand is moved from dune front to top. During low-water stages, broadened beach provides sand to new foredune ridge which is plastered against face of earlier dune; the innermost "ball" may be exposed to wind at this time, adding to supply of sand. Next high-water stage initiates destruction of foredune ridge and cycle repeats itself. Light, dashed line on high-water stage diagrams indicates sand removed from face of foredune (below line) and piled on top (above line). Vegetation encroaches progressively, stabilizing earlier dunes. (Modified from Olsen, 1958, in *Journal of Geology*, permission of University of Chicago Press.)

# X

# Petroleum and Natural Gas in Michigan

## Introduction

Oil and gas are not confined by state boundaries. Their geographic distribution and the geologic situations in which they occur involve conditions which are not unique to Michigan. If the larger, world-wide picture is presented first, then the occurrence of oil and gas in the state can be described more meaningfully.

Petroleum geology, in all its details, has occupied the lives of thousands of geologists for more than 50 years; it is an interesting and exacting scientific profession which has many intriguing problems left to solve. We can only introduce the subject, skipping the involved techniques and dwelling mainly on the question, "How do the geology and geologic history of an area, such as Michigan, control the distribution of oil and gas?" Additional details can be found in any good, modern textbook on economic or petroleum geology, such as that by Landes (1959).

## Origin and Source

Two initial problems are how oil and gas originated and what the conditions were in their place of origin. These questions have not been fully answered to the satisfaction of all specialists, but some highly probable suggestions can be considered (Hedberg, 1964).

First, it is known that oil and gas, although they vary in composition, are chemical mixtures. Oil is chemically more complex than gas. Asphalt is a common associate. The mixtures, in turn, consist of a variable number of chemical compounds called *hydrocarbons*. These consist primarily of the elements carbon (C) and hydrogen (H), put together in various ways. Some of the simplest, natural hydrocarbons, such as methane ("marsh gas," $CH_4$) have been reported from the vapors of volcanoes and thus in some cases might be of inorganic origin, but geochemists believe that the more complex hydrocarbons that occur in nature came from the dead bodies of plants and animals, just as methane and other hydrocarbons are released from dead organic matter today. Significantly, oil and gas nearly always occur in or near sedimentary rocks that originally were deposited in water. Rarely are they found in igneous or metamorphic rocks and then mainly when those rocks are close to a potential sedimentary source rock. In some cases the sediments were fresh-water deposits; more often they were of marine origin. Modern sediments, of types which could form sedimentary rocks, are accumulating today and many such deposits are known to be rich in dead organic matter of both plant and animal origin. This is particularly true of the clay muds and lime oozes which, upon turning to sedimentary rock, would become shale and limestone. A small fraction of hydrocarbons may have been carried into water from the land. There probably never were enough large animals and plants living in lakes and seas of the past to have produced even a small part of all the known petroleum and natural gas in the world, but even today there are tremendous numbers of tiny microorganisms, one-celled plants and animals and some higher forms of life, living in the seas and lakes. Some are surface floaters (plankton) that fall to the bottom upon death and are buried in sediment; others live on or in the bottom sediments. Although some oil companies use symbols such as shells or dinosaurs to advertise their products, the remains of such large animals probably were only very minor contributors to the world's petroleum reserves.

Environmental conditions which in the past would have favored an abundance of microorganisms probably were similar to those today, that is, warm, relatively shallow, sunlit waters rich in dissolved mineral and organic matter. These same conditions often prevailed in the epeiric seas that covered large portions of the interiors of continents in the geologic past (Chapter V).

Natural hydrocarbons are susceptible to destruction when in contact with air or with water that has a high oxygen content; that is to say, they are readily oxidized. Excessive bacterial action also can destroy them. Hence it is believed that the natural hydrocarbons accumulated under conditions that prevented this destruction. Theoretically, and also from actual observation of modern sediments, it seems that favorable conditions would have occurred either in bottom sediments in bodies of water where wave action or currents did not stir in a constant supply of new oxygen, or in areas where fine-grained sediment accumulated rapidly, thus burying the organic material and isolating it from oxygen additions before it could completely decompose. These conditions are found in isolated or semi-isolated basins of sedimentary deposition such as were common in the ancient epeiric seas, as well as along the margins of continents. The Michigan Basin during the Paleozoic Era was one such area (Chapter V). As sediments accumulate, bacterial action eventually ceases at depth in the sediment because oxygen becomes depleted and because the toxic waste products given off by the bacteria themselves inhibit further bacterial growth. Some initial bacterial action may be necessary, however, to break down the original organic matter and to release undesirable elements such as nitrogen and sulfur.

Oil and gas are chemically and physically different from the types of hydrocarbons that may have been their source. Thus, many complex changes must have occurred to transform the raw organic oils, gases, and solids into liquid petroleum or natural gas. Specialists still are in doubt about the causes and manner of change, but contributing factors may have been reactions of solutions in water, heat (not over 185°C), pressure, and reactions with other chemicals in the rocks, all going on over thousands and even millions of years. Natural gas is a common associate of oil and it is reasonable to believe that it is a product derived from organic materials similar to those that produced petroleum. Gas may be produced both during the evolution of petroleum and as an end product of that process. Although oil and gas may occur separately, this is probably because gas, being more mobile, can escape and migrate through the rocks that are impermeable to the passage of oil.

## Migration Through the Rocks

Mention of *migration* raises the next problem in the origin of oil and gas reserves. The likely sources of petroleum and natural gas, at least as they are represented in the organic-rich sediments of today, could not have produced enough oil and gas in their place of origin to have given rise to the enormous quantities of those materials found in commercial fields. Moreover, oil and gas usually are produced from "reservoir" rocks which are not of the type in which organic material would have been abundant originally. Hence, there must have been migration of oil and gas from their places of actual origin (the *source* rock) to certain concentration points in rocks that serve as the reservoirs.

It has been established that oil and gas *can* migrate through rocks because they are known to move toward wells that have been drilled in producing oil fields. Another proof of the ability of oil and gas to migrate comes from the presence of oil and gas in "seeps" at the earth's surface in many places in the world. Oil in a reservoir, deep underground, is at high temperature and contains much gas in solution. These two factors increase the fluidity and mobility of the oil, bringing them close to that of water, and we have seen in Chapter IX that ground water moves readily through porous and permeable rocks. Water is also present with the oil underground and forms a coating film on the walls of pore spaces in the reservoir rock. Thus, the oil itself does not come into direct contact with the rock. This reduces adhesion of the oil to the rock and tends to increase the mobility of the oil through the reservoir. The actual causes of movement may be several, probably working in combination. Compaction of sediment after it is deposited may squeeze some of the natural hydrocarbons out from the source rocks. The weight of overlying rock material would cause the underlying material to compact; so too would the forces causing compression, buckling, and folding of the earth's crust. Almost all rocks of the earth's crust are saturated with water to great depths below the surface. Oil and gas both are lighter than the heavily mineralized and salt-saturated underground waters with which they occur. The oil and gas obey the law of gravity and float upward through the water, thus migrating until they are stopped by some impermeable rock material above. Some less certain causes of migration include "water flushing" (the movement of water through rocks, carrying the oil and gas with it), and capillary attraction (the force that tends to pull fluids into and along narrow tubes and passageways).

Although oil and gas most probably originate in a source rock, as discussed above, it is usually very difficult or impossible to prove, in the case of a particular oil field, just what rock formation actually was the source. This is because oil and gas rarely are produced from the source rock itself, but usually come from what is called a *reservoir rock*, whose characteristics are not those one would expect to find in a good source rock.

## Reservoir Rocks

It is a common misconception that oil and gas occur in open "pools" underground. This is rarely true, the holding capacity of the reservoir normally being the result of the presence of myriads of small interconnecting pores and cavities. If a cavern dissolved out of carbonate rocks later comes to contain oil and gas, this is an exception rather than the rule.

On a worldwide scale the commonest and best reservoir rocks are sandstones. Carbonate rocks, such as limestone and dolomite, rank second as reservoirs, although in Michigan the oil reservoirs are dominantly carbonates. In any case the reservoir rock must be porous.

Openings and spaces, no matter what size, make a rock porous. Most sedimentary rocks, at least in the upper part of the earth's crust, have some porosity, but the percentage varies from a maximum of nearly 50 percent of the volume occupied by the rock to practically zero. Porosity may be *primary*, that is produced at the time of deposition of the sediments. For example, in sedimentary rocks composed of fragments or grains, such as conglomerate, sandstone, siltstone, or shale, the porosity consists of the original or primary spaces between the grains. Tight packing, poor size sorting with a wide variation in particle sizes (as for example a mixture of sand, silt, and clay), or the filling of pore spaces with some mineral cement, all may result in lower porosity.

Even carbonate rocks may have some primary porosity. The original structures of fossils composing limestone often were porous. Organic reefs, especially those formed by the limy skeletons of corals, commonly have primary porosity. If pieces of the reef tops, torn loose by wave action in the seas of their time, accumulated along the flanks of the reef, these fragmental, flank deposits have primary porosity too. However, most porosity in carbonate rocks is *secondary*. It may be the result of (1) fracturing due to crustal movements, (2) subsequent solution by ground waters of material along the fractures, or (3) a process called dolomitization to be discussed later. The majority of the oil in Michigan occurs in secondarily porous carbonate rock reservoirs. The fact that secondary porosity formed *after* the original rock had solidified (lithified) proves that in most cases the hydrocarbons found in a carbonate reservoir migrated

# Petroleum and Natural Gas

into the reservoir rock later, rather than having been deposited there originally.

Porosity alone does not make a good reservoir. A reservoir rock must also be *permeable* to allow the free passage of fluids and gases. In order that the oil and gas may migrate into and through the reservoir, the pore spaces must interconnect. If interconnecting passages are too small, the forces of adhesion can prevent the free flow of fluids or even gases. This often is the case in very fine-grained and densely compacted rocks such as shale or anhydrite. Formations are said to be *tight* if they are impermeable.

## Seals and Traps

We have seen that even in the best source rocks oil and gas apparently are not generated in sufficient quantities to provide economically valuable amounts "in place." Even if present there, they are not producible because the source rocks normally are too tight for the hydrocarbons to be gotten out. They must have migrated in response to causes discussed earlier, and become concentrated in local pockets, "fields" or "pools." And there must be something to stop the movement so that the oil will collect. This brings us to the terms "seal" or "seal rock" and "trap." *Seal rocks* generally are nonporous or at least relatively impermeable rock formations. Shale and anhydrite are common seal rocks, but dense limestones and even thoroughly cemented sandstones may also serve as seals. The best seal rocks, especially in Michigan, are anhydrites and anhydritic shales.

A full discussion of *traps* could fill a book by itself. A trap is a geometric condition, involving the shape of rock formations or their composition or both, which collects and holds oil or gas in quantities sufficient to make their recovery economically worthwhile. Traps are of many types, as will be shown, and not all known types are important in Michigan, but understanding their general nature will allow a better appreciation of the kinds of traps which are important here. After all, a layman, hoping for a bonanza, should be aware of all the possibilities as well as the special conditions in the area of his particular interest. It is important to know what *not* to expect in the way of oil and gas traps in the state. There always are exceptions, so we cannot positively rule out any kind of trap in the state, but it is easier to assess the probability of success of a proposal if you

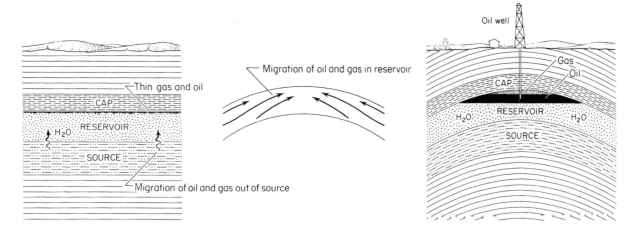

Figure x–1. Essentials of oil and gas accumulation.

*Left*—Migration from source rock into porous and permeable reservoir rock is halted beneath impermeable seal or cap.

*Center*—Inclination of reservoir boundaries favors movement toward higher regions of a structure, with consequent localization of oil and gas.

*Right*—Simple, anticlinal structure, showing separation of gas, oil, and water at crest of structure in porous reservoir, beneath seal.

know that unusually strong evidence may be needed to support it.

The classical type of trap, the first from which oil was produced in commercial quantities, and often the easiest for the geologist to find, is the *anticline* or *dome*. Anticlines and domes usually involve most of the factors that are important in the accumulation of oil or gas in the other types of traps so a detailed description of anticlinal trapping follows. Picture a series of sedimentary rock strata in their original horizontal position of deposition and including the following essentials: a source rock, reservoir rock, and seal rock, all as shown in Figure x–1a. Also imagine, as is the case in most sedimentary rocks, that the formations are saturated with water—the pore spaces are filled. Oil and gas, produced in the source rock and forced to migrate out by causes discussed earlier, find their way into the porous and permeable reservoir rock. There they rise to the top of the water and accumulate in the upper portion of the reservoir. They are held in the reservoir and prevented from rising farther by the impermeable seal rock above the reservoir (Fig. x–1b). But still they may not occur there in an economically worthwhile quantity. Suppose, however, that the strata including the reservoir and seal rocks are folded upward into a dome or anticline (Fig. x–1c). The oil and gas migrate upward along the inclined reservoir, beneath the seal rock, to the highest position possible at the crest of the anticline. There they are trapped and localized in greater concentration than would have been possible in a horizontal reservoir. A well might tap this trap and produce large quantities of oil or gas.

Gas and oil may occur singly or together in a trap. If both are present, the gas may be partially dissolved in the oil, thus building up pressure. The gas may also rise above the oil to form a "gas cap" which then exerts a downward pressure on the oil below. Water would lie below the oil and also down the flanks of the reservoir. If a well just penetrated the gas cap and stopped, then only gas would be produced. If the well missed both the oil and gas by passing down beyond them, or by being located too far out toward the edge of the trap, then only water might be produced. The well should penetrate the oil-rich zone within the trap. If a gas cap under pressure is present above the oil, expansion of that gas will force the oil through the reservoir toward the well and possibly out the top as a "gusher." The expansive force of gas dissolved in the oil would have a similar effect. As gas and oil are withdrawn from the reservoir, thus reducing the pressure therein, flanking water moves in behind, occupying the vacated pore spaces.

An essential feature of all traps is what is called *closure*. An anticline without closure might be likened to a fold in a rug where both ends of the fold are open. Oil or gas, migrating up the flanks of such a fold and reaching the crest could then move along the crest, possibly escaping entirely at the ends. However, if the ends of the fold were depressed or slanted, too, then the trap would be closed. The amount of closure is determined by the distance from the crest or high point on a trap down to the level of the highest point across which the oil or gas could escape. Picture an overturned bowl, held under water in a horizontal position and with air trapped beneath it. The bowl forms the seal for this structure. The closure beneath the bowl is the distance from the top down to the rim. However, if the bowl is tilted, some of the air will escape; lifting one edge of the rim has decreased the closure. Many geologic structures which otherwise might serve as oil or gas traps, fail to do so because they lack closure.

Now that the essential characters of a trap have been discussed, let us consider various types of traps. These may be divided into *structural* and *stratigraphic* groups, although they may occur in combination. The general mechanism of oil migration into the trap is the same for either group. Likewise, there must be reservoir, seal, and source rocks. Traps differ, however, in their mode of origin. The main types of traps within each group are illustrated in Figure x–2. A petroleum geologist must assemble data on past geologic events in order to make an intelligent appraisal or prediction of the nature, location, time of origin, size, and other aspects of earth structures which would favor oil and gas origin, migration, and trapping.

Folded or anticlinal traps may result from horizontal compression of the earth's crust (Fig. x–2a), or from differential compaction of sedimentary layers over a less compatible base such as a buried hill (Fig. x–2b). Some structural traps (Fig. x–2c) result from faulting (breaking with slippage) of the crust; tilted strata are sealed at one end where the reservoir formation is offset against one impermeable material and capped by another

**Petroleum and Natural Gas**

233

STRUCTURAL TRAPS

a. COMPRESSION FOLD   b. COMPACTION FOLD   c. FAULT   d. FISSURES   e. MULTIPLE LEVEL

STRATIGRAPHIC TRAPS

f. WEDGE-OUT   g. FACIES CHANGE   h. CHANNELS   i. REEF   j. UNCONFORMITY

Figure x–2. Major types of traps. In many fields these occur in some combination. Gas and oil, shown in black, occur in porous and permeable reservoir rocks below impermeable seal or cap rocks.

impermeable formation. This produces closure. The reservoir may also be sealed at its upper end by ground up material (fault gouge) in the zone of slippage along the fault. It is even possible that evaporation of volatile material may leave behind an asphaltic residue which then could seal off the upper end of the tilted reservoir. Cracks or fissures caused by movements of the earth's crust may produce porosity and thus form a reservoir in a formation which is included in a structure favorable for trapping oil and gas (Fig. x–2d). If reservoir and seal rocks occur at several levels within a favorable structure, then oil and gas may be trapped at several levels in one field (Fig. x–2e).

Stratigraphic traps result where there are changes in the porosity and permeability of rock formations. Picture, as in Figure x–2f, a series of sedimentary deposits laid down successively along the shore of an advancing or "transgressing" sea (Chapter XII). At any given time the layer of sediment that was being deposited might grade from porous and permeable shoreline sands, bars, or dunes which would make potential reservoirs, seaward into fine grained, organic-rich silts, clays, and lime oozes that could serve as source rocks. If the transgression of the sea resulted in a succession of such graded layers, laid down one on top of another, then in some places the near-shore, porous sands of a previous period, would be overlain by the offshore, impermeable shales or clays of a subsequent layer. Thus, there would be a reservoir (near-shore sand), sealed by impermeable material and in close association with hydrocarbon rich source rocks. All that would then be needed to close the trap would be tilting, which would cause the oil and gas to migrate upward along the tilted reservoir to the "pinch-out." Another possibility is that one of the potential reservoir layers might become impermeable "up dip" halting the oil and gas at that point (Fig. x–2g). Sandbars, shallow water sands, sand-filled tidal channels, stream channels, delta distributary channels, local dunes, all are common depositional features composed of sediments which might become porous reservoirs (Fig. x–2h). Another type of stratigraphic trap involves locally increased porosity and permeability as, for example, along the flanks of a buried organic reef (Fig. x–2i). Folded or tilted sedimentary layers, if uplifted, beveled by erosion, and then covered by an impermeable sealing sedimentary deposit, may provide a trap beneath the "unconformity" (Fig. x–2j).

Oil and gas traps frequently are combination or composite types which involve two or more of the simpler kinds described above. For example, movement of the earth's crust may tilt or fold any of the types of stratigraphic traps so as to produce local structures even more favorable for the accumulation of oil and gas. The difficulties of oil exploration are increased greatly when a combination of conditions and historical events must be explored. In actual practice, each trap has its own peculiarities and special features.

## Petroleum Exploration

Geologic exploration for oil and gas requires application of the principles of historical geology in a search for the right combinations of source rocks, seals, reservoirs, causes of migration, trap structure, and timing, all of which must be favorable.

Many exploratory tools and techniques are used by the petroleum geologist. We shall mention these only briefly, with emphasis on those most useful in the state of Michigan.

Sometimes there are surface indications that oil or gas occur in the subsurface. Oil and gas "seeps" and asphaltic deposits fall in this category. They are not invariably reliable because they may represent all that is left after most of the oil and gas have escaped; or the actual accumulations at depth may lie a great distance off, the seep material having migrated long distances up inclined reservoirs. Analysis of surface soils for organic hydrocarbons may reveal the presence of oil and gas in the subsurface. Mapping the distribution and shape of rock formations in surface outcrop, an exercise in 2-dimensional geometry, provides some information regarding the 3-dimensional subsurface shape of rock formations. This, in turn, could lead to the discovery of potential traps. Geologic surface mapping is one of the most basic techniques in petroleum exploration, but there must be sufficient surface rock exposure to provide a reliable picture. Unfortunately, the state of Michigan is mostly covered with glacial deposits which mask the bedrock, thus hindering surface mapping. Occasionally, a subsurface structure, potentially a trap, may have direct surface expression as an actual geographic feature. Examples would be hills uplifted above anticlines or salt domes. This is very rare, however. In most cases the hills and valleys at the surface are misleading as to the subsurface structure. A professional geologist is trained to recognize the bearing that surface topographic features have on subsurface structure, but the relationship is usually indirect. In the oil and gas fields of the Lower Peninsula of Michigan there is no relationship between the subsurface bedrock structure and the shapes of hills and valleys in the overlying glacial deposits at the surface. Drilling on top of a glacial hill in this state merely necessitates deeper drilling to a given zone in the bedrock below.

Geologic study of data from the subsurface, often from great depths, is sometimes called "subsurface geology." There are many ways of getting subsurface data. Direct sampling by drilling is one. Samples may be taken as part of an exploratory drilling program, or the study may involve chips, rock fragments or "cores" brought up from the depths during a drilling operation that was under way for other reasons. Wells drilled for brine or water might produce bedrock samples of significance to a petroleum geologist. Samples taken from wells drilled in known oil fields increase the fund of information already available concerning subsurface conditions. Wells drilled in extension outward from a producing field may turn out to be "dry holes" but provide valuable subsurface samples. The Subsurface Laboratory of the Department of Geology and Mineralogy at The University of Michigan in Ann Arbor is one repository for subsurface samples produced from some of the wells drilled in the state. Records from some wells drilled in surrounding states or in neighboring Canadian provinces also find their way into that collection. The samples are systematically stored so that they are available for study by geologists with a wide variety of interests (Chapter V, Fig. v-4). The Michigan Geological Survey in Lansing also keeps many samples from wells. According to Cohee (1965), over 26,000 oil and gas wells have been drilled in Michigan. Many others have been drilled in the search for coal and water. About 7500 sets of drill cuttings have been available as well as drillers' logs for many others.

Various geophysical methods also produce information about subsurface conditions. That information is indirect, reflecting some of the properties or conditions of the bedrock at depth, but not including an actual rock sample. Among the geophysical methods are magnetic, gravity, and seismograph surveys. These tell something about the subsurface shape and nature of bedrock without the necessity of drilling. When a well is drilled, instruments may be lowered down the hole to examine various properties of the rocks penetrated. These include electrical conductivity and resistivity, radioactive levels and types of radiation, and the resistance of the walls of the well to crumbling or caving (by caliper measurements). A "driller's log" may record the rate of penetration of the drill which often is a reflection of the differential hardness of the rock formations and this in turn of their composition and degree of cementation. Detailed descriptions of subsurface techniques are presented in "Petroleum Geology" by K. K. Landes (1959).

Subsurface and surface data may be analyzed in many ways, but maps and cross-section diagrams are among the most common ways of integrating the information. Paleogeographic and paleogeologic maps portray the geologist's interpretation of ancient geography and geology. Facies maps show the location, shape, and type of ancient sedimentary deposits. Isopach maps reveal thickness variations, and structure contour maps indicate the structural shapes of the surfaces of sedimentary formations. All are of great use in the search for petroleum and gas. Cross-sectional diagrams also show the subsurface shape and distribution of sedimentary rock layers. All of these maps and diagrams are based on great quantities of data. Much of this can now be analyzed and plotted by computers, but recognition of the geologic significance of the maps and diagrams in the search for oil and gas still requires the skill of a professional geologist.

Now a word of caution! A great deal is known regarding oil and gas origin, migration and accumulation. The prerequisites of timing, source, reservoir, seal and trap have been fairly well understood for many years, and the techniques of oil exploration constantly become more highly refined. *But* there is as yet *no certain means* to prove the presence of oil and gas in the subsurface without drilling. It is not yet possible to be sure in any case, even within the limits of a known field and certainly not beyond those limits, that *all* of the requirements for the accumulation of oil or gas have been exactly right throughout the long span of geologic time. Even the best and most modern geologic survey of the oil and gas possibilities of an area offers no guarantee of success. You may ask then, "Why bother with geology?" Statistics on discoveries for a typical year give a simple answer (Landes, 1951). During the year 1949 many "wildcat wells" were drilled for oil in the United States. (A wildcat well is one outside the known limits of a producing field.) Of these wells 80 percent were drilled with at least some professional advice, yet only 13 percent were successful discovery wells. This may look bad for the geological profession. But, of the 20 percent of remaining wildcat wells, those drilled without professional advice, only 3 percent were successful. It should be pointed out, too, that many of the latter wells were drilled by persons who selected areas already known to have produced some oil or gas, so they really were dependent on advice of an indirect sort. The search for oil is risky, the odds are strongly against success, and the whole business is therefore very costly. As can be seen from the figures given, geologic advice can improve the odds about four-fold, but still cannot guarantee success. Many wells must be drilled, at great cost in each case, for the few producers which then must support the whole cost and still yield some profit. The odds against the layman without professional advice are very great. True, an occasional individual or small group of investors may "wildcat" with spectacular success. These are the ones you hear of. The many failures (the other 97 percent) go unheralded. Improving the odds in their favor from 3 percent to 13 percent makes good economic sense to an oil company. This is why oil companies who actually find their own oil, rather than buying it from other producers, all maintain some kind of a geologic staff. Those of you who might consider investing in a wildcat venture should also remember that the quality of geologic professional advice varies. In most cases, just as with a legal or medical opinion, you get about what you pay for.

## Production of Oil and Gas

Once a geologic survey has shown that an area has good possibilities, the actual drilling of wells and recovery of what oil and gas may be found become the responsibility of the petroleum engineer, although a geologist may continue to "sit on the well," as it is being drilled, to gather information and make geologic recommendations. The final proof of the presence of oil or gas is the producing well.

The history of production and use of oil goes back hundreds of years. Surface seeps, or accidental discoveries in wells drilled for other reasons, accounted for most of that early production. The first discoveries in the United States, in the middle 1800's, were accidental in the course of drilling for brine. The first producing well intentionally drilled for oil in the United States was the famous Drake Well (1859) at Titusville, Pennsylvania. The real impetus for oil exploration was felt in the early 1920's with the rapid increase in the use of the gasoline engine, which created a need for machinery lubricants and fuel in large quantities.

Drilling methods basically are of two types, cable and rotary. Cable tool drilling, once most common, involves repeated raising and dropping of a rock bit on the end of a rope or cable. The shattered rock periodically is removed from the bottom of the hole by a bailer. Costly steel casing must be inserted as the drilling proceeds, to prevent collapse of the hole. Cable tool or "impact" drilling is relatively slow and is limited to shallow depths. Rotary drilling now is the more common type, especially for deep wells. In this method, a cylindrical drill bit of hardened steel, or one studded with diamond fragments, is turned at the end of a "string" of pipe. The diameter of the bit is slightly greater than that of the pipe. Added sections of pipe are screwed on at the top as the well deepens. A dense fluid, called "mud," is pumped down through the pipe to the bottom of the hole. The mud flows back to the surface along the outside of the pipe in the space between the drill stem and the rock walls of the hole. The flow of mud carries the drill cuttings up to the surface. This movement is aided by the fact that the cuttings are lighter than the mud and float in it. The circulation of mud removes the need for periodic bailing. The hydrostatic pressure of the mud against the walls of the hole helps to prevent caving of the rock walls or influx of subsurface water, and to some extent removes the need for casing. The drill cuttings that are carried to the surface are caught, washed to remove the drilling mud, placed in sacks labeled as to the level from which they were cut (as determined by the known depth of the drill bit), and saved for geologic study. Samples of this sort, mounted on cards and stored, provide a source of data of inestimable value for further oil and gas exploration. If a solid rock sample or "core" is desired from a certain level, then the drill stem is pulled up out of the hole, unscrewed section by section, and stacked inside the derrick. The regular drill bit is removed and replaced by a coring bit which has a hole in the center like a doughnut cutter. A core barrel is screwed onto the coring bit and the whole assembly, including the reassembled drill stem, is lowered back down the hole. The core drill cuts a cylindrical core of the bedrock which is forced up into the core barrel. The core can then be removed at the surface by "pulling out of the hole" again. Coring is an expensive procedure, but may be justified when the need for information is great; for example, when a drill begins to approach an anticipated oil or gas producing level, when the porosity and permeability of a reservoir rock must be tested, or when the physical characteristics of the bedrock must be known in order to assist in further exploration, the geologist "sitting on the hole" may make the decision to core.

When oil is encountered in a well it may flow unaided to the surface because of the pressure of gas in or above it. But sooner or later most wells must be pumped. The old practice of letting gas escape freely from an oil well was wasteful because it reduced the subterranean gas pressure, thus requiring earlier pumping. Moreover, dissolved gas makes oil more fluid and hence more likely to flow freely out of the porous reservoir rocks into the well. This in turn increases the ultimate yield of the well or field. In modern practice, oil wells may be "repressurized" by pumping gas or air back down into the reservoir through neighboring wells, some of which may be drilled for that very purpose. Modern practice also involves "prorating" the production or withdrawal rate of oil from producing wells. The reasons for this may be partly economic, to avoid depressing the market or to allow each well owner his fair share of the oil from a particular reservoir. However, a more important geologic reason for prorating is that slow withdrawal is a good conservation practice. It tends to increase the ultimate recovery of oil from underground because too rapid withdrawal leaves oil behind, isolated in the pores of the rock reservoir. If water moves into the reservoir too rapidly behind the withdrawing oil and gas, it tends to isolate pockets of the oil which may not be recovered.

If the porosity or permeability of the reservoir rock are inadequate for profitable recovery of oil

## Petroleum and Natural Gas

and gas, there are various means by which these properties may be artificially increased. Explosives set off at the bottom of the hole may fracture the reservoir rock and open it up. Large quantities of acid may be introduced through the well in order to dissolve additional and larger openings in calcareous reservoir rocks such as limestone, dolomite, or lime-cemented sandstones. Acidizing also increases the area of exposed reservoir rock surface in the vicinity of the well so that the oil can "bleed" more rapidly out of the reservoir rock pores into the well. "Hydrafracting" is a process whereby porosity and permeability of the reservoir are increased by pumping a soluble gel into the reservoir under tremendous pressure. The gel can then be dissolved and pumped out again. Water may be pumped down certain wells in or around a field in order to "flush" the oil toward other producing wells.

An interesting and relatively recent development in the case of depleted and partially depleted natural gas reservoirs in Michigan is their use as temporary storage for natural gas piped into the state from other areas. Imported gas is pumped back down underground into a porous and permeable reservoir rock from which natural gas had been removed earlier. Advantage is taken of the fact that the proper natural combination of reservoir rock, seal rock, and trap structure already is there. This practice allows gas to be acquired during summer periods of low use in order that it will be available in adequate quantity during winter months of high use, a time when the source of supply might otherwise be inadequate.

### Oil and Gas in Michigan

Now that we have surveyed the subject of petroleum geology, let us apply this knowledge to the special conditions that prevail here in this state. In 1962 Michigan accounted for about 0.5 percent of the total annual petroleum and natural gas production of the United States (Independent Petroleum Association of America). Obviously, this state is not a major producer. Nevertheless, the state's share turned out to be a sizeable amount in actual dollars and cents. In that same year the value of oil and gas produced in the state was nearly 57 million dollars, contributing 14 percent of the total value of mineral products from the state, and ranking third in the state after iron and cement in importance. In 1964 the state produced nearly 52 million dollars worth of oil and gas, again third place in the state's total mineral production.

Sorenson and Carlson (1959) have shown that in 1958, of the 152 exploratory (wildcat) wells drilled in Michigan only 11 produced oil, only seven produced gas, and 134 were dry holes! Only about 12 percent were productive, and you will recall this is very close to the national average for wells drilled with professional advice. The Michigan figures do not, however, indicate what proportion of the holes was drilled with such advice. Even wells drilled in producing Michigan fields have a high percentage of failure (118 dry out of 216 completed in 1967). In 1967 only 1 out of every 20 exploratory wells was productive, about 5.0 percent. Thus, we see again that the risk inherent in oil and gas exploration is very great, especially in wildcat operations.

### History of Michigan Oil and Gas Production

Oil was first discovered in Michigan at Port Huron in 1886, but extensive production did not begin until after the discovery of oil in the Saginaw area in 1925. Oil was found in the important Mount Pleasant field in 1928. Many additional fields have been located and developed since then. The locations and names of some of these are shown on Figure x–3. The most recent developments have been those of the Albion-Scipio "Trend" in Hillsdale, Jackson, and Calhoun counties and the prolific reef accumulation in St. Clair County.

### Geology of Michigan Oil and Gas

The nature of the sedimentary bedrocks of the state, and their history, are discussed in detail in Chapter V. The names and ages of the important rock formations are shown on Figure v–2. In brief review, most of the surface of the state is covered by unconsolidated glacial deposits of the relatively recent Pleistocene or Ice Age. The thickness of the glacial deposits varies from zero in a few areas to several hundreds of feet in most places in the Lower Peninsula. The surface glacial deposits are not significant oil or gas reservoirs in this state.

# Geology of Michigan

Beneath the glacial deposits lies a thick series of Paleozoic age sedimentary rock formations of many types. The thickness of the Paleozoic sedimentary rock series varies. Some of the individual formations thicken toward the center of the Michigan Basin; others thin out or even disappear. Some rocks change character over a short distance. The Paleozoic deposits are largely marine in origin and include near-shore, beach, and dune sands, near-shore to offshore shales and limestones, fossiliferous reef deposits, and marine evaporites such as salt and anhydrite.

The following parts of geologic time are not represented by rocks in Michigan: Permian, all of the Mesozoic Era except the Late Jurassic, and all of the Cenozoic Era except the latest Pleistocene (Chapter VI). The virtual absence of rocks of Mesozoic and Cenozoic age age is significant because elsewhere in the United States, and in other parts of the world, oil and gas occur in great quantities in rocks of those ages. The greater part of the world's oil production comes, in fact, from Cenozoic (or Tertiary) rocks. Beneath the Paleozoic Era rocks of Michigan lie rocks of Precambrian age (Chapter IV). In the center of the Lower Peninsula, the Precambrian rocks are about 14,000 feet below the surface, but they are closer to the surface near the southern edge of the state along the flanks of the Cincinnati, Findlay, and Kankakee arches. Precambrian rocks actually crop out at the surface in the western half of the Upper Peninsula. No oil or gas have been discovered in the Precambrian rocks beneath the state; wells seldom are drilled that deep because the chances of finding oil there are judged to be unfavorable. It is true that a very minor amount of oil has been produced from Precambrian rocks elsewhere in the world, but only where, due to some peculiarity of subsurface structure, it has migrated out of much younger rocks to a secondary position in cracks or fissures in the Precambrian. In most cases Precambrian rocks are not likely source rocks for petroleum or gas, either because they are igneous and formed from a molten condition that would have destroyed the hydrocarbons, or because they have been so deformed and altered (metamorphosed) during the long course of geologic time that any hydrocarbons originally in them have long since been destroyed. Thus, it is in the Paleozoic rocks of the state that we search for oil and gas.

The general geologic structure of Michigan was described in Chapter III and Figure III-2 as being somewhat like a series of nesting mixing bowls. This comparison may be somewhat misleading because the sides of the "rock bowls" in the Michigan Basin dip so gently that when seen in a quarry or road cut they may appear horizontal. Nevertheless the dip is sufficient over a long distance to carry formations from their surface outcrop down to great depths. Near the center of the Lower Peninsula the oldest Paleozoic rocks (Cambrian) are deep below, resting on the Precambrian. In the same area, the Paleozoic formations directly beneath the glacial deposits are much younger. Outward in all directions from the center of the basin the rocks become progressively older so that around the margins of the Michigan Basin the bedrock beneath the glacial deposits is Early Paleozoic in age. The pattern formed on a geologic map consists of a series of roughly concentric rings of bedrock ranging from young at the center to old around the edge. The map in Figure III-2 shows this. The concentric bands of bedrock of different ages which one sees on such a map are the edges of the nesting bedrock bowls in the subsurface. When originally deposited, the bedrock formations of Paleozoic age were more nearly horizontal and in most cases extended far out beyond the present margins of the Michigan Basin. Some of the crustal downwarp which produced the Michigan Basin occurred as the layers of rock accumulated so that the older ones on the bottom are tilted more steeply than the younger overlying layers. After the basin structure had formed, the earth's crust was uplifted over an area much wider than that of the basin itself and subsequent erosion removed the higher parts of the formations. Thus, the edges of the "nesting bowls" are the erosion-beveled edges of those formations. This picture is essential to an understanding of the distribution of oil and gas in Michigan. The tilt of the formations encourages up-dip migration of oil and gas, and migration is essential if widely dispersed quantities of gas and oil are to accumulate in economically valuable quantities in reservoirs and traps. One would not drill for oil or gas in an area where the bedrock formation expected to contain them lay exposed directly below the glacial deposits. No seal rock would be present over the potential reservoir there, and any oil and gas that had reached that point would have escaped

# Petroleum and Natural Gas

at the surface long ago. A potential reservoir in the Paleozoic rocks must be investigated beneath an area that lies somewhere toward the center of the basin, away from the beveled, outcropping edge of the reservoir. The reservoir must be sealed by overlying deposits. Perhaps an example would be useful here. Over the years, the most productive reservoir rock for oil in the state has been the Devonian Dundee Limestone (although recently, the Ordovician Trenton and Black River and Silurian rocks have become major petroleum producers). Looking at the map in Figure III–2, you will note that there are extensive areas in southeastern Michigan and around the northern edge of the Lower Peninsula where Devonian age rocks come up to the surface beneath the thin glacial deposits. However, the oil fields that produce from the Dundee Limestone of Devonian age are located mainly in the center of the Lower Peninsula, where the Dundee lies far below the surface, beneath potential seal rocks.

The prevalence of marine environments in the state during Paleozoic times accounts for the origin of the hydrocarbon-rich source rocks that initially produced the oil and gas, the seal rocks, and the potential reservoir rocks. The general basin structure promoted migration of the oil and gas up the dipping flanks of the Michigan Basin, into and along reservoir rock formations, and toward the exposed edges of the formations near the outer margins of the basin. However, this situation alone would have resulted in the eventual escape of all the oil and gas at the surface outcrop. There had to be local traps to catch and hold the oil and gas before they actually reached the outcrop. Fortunately, these were available. Two types of traps prevail, (1) anticlinal, and (2) variable porosity-stratigraphic. The latter is the most important in Michigan today. The anticlinal type of traps occur because the basin structure is complicated by a number of minor folds or groups of folds, as well as a few faults, superimposed on the major basin structure. Picture a thick pile of big sheets of paper. Push the center of the pile down. The depressed center is the Michigan Basin and the sheets represent the tilted Paleozoic rock layers. Landes (oral communication) has described the minor anticlines as a series of "capsized canoes" distributed irregularly among the sheets. The canoe-like ends of those anticlines produce the necessary structural closure (see earlier discussion).

If the proper combination of source, reservoir, and seal rocks is available, petroleum may be present. The anticlines occur in groups which determine the locations of some of the oil fields as shown on Figure X–3. The folds are elongated in a northwest-southeast direction in the southern and central parts of the Lower Peninsula, but trend northeast-southwest in the northern part of the Lower Peninsula. The small, individual folds range from 4 to 6 miles in length and from 1 to 1.5 miles in width. Dips rarely exceed 200 feet per mile. Closure ranges from 50 to 100 feet. The anticlines tend to be asymmetrical with their steeper flanks facing the center of the Michigan Basin. These folds are small as anticlinal oil traps go. Therefore, they hold relatively small quantities of oil and gas compared with some of the really big fields elsewhere in the United States and the world. The elongate shapes of the fields at the surface reflect the shapes of the traps at depth. Not all the minor anticlines or groups of anticlines that have been found in the state are productive. Possible reasons for this were discussed earlier. Many dry holes have been drilled into them simply because even the most advanced techniques for oil exploration cannot guarantee that all necessary factors are favorable in each potential trap. We have seen that any oil well involves an economic risk. A large company, drilling many wells, has sufficient resources to "play the percentages" and to increase the odds in its favor by expensive research; the private individual who sinks all of his savings in one well cannot hope to do this.

## Reservoir Rocks in Michigan

Figure V–2 in Chapter V has a column showing the oil and gas producing reservoir rocks of Michigan. Carbonate rocks (limestone and dolomite) constitute over 90 percent of the proven oil reservoirs. This is unusual. Although carbonate rocks do contain oil elsewhere in the nation and the world, sandstones are the most productive reservoirs outside the state. Sandstones with adequate porosity and permeability do occur in Michigan Paleozoic rocks, but they do not match the carbonates in total thickness, number, or extent. Until recently, Devonian age carbonate rocks of the Traverse Group, the Rogers City Limestone, the Dundee Limestone, and the Lucas Formation of the Detroit River Group were the principal Michigan petroleum reservoirs. These formations still hold the record for the largest cumulative production over the years, but now they account for only about 25 percent of the total annual production. Porosity in the carbonate reservoirs is due mainly to secondary solution and dolomitization. By 1962 over 50 percent of the annual petroleum production was from Ordovician age (Trenton and Black River) and Silurian age (Niagaran) rock reservoirs; these now yield about 30 percent of the state's gas as well. Porosity in the Ordovician carbonate rocks occurs in locally restricted, secondarily dolomitized fracture and fault zones in limestones. The major oil and gas reservoirs in Silurian rocks occur in the following: the A–1, A–2, and E (Kintigh) carbonates within the Salina evaporite series (Fig. v–2); in porous Niagaran limestones and dolomites; and in Niagaran organic reefs sealed above by evaporites (Chapter V). The Mississippian Michigan Stray Sand and Upper Marshall Sandstone, once major gas reservoirs in the central part of the Lower Peninsula, now are largely depleted. The Mississippian Berea Sandstone also has produced some gas and oil in the Saginaw Bay area and in the central western Lower Peninsula along Lake Michigan (Pentwater Field). A minor amount of gas has been recovered from the Late Devonian Antrim Shale and from the Michigan and Coldwater formations of Mississippian age.

Source rocks which could have generated oil and gas are abundant in Michigan. The oil and gas reservoirs originated as marine sediments. In each case they are associated with many thick marine deposits which are fossiliferous now (Chapter XIII), which probably were rich in organic matter at the time of deposition, and which could well have provided the necessary hydrocarbons. There is no problem in explaining the origin of the porosity and permeability of the sandstone reservoirs of the state. Those rocks are composed of sedimentary grains or fragments which had interjoined pore spaces between the grains at the time they were deposited. On the other hand, there has been some controversy regarding the origin of porosity and permeability in the more important carbonate rock reservoirs of the state. We have already seen that some primary porosity does occur in fossiliferous carbonates, especially in organic reefs and reef flank deposits.

However, all these sources of primary porosity and permeability are insufficient to explain the total reservoir space available for the tremendous quantities of oil and gas contained in Michigan carbonate reservoirs. Many of the Michigan carbonates originally were nonporous, chemical precipitates consisting of intergrown crystals and tending to be rather dense and compact. Postdepositional or secondary porosity and permeability are more important in these. Some secondary porosity and permeability in carbonates resulted from earth crustal movements which cracked and fissured the rocks. In other instances a limestone rock at one time lay near or above the ground water table (Chapter IX). Circulation of ground water through relatively few primary pores, cracks, and fissures dissolved away some of the limestone, thus increasing the porosity and permeability. In the case of carbonate rocks deposited in ancient seas, this would mean that at some time after deposition of the sediments the seas must have withdrawn and the land risen, allowing water to circulate downward. It has been suggested that one of the principal oil reservoirs in Michigan, the Dundee Limestone, in places was involved in such a process, the upper part of that formation having been subject to considerable ground water solution during a period of uplift prior to deposition of the Rogers City Limestone that now overlies the Dundee.

A process called *dolomitization* has received much attention as an important cause of porosity in carbonate rocks. Many porous zones in carbonate rocks occur in parts of the formations that are largely dolomite, calcium magnesium carbonate, instead of limestone, calcium carbonate. The hypothesis of secondary dolomitization, which now receives strong support, maintains that limestone was locally dissolved by groundwater and partially replaced by dolomite in lesser volume than had been dissolved, thus leaving some open pore space. The fact that the dolomitized portions occur in locally restricted zones within limestones may support this hypothesis. For example, in the Deep River Field and the Albion–Scipio "Trend" the dolomitized zones stand vertically, possibly occupying fractured zones above faults that extend upward from the Precambrian basement.

# Petroleum and Natural Gas

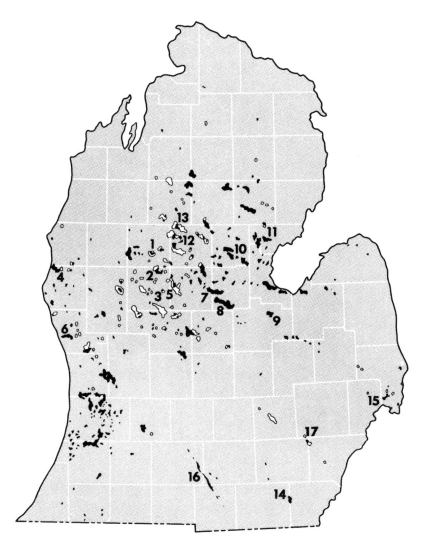

Figure x-3. Oil (black) and gas (white) fields of Michigan. (Modified with authors' additions from *The Michigan Story* published by Mich. Assoc. of Petroleum Landmen.) Field names as follows:
1—Evart
2—Fork
3—Coldwater
4—Pentwater
5—Sherman
6—Muskegon
7—Mt. Pleasant
8—Porter
9—Saginaw
10—Buckeye
11—Deep River
12—Temple
13—Winterfield
14—Deerfield
15—St. Clair
16—Albion-Scipio "trend"
17—Northville Field

## Oil and Gas Fields in Michigan

The production statistics for oil and gas change rapidly over the course of years so that any information provided now may soon be out of date. But we can use current statistics to get some idea of the importance of petroleum and gas in the economy of the state and the nature of some currently producing fields.

In 1967, oil was being produced from 44 of the 83 counties of the state, and natural gas from 25 all in the Lower Peninsula *(Michigan Geological Survey Annual Statistical Summary)*. Just over half of the gas was produced from wells that also yielded oil. The organic reefs of St. Clair County provided 45 percent of the gas, and together with the Albion–Scipio "Trend" and gas fields in Macomb, Missaukee, and Roscommon counties accounted for 87 percent of the total gas produced in the state. The Albion–Scipio "Trend" produced the largest share of oil in the state during 1967. Exploratory wells drilled outside known fields and developmental wells drilled within known fields together totaled 393 in 1967. Of the exploratory or "wildcat" wells only 1 out of 20 was productive. Again we see the risk involved in oil and gas exploration!

The producing fields fall roughly into the following "districts": Central Basin, Western, Southwestern, and Eastern (or Southeastern).

In the past, most of the oil and gas in the state was produced from anticlinal or dome-shaped

**Geology of Michigan**

Figure x–4. Gas and oil mixture, under pressure in subsurface, blows temporarily out of control—*a very rare accident* thanks to the precautions taken by drilling companies and to the safeguards stringently enforced by the Michigan Department of Conservation. Well near Albion. (Michigan Department of Conservation, photo by Clyde Allison, 1959.)

traps of small size (see earlier discussion on traps). In most but not all of these the oil and gas were found in the tops of the structures (Fig. x–1). Oil and gas commonly occurred in more than one horizon or reservoir within these anticlinal traps.

Let us look at a few of the Michigan fields as examples of oil and gas occurrence. The Porter, Buckeye, and Muskegon fields, at least in part, produce from typical anticlinal traps. The Muskegon Field is a dome with 60 to 70 feet of closure, producing from a porous Devonian limestone reservoir. In the Porter Field the porous upper part of the Devonian Dundee Limestone is the reservoir; porosity there was caused by solution of openings into the limestone during a time of land uplift and sea withdrawal after deposition of the Dundee Limestone and before the deposition of the overlying Rogers City Limestone (Chapter V, Fig. v–2). Trapping in the Porter Field also results from local changes in the reservoir rock permeability, thus that field involves both structural and stratigraphic trapping.

The Northville Field in southeastern Michigan, lies on the southeastward extension of the Howell Anticline and associated fault. Production there is from several reservoir horizons including the Dundee Limestone (gas), Salina–Niagara rocks (gas), and dolomitized zones in the Trenton–Black River rocks (oil and gas). These zones are shown on Figure v–2 in Chapter V. This is another compound, structural-stratigraphic trap because production from the Trenton–Black River reservoir is not only in a dolomitized zone but is localized along an anticline.

The Albion–Scipio "Trend," discovered in 1955, is another interesting example of a Michigan field. By 1960 this field accounted for over half of the state's oil production. The reservoir is a nearly vertical zone in dolomitized upper parts of the Ordovician Trenton Limestone. Dolomitization took place in a fractured zone, probably above a more deep-seated fault zone which may extend down into the Precambrian basement rocks (Chapter V). The "Trend" is a narrow belt, with some gaps, extending at least 35 miles in a northwestward direction across Calhoun, Jackson, and Hillsdale counties (Fig. x–3).

In Silurian rocks in St. Clair County, oil and gas are trapped in porous organic reefs (Chapter V) which are overlain by seal rocks consisting of Salina evaporites. The reef structures occur as mounds, resembling anticlines, and serving a similar function as traps. Although reef production now is mainly from St. Clair County, similar reefs are known to occur in many other places throughout the state. Some of these, as well as others yet to be discovered, may become producing fields in the future. In the present area of reef production, 9 out of 10 reefs have been found to contain oil, gas, or both.

Stratigraphic traps, owing their origin largely to local dolomitization, have become increasingly important in recent years until they now surpass anticlinal traps as oil and gas producers in the Michigan Basin.

## Summary

The origin of oil and gas in Michigan was a direct consequence of the Paleozoic marine history of the state. The source, reservoir, and seal rocks were deposited as marine sediments. Carbonates (limestone and dolomite) are the most productive reservoir rocks. Traps are of both the structural and stratigraphic type. Data on geologic conditions at the surface, combined with far greater information from the subsurface, are essential to an understanding of the geology and geologic history of the state. Interpretation of that history in terms of its effects on rock strata and their structures aids in the discovery of oil and gas.

# XI

# Minerals in Michigan

## Introduction

The geologic history of Michigan is written in the language of mineralogy (study of minerals), petrology (study of rocks), and paleontology (study of fossils). Although these are branches of geology, many hobbyists find one of these fields intriguing for its own sake. These are fields of knowledge and research in which professional geologists receive intensive training. The brief review of mineralogy presented in this chapter emphasizes things found in Michigan and uses examples drawn from this state. Nevertheless, the basic principles presented are applicable anywhere; materials similar to or identical with those found in Michigan occur in many other places.

This chapter is intended to serve mainly as a reference section rather than to be read from beginning to end. It includes identification charts and tables with instructions for their use, a list of Michigan minerals, and a list of mineral exhibits and collecting places in the state. The latter may be of use in starting a collection or in locating places of geologic interest near your home. Reference books mentioned later would be useful for those who might wish to extend their knowledge beyond the material presented here.

## Nature of Minerals

A *mineral* is a naturally occurring inorganic substance whose chemical composition and physical properties vary within definite limits. The great majority of minerals have an orderly internal structure, that is, they contain atoms of specific elements in definite proportions and arranged in a regular geometric pattern. This invisible array of atoms is sometimes referred to as a *lattice*. This "internal" order is responsible for the characteristic and constant "external" physical properties that are useful in identification of minerals. It is both the kind and arrangement of different atoms that establish the uniqueness of a mineral. Two different minerals may have exactly the same chemical composition (same kinds and amounts of atoms), but are unique in that the arrangements of the atoms are different; such minerals are *polymorphs*. Conversely, two different minerals may contain entirely different kinds of atoms which are arranged in the same regular pattern; such minerals are called *isomorphs*.

A brief visit to any mineralogy museum (see list at end of chapter) should readily impress you with the number and variety of mineral species. Over 1500 minerals have so far been discovered and described, but even a well-trained mineralogist can recognize only a small fraction of these at first glance. Fortunately, this complexity is not critical for a student of elementary geology because only about 1 percent of all the mineral species are common in the earth's crust. In fact, only a dozen or so minerals are major rock formers, and that small number makes up over 95 percent (by volume) of all the rocks exposed in the earth's crust. This limitation to the number of common minerals is due to the fact that there are a limited number of common chemical elements in the crust. Although over 100 chemical elements are now known, only 8 are common in nature (Figs. xi–3, 5). These constitute over 98 percent of the mass of the crust and include, in order of decreasing abundance, oxygen, silicon, aluminum, iron, calcium, sodium, potassium, and magnesium. These 8, in various combinations and proportions make up the common, rock-forming minerals. Oxygen and silicon, the most abundant by far, occur very commonly, forming a large variety of minerals called silicates.

## Methods of Mineral Study

Schemes of mineral identification emphasizing simple observations of physical properties are rapid and practical for field study and do not require the use of involved and expensive equipment. However, even the experienced mineralogist must occasionally throw up his hands at a difficult unknown and turn to more elaborate procedures of mineral study. Indeed, there are many minerals such as the fine-grained clays that simply cannot be distinguished from one another without resort to other laboratory procedures. Because of their restricted chemical composition, one method of identifying minerals is through chemical analysis. A variety of techniques are used to determine which of the chemical elements are present in a mineral, and thus determine the qualitative composition. Such tests generally reveal which elements are present but not the precise quantities of each. Even such an incomplete chemical analysis may require considerable time and equipment.

Also, in some cases, more than one mineral species may have the same chemical composition (e.g., calcite and aragonite) and the composition is not definitive. It is difficult, if not impractical, to use chemical analysis in field identification.

The professional mineralogist frequently identifies unknown minerals by determining their optical properties, which vary in accord with differences in internal structure. For this purpose, extremely thin slices of a mineral, or rock, specimen are cut, and these are studied in transmitted light under the microscope much as one would use a biological section. The opaque minerals cannot be studied in this fashion since they do not transmit light; they are prepared by sawing the sample and polishing the cut surface to a high finish. The polished surface is then examined in reflected light under a special ore microscope. These methods, the study of *thin* and *polished sections,* are far from suitable for use in the field.

Still another method of identification of unknown minerals is through use of x-ray methods, which indirectly take a picture of the arrangement of atoms in the internal structure of the mineral. Since the orderly arrangement of atoms is a feature characteristic of many particular minerals, the *x-ray pattern* provides one of the most reliable means of identifying mineral unknowns.

Another technique, commonly used in identification of many fine-grained mineral aggregates is called *differential thermal analysis.* Here a sample of the mineral is heated at a constant rate while a record is kept of the heat absorbed or given off by the sample at any specific temperature. Loosely speaking, the mineral is forced to write its thermal signature, a signature in which the peaks and troughs are caused by thermal reactions characteristic of the mineral tested. This technique is especially useful in identifying clays, carbonates and other thermally sensitive minerals. Although they provide very definitive answers, the equipment and experience required for application of these methods are considerable.

Field geologists and most hobbyists learn to identify minerals by means of their various physical properties. Although this method is not always exact or certain, it is by far the simplest and speediest method and is accurate enough for identification of the major rock-forming minerals in most of the cases where they are encountered in rocks of the earth's crust.

# Minerals

Figure XI-1. Physical properties of minerals.
1—Double refraction, calcite.
2—Irregular fracture, quartz.
3—Conchoidal fracture, obsidian.
4—Earthy fracture, kaolinite.
5, 6—Rhombohedral cleavage, calcite.
7—Platy basal cleavage, mica.
8—Octahedral cleavage, fluorite.
9, 10—Cubic cleavage, galena and halite.
11—Cubic crystals, galena.
12—Octahedral crystal, fluorite.
13—Hexagonal crystals, quartz.
14—Monoclinic tabular crystal, gypsum.
15, 16, 17—"Dog Tooth spar" hexagonal scalenohedrons, calcite crystals.
18, 19—Monoclinic prismatic, orthoclase crystals.
20, 21, 22—Hexoctahedral crystals, garnet.
23—Elongate monoclinic prismatic crystals, hornblende.
24—Short monoclinic prismatic crystals, augite.

## Physical Properties of Minerals

The important physical properties useful in the identification of minerals are: (1) hardness; (2) specific gravity; (3) color, particularly that of the powdered mineral, or streak; (4) crystal form, cleavage, and fracture, which determine the mineral's shape; (5) some of the optical properties of minerals such as luster, diaphaneity ("transparency"), and refraction; and (6) a group of rather special properties, such as magnetism, taste, odor, and feel.

*Hardness:* The hardness of a mineral may be defined as a measure of its relative resistance to abrasion. Hardness is determined by scratching the smooth surface of one mineral with the sharp edge of another. To be sure that you recognize which mineral actually does the scratching, try wiping off the "scratch" to see if it is truly a scratch or whether it consists of small particles rubbed off the other mineral. Be sure that you are actually scratching individual grains of the unknown mineral and not merely producing a scratch by tearing apart hard grains loosely cemented in an aggregate sample. Even quartz, which is quite hard, may be scratched by a soft fingernail if the grains in a fine-grained quartz sandstone are easily broken apart. This, however, is not a true abrasion scratch of the type you will wish to make. On many mineral identification keys you will find the hardness of a mineral indicated by a number; this refers to the abrasion hardness of the mineral relative to that of 10 standard minerals arranged in order of increasing hardness (1, softest; 10, hardest). Each of the minerals will scratch those with a lower number on the scale and will, in turn, be scratched by minerals with a higher hardness number. This set of standards, called *Mohs' scale of hardness,* is truly a relative scale—for example, the difference in hardness between numbers 9 and 10 is much greater than that between 1 and 9.

MOHS' SCALE OF HARDNESS

1. Talc (softest)
2. Gypsum
    2½ Fingernail
3. Calcite             About 3 Copper coin
4. Fluorite
5. Apatite             5–6½ Knife blade or plate glass
6. Orthoclase feldspar   About 6 Steel file
7. Quartz
8. Topaz
9. Corundum (gem—ruby or sapphire)
10. Diamond (hardest)

Minerals may be grouped in two hardness classes based on the common test substances listed above; minerals softer than a knife blade or glass plate are "soft"; those harder than a knife or glass plate are "hard." These terms are used in the mineral identification key in this chapter.

Abrasion hardness may vary with direction in crystal faces tested. Test good crystals of calcite to see if you can detect any such variation.

*Specific gravity:* By specific gravity is meant the weight of the substance relative to that of an equal volume of water. It is stated as a number indicating the ratio of the two weights. Measurement of specific gravity, therefore, provides a means of comparing the weight per volume of different substances.

Specific gravity may be determined by a series of weighings using different types of balances. With a little practice it is also possible to detect relatively small differences in specific gravity merely by the "heft" of a specimen of moderate size. Specimens of about the same size or volume should, of course, be used when attempting to judge the specific gravity by hand. The silicate minerals, the major group of rock-forming minerals, have an average specific gravity in the range 2.6–3.7. The major metallic ore minerals, on the other hand, range in specific gravity from about 4.5 to 8.0. As minerals go, this latter group is "very dense."

*Color:* The color of a mineral is a property depending on many factors such as the internal structure of the mineral, the presence of minor amounts of impurities, the habit and arrangement of individual mineral grains, and many others. Some minerals, like fluorite and quartz, may change color in time, if exposed to natural radioactivity, and may lose their radiation colors if heated. The reliability of color as a guide in identification varies to the extent that these other variables may change. For some minerals, such as yellow pyrite or green malachite, the color is constant and characteristic, but with many other minerals the color is quite variable and of little assistance in the identification process. Quartz, for example, may display many different colors, and, in fact, many of the semiprecious gem stones are merely varieties of quartz that differ in color (e.g., amethyst, citrine, onyx, chrysoprase). It is thus necessary to use the property of color with care in the identification of minerals.

The color of the powdered mineral, called the *streak,* is, however, much more diagnostic. Streak is determined by rubbing a mineral on a piece of unglazed porcelain. The streak of a mineral may be similar to its color in a hand specimen, or it may be quite different; it is a property well worth testing.

*Shape:* The shape of a mineral specimen is dependent on many things. As used here, however, shape will refer to any regularity of external form displayed in individual grains of a mineral. The crystal form (habit), cleavage, or fracture of a mineral may control to a greater or lesser extent its form and may contribute regularity to the shape. Minerals, because of their orderly internal arrangement, tend to crystallize into definite *crystal forms* with a characteristic shape. These are bounded by plane surfaces called crystal faces. Such faces mark the outer limit of growth of the crystal in one particular direction. The shapes of the various crystal faces and the overall form of the crystal are diagnostic and regular, but more importantly, the angle between any two particular crystal faces is uniform (the constancy of interfacial angles as discovered by Nicholas Steno in 1669). Regardless of the size or other properties of the crystal, the interfacial angles are constant from specimen to specimen of a given species. Each mineral, then, is characterized by a certain crystal habit or a certain few crystal habits which provide an important tool for identification. Not all specimens may display the form well, but where good crystal form is present, it may be helpful in identifying mineral unknowns.

Many minerals *cleave,* or break, along smooth planes whose orientation is also controlled by the orderly internal atomic arrangement. The number and orientation of the *cleavage* directions of many minerals determine the shape of most of the mineral fragments. Some minerals (e.g., the micas) have only 1 direction of cleavage, resulting in a platy shape for all specimens; others have 2, 3, 4, or even 6 directions of cleavage. When cleavage is well developed, it is difficult for a mineral to break other than along the cleavage; galena and halite fragments rarely occur except as cubes, because each of these minerals is characterized by 3 directions of nearly perfect cleavage intersecting one another at right angles. Calcite fragments typically occur as rhombohedrons, with the shape defined by three planes of perfect cleavage intersecting at other than right angles. In some cases (e.g., augite and hornblende, a pyroxene and amphibole, respectively) cleavage is practically the only physical property which is diagnostic; augite has two directions of cleavage which intersect nearly at right angles, and hornblende has two directions of cleavage which intersect, roughly, at angles of 60° and 120°.

Avoid confusing good *crystal faces* with *cleavage planes* or *crystal forms* with *cleavage fragments*. The crystal form of a mineral and the faces defining that form are external properties of a crystal. Crystal faces exist only at the exterior of a crystal. Planes of cleavage, on the other hand, are planes along which the mineral may easily fracture due to internal weaknesses in the crystal structure. Unlike crystal faces, there is a myriad of potential cleavage planes within a crystal that possesses cleavage. These planes are parallel because the directions of weakness through a given crystal structure are constant. Some minerals having good crystal form may lack cleavage altogether; others having good cleavage may lack crystal form, and so the properties are entirely separate features. To clarify the distinction, in some cases, a crystal face may parallel another crystal face only at the opposite side of the mineral but not along many repeated planes or cracks passing throughout the entire specimen. However, it is not uncommon to see a crystal face paralleled by a prominent set of cleavage planes and in this case, where both are parallel, it may be difficult to distinguish the contributions of form and cleavage to the overall mineral shape. Cleavage may often be detected as many parallel visible cracks in a specimen or by a miniature "stair step" profile on the specimen. It is often helpful to break the mineral specimen and to examine the resulting "cleavage" fragments under a binocular microscope or hand lens. These forms may look like crystal forms but are of course bounded by cleavage surfaces, not crystal faces. If a mineral specimen having good cleavage is turned under a lamp or other source of nondiffuse light, numerous parallel cleavage surfaces will simultaneously reflect the light in certain positions. Such flashes of light are useful in detecting the presence of cleavage planes and determining their orientation. With practice you can estimate the angle between cleavages by sensing the angle through which you rotated the sample between successive flashes.

Minerals may contain up to several good cleavages; their perfection and orientation are dependent upon the internal structure. The number orientation and perfection of such cleavages are a valuable tool in identification, since they are relatively constant and dependent upon the internal structure of the mineral.

In tiny mineral fragments and partial crystals, such as usually are found in rocks, the presence of cleavage is revealed by tiny flashes (sparkles) of light which appear when the specimen is rotated quickly. The light flashes are produced by light reflecting from the flat cleavage faces in one particular position. Often in mineral identification it is sufficient to determine that cleavage is present in a mineral without being able to determine the type or nature of cleavage. Note on the following mineral key that in most cases you must simply decide whether cleavage is present or absent.

Minerals which do not cleave along smooth planes may *fracture* (break) irregularly, and even cleavable minerals may fracture in directions other than the cleavage directions. Some minerals have a typical kind of fracture. Quartz, particularly the finely granular varieties (chert and flint), frequently breaks with a conchoidal (shell-shaped or curved) fracture, a property utilized by primitive man in the manufacture of stone tools and weapons. Other minerals have a fibrous fracture (Fig. XI–1).

*Light response:* The three physical properties describing the response of a mineral to light are luster, diaphaneity, and refraction. By luster is meant the character of the light reflected from the fresh surface of the mineral. One major type of luster is *metallic* (opaque, and reflecting light like a metal); the other major luster type is *nonmetallic,* including vitreous (like that of broken glass), pearly, greasy, resinous, silky, and earthy (like that of a lump of dirt). There are all gradations between minerals that are metallic and those which are clearly nonmetallic, and the division is really an artificial one. Many of the metallic minerals do not always display a metallic luster, but instead appear somewhat dull and tarnished. Luster, like color, is not always a diagnostic property and is very dependent on the physical arrangement of mineral grains.

*Diaphaneity* is a sophisticated term for the transparency of a mineral or its capacity to transmit light. Most of the minerals you will be dealing with are clearly *opaque* (do not transmit light), or *transparent* (do transmit light), but there are many intermediate minerals. In questionable cases you should examine thin edges of the specimen most apt to transmit light or actually break off minute chips and examine them in strong light under the binocular microscope. In the laboratory you should be able to differentiate among opaque, translucent, and transparent materials. Transparent minerals transmit optical images, whereas translucent minerals transmit light but not clear images.

All the minerals which transmit light bend (refract) the light rays as they pass through because the velocity of the light is different through the mineral than it is through air. Many minerals also split the light into differently polarized rays, each of which is transmitted at different velocity. In most cases the difference in the ray velocities is too slight to be detected by the naked eye, but in the case of calcite the two rays travel at very different rates. As a result, one actually "sees double" when looking through a piece of transparent calcite; we say that calcite is characterized by double refraction.

*Special properties:* The other, more special physical properties of minerals, such as odor, taste, feel, and magnetism, need little further comment. It should be mentioned, however, that by magnetism we do not mean that the mineral itself acts as a magnet; we merely mean that the mineral will affect a magnet. To test for magnetism, powder some of the mineral and try to pick it up with a magnet, or balance a magnet on the edge of a box or table and see if the specimen will "pull" the end of the magnet down when it is brought near the magnet from below.

The *reaction* of a mineral with dilute hydrochloric *acid* is another important test. Calcite and dolomite react with acid; both are common, rock-forming minerals producing limestone and the rock, dolomite. Calcite "fizzes" or bubbles when a drop of such acid is placed on it. Dolomite does not react so readily, but will fizz when acid is placed on powdered dolomite. The powder may be a small quantity scratched loose with a knife point.

These, then, are the principal properties by means of which field identification of minerals is possible; they are the properties used as distinctions on the mineral key (Fig. XI-2). Mineral identification requires skill which comes with practice. At first only the most obvious and perfect specimens may be easy to identify. Practice on an already identified set of minerals. To acquire a better understanding of what is meant by the various mineral properties a set of minerals showing each property would be useful. Identification of unknowns can come later. Do not expect to be an expert at first. Get help from someone.

A *mineral and rock identification kit* will be necessary for testing minerals for their physical properties. Kits may be purchased from science supply companies or may easily be assembled from local stores. The items needed are:

1. A hand lens or magnifying glass (about 10 power).
2. Acid bottle and acid. Get dilute hydrochloric acid (5 percent) at a drugstore. Keep in a nose-drop bottle, using the dropper to place a small drop of acid on the mineral to be tested. Wrap the bottle with masking or friction tape and carry carefully in an upright position. The dilute acid is not dangerous but will discolor skin and deteriorate fabrics.
3. Small pocket knife for hardness test.
4. Small piece of plate glass for hardness test.
5. Copper penny for hardness test.
6. Streak plate. Any piece of unglazed white tile or fragment of unglazed dishware will do. May be cleaned for reuse by soaking in dilute hydrochloric acid and brushing with soap and water.
7. Small magnet. A magnetized knife blade will work if strong.
8. You supply the senses of taste, odor and specific gravity (relative weight); and the fingernail for testing hardness.

A *special note* about mineral identification is important. If you wish to learn to identify rocks (which are composed of minerals) you must learn to identify minerals first. Although a very few rock types are so distinctive that they can be identified without identifying the minerals that compose them, these are the very rare exceptions. The following chapter on rocks should be studied only after you have worked on minerals.

In one respect, identification of minerals in rocks is easier because only certain combinations of minerals are to be expected in nature. For example, igneous rocks that form from cooling of rock melts initially contain only minerals stable at high temperature and never contain minerals like gypsum which cannot form or survive in a high-temperature environment. Such mineral *associations* are very helpful in mineral identification but cannot be put to full use until you have gained some knowledge of rock-forming environments.

The *mineral identification key* that follows is simple to use and will aid in the recognition of a few of the many thousands of known minerals. Those few shown on the key include all the important rock-forming minerals, a few "showy" types, and some that are common in hobby collections; the key includes all common Michigan minerals. The rock-forming minerals are important for development of a greater understanding of geology and geologic history. It is unfortunate that many of the mineral "sets" that are sold commercially to tourists as souvenirs include minerals which are rare, found only at very special places, and not very important geologically. These cannot be included in a common mineral key. For further reference and more detail on minerals see a mineralogy textbook such as *Mineralogy*, fifth edition, by Kraus, Hunt and Ramsdell (McGraw-Hill Co.), *Dana's Manual of Mineralogy*, by C. S. Hurlbut (J. Wiley & Sons), now in its seventeenth edition, and *Mineralogy for Amateurs*, by J. Sinkankas (Van Nostrand, 1964), or one of the popular field guides available at most local bookshops. *A Field Guide to Rocks and Minerals,* by F. H. Pough (Houghton-Mifflin, 1951) is good, as is *Rocks and Minerals,* by H. S. Zim and P. R. Shaffer (Golden Press, 1957).

To use the mineral key you apply successive "tests" and make a decision after each which leads you on to the next test (Fig. XI-2). First examine luster and color and decide if the specimen is *nonmetallic-dark, nonmetallic-light,* or *metallic* in luster. Be sure you are examining a pure mineral, not a rock with a mixture of several minerals. If the mineral has a metallic luster, then test for streak. If it is nonmetallic the next tests are for hardness and then cleavage. Finally, compare the specimen with the description of special properties and pick the best fit. Remember that no artificial key can be perfect, especially not one restricted to a very few minerals. Do not expect to be right all the time, nor that all specimens will be equally easy to identify. Field geologists and mineralogists may be unable to identify certain specimens that are rare types or abnormal; they often suspend judgment until they can apply more exact laboratory tests. Mineral identification requires some "flair" or intuition which comes with practice. The abbreviations on the key are: H, for hardness; C, for cleavage; F, for fracture; L, for luster; S, for streak; and SpG, for specific gravity.

## Mineral Collecting in Michigan

The list and maps that follow give the locations of mineral collections which are on public display and collecting localities which have been mentioned in the literature and which have yielded some mineral specimens to amateur collectors.

To save space, the locations are given only in a very general way; in all cases one should inquire locally after he has come as close to the indicated locality as possible. It will be a help to the collector of minerals (as well as fossils) to obtain copies of the county road maps of Michigan which are available from the Michigan Department of Natural Resources, Lansing, Michigan. All localities listed here may be located more closely on the county road maps, many of which indicate quarries and mines by means of appropriate symbols. Some of the collecting localities, particularly old mines and quarries, are "worked out" as far as their economic value is concerned and may be closed. Many of the minerals listed for certain localities are rare and hard to find; this list offers no guarantee that the amateur actually will find such minerals. Many localities have simply been assembled from reports and the authors of this book have been unable to visit personally and collect at all of them.

*Precautions* and *rules* to be observed follow:

The publication of this list does not imply that permission is automatically granted by property owners for collectors to enter on the land where the locality is situated. In all cases, obtain permission first.

Various safety precautions should be taken. Wear goggles if you hammer on rocks. Use the proper kinds of tools, such as a rock hammer and rock chisel. Avoid steep cliffs, loose rock, and water-filled portions of quarries. Look out for people above and below you. *Do not,* under any circumstances, *roll or throw rocks.* Don't be a litterbug or a vandal. Stay away from buildings and equipment located in the area. The willingness of property owners to permit collectors on their land decreases rapidly with time because some collectors are thoughtless.

To increase the likelihood of finding minerals, it is well to prepare oneself in advance. Actually, experience is the best teacher; however, one may join mineral clubs, subscribe to mineral journals, and visit museums and local rock shops. If possible, arrange to start collecting activities in the company of an experienced mineral collector.

The following popular sources of information on mineral collecting may be found in most libraries, and the librarian can usually provide current subscription data. This list of literature sources does not pertain exclusively to Michigan but each of the publications may include some information on the state, including localities, types of minerals, clubs to join, rock shops, interested laymen, equipment, and so forth. The list is as follows:

*Earth Science,* P. O. Box 550, Downers Grove, Illinois

*Gems and Minerals,* Box 687, Mentone, California

*Lapidary Journal,* P. O. Box 518, Del Mar, California

*Rocks and Minerals,* P. O. Box 29, Peekskill, New York.

### Figure XI–2
### Michigan Common Mineral Identification Key
(Variety names in parentheses if found in Michigan)

| | | | | |
|---|---|---|---|---|
| LUSTER NON-METALLIC, LIGHT COLORED | HARD harder than glass or knife | CLEAVAGE PRESENT | Flesh colored or white; C, 2 planes nearly at right angles, no striations on cleavage faces; H, 6. | ORTHOCLASE feldspar, $KAlSi_3O_8$ (adularia) |
| | | | White to pale yellow or green; H, 6; 2 planes nearly at right angles; frequently has irregular and discontinuous bands crossing cleavage planes. | MICROCLINE, $KAlSi_3O_8$ |
| | | | Gray, gray-green, green-white or white (rarely pink); C, 2 planes nearly at right angles, faces show striations; H, 6. | PLAGIOCLASE feldspar, $NaAlSi_3O_8$ and/or $CaAl_2Si_2O_8$ (albite including peristerite) |
| | | NO CLEAVAGE | L, vitreous; transparent to translucent; may be in hexagonal (6-sided) crystals; H, 7; F, conchoidal. Many colors. | QUARTZ $SiO_2$ (rock crystal, agate, amethyst, sardonyx, rose carnelian, chalcedony, milky) |
| | | | L, dull; light gray, yellow, light brown or white; H, 7; F, conchoidal; a noncrystalline or micro-crystalline form of quartz. | CHERT (or Flint), $SiO_2$ |
| | | | Light green to white; H, 6-6½. | PREHNITE, hydrous Ca, Al silicate |
| | | | L, vitreous; olive green or green-brown; transparent to translucent; H, 6. | OLIVINE, $(Mg$ or $Fe)_2SiO_4$ |
| | SOFT softer than glass or knife | CLEAVAGE PRESENT | Colorless to white; salty taste; C, cubic. | HALITE (or rock salt), NaCl |
| | | | White, yellow or colorless; H, 3; C, rhombohedral; crystal faces flat if present; fizzes readily with dilute hydrochloric acid. | CALCITE, $CaCO_3$ (dog tooth spar, anthraconite, onyx, satin spar, limestone, marble) |
| | | | White, yellow, red, brown, or black; H, 3.5; C, rhombohedral; crystal faces may be curved if present; fizzes slowly or not at all with dilute hydrochloric except in powdered form. | DOLOMITE, $CaMg(CO_3)_2$ (some marble) |
| | | | White to transparent; H, 2; occurs as flexible plates or finely crystalline or as earthy masses. Soft enough to scratch with fingernail in most cases. | GYPSUM, $CaSO_4 \cdot 2H_2O$ (alabaster, satin spar, selenite) ANHYDRITE, $CaSO_4$ |
| | | | Green to white; soapy feel; H, 1; very soft. | TALC, $Mg_3(OH)_2Si_4O_{10}$ |
| | | | Colorless to light yellow; transparent when in thin sheets; C, 1 good plane; occurs as flexible plates. | MUSCOVITE mica, approx. $KAl_3Si_3O_{10}(OH)_2$ (sericite) |
| | | | Faint bluish white, or yellow (rarely green or red); transparent to translucent; C, perfect basal and prismatic; F, uneven; H, 3–3.5; L, vitreous to pearly. | CELESTITE, $SrSO_4$ |
| | | | White; yellow, purple, or green; C, octahedral; H, 4. | FLUORITE, $CaF_2$ |
| | | NO CLEAVAGE | White, earthy odor when damp; F, earthy; H, 2. | KAOLINITE, $H_4Al_2Si_2O_9$ |
| | | | Green to white; soapy feel; H, 1. | TALC, $Mg_3(OH)_2Si_4O_{10}$ (soapstone) |
| LUSTER NON-METALLIC, DARK COLORED (cont. top of next page) | HARD | CLEAVAGE PRESENT | Black, greenish-brown, greenish-black; H, 5 to 7; C, 2 planes nearly at right angles; short, thick, dull crystals. | AUGITE, pyroxene. Variable and complex chemically, Ca, Mg, and Fe silicate. |
| | | | Black, greenish-black, greenish-brown; H, 5 to 7; C, 2 planes at about 60 and 120 degrees to one another; long, shiny, slender crystals. | HORNBLENDE, amphibole. Variable and complex chemically, Ca, Mg, and Fe silicate. |
| | | | Gray to blue-gray; H, 6; C, 2 planes nearly at right angles; striations on cleavage faces. | PLAGIOCLASE feldspar, many varieties ranging from $NaAlSi_3O_8$ to $CaAl_2Si_2O_8$ |
| | | | L, vitreous to resinous; color, yellowish to blackish-green; F, uneven; H, 6 to 7; transparent to opaque. | EPIDOTE, $Ca_2(Al$ or $Fe)_2(AlOH)(SiO_4)_3$ |

# Minerals

**Figure XI-2—Cont.**
MICHIGAN COMMON MINERAL IDENTIFICATION KEY
(Variety names in parentheses if found in Michigan)

| | | | | |
|---|---|---|---|---|
| *(cont. from preceding page)* | HARD | NO CLEAVAGE | Typically, red to red-brown, but color varies widely; H, 6.5 to 7.5; fracture resembles poor cleavage; brittle; often in crystals that are nearly round. | GARNET, a very complex Fe, Ca, Al silicate. |
| | | | Gray to gray-black; L, vitreous; H 7. | SMOKY QUARTZ, $SiO_2$ |
| | | | Red to red-brown, dark gray or black; L, dull; H, 7; F, conchoidal. | JASPER (red) and FLINT (black) forms of quartz, $SiO_2$ |
| LUSTER NON-METALLIC, DARK COLORED | SOFT | CLEAVAGE PRESENT | Brown to black; C, 1 very good plane; thin, elastic plates. | BIOTITE mica, $K(Mg, Fe)_3AlSi_3O_{10}(OH)_2$ |
| | | | Dark green to greenish-black; H, 2 to 2.5; S, white; slippery feel; C, 1 plane; flexible plates. | CHLORITE, $(Mg, Fe)_5(Al, Fe)_2Si_3O_{10}(OH)_8$ (aphrosiderite, bowlingite, delessite, masonite, thuringite) |
| | | | Brownish to nearly black; L, vitreous to pearly; F, conchoidal; C, perfect rhombohedral; H, 3.5 to 4.5; S, pale yellow or white. | SIDERITE, $FeCO_3$ |
| | | | C, fibrous, or splintery or conchoidal; green (or yellow, gray, red, brown, or black); may be spotted or varicolored; L, dull, resinous, greasy or waxy; H, 2.5 to 4. | SERPENTINE (asbestos), varied complex $H_4Mg_3Si_2O_9$ (antigorite, picrolite, "verde antique") |
| | | NO CLEAVAGE | Blue; S, blue; H, 3.5 to 4. | AZURITE, $Cu_3(CO_3)_2(OH)_2$ |
| | | | Bright green; S, pale green; H, 3.5 to 4. | MALACHITE, $Cu_2CO_3(OH)_2$ |
| | | | Green, brown, blue, purple; H, 5; C poor basal. | APATITE, $Ca_5(F, Cl, OH)(PO_4)_3$ |
| METALLIC LUSTER (Note. metallic lusters may tarnish, become dull, or disappear. If so, then try streak.) | | LEAD-GRAY SHINY STREAK | Dark lead gray; metallic luster on fresh surface may dull to black; F, conchoidal; C, poor or none; heavy, specific gravity 5.5 to 5.8; H, 2.5 to 3. | CHALCOCITE, $Cu_2S$ |
| | | BLACK, GREENISH-BLACK OR DARK GREEN STREAK | Lead-pencil black; smudges fingers when handled; greasy feel; H, 1. | GRAPHITE, pure C (carbon) |
| | | | Pale brass-yellow; H, 6 to 6.5; often in cubic crystal. (Marcasite lighter colored. Crystals rare.) | PYRITE, "fool's gold," MARCASITE, $FeS_2$ |
| | | | Dark brass-yellow; H, 3.5 to 4; tarnishes purple. | CHALCOPYRITE, "fool's gold," $CuFeS_2$ |
| | | | Shiny gray; very heavy; C, perfect cubic; H, 2.5. | GALENA, PbS |
| | | BLACK COLOR BLACK STREAK | Iron black; H, 6; strongly magnetic. | MAGNETITE, $Fe_3O_4$ |
| | | RED, REDDISH-BROWN STREAK | Red-brown to nearly black; may appear quite soft. | HEMATITE, $Fe_2O_3$ (martite) |
| | | YELLOW OR YELLOWISH-BROWN STREAK | Yellow-brown to dark brown; may be almost black; H, 6. | GOETHITE, $FeO(OH)–nH_2O + Fe_2O_3–nH_2O$ (limonite, "sammet blende") |
| | | PALE YELLOW OR WHITE STREAK | Yellow-brown, pale yellow (or rarely white); L, resinous; cleavage faces common; H, 3.5 to 4. | SPHALERITE, ZnS |
| | | COPPER-RED STREAK | Copper red color when fresh, but tarnishing to black, red, green, or blue; H, 2.5 to 3; heavy SpG, 8.5 to 9; malleable (can be flattened by pounding). | COPPER, "native copper," pure metallic Cu |

# Geology of Michigan

Figure XI-3
COMMON MINERAL-FORMING CHEMICAL ELEMENTS

| Element | Approximate Percent by Weight in Earth's Crust | Chemical Symbol |
|---|---|---|
| Aluminum | 8.1 | Al |
| Calcium | 3.6 | Ca |
| Carbon | | C |
| Chlorine | | Cl |
| Copper | | Cu |
| Fluorine | | F |
| Gold | | Au |
| Hydrogen | | H |
| Iron | 5.0 | Fe |
| Lead | | Pb |
| Magnesium | 2.1 | Mg |
| Manganese | | Mn |
| Mercury | | Hg |
| Nitrogen | | N |
| Oxygen | 46.6 | O |
| Phosphorus | | P |
| Potassium | 2.6 | K |
| Silicon | 27.7 | Si |
| Silver | | Ag |
| Sodium | 2.8 | Na |
| Strontium | | Sr |
| Sulphur | | S |
| Uranium | | U |
| Zinc | | Zn |

Total: 98.5 percent for those indicated. All other elements including those listed above constitute the remaining 1.5 percent.

The list of fossil-collecting localities in Chapter XIII will also serve as a useful guide to other mineral-collecting localities because at many of these fossil-collecting sites, especially in geodes and cavities within the rocks, one may find the following minerals: pyrite, celestite, calcite, marcasite, sphalerite, galena, gypsum (including selenite), and many varieties of quartz including chert, flint, and chalcedony.

The list of localities that follows is numbered according to the locations shown on the accompanying maps. These numbers are arranged along the routes which a traveler would be likely to follow. Localities 1 through 48 are in the Lower Peninsula (Fig. XI-4); localities 49 through 115 in the Upper Peninsula (Fig. XI-6).

Figure XI-4. Map of Lower Peninsula mineral collecting localities and exhibits to go with locality list in text.

## Lower Peninsula Mineral Localities and Exhibits

1. Wayne State University, Detroit. Mineral exhibits.
2. Ojibway, Ontario, Canada, south of Windsor, owned by International Salt Company. Although this locality is in Canada, it is near Detroit and open to the public upon special arrangement with the company. It is an underground salt mine. The mineral halite occurs in massive form (rock salt), and as large, clear crystals.
3. International Salt Company Mine, Wyandotte, Michigan. *Not open to public,* but see Ojibway, Ontario, above in this list. Mineral is halite.
4. Rockwood Quarry, Rockwood, Michigan. In Sylvania Sandstone of Devonian age. *Not open to public.* Minerals include: calcite, yellow crystalline variety in geodes; celestite, crystals, in geodes.
5. Maumee Stone Co. Quarry, dolomite quarry. Scofield Quarry at Scofield, about 2 miles NE of Maybee. Minerals are: sulfur crystals (yellow), celestite crystals, calcite crystals.
6. France Stone Company quarry, Monroe, Michigan. *Not open to public.* Minerals are: calcite (both yellow and colorless), marcasite.

# Minerals

7. Ida, old quarry, inquire locally. According to old reports, once furnished following minerals: "gashed" dolomite; strontianite (massive, white, and minute fibrous crystals, in cavities).
8. Adrian College, mineral collection.
9. Hillsdale College, Hillsdale, Michigan. Mineral collection.
10. Just southeast of Coldwater. Abandoned shale quarries in the Mississippian age Coldwater shale; inquire locally. Crack open clay-ironstone concretions on old dumps and look for the following minerals: galena, sphalerite, pyrite; also fossils. Detailed directions to quarries in this area are as follows: (a) 1 mile south from junction U.S. 112–M27 in Coldwater; right (west) about 1.25 miles (along south edge of Coldwater) to second road south; left (south) about 1.25 miles (to just past first road from left); old quarry to east of road about 0.25 mile. *Get permission locally to enter.* (b) 1 mile south from junction U.S. 112–M27 in Coldwater; right (west) 2.3 miles; left (south) about 0.25 mile to old quarries on right (west) side of road.
11. Albion College, Albion, Michigan. Mineral exhibits.
12. Several quarries near Bellevue, Eaton County, Michigan, just northeast of Battle Creek, in Mississippian age Bayport Limestone. Inquire locally for the following localities: (a) Holden quarry, 1 mile north and 1 mile east of center of town. (b) Cheney quarry, 1.25 miles west of center of town on south side of M-78. (c) Three abandoned quarries about 0.5 mile west and 1.25 miles southwest of center of town, along road south of River Road. (d) Abandoned quarry on White Farm, about 1.25 miles due south southeast of center of town.

    Minerals to look for include: pyrite, calcite, marcasite, and occasionally sphalerite, celestite.
13. Battle Creek, Michigan, Leila Arboretum, Kingman Museum of Natural History. Mineral, fossil, and other exhibits.
14. Kalamazoo, Michigan, Nature Center north of town. Mineral collections.
15. Western Michigan University, Geology Department, Science Building, Kalamazoo, Michigan. Collections of rocks, minerals, and fossils.
16. Chamberlain Memorial Museum, Three Oaks, Michigan. Mineral collection.
17. The University of Michigan, Ann Arbor, Michigan, Department of Geology and Mineralogy. Extensive mineral and rock collections and displays.
18. The University of Michigan, Ann Arbor, Michigan, Exhibit Museum, Museums Building. Many exhibits include minerals, rocks, astronomy, planetarium, anthropology, zoology, botany, and fossils. Guided tours available throughout the week for school groups upon advance arrangement.
19. Nature Center, Kensington Metropolitan Park, Milford, Michigan. Mineral exhibits.
20. Michigan State University, East Lansing, Michigan, Geology Department, Natural Science Building. Minerals and fossils.
21. Michigan State University, East Lansing, Michigan, The Museum. Displays of minerals, fossils, and other natural history materials.
22. Lansing, Michigan, 505 N. Washington Street, Michigan Historical Museum. Mineral and fossil exhibits.
23. Grand Ledge Clay Products Company quarries, Grand Ledge, Michigan. Fossils and minerals in Pennsylvanian age Saginaw Formation. Inquire locally. Look for pyrite, sphalerite, marcasite, siderite, and galena, all in ironstone concretions which must be cracked open.
24. Grand Rapids Public Museum, Grand Rapids, Michigan, 54 Jefferson Avenue SE. Displays of minerals, fossils, and zoological and anthropological materials.
25. Grand Rapids Gypsum Company mine, Grand Rapids, Michigan. Inquire locally. Gypsum, including the varieties: rock gypsum, "pencil," and selenite (clear crystal). Also small celestite crystals.
26. Quarries in the vicinity of Grandville near Grand Rapids; inquire locally. Gypsum in the following varieties: massive white; elongate crystals of yellow, brown, and pink selenite, formerly brown calcite crystals.
27. Lake Michigan sand beaches. Look for quartz sand, magnetite in dark grains which stick to a magnet, and many rocks and minerals in the beach pebbles and cobbles.
28. Cranbrook Institute of Science Museum, Bloomfield Hills, Michigan. Many superb displays of minerals and fossils.

Figure XI-5

COMMON MINERALS GROUPED ACCORDING TO CHEMICAL COMPOSITION, WITH USES AND COMMON OCCURRENCE
(All the minerals listed occur in Michigan but not all are abundant or of economic importance here.)

| Chemical Classes | Mineral | Chemical Composition | Principal Uses and Occurrence |
|---|---|---|---|
| Pure or "native" chemical elements | Native copper | Cu | Keweenaw Peninsula; wire, hardware, brass, bronze, electrical, coins, chemical, jewelry. |
| | Graphite | C | In igneous dikes and veins and in metamorphic rocks; near L'Anse in Upper Peninsula; crucibles, electrical brushes, foundry facings, pencils, paint, lubricants, stove polish. |
| Oxides and hydrated oxides | Quartz, including several varieties as well as flint, chert and jasper. | $SiO_2$ | Common in all rock types and in most areas; jewelry, dishes, optical and electronic instruments, mortar, glass, grindstones, building stones, scouring, and abrasives. |
| | Hematite | $Fe_2O_3$ | Sedimentary deposits (often metamorphosed), in igneous rocks, cracks, and fissures; Lake Superior district of Michigan; iron ore, paint pigment, jeweler's rouge, jewelry. |
| | Magnetite | $Fe_3O_4$ | In basic igneous rocks, metamorphic rocks, sedimentary "black sands"; Pre-Cambrian rocks of the Lake Superior region, sands of Great Lakes; iron ore. |
| | Limonite | $Fe_2O_3 - nH_2O$ | Very common in minor amounts in most regions; common in weathered rocks; iron ore, pigments. |
| Sulphides | Chalcopyrite | $CuFeS_2$ | In some regions a very important copper ore. |
| | Galena | PbS | Important in some regions. Main lead ore. |
| | Marcasite | $FeS_2$ (orthorhombic) | Manufacture of sulphuric acid. |
| | Pyrite | $FeS_2$ (cubic) | Manufacture of sulphuric acid. |
| | Sphalerite | ZnS | Main source of zinc, cadmium, and thallium. |
| Sulphates | Anhydrite | $CaSO_4$ | Little used commercially. Occurs in sedimentary rocks in association with salt, gypsum, limestone. |
| | Gypsum | $CaSO_4 \cdot 2H_2O$ | Occurs in sedimentary rocks with salt and limestone. Principal use is in plaster. |
| | Celestite | $SrSO_4$ | In sedimentary rocks; Maybee, Michigan; source of strontium. |
| Carbonates | Calcite | $CaCO_3$ | Sedimentary rocks; widespread; limestones; building stone, cement, steel flux, ballast, road surfacing. |
| | Dolomite | $CaMg(CO_3)_2$ | Abundant in sedimentary rocks in many areas; building stone, source of magnesium. |
| | Azurite | $Cu_3(OH)_2(CO_3)_2$ | In copper mines; copper ore; ornaments. |
| | Malachite | $Cu_2(OH)_2CO_3$ | In copper mines; copper ore; ornaments. |
| | Siderite | $FeCO_3$ | In ore deposits, in sedimentary rocks; minor source of iron ore. |

Figure XI-5—*Cont.*

COMMON MINERALS GROUPED ACCORDING TO CHEMICAL COMPOSITION, WITH USES AND COMMON OCCURRENCE

(All the minerals listed occur in Michigan but not all are abundant or of economic importance here.)

| Chemical Classes | Mineral | Chemical Composition | Principal Uses and Occurrence |
|---|---|---|---|
| Phosphate | Apatite | $Ca_5F(PO_4)_3$ | Igneous rocks, limestones, guano; source of fertilizer, phosphorus, gem stone. |
| Halides | Halite | $NaCl$ | Sedimentary rocks; salt for many uses. |
| Halides | Fluorite | $CaF_2$ | Veins in limestone and dolomite, also with galena; steel flux, glass, ornamental, fluorine. |
| SILICATES — Olivine group | Olivine | $(Mg, Fe)_2SiO_4$ | In basic igneous rocks; gemstones. |
| SILICATES — Amphibole group | Hornblende | $Ca_2(Mg, Fe)_5(OH)_2(Al, Si)_8O_{22}$ | Common constituent of igneous rocks. |
| SILICATES — Pyroxene group | Augite | $Ca(Mg, Fe, Al)(Al, Si)_2O_6$ | Igneous and metamorphic rocks. |
| SILICATES — Mica group | Muscovite | $KAl_2(OH)_2AlSi_3O_{10}$ | Common in all rocks; electrical apparatus, paper, lubricants, rubber. |
| SILICATES — Mica group | Biotite | $K(Mg, Fe)_3(OH)_2AlSi_3O_{10}$ | Common in all rocks; little used. |
| SILICATES — Mica group | Chlorite | $Mg_5Al(OH)_8AlSi_3O_{10}$ | Metamorphic rocks. |
| SILICATES — Mica group | Kaolinite | $Al_4(OH)_8Si_4O_{10}$ | Clays, weathered rocks; soil; ceramics, paper filler. |
| SILICATES — Mica group | Serpentine (asbestos) | $Mg_6(OH)_8Si_4O_{10}$ | Metamorphic rocks; ornamental, insulation. |
| SILICATES — Mica group | Talc | $Mg_3(OH)_2Si_4O_{10}$ | Metamorphic rocks; soapstone, powder, building, ornamental, paper filler. |
| SILICATES — Feldspar group | Orthoclase | $KAlSi_3O_8$ | Common in granite and sedimentary rocks; pottery, enamelware, abrasives. |
| SILICATES — Feldspar group | Plagioclase | Mixture of albite, anorthite | Common in igneous rocks. |
| SILICATES — Feldspar group | Albite | $NaAlSi_3O_8$ | Common in igneous rocks. |
| SILICATES — Feldspar group | Anorthite | $CaAl_2Si_2O_8$ | Common in igneous rocks. |
| SILICATES — Feldspathoid group | Nepheline | $Na(AlSiO_4)$ | Igneous rocks; glass and ceramics. |
| SILICATES — Garnet group | Garnet (many varieties, all complex silicates including several of Ca, Mg, Mn, Fe, Al, Cr) | | Common in igneous and metamorphic rocks; gemstones, abrasives. |

Data mainly from *Mineralogy*, by E. H. Kraus, W. F. Hunt, and L. S. Ramsdell; McGraw-Hill, 1959, 5th edition.

29. Port Huron Public Library Museum, Port Huron, Michigan, 1115 Sixth Street. Exhibits of minerals and fossils.
30. Quarries in the vicinity of Pigeon, east southeast of Bayport. Mississippian age Bayport limestone. Inquire locally. Go 1.5 miles south on M-25 from Bayport, 5.5 miles east on M-142. Look for chert in nodules, and calcite in nodules and geodes.
31. Marina 1.5 miles northeast of Bayport. Look in limestone fragments and blocks for brown calcite and chert in nodules and geodes.
32. Saginaw Museum, Bristol and N. Michigan Streets, Saginaw, Michigan. Mineral displays.
33. Jenison Trailside Museum, 5 miles north of Bay City at Bay City State Park. Mineral exhibits.
34. Alma College, Alma, Michigan. Mineral collection.
35. Griffin and Burt quarries in the vicinity of Omer, Michigan. About 2 miles due northeast of Omer, 1 mile east of M-65. Inquire locally. Look for brown calcite and chert in concretions.
36. Old quarries in vicinity of Au Gres, in the Mississippian age Bayport limestone. Inquire locally. Chert and calcite in nodules and geodes.

## Geology of Michigan

37. Gypsum pit 1 mile east of Turner, Arenac County. Inquire locally. Gypsum.
38. U.S. Gypsum Company quarries along highway at Alabaster. Gypsum occurs in the following varieties: massive rock gypsum in colors ranging from white to pink or red, white "pencil" gypsum, and crystalline selenite (in nodules).
39. National Gypsum Company quarries 1.5 miles south of National City in Mississippian age Michigan formation. Gypsum occurs as massive "rock gypsum" and crystalline selenite varieties.
40. North and east of Tawas. Look in roadside rock piles for glacial cobbles of jasper conglomerate or "puddingstone" which is a light colored quartzite with red jasper pebbles in it.
41. Shoreline of Squaw Bay, south of Alpena. Black shale bedrock outcrop of Devonian age Antrim Shale. Crack concretions for pyrite, marcasite, and anthraconite (a variety of calcite).
42. Calcite Quarry View. Maintained by Michigan Limestone Division, U.S. Steel Corporation, Rogers City. Exhibit overlooks quarry. Displays show fossils, limestone use. Just off right (east) side of U.S. 23A entering Rogers City from south.
43. Roadside ditches and road cuts along U.S. 23 south of Rogers City and along M-68 about 2.5 miles west of town. Dark colored, Devonian age Bell Shale outcrops. Pyrite.
44. Michigan in general. Sand and gravel pits in glacial deposits throughout the state. Look for many varieties of rocks and minerals, especially including quartz (many varieties), agate, chert (including jasper and flint), and minerals in igneous and metamorphic rock cobbles.
45. Con Foster Museum, Clinch Park, Traverse City. Exhibits of minerals and fossils.
46. About 0.5 mile south of Norwood. Bedrock along Lake Michigan shore and in road cuts in area. Outcrops of the Gravel Point Formation of the Traverse Group. Also concretions weathered from Antrim Shale. Look for calcite, anthraconite (a variety of calcite), pyrite, marcasite, and chert.
47. Lake Michigan shoreline bedrock outcrops north of Norwood, in the Devonian age–Traverse Limestone. Look for chert.
48. Shoreline of Lake Michigan and local inland lakes in vicinity of Petoskey. Especially in old quarries and beach at roadside–lakeshore park along U.S. 31 on east edge of Petoskey. Look in beach pebbles for "Petoskey stones," which are partially or fully wave polished and rounded pebbles and cobbles composed of the calcite skeletons of colonial corals and other marine invertebrate animals.

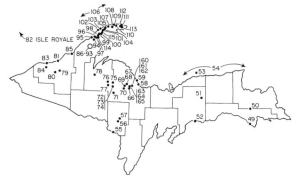

Figure XI–6. Map of Upper Peninsula mineral collecting localities and exhibits to go with locality list in text.

### Upper Peninsula Mineral Localities

49. Road cuts in the vicinity of Pointe Aux Chene near St. Ignace. Outcrops of the Pointe Aux Chene formation. Look for gypsum.
50. Scotts Quarry, near town of Trout Lake. Nine miles east of town on M-48 and about 0.5 mile south on side road off M-48. Inquire locally. Look for chert and flint.
51. About 0.75 mile south of Seney, and also west of town in swamps along highway. Soft, yellow, and red "Bog iron ore" (a variety of limonite).
52. Old quarries on east side of Manistique, near football field. Look for chert and flint in nodules.
53. Lake Superior beaches in vicinity of Grand Marais. Search among pebbles and cobbles on beach. Look for color banded, translucent agates and thomsonite.
54. Lake Superior beaches in Alger, Luce, and Chippewa counties. Look in beach pebbles for occasional wave-worn agates. Not common.
55. Old iron mines in vicinity of Loretto, Vulcan, Norway, Quinnesec, and Iron Mountain. Inquire locally. Iron ore and "iron formation" specimens on old iron mine dumps.

56. Rian Quarry, about 0.5 mile east of Metropolitan. Inquire locally. *Seek permission.* Look for following: Randville dolomite (altered by hydrothermal activity) is the rock being quarried here; also quartzite; diopside, medium to dark green, 90° cleavage angles, masses up to 4 inches; hornblende, black, lustrous, thin prismatic crystals up to 1 inch long; quartz; wollastonite.
57. Metro-Nite Quarry, about 2.6 miles east of Felch, along side road forking north from M-69. Inquire locally. *Seek permission.* Look for Randville dolomite intruded and metamorphosed by granite dikes; dolomite, white to pink, coarse to sugary crystals, good crystals rare; tremolite (a variety of amphibole) in light green, and white to dark green masses, satiny, bladed crystal clusters, some asbestiform masses; quartz, glassy, near dike margins; fluorite, rare, flesh-colored cubes in cavities; pyroxene (varieties pigeonite and salite) rare; diopside (another variety of pyroxene), light colored; phlogopite, small crystals up to 0.5 inch long, near dikes; scapolite, violet-colored, coarse, bladed crystals in clusters or "pods"; serpentine along dike borders; chalcopyrite; pyrite; galena.
58. Marquette, granite outcrop at Dead River Bridge near ore dock. Look in rocks for greenish epidote.
59. Sugar Loaf Mountain, several miles northwest of Marquette. Epidote, green, in rock. Poor mineral locality, but interesting country.
60. Jasper Hill or Jasper "Knob," Ishpeming; inquire locally in town. Look for jaspilite (iron formation with bands of specular hematite and jasper), hematite (specular variety).
61. South Jackson Mine dumps, Negaunee. Iron ore, manganite, barite.
62. Athens Mine dumps, southeast side of Negaunee; inquire locally. Hematite, red, including "pencil ore" and "kidney ore"; goethite, in cavities in ore, as prismatic crystals with cleavage; limonite, dark brown, noncrystalline, compact masses.
63. National Ski Hall of Fame, Ishpeming. Mineral exhibits.
64. Sellwood Mine dumps, Ishpeming. Goethite, limonite crystals pseudomorph after pyrite cubes.
65. Vicinity of Ishpeming, road cuts. Pyrite, hematite (several varieties), magnetite, barite, jasper, banded iron formation (jaspilite).
66. National Mine dump, about 0.5 mile west and 0.25 mile south of town of National Mine, about 3.5 miles south of Ishpeming. Goethite, quartz (crystals stained with iron).
67. Ropes Gold Mine, northwest of Ishpeming, about 3.5 to 4 miles from town on side road leading west from west end of Deer Lake. Inquire in Ishpeming for road to Deer Lake; at lake inquire for mine. Look for asbestos, low grade in veins, light green talc, gold (very rare but specimens have been found).
68. Verde Antique quarry a few hundred feet from Ropes Gold Mine (see 67 above). Look for verde antique "marble," which actually is serpentine streaked with calcite and dolomite.
69. Michigan Gold Mine. Go about 6 miles west of Ishpeming on U.S. 41 and about 1.5 miles north on Gold Mine Lake Road. Then take side road off southeast side of Gold Mine Lake Road. Inquire locally at Gold Mine Lake for directions. Look for quartz in veins; tourmaline, black, needle-like crystals in the quartz; molybdenite, rare, foliated (flaky), soft, lead colored; gold (very rare).
70. Mine dumps in vicinity of Greenwood, at Greenwood Mine 1 mile southeast of town, and Old Furnace Mine 0.5 mile north of town. Inquire locally. Look for specularite (variety of hematite).
71. Republic, iron mine dumps in vicinity. Inquire locally. Look for garnets, feldspars (orthoclase and microcline), jaspilite, siderite, tourmaline, quartz, muscovite, specularite, magnetite, quartz crystals in cavities in specularite, staurolite crystals in light colored mica schist, molybdenite (rare, soft, lead gray, scaly, in quartz veins), beryl, topaz.
72. Beacon area; see Beacon Mine and Phenix Pits in this list; also see reference in Bibliography to Mandarino (1950) for details and locations of Northhampton Mine, Marine Mine, North Phenix Pit, Pascoe Mine, and Hortense Mine in same general area.
73. Beacon Mine on east side of Beacon, North Range Mining Company; see Mandarino, 1950. Look for many minerals as follows: pyrite, usually badly oxidized, cubes (some dodecahedrons) associated with quartz, tour-

maline, chlorite, hematite, and garnet; quartz, in veins and masses, milky or yellow; look in it for tourmaline, chlorite, and garnet; siderite, in cracks and crevices and in quartz, as cleavage plates or in masses, cream colored or reddish brown with a slightly metallic oxidized coating; gypsum (rare), reddish-orange platy coating on rock; hematite, variety specularite, black, micaceous masses, often with quartz and/or pyrite; magnetite, crystals or masses in quartzitic parts of rocks, crystals are octahedrons (8 sided) in chlorite schist, associated with garnet, aphrosiderite, martite, magnet will attract; martite, a variety of hematite replacing original magnetite in octahedrons, red streak, nonmagnetic; tourmaline, long, slender, dark green, brown, or black crystals up to 4 inches long, mainly occurs in chlorite schist, associated with pyrite and garnet; garnet, altered to aphrosiderite, associated with quartz and chlorite; staurolite, abundant, long, slender crystals intertwined in schist; chlorite, in form of aphrosiderite altered from garnet, also in "rosettes"; apatite (rare), long hexagonal (6 sided), prismatic crystals, 0.5 inch to 2 inches long, pink or yellow. Additional but rare minerals include: goethite, rhodochrosite, adularia, barite, molybdenite, marcasite, chalcopyrite, uralite, sericite, grunerite, cummingtonite, chloritoid, clinochlore, muscovite, biotite, bornite, anhydrite, bismuthinite, andalusite, corundum.

74. Champion Mine dumps about 0.5 mile west of center of Beacon. Look for specularite (variety of hematite), magnetite, tourmaline in quartz, pyrite, masonite, garnets (rare).

75. Phenix Pits about 1 mile north of Champion on north side of U.S. 41. Inquire locally. Three pits along path leading west from gravel road in SW ¼, Sec. 29, T48N–R29W; see Mandarino, 1950, in Bibliography for details. About 1 mile due east of Van Riper State Park. About 0.5 mile west of Hortense and Pascoe Pits. Look for the following minerals: calcite, poor, pinkish crystals on goethite; siderite, small, buff, rhombohedra with goethite; hematite, rare, some with apatite; goethite, stalactitic, occasionally iridescent, abundant; apatite, small, red, hexagonal (6-sided) plates in cavities in hematite; quartz, smoky variety, small crystals with goethite.

76. Old iron mine dumps about 0.5 mile east of town of Michigamme, north of old main highway. Look for the following minerals: hematite replacing garnet, 12-sided crystals; tourmaline, black, slender crystals in quartz; chlorite schist (a rock), dark green, soft, common, making up much of the rock; martite, 8-sided crystals in the chlorite schist; garnet, 12-sided crystals up to 2 inches in diameter, badly weathered, dark red, in chlorite; magnetite; grunerite, with a silky texture; aphrosiderite replacing garnets in chlorite; staurolite.

77. South side of Lake Michigamme. Go south out of Michigamme and follow road along south side of lake. Outcrops in fields along road and in road cut about 1.5 miles from town of Michigamme. Look for staurolite in form of X-shaped (twinned) crystals, more resistant to weathering, standing out in relief on softer, dark, schist rock.

78. Near Alberta. About 2 miles north of town and 7 miles south of L'Anse. Old graphite mine concealed along U.S. 41. Difficult to find. Inquire locally. Look for the following: graphite, once mined for paint pigments for old Detroit Graphite Paint Company of Detroit; pyrite; also look in road cuts in vicinity for rare specimens of pyrolusite pseudomorphs after manganite, jaspilite, hematite, quartz.

79. Old copper mine dumps in the vicinity of Mass. Look for tenorite with chrysocolla; datolite; native copper, adularia, calcite enclosing copper.

80. Old copper mine dumps near Rockland. Inquire locally. Native copper, datolite.

81. Lake Superior beaches in vicinity of Ontonagon. Look in beach pebbles for agate, adularia, prehnite, copper, jasper.

82. Isle Royale. *Special note.* The following is a list of minerals reported to have been found on the island, mainly among the beach pebbles, and on neighboring islands. The list is included here for sake of completeness, but the *reader is cautioned* that Isle Royale is a national park, and that mineral and rock *collecting* is now *prohibited* there. Reported are: chlorastrolite, ? lintonite, thomsonite, ? mesolite, carnelian (or sard, a red variety of chalcedony), agate, prehnite, amethyst quartz, rose quartz, chalcedony, jasper, opal, datolite,

epidote, wollastonite, obsidian, calcite, barite, and onyx. For details refer to Dustin, 1932, in Bibliography.

83. Gull Point on M-64 in Ontonogan County about 1 mile east of Silver City. Look in beach pebbles for agates.

84. White Pine Mine at White Pine, southwest of Ontonogan. Chalcocite, minutely dispersed in the Late Precambrian Nonesuch shale; copper, greenockite. Shows on cut and polished surfaces.

85. Agate Beach (or Misery Bay) about 10 miles west of Toivola. Look in beach pebbles for agates; peristerite (variety of albite with "play of colors"), not common.

86. Beaches in vicinity of Chassell. Look in shoreline pebbles for agates.

87. Baltic Mine No. 2, near South Range, about 0.25 mile southeast of M-26, southeast of town of South Range. Minerals reported from this vicinity include: native copper, irregular masses, tarnished black, gray, or green on surface, but showing metallic red when scratched, heavy, thin masses bend; chalcocite, bluish black to lead gray, metallic veins in rock; bornite, peacock-hued or copper red to brown, bronze on fresh surface, purple or red when weathered; chalcopyrite, brassy color, metallic, soft; ankerite, white, with good cleavage, rare; rhodochrosite, rare, pink; calcite, also pink but fizzes with acid; prehnite in association with copper; epidote, mottled with finely divided copper; saussurite, laumontite, domeykite.

88. South Range Quarry, west side of M-26, 0.5 mile north of South Range. Epidote, thomsonite, saussurite, and good exposures of Keweenawan–age felsitic lava flows and conglomerates.

89. Champion Copper Mine dumps in vicinity of Painesdale, about 0.5 mile due south of center of town. Inquire locally. Look for calcite, copper, epidote, chalcocite.

90. Old Isle Royale Mine, copper mine dumps, about 1 mile south of Houghton or 0.5 mile north of Dodgeville off M-26. Inquire locally. Look for epidote (massive variety); quartz (crystals up to 0.5 inch in length as amygdules or in geodes in the lava); chlorite, dark green, BB-sized particles in the basalt rock; calcite, white, cleavage faces form parallelograms; prehnite, vitreous or glassy luster, light green, in veins and cavities; native copper. Rare barite crystals, massive pink anhydrite, laumontite, datolite, analcite.

91. Huron Mine near Atlantic. Old copper mine dumps. Look for quartz crystals in cavities; soft, pink laumontite; calcite, white, with parallelogram cleavage faces; prehnite, glassy (vitreous) luster, light green, in veins and cavities; native copper.

92. A. E. Seaman Mineralogical Museum, Michigan College of Mining and Technology, Houghton. Very extensive mineral exhibits and mine models.

93. Quincy Mine on northern outskirts of Hancock, near crest of river bluff above town going north on U.S. 41. Datolite, rare, porcelainic; copper in amygdules in rock; quartz crystals of the white and smoky varieties; gypsum, calcite enclosing copper.

94. Osceola Mine. Old copper mine dumps. Inquire locally. Quartz and epidote.

95. Calumet Water Works. Look in beach pebbles for thomsonite in basalt.

96. Old Tamarack Mine dumps, about 0.5 mile north of Tamarack, about 1 mile northeast of Calumet. Inquire locally. Copper bearing conglomerate.

97. Wolverine Mine in vicinity of Kearsarge. Inquire locally. Minerals reported include: microcline, a red feldspar in cavities in lava often with white calcite in center of cavity; epidote, green, in cavities; chrysocolla, bluish-green; cuprite, brick red; tenorite, black; quartz; copper; silver.

98. Allouez No. 3 Shaft and others in vicinity. Old copper mine dumps. Mines were in conglomerates. Turn northwest off U.S. 41 at Allouez, and go about 0.25 mile on side road. Look for: native copper, bluish-green chrysocolla, cuprite, tenorite, quartz, datolite, and associations of these.

99. Old copper mine dumps on right and left sides of U.S. 41 at Ahmeek. Minerals reported include: domeykite, metallic, silvery on fresh break, bronze on weathered surface; associated with quartz, copper, epidote, calcite, algodonite, "whitneyite," and feldspars (greenish crystals in basalt).

100. Mine dumps in the vicinity of Mowhawk. About 0.5 mile north of Mowhawk on south-

east side of U.S. 41. Also the Seneca Mine about 0.5 mile north of center of town on Seneca Lake Road. Also Calumet and Hecla Mine on north side of U.S. 41 about 0.5 mile east of center of town. Minerals reported from this general area include: domeykite, metallic luster, silvery on fresh break, bronze on weathered surfaces, associated frequently with white quartz; algodonite, "whitneyite," native silver, silver and copper intermixed, native copper, prehnite, calcite, quartz, green feldspar crystals in basalt.

101. Old Cliff Copper Mine at town of Cliff on old highway about 2 miles southwest of Phoenix. Going south on M-26 bear right off highway about 1.5 miles southeast of Phoenix. Apophyllite, clear, glassy.
102. Seven Mile Point north of Ahmeek, about 3 miles by foot along beach southwest of Five Mile Point and the Sand Hills lighthouse. Must walk from Five Mile Point. Look in beach pebbles along Lake Superior shore for: agate, prehnite (streaks in pebbles), copper (streaks in pebbles), peristerite (variety of albite).
103. Five Mile Point and Sand Hills Lighthouse, about 8 miles north of Ahmeek along Lake Superior shore. Look in beach pebbles for agate, prehnite, copper, peristerite, thomsonite.
104. Phoenix Mine, old copper mine dump on northwest edge of Phoenix. Inquire locally. Look for prehnite in seams and veins, native copper, apophyllite, and chlorite.
105. Ash Bed Mine, southeast side of road between Eagle River and Phoenix. Look for native copper, columnar basalt.
106. Lake Superior shore between Eagle River and Copper Harbor. Look in beach pebbles, especially at Great Sand Bay between Eagle Harbor and Eagle River, for agates, and thomsonite.
107. Copper Falls Mine, old copper mine dumps, walk in about 0.25 mile south of road from point about 3 miles southwest of Eagle Harbor between Eagle Harbor and Phoenix. Look for the following: calcite, large, white cleavage rhombs; zeolites including analcime (or analcite) as small red crystals in basaltic lava rock, and natrolite in the form of long, needle-shaped crystals; datolite, finely crystalline and massive varieties, crystals include colors from white to pink, yellow, brown, and green; prehnite, faujasite.
108. Delaware Copper Mine, old dumps, about 0.5 mile north of Delaware on north side of county road 586. Go north off U.S. 41 at Delaware to county road and then left a short distance to old mine dump on right. Look for: chlorastrolite; datolite, massive variety in nodules; native copper; prehnite.
109. Essey Park, along Lake Superior beach about 0.75 mile east of village of Agate Harbor. Look in pebbles on Lake Superior beach for agates.
110. Mandan, old copper mine dump. Look for prehnite, and native copper.
111. Fort Wilkins Museum, Fort Wilkins State Park, Copper Harbor. Exhibits of Lake Superior Region ores and minerals.
112. Lake Manganese, south of Copper Harbor, along stream flowing out of lake. Stream cuts over vein of manganese ore, braunite.
113. Clark Mine, about 0.5 mile southeast of Lake Manganese. Inquire at Fort Wilkins or Copper Harbor. Walk in along woods road about 1.5 or 2 miles and off to the left side of woods road. Look for: analcime (or analcite); datolite in nodules, chlorite, copper in quartz, hornblende, badly weathered manganese ore, prehnite enclosing copper.
114. Centennial Mine, about 1.5 miles northeast of Calumet or about 1 mile south of Kearsarge on side road about 0.25 mile east of U.S. 41. Look for chlorite, epidote, laumontite, calumetite.
115. Ojibway, old copper mine dumps, about 3 miles northeast of Mohawk on U.S. 41 and M-26 and about 0.5 mile east from there on side road off M-26. Look for quartz and epidote.

## List of Michigan Minerals

Most of the minerals that amateur and professional collectors have reported finding in the state are listed here. Some species are very common, others less so. Some are no longer obtainable except by purchase or trade. An asterisk (*) in front of the more common species and some of their varieties indicates that these species appear in the

# Minerals

mineral identification key earlier in this chapter. Persons interested in the properties and identification of the less common types should obtain a standard mineralogy textbook, such as *Mineralogy* by Kraus, Hunt and Ramsdell (1959; see Bibliography). Some of the names below are applied to varieties, special forms, of common mineral species. In such cases, reference is made to the common alternative name. The names include all those minerals reported from one or more of the Michigan collecting localities listed earlier in this chapter. Some names of especially interesting Michigan rock types also are included. Certain of the names are used only by nonprofessional hobbyists for special Michigan varieties that may not be described or even listed in standard mineralogy texts.

The letters following each name roughly indicate the relative abundance of the species or variety as follows:

VC—very common; C—common; U—unusual; R—rare; VR—very rare.

actinolite—U
*adularia—C (var. of orthoclase)
*agate—U (var. of quartz)
*alabaster—C (var. of gypsum)
*albite—C (rock forming feldspar, a variety of plagioclase)
algodonite—R
allanite—VR
*amethyst—R (var. of quartz)
*amphibole—VC (as rock forming mineral; also see hornblende)
analcime (analcite)—U
analcite (is analcime)
anatase—U (as rock forming mineral)
andalusite—R
anglesite—VR
*anhydrite—U
ankerite—U
annabergite—R
anthonyite—VR
*anthraconite—U (var. oil bearing calcite)
*antigorite—U (var. of serpentine)

*apatite—U
*aphrosiderite—U (var. of chlorite in pseudomorphs after garnet)
apophyllite—R
aragonite—R
argentite—VR
*asbestos—R (var. of serpentine and of tremolite)
atacamite—VR
*augite—VC (as rock forming mineral, var. of pyroxene)
*azurite—R

babingtonite—VR
banded iron formation—VC (rock name, also see jaspilite)
barite—U
bassetite—VR
beryl—R
*biotite—U (var. of mica)
bismuthinite—VR
blumstrandine—VR
*bog iron ore—VC (var. of goethite or hematite)
bornite—R

*botryoidal ore—R (var. of hematite)
*bowlingite—R (var. of chlorite)
braunite—R
brucite—R
buttgenbachite—VR

*calcite—VC
calumetite—R
*carnelian—R (var. of quartz)
*celestite—U
chalcanthite—VR
*chalcedony—C (var. of quartz)
*chalcocite—U
*chalcopyrite—U
*chert—VC (var. of quartz)
chlorastorite—R (var. of pumpellyite)
*chlorite—C
chloritoid—R
chromite—R
chrysocolla—R
*chrysoprase
cinnabar—VR
clinochlore—R
*copper—U
corundum—R
covellite—R
cummingtonite—U
cuprite—R

datolite—R
*delessite—U (var. of chlorite)
diopside—R (var. of pyroxene)
dioptase—VR
*dog tooth spar—C (var. of calcite)
*dolomite—VC (as rock forming mineral)
domeykite—R (both alpha and beta)

*epidote—C
epsomite—U
erythrite—VR

faujasite—VR
*feldspar—VC (as rock forming minerals; see orthoclase and plagioclase)
*flint—C (var. of quartz)
*fluorite—R

*galena—R
garnerite—VR
*garnet—U
glauconite—U
*goethite—C
gold—VR
*grape ore—R (var. of hematite)
*graphite—U
greenockite—R
grunerite—R (an amphibole)
*gypsum—VC

*halite—VC
hausmannite—R
halloysite—R
harmotome—VR
*hematite—VC
heulandite—R
*hornblende—C (as rock forming mineral)

*iddingsite—R (mixture of goethite and chlorite)
ilmenite—R

*jasper—C (var. of quartz)
jasper conglomerate—R (rock name)
jaspilite—C (rock name for iron rich cherty iron formation)

*kaolinite—R
keweenawite—VR (mixture of copper and copper arsenides)
*kidney ore—R (var. of hematite)
*Kona dolomite—U (rock from near Marquette)

laumontite—U
leonhardite—U
lepidochrosite—R

# Geology of Michigan

*limonite—C (field name for fine grained goethite)
lintonite—R (var. of thomsonite)
lotrite—VR

*magnetite—VC
*malachite—R
manganite—R
*marble—U (rock name for metamorphosed limestone or massive "verde antique" serpentine)
*marcasite—U
*martite—U (hematite pseudomorphs after magnetite)
*masonite—R (var. of chlorite)
*mass copper—U (large pieces of native copper)
maucherite—VR
melanterite—R
melilite—R (artificial in slag)
meta-autunite—VR
meta-torbernite—VR
meta-tyuyamunite—VR
*mica—VC
*microcline—VC
*millerite—VR
*milky quartz—R (var. of quartz)
minnesotaite—R
mohawkite—R (mixture of copper arsenides)
molybdenite—R
monazite—R
*muscovite—C (var. of mica)

nantokite—VR
*native copper—VC (copper occurring naturally in the metallic state)
natrojarosite—R
natrolite—R
*needle ore (var. of hematite)
niccolite—R

nontronite—R

*olivine—C
*onyx—R (var. of quartz or calcite)
opal—VR
*orthoclase—C (var. of feldspar)
ottrelite—R (var. of chloritoid)

*pencil ore—R (var. of hematite)
paramelaconite—VR
pararammelsbergite—VR
pectolite—VR (if found at all)
pennine—R (var. of chlorite)
*peristerite—VR (var. of albite showing play of colors)
petoskey stone—C (see corals under chapter on fossils)
pharmacolite—R
phlogopite—R (var. of mica)
*picrolite—R (var. of serpentine)
picromacolite—VR
pigeonite—R (rock forming mineral, var. of pyroxene)
*plagioclase—VC (rock forming mineral, var. of feldspar)
plancheite—R
powellite—VR
*prehnite—C
psilomelane—R
pitchblende—VR (var. of uraninite)
pumpellyite—R
*pyrite—U
pyrolusite—R
*pyroxene—VC (also see augite)
pyrrhotite—R

*quartz—VC

rammelsbergite—R
rhodochrosite—R

*rock crystal—R (var. of quartz)
*rock gypsum—VC (var. of gypsum)
*rock salt—VC (halite)
*rose quartz—C
rutile—R

salite—R (var. of pyroxene)
*sammet blende—R (var. of goethite)
saponite—U
*sardonyx—R (var. of quartz)
*satin spar—U (var. of gypsum and calcite)
saussurite—R (mixture of several minerals)
scapolite—R
scheelite—VR
scorzalite—VR
seamanite—VR
*selenite—U (var. of gypsum)
*sericite—R (fine grained mica)
*serpentine—C
*shot copper—R (native copper in small shot shaped masses in rock)
*siderite—U
silver—R
*slate ore—U (var. of hematite)
*smoky quartz—R (var. of quartz)
*soapstone—R (var. of talc)
*specular hematite—VC (var. of hematite)
*specularite—VC (var. of hematite)
*sphalerite—R
sphene—R
staurolite—U
stellerite—VR
stilbite—VR (if found at all)

stilpnomelane—R
strontianite—R
sulfur—R
sussexite—VR

talc—R
tenorite—R
thomsonite—R
*thuringite—R (var. of chlorite)
titaniferous hematite—R (var. of ilmenite)
titanite—R (is sphene)
topaz—VR
tourmaline—U
tremolite—U (var. of amphibole)
trichalcite—VR (is tyrolite)
tyrolite—VR

uralite—R (pyroxene altering to amphibole)
uraninite—VR

*verde antique marble—U (serpentine rock streaked with calcite and dolomite)

whitneyite—R (arsenic rich copper)
wolframite—VR
wollastonite—VR (if found at all)

xonotlite—R

zeolites—U (a group of silicates including analcite, natrolite, stilbite, heulandite, thomsonite, laumontite, leonhardite)
zircon—R
zoisite—R

# XII
# Rocks

## Introduction

Rocks are aggregates of minerals. Some rocks are made up entirely, or almost entirely, of a single mineral; limestone made of calcite and rock salt made of halite are two examples. More often, however, rocks are composed of several kinds of minerals. The rock types are variable in their mineral and chemical composition, whereas minerals are chemically homogeneous.

## Main Rock Types

The three main types of rock are *igneous, sedimentary* and *metamorphic*. Each of these terms gives a general idea of the way in which that particular rock type forms. Igneous rocks solidify from a molten magma. Sedimentary rocks are deposited either as fragments from some preexisting rock or as chemical precipitates, and under normal earth surface temperatures and pressures. Metamorphic rocks are formed by alteration under great heat and pressure from preexisting rocks of any of the three types. The rock classifications presented in this chapter are field classifications. They can be used without the aid of a microscope or other expensive equipment. Much more elaborate classifications have been made, but these mainly involve further subdivision of the broad divisions of the field classifications. Field classifications of rocks are based primarily on the texture of the rock and upon its mineral composition. Since both texture and composition are gradational things, the various rock types named on the following charts are merely arbitrary subdivisions of actually continuous ranges. For example, there are igneous rocks that are almost exactly intermediate in mineral composition between a normal diorite and a normal gabbro; or intermediate in texture between a diorite and an andesite.

## Rock Textures

The term *texture*, as applied to rocks, means the size, shape, and arrangement of the constituent mineral particles; the term is used in slightly different ways for each of the three main rock types, as will be explained later.

Differences in the degree of crystallinity (crystal

265

**Geology of Michigan**

Figure XII-1. Igneous rock textures as seen in hand specimens.

A—Equigranular (even grained), granite.
B—Porphyritic (inequigrained), basalt porphyry.
C—Porphyritic (inequigrained), felsite porphyry.
D—Aphanitic (dense, fine-grained), basalt.
E—Glassy, obsidian.
F—Vesicular, basalt.

faces well formed and obvious, or not so) and in the size of the crystals determine the texture of an igneous rock. Both characteristics depend largely on the rate of crystal growth and solidification in the molten magma. These, in turn, are chiefly determined by the rate of cooling, but may also be affected by the chemical composition and volatile gas and fluid content of the magma. The principal textures of igneous rocks (Fig. XII-1) are: *granular* (grained with individual crystal grains large enough to be seen with the naked eye), *aphanitic* or *dense* (with mineral grains too small to be seen by the naked eye), *glassy* (noncrystalline or microcrystalline), and *pyroclastic* (composed of volcanic fragments). Pyroclastic rocks sometimes are considered as sedimentary rocks of a rather special sort. Many other textural terms of lesser importance may be applied to igneous rocks. *Por-*

Figure XII-2. Sedimentary rock textures as seen in hand specimens.

A—Exogenous (clastic, detrital), rounded and sorted grains in a matrix or cement. Conglomerate specimen.
B—Exogenous; broken, abraded, size-sorted shell fragments, in coquina.
C—Endogenous (chemical), fine-grained calcite crystals (not visible) in a marine limestone. Shell is a brachiopod.
D—Endogenous, coarse-grained halite (rock salt) intergrown crystals.

*phyritic* refers to a "2-textured" rock containing some large crystals in a background of smaller ones which results from 2 periods of cooling at different rates during the time of solidification of a magma. *Vesicular* rock texture results from the trapping of gas bubbles in a rapidly solidifying magma; this texture usually is restricted to the upper parts of lava flows. *Amygdaloidal* is a texture resulting from the filling in of vesicles with mineral matter.

The texture of a sedimentary rock depends upon the nature of the particles (pieces of shells, differently sized fragments of rocks and minerals) and the manner in which the particles are bound together. The binding agent may be a cement carried into the sediment and deposited around the particles, or the rock may be held together by the intergrowth of the constituent mineral grains in the case of mineral chemical precipitates. The 2 main sedimentary rock textures (Fig. XII-2) are *clastic* (fragmental, detrital) and *nonclastic* (chemical precipitates).

The various textures of metamorphic rocks (Fig. XII-3) are dependent on differences in orientation or alignment of the crystals and on the size

Figure XII–3. Metamorphic rock textures as seen in hand specimens.

A, B, and C—Foliated (gneissic).
D and E—Foliated (schistose).
F—Foliated (slaty).

G—Nonfoliated, specimen is quartzite.
H and I—Nonfoliated, specimens are marble.

of the crystals. The two general textural types are *foliated* (in which the constituent crystals are aligned parallel with one another so that the rock tends to split into folia or plates) and *nonfoliated*. The foliated metamorphic rocks are further subdivided into: *gneissic* (adjacent folia of different mineral composition); *schistose* (dominantly comprised of platy or rod-like minerals with their long dimensions oriented in a single direction, with adjacent bands of similar mineral composition, and tending to split along irregular surfaces); and *slaty* (very finely foliated, crystals microscopic in size, tending to split readily along almost perfectly flat, parallel planes).

In nearly all field classifications of rocks, the second major variable is chemical (mineral) composition, which may be revealed either by the mineral composition of the rock or, to some extent, by its general color.

## Rock Identification

With this brief introduction to rocks and with the introduction to mineral identification in Chapter XI you are ready to try your hand at identifying rocks using the tables that follow. Before using a particular table you should ask yourself several questions:

1. Is the rock igneous, sedimentary, or metamorphic?
2. What is the texture of the rock?
3. What is the mineral composition of the rock? A hand lens and the mineral identification kit and key will be useful here.

If you can correctly answer these questions you are then ready to refer to the rock identification charts that follow and to continue the identification. However, you should not expect to be able to identify rocks skillfully until you have become familiar with mineral identification. An identified set of standard rock samples and a set of rocks illustrating textures will be of help initially. Such sets may be purchased from scientific supply companies. When you develop confidence using the identified rock set that you have purchased, then take off the labels and shuffle the specimens around, or get another, unlabeled set. When you are satisfied that you can identify the unlabeled specimens, go to a gravel pit where rocks of many different types usually occur and try your skill there. Remember to take a hammer and to crack open the cobbles and pebbles. Smoothed and weathered rocks such as those found in gravel pits are difficult to study; freshly broken surfaces are necessary. Do not be discouraged if at first you can identify with certainty only a few specimens; skill will come with practice and as you learn from experience the variation that is to be expected in the main rock types. If possible, have someone more experienced than yourself check your identifications, or, better yet, have him go with you the first few times.

## Igneous Rocks

The *Chart for Igneous Rocks* (Fig. xii–4) is nearly self-explanatory. One major criterion, composition, is easily applied to the coarsely textured rocks but you may wonder how this can be of use in the case of the aphanitic rocks in which the individual mineral grains are too small to be seen. In these finer grained rocks the general color of the rock is used as an indication of the major mineral composition. For example, a dark-colored, aphanitic rock is likely a basalt; it is dark colored because it is composed of a relatively high proportion of iron-rich (ferromagnesian), dark-colored amphibole or pyroxene minerals. On the other hand, a light-colored, fine-grained igneous rock is most likely a felsite, composed of feldspar and quartz, both of which tend to be relatively light colored. In igneous rocks, the term "dark" is applied only to rocks that are dark gray, dark green, or black. All others, including dark purple and red, are considered "light" colored. Color cannot be used to identify volcanic glasses. Obsidian, which is silica-rich is usually dark. It is difficult or impossible to tell different kinds of volcanic glass apart by simple hand specimen identification techniques.

The distinction between diorite and gabbro needs special mention. Diorite usually contains a smaller percentage of ferromagnesian minerals, and hence is lighter in color than gabbro, but this is not always the case. The real distinction is based on mineral composition. Gabbro nearly always contains a pyroxene, whereas diorite usually contains an amphibole; but even more importantly, the distinction between diorite and gabbro is based upon the chemical composition of the plagioclase feldspar, which cannot easily be determined by simple visual inspection. Thus, it is difficult definitely to distinguish diorite from gabbro by simple field identification tests. Your best guess, without laboratory study, might be wrong if based upon color, but so might the guess of a professional geologist.

Another problem involves the distinction between rhyolite and dacite. In the field rhyolite can be distinguished from dacite (the fine-grained equivalent of granodiorite) only with difficulty; such distinctions may be avoided by simply calling both kinds of rocks by the more inclusive term, *felsite*.

## Metamorphic Rocks

The *Metamorphic Rock Chart* (Fig. xii–5) presents an extremely simplified key. The major subdivisions, foliated and nonfoliated, and the special

### Figure XII-4

### Igneous Rock Chart

(\* Indicates common in Michigan)

| TEXTURES (read down) | PREDOMINANT MINERALS, REFLECTING CHEMICAL COMPOSITION (read across) →→→ Decreasing silica →→→ | | | |
|---|---|---|---|---|
| | *Feldspar and Quartz* | *Feldspar Predominant (No Quartz)* | *Ferromagnesian Minerals (Biotite, Hornblende, Augite) Predominate. Plagioclase Feldspar (No Quartz)* | *Ferromagnesian Minerals Only (No Quartz, No Feldspar)* |
| *Granular* Mineral crystals clearly visible. May be porphyritic. | *GRANITE (potassium feldspars such as orthoclase and microcline predominate) *GRANODIORITE (plagioclase feldspars predominate) | *DIORITE | *GABBRO | ULTRABASIC ROCKS — PERIDOTITE (with olivine and a pyroxene mineral) PYROXENITE (pyroxene alone) *SERPENTINE (altered olivine and pyroxene minerals) |
| *Aphanitic* Fine-grained. Crystals too small to see. Some porphyritic | *FELSITE (including rhyolite and dacite) | *ANDESITE | *BASALT | |
| *Glassy* | OBSIDIAN—if dense or massive PUMICE—if frothy | | BASALT GLASS (quite rare) | |
| *Pyroclastic* or fragmental; often classed as sedimentary | VOLCANIC BRECCIA or CONGLOMERATE—fragments over 4 millimeters in diameter VOLCANIC TUFF or ASH—fragments less than 4 millimeters in diameter | | | |

↑ Increasing grain size—Slower cooling rates

Additional notes: SCORIA is a vesicular igneous rock of fine or glassy texture, usually of basaltic composition.

AMYGDALOIDAL STRUCTURE is that produced in a vesicular rock by filling of vesicles with mineral matter.

If a rock falls in one of the above categories on the chart but is porphyritic, use a compound name. For example, granite porphyry (or porphyritic granite), and basalt porphyry.

---

types of the foliated metamorphics, such as schist and gneiss, are distinguished chiefly by their texture. Beyond this metamorphic rocks are further subdivided largely on the basis of their major constituent minerals. The varieties of schist, for example, include such possibilities as quartz-mica-schist, hornblende-schist, garnetiferous-biotite-schist, staurolite-schist, and others, with the name in each case including one or more of the minerals important in the rock. In each of the examples just given, the term schist alone would have been correct but simply not as exact. The major subdivisions of the nonfoliated metamorphic rocks also are based upon mineral composition.

Some of the differences in metamorphic rocks are the result of different compositions of the original rocks that subsequently were metamorphosed. However, other differences result from the nature of the metamorphic process to which the original rocks were subjected. This latter type of difference is reflected in the lower half of the chart in what is termed a "genetic" classification. By genetic we mean how, beginning with an original rock of a certain type, different kinds of metamorphic rock products may be formed depending upon the degree of metamorphism. Heat and pressure are the principal agents of change. With increasingly greater heat and pres-

## Figure XII–5

### METAMORPHIC ROCK CHART

(All types occur in Michigan)

| IDENTIFICATION KEY | | |
|---|---|---|
| Texture | Diagnostic Features | Rock Name |
| FOLIATED | Fine grained; most or all mineral grains invisible to the naked eye; similar composition in adjacent folia (bands); smooth, even slaty cleavage. | SLATE |
| FOLIATED | Fine grained; mineral grains barely or not visible; similar composition in adjacent folia; folia minutely wavy. | PHYLLITE |
| FOLIATED | Medium grained; many of mineral grains visible to the naked eye; relatively uniform and similar mineral composition in adjacent folia; folia irregular and discontinuous; often rich in mica. | SCHIST |
| FOLIATED | Coarse to medium grained; mineral grains visible to naked eye; adjacent folia of different mineral composition; contains abundant feldspar; folia irregular and discontinuous. | GNEISS |
| NONFOLIATED (but may be faintly banded due to presence of original stratification) | Chiefly composed of quartz; if original sedimentary quartz grains are distinguishable, note that rock breaks through the grains rather than along grain boundaries. May be banded. | QUARTZITE |
| NONFOLIATED | Chiefly calcite ($CaCO_3$) or dolomite ($CaMg(CO_3)_2$). | MARBLE |

*Note:* The many varieties of schists and gneisses are subdivided on the basis of their mineral composition, which is determined largely by (a) the composition of the original rock, (b) the "grade" or intensity of metamorphism, and (c) the kinds of chemical substances either removed or introduced during metamorphism. Examples are: garnetiferous schist, biotite schist, etc.

### GENETIC RELATIONS OF METAMORPHIC ROCKS

| Original Material | Products Resulting from Increasing Metamorphic Intensity (Low intensity to left—high intensity to right) |
|---|---|
| Limestone and Dolomite ⟶ | MARBLE (alteration occurs at low intensity. Little or no change thereafter) |
| Quartz sandstone ⟶ | QUARTZITE (alteration occurs at moderate to high intensities. No mineral change because original composition simple—$SiO_2$) |
| Shale ⟶ | SLATE→PHYLLITE→SCHIST→GNEISS→GRANITE |
| Peat ⟶ | LIGNITE→BITUMINOUS→ANTHRACITE→GRAPHITE |
| Igneous rocks (many types) ⟶ | These become schists and gneisses. In general, the iron and magnesium rich (ferromagnesian or simatic) rocks are altered to schists and amphibolites, whereas the silica and aluminum rich (sialic) rocks form gneisses, but almost any variation is possible because of variations in intensity of metamorphism and also because of the opportunity for the addition or removal of elements. |

sure a series of metamorphic types is produced by rearrangement of atoms and recombination of the chemical elements present in the original rock. For example, a shale (one possible original rock type) alters first to slate and then to schist or possibly even to "granite." When such changes occur, the chemical composition may remain much the same, but the chemical elements combine in new ways, thus forming new minerals.

Frequently, it is found that over a large area in a region of metamorphic rocks there is a zonation or "banding" of metamorphic rock types, the zones being arranged around a center of maximum metamorphic intensity. The zonation usually does not parallel or correspond with any changes in the bulk chemical composition of the rocks of the region. This arrangement into zones of different "grade" resulted from differences in the intensity of metamorphism and is revealed by the first appearance in the rocks of certain key minerals which record a definite temperature of formation. A common set of mineral zones in a schist (formed from shale), for example, would include the following key minerals listed in order of increasing metamorphic intensity: chlorite, biotite, red garnet, staurolite, kyanite, and sillimanite.

A study of the relationship between the composition and grade of metamorphism of rocks derived originally from basalt has given rise to the concept of *metamorphic facies,* each facies characterized by a unique mineral assemblage. Each metamorphic facies represents a particular metamorphic environment (range of temperature and pressure) in which the mineral assemblage is stable and hence likely to form. With a change in environment, particularly with an increase in temperature and pressure, the mineral assemblage becomes unstable and a new assemblage, stable under the new set of conditions, is formed in its place. Eskola, a Finnish geologist who conducted much of the early work that led to the idea of metamorphic facies, recognized 4 facies equivalent to basalt in bulk chemical composition, as shown below:

METAMORPHIC FACIES EQUIVALENT TO BASALT
IN BULK CHEMICAL COMPOSITION

| ORIGINAL IGNEOUS ROCK | KEY MINERAL COMPOSITION |
|---|---|
| Basalt | *Pyroxene* and *plagioclase* |
| METAMORPHIC FACIES | METAMORPHIC MINERALS |
| *Low metamorphism* | |
| Greenschist facies | Chlorite, epidote, albite |
| Amphibolite facies | Amphibole and calcium- and sodium-rich plagioclase |
| Pyroxene granulite facies | Calcium, magnesium or iron pyroxene, and calcium plagioclase |
| *High metamorphism* | |
| Eclogite facies | Sodium pyroxene and magnesium garnet |

We have seen that rocks may be metamorphosed by the application of heat and pressure alone. However, in many cases the events leading to metamorphism are such that chemical elements not present in the original rocks also are carried in as metamorphic change progresses. This, of course, changes the bulk chemical composition of the original rocks and adds greatly to the variety of new mineral and rock types that may form. When this happens the situation becomes more complicated, and it is often very difficult or even impossible to determine from the metamorphic end product, exactly what was the nature of the original rock.

Bedrock exposures of metamorphic rocks in Michigan generally are restricted to the areas of Precambrian age rock outcrops in the western half of the Upper Peninsula. The reasons for this are discussed in earlier chapters. However, fragments of many types of metamorphic rocks may be found in almost any gravel pit throughout the state. These were torn loose from bedrock over which Pleistocene glaciers rode, carried southward by the ice, and dumped in the many glacial deposits left covering the state when the ice melted. Discussions of particular types and ages of metamorphic rocks in Michigan will be found in Chapter IV dealing with the Precambrian of Michigan in the sections describing the geology of the Iron Country and Copper Country.

## Sedimentary Rocks

*Sedimentary rocks* provide a more detailed, more easily read, and more meaningful record of the earth surface conditions during the geologic past than do igneous and metamorphic rocks. The reasons for this are several. First, sedimentary rocks form at the earth's surface and, unless affected by subsequent geologic events, tend to remain at or near the surface; therefore they tend to be more accessible. Second, rocks of this type originate under normal earth surface conditions of the sort we are accustomed to see and with which we are familiar. Finally, they preserve various structures, textures, fossils, and other easily interpreted clues to the conditions and events which produced them in the geologic past. Therefore, this section will be somewhat more detailed than those preceding sections devoted to igneous and metamorphic rocks. Nearly all of the sedimentary rock types, structures, fossils, and other features discussed in the following paragraphs occur at one or more places in Michigan.

The *classification of sedimentary rocks* presented in the next 2 charts recognizes 2 major categories, *endogenous* (synonyms are *chemical* or *biochemical*), and *exogenous* (synonyms include *detrital, fragmental,* and *clastic*). Although all sedimentary rocks are composed of materials derived from preexisting rock materials, the manner of their origin varies.

*Endogenous sediments* are those produced by either organic or inorganic mineral precipitation and crystal growth. In this group the chemical elements that precipitate were dissolved from some preexisting rock, perhaps at a great distance from the site of deposition. *Organic precipitation* of mineral matter to form sediment occurs when some organism takes chemicals out of the environment and in the course of its normal life processes forms a shell or skeleton which is the "precipitate." An example of an *inorganic* sedimentary mineral *precipitate* of the endogenous type would be inorganic limestone formed from lime oozes precipitated directly out of sea water without the aid of organic activity.

*Exogenous sediments* are formed of rock and mineral fragments eroded from some source rock area and carried (transported), without going into solution, to an area of deposition. Examples would be sand, silt, and clay materials deposited along the margins of the sea. The transporting agent may be moving water, wind, or ice. Organisms may be important contributors to either major sediment type. Fossilized remains often make up major or minor portions of a sediment.

Endogenous sedimentary rocks are classified on the basis of two criteria in the following chart. These are the abundance or relative proportion of fossils making up the rock, and the principal minerals composing the rock. The fossil remains often found in endogenous rocks are characterized by being little broken or even unbroken, indicating that the fossil materials were not transported from a distance but accumulated near where the organisms themselves lived. Hence, they are truly "in place" organic precipitates.

The exogenous sediments are composed largely of silicate mineral fragments, but also included in this group are the clastic carbonate sediments and (by some) even the pyroclastic volcanic rocks such as ash and tuff. The exogenous group is characterized by rounded to subangular grains set in a "matrix." By matrix is meant the fine-grained background material or groundmass which surrounds the sedimentary grains. The grains of clastic sedimentary rocks commonly are quartz, feldspar, mica, chert, limestone or fossil fragments (the latter as opposed to whole or nearly whole shells). The matrix ordinarily consists of finer particles of sand, silt, and clay, or a mineral cement such as calcite, opal, or limonite. Only a few combinations of grain types and matrix are common, and these make up the more typical clastic (exogenous) sedimentary rock types. The exogenous sedimentary rocks are classified on the basis of their texture (grain size) and their principal minerals. The fragmentary nature of any fossil material that may occur in clastic sediments is important because it indicates that the fossil material was transported, by waves or currents, thus distinguishing it from the "in place" deposits of whole or nearly whole shells of endogenous deposits that accumulated close to the place where the contributing organisms actually lived.

The two charts that follow will aid in identification of sedimentary rocks (Fig. xii–6, 7). First decide if the rock in question is sedimentary. Then determine if its characteristics are those of exogenous or endogenous material as described above. As with identification of minerals and the other rock types, recognition of sedimentary rocks involves skill that comes with practice. If possible, visit and study collections of sedimentary rocks in a museum or private collection for help at the start.

If the rock is exogenous then determine the mineral composition (listed for each vertical column across the top of the chart) and the grain size (listed for each horizontal row along the left side of the chart). The particular combination of grain size and mineral composition determines the name on the chart. The grain size should be estimated on the basis of the majority composition; that is, if the majority of grains are of sand size, consider the rock a sandstone, even though grains of other sizes may be present in smaller proportions.

If the rock is endogenous, determine the combination of mineral composition (top of chart) and proportion of fossils that fits the specimen best. This involves personal judgment in field identifications.

*Sedimentary rock textures* and *structures* are particularly useful in historical geology because they often reveal much concerning the conditions of transport and environment of deposition of those rocks. (See Fig. xii–9 for summary of following section). Grain size and degree of sorting (Fig.

## Figure XII–6

### Exogenous (Detrital, Clastic) Sedimentary Rocks

| Texture (Grain Size) | ←——Principal Minerals——→ | | | | | |
|---|---|---|---|---|---|---|
| | Quartz (± cement) | Quartz + Feldspar (± cement) | Quartz + Clay Matrix (± cement) | Calcite Grains (Not Inter-Grown Crystals) + calcite cement | | Calcareous Fossil Fragments + calcite cement (± calcite grains) |
| **Gravel** Coarse grained Majority of grains over 2 mm. in diameter | Quartzose Conglomerate | Arkosic Conglomerate | Graywacke Conglomerate | Limestone Conglomerate | Coquina (if fossil fragments dominant or abundant) | Bioclastic Conglomerate (if fossil fragments common to few) |
| **Sand** Medium grained Majority of grains 2 mm.–1/16 mm. in diameter Visible to naked eye | Quartzose Sandstone | Arkose (or Arkosic sandstone) | Graywacke (or graywacke sandstone) | Calcarenite (or detrital limestone, or clastic limestone) | Coquina | Bioclastic Calcarenite |
| **Silt** Fine grained Majority of grains 1/16 mm.–1/256 mm. Invisible to naked eye but feels gritty when scratched with fingernail | SILTSTONE (Very difficult to distinguish different types of mineralogy without microscope or laboratory analysis) | | | LIMESTONE (Difficult to distinguish from endogenous types of limestone) | | |
| **Clay** Very fine grained Majority of grains less than 1/256 mm. Feels smooth when scratched with fingernail Earthy or clayey odor when moist | SHALE | | | LITHOGRAPHIC LIMESTONE | | |

Modified from *Historical Geology Manual and Exercises*, by J. A. Dorr, Jr., and L. I. Briggs, Geo. Wahr Publ., Co., Ann Arbor, Michigan.

XII–8) are two such characters. A coarse grain size is related to a high competence of the transporting agent, such as strong currents or waves. A fine grain size indicates a low transportation competence, in quiet waters, for example. Deposition occurs where the transporting agent slows down. Sorting is a measure of the mixture of grain sizes. A perfectly sorted sediment would contain grains of only one size, a poorly sorted sediment grains of many sizes. Good sorting commonly indicates continual reworking by a transporting agent, for example, by waves along the shore or winds in dune areas. Poor sorting indicates rapid deposition without reworking, as in the morainal deposits laid down by melting glacial ice, or in flash flood deposits on alluvial fans. The following combinations have been shown, by observation of modern conditions, to be important and historically significant when found in ancient sediments. The listing will serve not only as a source of information for interpretation of Michigan sedimentary rock history, but also as an example of the Principle of Uniformitarianism:

## Figure XII–7

### ENDOGENOUS (CHEMICAL, BIOCHEMICAL) SEDIMENTARY ROCKS

←———Principal Minerals———→

| Abundance of Fossils | Quartz (or opal) | Calcite (or Dolomite) | Carbonaceous (Plant Remains) | Hematite (Possibly Some Silica and/or Calcite) | Halite (Possibly Some Anhydrite) | Gypsum |
|---|---|---|---|---|---|---|
| Fossils<br><br>Predominant | Diatomite (if diatoms)<br><br>Radiolarite (if radiolarians) | Fossiliferous limestone*<br>or<br>Fossiliferous dolomite*<br>May be named for the predominant fossil type.<br>Ex: Coralline limestone*<br>Crinoidal dolomite | Peat*<br><br>Lignite* | (rare) | (rare) | (rare) |
| Fossils<br><br>Subordinate | Diatomaceous or radiolarian Chert | Fossiliferous limestone*<br>or<br>Fossiliferous dolomite* | Lignite*<br>Bituminous Coal* | "Clinton" type iron ore | (rare) | (rare) |
| Fossils rare or absent<br>Minerals in a chemically precipitated crystalline intergrowth | Chert*<br>Flint*<br>Geyserite | Limestone*<br>Dolomite*<br>(Travertine and caliche more rarely) | Anthracite | Oolitic Iron ore (fossils rare)<br>Cherty iron ore (fossils absent) | Rock salt*<br><br>EVAPORITES | Rock gypsum* |

Between the Carbonaceous and Hematite columns: *Increasing alteration due to heat and pressure* / *Decreasingly recognizable fossil plant content*

\* Common in Michigan.

Modified from *Historical Geology Manual and Exercises*, by J. A. Dorr, Jr., and L. I. Briggs, Geo. Wahr Publ. Co., Ann Arbor, Michigan.

A. *Coarse grain size* and *poor sorting*. Relatively rare. Restricted largely to boulder clay (glacial till) and some alluvial fans.

B. *Coarse grain size* and *good sorting*. Common in the youthful (upstream) parts of streams, on small deltas, and along lake and ocean shores in pebble or gravel beaches.

C. *Medium grain size* and *poor sorting*. Common in alluvial fans, marshes, swamps, and marine turbidity current deposits (as graywacke).

D. *Medium grain size* and *good sorting*. Most streams, beaches, and sand dunes.

E. *Fine* and *very fine grain sizes* and *poor sorting*. In quiet waters offshore, wind blown dust (loess), river flood plains, large deltas, swamps and marshes.

F. *Fine* and *very fine grain sizes* and *good sorting*. Very rare. Generally in residual soils such as laterites, bauxites, and underclays.

Figure xii–8. Some sedimentary rock structures of historical significance.

A—Current ripples.
B—Wave ripples.
C—Incipient mudcracks and raindrop impressions.
D—Raindrop impressions.
E—Flow casts.
F—Mudcracks.
G, H—Faceted and striated glacial cobbles.
I, J, K—Cross-bedding and channel lensing.
L—Unsorted (different sizes), pebbles and cobbles in clayey glacial till.
M—Poorly sorted, dark jasper pebbles in finer-grained sand matrix.

## Figure XII-9

### Historical Significance of Some Physical Features of Sedimentary Rocks

| | |
|---|---|
| Aligned particles (tend to parallel one another) | Moving currents as in rivers, shore-line areas, glacial ice |
| Alternating (or nonalternating) lithology | Environments of deposition alternated (or did not) in repeated cycles |
| Changing lithology | Environments of deposition changed |
| Cross-bedding | Moving currents in rivers, shore-line areas, turbidity currents, or wind-blown sand |
| Faceted cobbles or planed bedrock | Glacial ice abrasion |
| Flow casts | Slumping, sliding, squeezing of soft, water-saturated sediment |
| Frosted grain surfaces | Wind abrasion |
| Mudcracks | Drying of wet, fine-grained sediment, usually during exposure to air, as on tidal flats, river floodplains, or fluctuating lakes |
| Raindrop impressions | Impact of water drops on wet, fine-grained sediment during exposure to air, as on tidal flats, river floodplains, or dry lake beds |
| Ripplemarks, asymmetrical | Moving currents of water or air, as in rivers, shore-line areas, dunes, turbidity currents |
| Ripplemarks, symmetrical | Oscillation of waves in relatively shallow water (depths less than one-half maximum wave length) |
| Rounded grains | Long continued wave or current abrasion, usually during distant transport |
| Size of grains—large (or small) | Powerful (or weak) transporting agent, nearby (or distant) source area, shallow (or deep) water |
| Sorting of grains good—all close to same size | Transport and deposition by effective sorting agent such as water or wind |
| Sorting of grains poor or absent—mixture of many sizes | Transport and deposition by poor sorting agent such as glacial ice or turbidity currents |
| Striated surfaces (scratched, grooved) | Glacial ice abrasion (or landslide, or mud flow) |
| Thick (or thin) bedded | Long (or short) period of deposition, in stable (or unstable) environment |

Another significant character of sedimentary rocks is the *roundness of the grains* (Fig. XII-8). This may serve as an index to the distance of transport, but it must be so used with caution. In general, the farther a sediment particle is carried the more the corners are abraded and the rounder it becomes. However, pebbles and boulders (coarse grains) round more rapidly than sand grains because the smaller grains are often crushed by the larger ones, thus to become angular again. Moreover, softer minerals, such as calcite, round much more rapidly than harder minerals, such as quartz. Also, minerals with good cleavage tend to break along the cleavage planes while those without cleavage soon become rounded. Wind as a transportation agent selectively sorts out already rounded grains in preference to angular grains, because the rounder grains move more easily and roll farther than the angular ones; in contrast, streams selectively move angular particles faster than rounded ones largely because the angular particles, once in the moving stream, fall to the bottom more slowly and therefore move farther than the rounded grains. Consequently, round sand grains are typical of sand dunes, but atypical of stream deposits; and stream conglomerates char-

# Geology of Michigan

acteristically contain round pebbles and boulders with angular sand grains. Lastly, rounded grains are most likely to be produced by erosion and redeposition of an older sedimentary rock, thus being the product of 2 or more cycles of erosion and deposition.

The thickness of the layers (strata) in sedimentary rocks is related to the length of time a particular episode of deposition in a particular environment lasted. Thick bedding suggests long uninterrupted deposition, thin bedding indicates frequently interrupted, short periods of deposition. Remember, however, that the coarse-grained sediments accumulate faster than do the finer materials. For example, a thick gravel bed may build up in the same length of time as would a thinner clay layer. Comparisons of length of deposition should always be made between sediments with equivalent grains sizes (e.g., a thick as contrasted with a thin clay).

An *alternation of rock types* in the layers of a sedimentary rock sequence implies alternation or variation in the environment of deposition. If the environments of the rock types can be interpreted from the clues in the rocks themselves one may tell what variations in conditions of the area of deposition occurred in the past. For example, alternating layers of marine shale and freshwater sandstone would indicate fluctuations of a marine body of water in and out of an area.

*Cross-bedding* is a sedimentary rock structure commonly encountered in medium and coarse grained deposits (Fig. XII–8). It results from current flow, as on deltas, floodplains of rivers, alluvial fans, and sand dunes. A shift in current often results in erosion of the upper parts of one set of inclined, cross strata so that when deposition occurs again the strata in the lower set terminate abruptly against the bottom of the next set above. The concave faces of cross-bedded strata face in the direction of current flow, hence can be used as a clue to ancient current directions and thus to the paleoslopes and geography of the region. Variations in crossbed direction imply variations in current direction. (See Chapter V for actual examples in the Jacobsville and Munising sandstones in the Upper Peninsula of Michigan).

Two types of *ripple marks* are fairly common in sedimentary rocks. *Current ripples* are asymmetrical (like a miniature sand dune) having a steep side facing in the down-current direction (Fig. XII–8). Water or wind currents are implied.

Figure XII–10

NATURAL ENVIRONMENTS OF SOME FOSSIL INVERTEBRATE ANIMALS AND PLANTS COMMON IN ROCKS OF MICHIGAN

| FOSSIL TYPE | Certainly Marine | Probably Marine | Probable Water Conditions | | |
|---|---|---|---|---|---|
| | | | Shallow | Clear | Warm |
| Foraminifera (Phylum Protozoa) | | X | | | |
| Colonial corals (Phylum Coelenterata Class Anthozoa) Reef builders | X | | X (not over 200 feet) | X | Mean annual temperature (about 68°F) |
| All corals | X | | X | X | |
| Stromatoporoidea | X | | | X | |
| Bryozoans | | X. Most types marine | | X | |
| Brachiopods | X | | | | |
| Crinoids (Phylum Echinoderma) | X | | | X | |
| All echinoderms | X | | | | |
| Cephalopoda (squid, cuttlefish, nautiloids, ammonoids, belemnoids, Phylum Mollusca) | X | | | | |
| Trilobites (Phylum Arthropoda) | X | | | | |
| Graptozoa or graptolites (Phylum ?Chordata) | X | | | | |
| Lime-secreting algae (plants) | | X | | | |

*Wave* or *oscillation ripples* (Fig. xii–8) are symmetrical, with sharp crests and rounded troughs, or with rounded crests and troughs; their historical implication is that standing water (ponds, lakes, sea) was the general environment of deposition because such ripples are caused by the oscillation of water in wind-produced water waves acting on bottom sediments.

Some sedimentary rocks exhibit *lensing bedding* (Fig. v–5b). The individual strata are lens-shaped in transverse cross section, thickening and thinning rather abruptly, terminating against other lenses by "pinching out," and, in general, being irregular and discontinuous in shape. The lenses may vary from a few inches in width and depth to many feet or hundreds of feet. Lensing-bedding results from rapid lateral shifting of stream channels or other current courses and commonly is associated with cross-bedding. Water-laid, fine-grained sediments which are intermittently exposed to the drying effect of the atmosphere (as on tidal flats or river flood plains) commonly develop *mudcracks* (Fig. xii–8). The impact of raindrops on such sediments while they are exposed to the air may produce *raindrop impressions*. Both features imply deposition of sediments in relatively shallow areas that were intermittently water-covered and then exposed.

Major breaks *(unconformities)* in sedimentary rock sequences are historically very instructive. Two main types are recognized. One type of break, called a *disconformity* (Fig. i–5b, and Fig. v–30), is produced during a period of interrupted sedimentation accompanied by erosion and commonly is caused by simple uplift of the area. A "time gap" in the sedimentary sequence results, but the strata above and below the break remain essentially parallel with one another because no angular deformation of the earth's crust was involved. *Angular unconformities* (or *nonconformities*), the other main type, are produced by crustal deformation (folding or tilting) and erosion of earlier deposits followed by a subsequent period of deposition. The strata above and below the unconformity of this type are nonparallel and the lower (older) set terminates against the overlying (younger) set (Fig. i–2, and Fig. v–14). Such features imply a period of earth crustal disturbance, uplift, erosion, later subsidence, and finally a resumption of sedimentary deposition.

The color of a sedimentary rock sometimes offers important historical clues as to the mode and environment of origin of the rock, although color must be used with extreme caution, and the conclusions you reach must be buttressed, if at all possible, by other lines of evidence. Especially, remember to examine the color on a freshly broken rock surface; crack the rock with your hammer. This is necessary because the effect of various weathering processes on the surfaces of exposed rocks, as well as the introduction of minerals by groundwater along cracks and fissures in bedrock, tends to alter the original color of the rock. *Red* may be due to the color of the original sedimentary materials of which the rock is composed (e.g., flesh-colored or red orthoclase feldspar), or it may be due to oxidizing conditions at the site of deposition, and hence a clue to environment. Red commonly is produced by relatively small amounts of the red iron oxide mineral, hematite. Most red sediments are the products of accumulation in warm, well-drained continental upland areas, or in other oxygen-rich environments. River floodplains, delta surfaces, tidal flats, water-covered areas where circulation introduces oxygen, all are places where red due to oxidized iron may develop in sediments. *Yellow, orange,* and *brown* result from the presence of hydrated iron oxides (limonite) in small quantities in the sedimentary rock. These colors indicate both oxidizing and hydrating conditions of sedimentation, possibly a moist well-drained area where vegetation did not accumulate. *Dark green, dark gray,* and *black* commonly indicate reducing (oxygen-poor, acidic) conditions. The ferrous iron minerals are green, and a black coloration is commonly due to incompletely decomposed organic (carbonaceous) material. Reducing conditions occur in marshes and swamps and in marine basins having restricted circulation with the open sea. They are known to exist today in the Black Sea, some fjords, and many marshes and swamps. *Color banding,* that is, alternations of reds, grays, purples, and greens may indicate fluctuating conditions at the time of deposition, as between well-drained and marshy. Many Cenozoic continental deposits are of this sort.

*Fossils* are another guide to ancient sedimentary *environments*. Some animals and plants of the past are thought to have had environmental preferences similar to those of closely related living types. These preferences of the extinct forms may be discovered from their association

Figure XII–11

CHART OF PHYSIOGRAPHIC (GEOGRAPHIC) ENVIRONMENTS OF SEDIMENTARY DEPOSITION

| | | | |
|---|---|---|---|
| Terrestrial (laid on land) | Sand dune | | Rounded, well-sorted, cross-bedded, quartzose sands |
| | Talus | | Unsorted, coarse, angular, exogenous sediment |
| | Mantle rock | | Largely soils |
| | Gravity slide and flow | | Coarse to fine, unsorted, angular, exogenous |
| | Glacial till or boulder clay | | Unsorted; larger particles may show smoothing or polishing, facets, striations; unstratified |
| Fluvial (running water) | Stream channels | | Coarse and medium grained, well sorted, detrital sediment; cross-bedding and current ripple marks common. Lensing common |
| | Floodplain | | Fine to very fine, thin-bedded, detrital sediment; cross-bedding common. Lensing |
| | Alluvial fan and apron | | Fair to poor sorting; cross-bedding and current ripple marks common |
| | Glacio-fluvial (melt-water streams) | | Outwash channels and aprons, like stream channels and alluvial fans. Often occur within or around margins of glacial till deposits |
| Lacustrine (lakes) | Beaches | *Foreshore*— (between low and high-water levels) | Medium to very fine exogenous and endogenous sediment; cross-bedding, current and wave ripple marks common; well sorted |
| | | *Backshore*— (above high-water level) | Coarse to medium-grained detrital sediment; dune sands commonly associated; cross-bedded |
| | Offshore—(Below low-water level) | | Typically fine-grained exogenous and endogenous sediment; thinly stratified; ripple marks common, particularly wave ripples |
| | Delta | | Coarse to fine-grained detrital sediment; topset, foreset, and bottomset bedding characteristic |
| | Marsh and lagoon | | Very fine exogenous and endogenous sediment (often mixed); carbonaceous material from plants common; thin, regular bedding |
| | Playa | | Medium to fine-grained exogenous sediment mixed with evaporites including halite, thenardite, borax, trona |
| Marginal Marine | Lagoon | | Like lacustrine lagoons, see above |
| | Tidal swamp | | Fine to very fine grained; well and thinly stratified (laminated); locally lensing and often carbonaceous (coaly, with or without well-preserved and distinct plant remains visible) |
| | Estuary | Open circulation | As in mixed delta and foreshore beach (see above) |
| | | Restricted circulation | Black, carbonaceous muds rich in pyrite |
| | Delta | | Medium to fine-grained detrital sediment; cross-bedding and ripple marks common |
| | Beaches | Foreshore | Like lacustrine foreshore (above) |
| | | Backshore | Like lacustrine backshore (see above) |
| Marine offshore (Below low tide level) | Open circulation | | Medium to very fine exogenous and endogenous sediment; ripple marks common in neritic zone. Includes following zones:<br>*Neritic*—Low tide level to 600 foot depths; the continental shelves<br>*Bathyal*—600–6000 feet. Continental slope<br>*Abyssal*—over 6000 feet. True ocean basins |
| | Restricted circulation | | Black, pyritic muds; evaporites (halite, anhydrite, gypsum, dolomite). Found in both neritic and abyssal zones although most common in neritic |
| | Organic reefs | | Endogenous fossiliferous limestone and dolomite, and exogenous calcarenites, bioclastic limestone, and coquina are characteristic |
| | Gravity slide and flow turbidity currents | | Exogenous sediment deposited from density currents; graded bedding common; graywacke type sediments common |

Figure XII–12

CHART OF TECTONIC ENVIRONMENTS OF SEDIMENT DEPOSITION

(*not* all represented in the geologic history of Michigan)

| | | | |
|---|---|---|---|
| Cratonic basins of deposition | The relatively stable, central platforms of the continental regions. Condition of earth's crust may vary here from slightly positive (rising) to neutral to slightly negative (sinking) with reference to sea level and surrounding regions. Sedimentary deposits and strata tend to be thin and discontinuous. Typical sedimentary rocks are: quartzose sandstones, calcarenites, and bioclastic limestones | Intracratonic epeirogenic basins (This was the tectonic condition in *Michigan during* most of the *Paleozoic Era*) | Negative or subsiding areas of moderate downwarp over long periods of time. Lie within the craton. Roughly circular. Total thickness of sediment deposited during an interval is greater than elsewhere on craton. Sedimentary rock sequences are more complete with few unconformities. Sediments like craton in general, but with evaporites added (halite, gypsum, dolomite). Fossils abundant |
| | | Faulted intracratonic basins | Typically long, narrow downfaulted troughs (grabens) bordered by granitic upfaulted highlands. Sediment very thick. Marine deposits few. Red arkose, arkosic conglomerate and shale most common. Alluvial fan deposition dominant. Fossils common |
| Geosynclinal basins of deposition | Strongly negative, linear belts marginal to the craton and bordered by uplifted volcanic and metamorphic highlands on the seaward side. Sediment accumulation very thick. In later stages geosynclines commonly become involved in crustal compression, intense deformation, and mountain building of the "alpine" (folded and overthrusted) type<br>Some of the Precambrian rocks of the Upper Peninsula of Michigan may originally have formed in this type of tectonic environment | Volcanic geosyncline (Eugeosyncline) | Very strongly negative. Bordered by volcanic highlands or island arcs. Volcanic debris accumulates in deep waters as *graywacke* sediment. Usually, the outer portion of the geosyncline as a whole. Fossils rare |
| | | Continental geosyncline (Miogeosyncline) | Moderately negative. Bordered by nonvolcanic uplands or craton of the continent. Sediments like those of craton but thicker, with fewer time gaps (unconformities). Endogenous limestone and dolomite common. Fossils abundant |

with various types still living. For example, the extinct trilobites, when found with other types of fossils, always occur with marine types; hence trilobites alone are considered indices to a marine environment of deposition for the rock containing them. Some degree of caution must be exercised when using fossils to interpret past environments. It is not always certain that the environmental preferences of all ancient animals were exactly the same as the living types. Furthermore, organisms were not always buried exactly where they lived and died. They may have been transported after death to their final resting place in sediment, although in such cases their hard skeletal parts usually show some evidence of wear (breakage, rounding, scratching) during transport. Generally, the land-dwelling and freshwater-dwelling fossil organisms such as the land vertebrates and plants are considered indices of a continental, as opposed to marine, environment of deposition. Many other types of fossils indicate marine environments and are common in the fossil record of Michigan and elsewhere in areas which are now dry land. Figure XII–10 lists a few of the main environmentally significant types of marine invertebrates. For additional discussion of fossils and help in their identification see Chapters XIII, XIV, and XV.

Figure XII–11 lists the types of sedimentary deposits that accumulate in many of the main geographic situations found at the earth's surface today. Here, we are concerned with recognition of

# Geology of Michigan

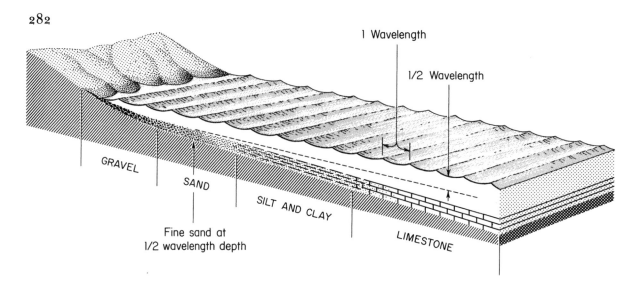

Figure xii–13. Sedimentary facies gradation, from coarse to fine grained, in an offshore direction. Below depths of about one half of maximum wavelength, wave action does not put fine sand or finer-grained material back into suspension after it has settled to the bottom. A highly diagrammatic illustration of the concept that at *any given time* sediments of *different types* may be accumulating in different physical environments, thus becoming parts of a continuous sedimentary unit which represents the same part of geologic time throughout its extent.

similar conditions in the past by analysis of the sedimentary rock record.

Figure xii–12 arranges the main types of sedimentary deposits according to the tectonic (as opposed to geographic) setting in which they commonly form. By tectonic environments we mean situations, present or past, in which certain types of subcrustal processes and forces caused long-term changes in both the earth's surface and subsurface. Tectonic forces and processes and the terminology used by geologists to describe them may be unfamiliar to the layman. If you are in doubt, look in the index for reference to descriptive sections included elsewhere in this book.

Overlap and offlap deposits, the marks of shifting seas, are another important source of historical information in the sedimentary rock record. The Paleozoic seas (Chapter V) left superposed sedimentary rock deposits of several different ages in the Michigan Basin. In those deposits there is an interesting way to recognize the ancient marine advances and retreats. As can be observed today, much sediment and dissolved mineral matter is carried by streams into standing water bodies, and more rock particles are added by the erosive action of waves along the shorelines. In shallow water (usually near the shore), the action of waves stirs up that sediment, lifting much of it into suspension. Currents moving away from shore tend to carry the suspended sediment toward deeper offshore water. The turmoil of breakers is greatest in shallow water along the beach, and only the coarsest and heaviest particles of gravel and sand can settle permanently to the bottom there. The finer material stays suspended longer and is carried farther out from shore, where it settles to rest on the bottom in deeper water below the reach of wave action. Thus, there is a tendency for sediments deposited in standing bodies of water, such as the sea and lakes, to range from coarser-grained near shore to finer-grained offshore; in areas far out beyond the normal reach of even the finest grains of sediment, chemical precipitates (endogenous sediments), such as limestone, predominate. This relationship is shown diagrammatically in Figure xii–13. Many factors, such as shoreline irregularities, growth of offshore sand-bars and lagoons, and entry of major streams with their deltas, mar the idealized picture, but the tendency toward such an arrangement remains. Now suppose that as sediment accumulates in a graded series from coarse near shore to fine offshore, the margin of the sea moves inland or *transgresses* (Fig. xii–14). In such a situation, the relationship of coarse and fine sediments with respect to the shoreline will remain the same but the *place* of deposition of each grain size will change with the changing shoreline position. A point formerly near shore or part of the beach may come to lie farther and farther offshore, and the sediment deposited above that point becomes progressively finer and finer grained or may even change to a chemical precipitate. Thus, in a *transgressive sea* deposit the sediments accumulating above any given point tend to vary from *coarse below to fine above*. Each successive blanket of sediment deposited in a transgressing sea would overlap the preceding one in a landward direction and a series of such overlapping deposits, representing an invading or transgressing sea is called an *overlap deposit*. If a series of such deposits, grading in grain size from coarse below to fine above, were penetrated by an oil well drill, exposed by erosion in a canyon, or revealed in a quarry, one could recognize it as an advancing sea deposit. Now suppose the sea retreats or *regresses*. Then the process is reversed. A selected point becomes progressively nearer shore and the grain sizes of the sediments deposited above that point become coarser in successive overlying layers. The successive layers deposited in a retreating sea *offlap* one another. An offlap sequence deposited in a regressing sea tends to be represented, at any place that lay within the basin of deposition, by a gradation of particle sizes ranging from *fine below to coarse above*. If deposits of a whole cycle of marine advance and retreat were represented in the sedimentary rock record, they ideally would include a vertical gradation of grain sizes from coarse to fine to coarse. However, if the sea moved off the land far enough,

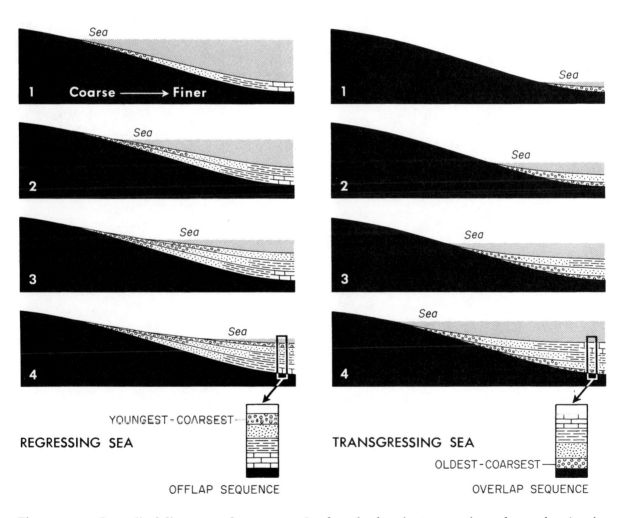

Figure XII–14. Generalized diagrams to show some results of a cycle of marine transgression and regression. A series of *offlap* and *overlap* deposits are produced in which sedimentary grain sizes tend to vary in a regular and predictable manner both vertically (through time) and laterally.

**Geology of Michigan**

the sediment deposited at a point formerly beneath the sea might even change to continental or freshwater type. Or erosion might begin and much of the preceding record of the transgressive-regressive cycle might be destroyed. In this last case, the deposits of a following transgressive cycle might be deposited on sediments representing an earlier part of a preceding cycle and the erosion surface representing a gap in the historical record would be called an *unconformity*. Thus, it is possible in theory to examine the sedimentary rock sequence and recognize ancient fluctuations in sea level, advances and retreats of former seas, and materials that were deposited at varying distances from shore. In practice, the record is not always so simple or easily interpreted, but the basic principles described above are applicable to the Paleozoic rock record in Michigan that is shown in Figure v-2. The account of the Paleozoic history of Michigan in Chapter V is based in part upon an analysis of such information. The evidence for fluctuations of land and sea becomes stronger if the sediments clearly alternate between continental types such as sand and gravel stream channel deposits or coal swamp deposits with continental types of fossil animals, and marine sediments such as limestone containing marine fossil corals or brachiopods. Cases of this particular type are described in Chapter V in the section dealing with the Pennsylvanian Period in Michigan.

# XIII

# Fossil Invertebrates

## Introduction

Studies of the animals and plants of the geologic past make important contributions to historical geology. This chapter will serve as an introduction to the fascinating subject of invertebrate paleontology. Chapters XIV and XV deal with fossil vertebrate animals and plants. The emphasis in the illustrations is on Michigan fossils.

## Definition of a Fossil

A *fossil* may be *defined* as any evidence of past life naturally preserved on or in the earth's crust. The geologic age of a fossil may be extremely great, or the organism whose remains are preserved may have died very recently in a geologic sense. Remains of still living types of organisms, particularly domestic animals, buried during historic times are not usually considered as fossils, but the necessary degree of antiquity is difficult to specify. Generally speaking, buried organic remains are paleontologically important if they provide evidence of the nature of past life of a sort that would not be available either through human historic records or by study of organisms that live today. Fossils are not restricted in occurrence to "hard" rocks. Poorly compacted, unconsolidated sediments and even ice can be of considerable age and may contain fossils of great interest.

## Types of Fossil Preservation

There are many *modes of preservation of fossils* (Fig. XIII–1). Although it usually is some hard skeletal part which withstands the ravages of time to bring us a record of past life, this is not necessarily so. Part, or all of a hard part (bone or shell), may be preserved unaltered as it was in life. More often, some portion may be decayed or dissolved either before or after burial in sediment. The remains may have been incomplete at the time of burial. Weather, the processes of erosion, and scavengers and bacterial decay destroy the vast majority of dead animal and plant remains shortly after death; it is only the rare individual that is preserved even in part. If the open spaces in porous hard parts become filled in with mineral matter (calcite, silica, hematite, or other), the remains

Figure XIII-1. Some common modes of preservation illustrated by Michigan fossils.

*Top left*—Original hard parts, in this case the calcium carbonate shells of Devonian age brachiopods.

*Top right*—Mineral replacement. Here quartz has replaced original calcium carbonate exoskeleton of a colony of Devonian corals. Bright specks are light reflections from quartz crystal faces.

*Bottom left*—Carbonization by compression. Carbon film faithfully preserves general form of an extinct, Pennsylvanian age "seed-fern" of the genus *Neuropteris*.

*Bottom right*—Natural molds (impressions or fillings). Original hard parts are gone. Clam shells in Mississippian age marine sandstone. Scale for all is in millimeters and centimeters.

are said to be petrified by *permineralization*. Often the entire hard part decays or is dissolved little by little by water and completely replaced by mineral matter; it is then petrified by *replacement*. If an organism leaves an impression in soft sediment and that impression *(mold)* is later filled with more sediment, the filling may produce a *cast* which more or less faithfully reproduces the original form. Casts and molds are common forms of fossil preservation; they may sometimes give a clue as to the shape of the soft structures of an organism, or even of organisms that had no hard parts at all, e.g., a jellyfish. Casts and molds may be produced from both the convex outer surfaces or the concave inner surfaces of shells. Thus, a shell might be preserved in the following ways: unaltered, permineralized, replaced, outside (external) mold or cast, or internal mold or cast. Plant tissues often are preserved as carbon residues. These may be mere dark films in the rock, or, if sufficient tissue was present to begin with, even a coaly layer. Coal is an accumulation of the carbon residues of many plants pressed together with little or no intermixed sediment. More rarely, soft animal tissues may be reduced to carbon films, outlining the body form. In very rare instances even the unaltered soft tissues of animals may be preserved for hundreds or even thousands of years. Michigan is not noted for occurrences of this sort, but specimens found elsewhere include the frozen woolly mammoths (Ice Age elephants) of Siberia and Alaska, an ancient woolly rhinoceros "pickled" in an oil seep in Poland, and the hair and skin of an extinct ground sloth found dehydrated in a dry cave in Patagonia. Recalling our definition of a fossil as including "any evidence of past life," even the molds or casts of tracks, trails, and burrows of organisms are considered to be fossils. No individual organism stands a very good chance of becoming fossilized, and very few of the many billions that must have lived in the past were preserved, but the chances are best if the

# Fossil Invertebrates

remains are buried in sediment soon after death. There they are protected from the atmospheric elements, bacteria, and scavengers. Places of rapid sedimentation, such as river flood plains and deltas, are especially favorable. If wave or current action is too strong, the organic remains may be ground up and destroyed by the coarse sediment (gravel or sand) in motion, so quiet water conditions also favor preservation. Because so few of the many organisms that live stand any chance of fossilization, preservation of a record of abundant kinds of organisms in environments favorable to life in general is most common. The warm marine waters of the Paleozoic seas in Michigan often were ideal sites for preservation of fossils because life there was abundant, sediment accumulated rapidly for burial, and the waters were sufficiently quiet in offshore areas to avoid destruction of the remains. Naturally, the kinds of environments present in the past also determined the kinds of fossils that would be preserved. Moreover, the types to be found in a particular area depend upon the ages of the sedimentary rocks present. One may expect to find marine invertebrates of Paleozoic age in the Paleozoic marine deposits in Michigan, but fossilized land dwelling mammals of the Early Cenozoic (Chapter VI) do not occur in this state because no rocks of that age are found here.

## Naming and Studying Fossils

In most cases one cannot learn all he would like to know about fossil organisms from their fossil remains alone. Therefore, fossils are studied by comparing them with their closest living relatives if any exist. The *Law of Correlation of Parts* is employed. For example, a fossil shell with all the structures of a snail shell is assumed to have been occupied by an animal whose soft anatomy and habits were those of a snail. Whenever possible, fossil organisms are classified according to the same system used by biologists to classify living organisms. Of course, there are limits to the validity of this procedure. Some fossils belong to groups long extinct, with no close living relatives. Others belong to groups which, through evolution, have changed so much from their ancestors that the anatomy or habits of the ancient and modern ends of the lineage are no longer closely comparable. In still other cases, so little is known about a fossil type that it cannot yet be classified accurately with reference to any living types. Nevertheless, much of what we know about the history of living organisms is derived from studies of their fossil record. Paleontology provides the time dimension for biology.

Fossil and living organisms must have names that are commonly agreed upon so that we can relay information about them. Moreover, they must be classified so that the myriad facts known about them are organized into some kind of a meaningful framework whose significance can be recognized. If ancestor-descendant relationships are found to exist, then a "phylogenetic" classification provides the means for expressing these.

The system of zoological nomenclature (naming) now in use is called the Linnaean system, after the Swedish taxonomist, Carolus Linnaeus. The tenth edition, 1758, of his book, *Systema Naturae*, is accepted by international agreement as the starting point for modern zoological nomenclature. All names of animals, to be considered valid, must have been proposed in that edition, or elsewhere after that time and in accordance with procedures which are acceptable under a uniform code of rules established by the Committee on Zoological Nomenclature of the International Zoological Congress. Botanical taxonomy is similar in many respects but is governed by a separate code of rules.

In all countries, well-known animals and plants have common names which are understood by those who speak the language of the country, but those names are not universally intelligible. In some cases, too, a common name in one language is not applied to the same kind of animal by everyone who speaks the language. To an American, the word "elk" refers to an animal which zoologists classify in the genus *Cervus*. This genus includes the European red deer, stags, and the American elk or wapiti. In Europe, the word "elk" is applied to animals of the genus *Alce* which includes the Asiatic elk and the moose of both continents (Simpson, 1945). What is more, the animals and plants that have common names constitute only a tiny fraction of the millions of living and extinct animals and plants that are known to biologists and paleontologists. For these reasons, scientists give all organisms scientific names. Except when it is necessary to be even more exact, the name of an animal or plant consists of two parts, the *generic* name, and the *trivial*. These names are formed from Latin or Greek roots and together comprise the *specific*, or *species, name*. The names of persons are somewhat comparable in reverse. Both parts are underlined or italicized and the generic name is capitalized.

EXAMPLE:

| | |
|---|---|
| Generic name—*Felis* (certain cats) | *Doe* (certain people, named Doe) |
| Trivial name—*domesticus* (domesticated) | *John* (John kinds) |
| Specific name—*Felis domesticus* (the common domesticated cat) | *Doe, John* (one or more Does named John) |

Names often are selected to describe some feature supposedly characteristic of the animal, but this is not always the case; some refer to the places where the animals occur, others are given in honor of someone, and still others are made from arbitrary combinations of letters. The intended meanings of the names are not as important as the fact that the names are all distinctive. No two genera have the same generic name, and no two species in the same genus have the same trivial name.

In the modern system of biological classification, organisms are grouped into categories of different ranks according to degree of phylogenetic (ancestor-descendant) relationship as well as our understanding permits. Over 40 categories are possible, but those shown below are most commonly used. Each higher category is more inclusive than any of those below it. A single family, for example, may include one or more genera. The several species included in one genus of a family are so placed because they are believed, on the basis of many lines of evidence, to be more closely related to one another than to species placed in any other genus of the same family. The common categories, in descending order of inclusiveness, are:

CATEGORY:         EXAMPLE:

Kingdom......*Animalia* (animals as distinguished from plants)
Phylum.......*Chordata* (animals with a dorsal axial support called a notochord at some stage in development)
Subphylum....*Vertebrata* (Chordates with vertebrae of bone or cartilage partly or completely replacing the notochord)
Class........*Mammalia* (Chordates with hair, mammary glands and other distinctive features)
Order........*Carnivora* (placental Mammalia with certain structures *uniquely* adapted for a carnivorous diet (but not all carnivorous mammals are Carnivora)
Family.......*Felidae* (Carnivora with special features common to the cat family)
Genus........*Felis* (Cats of many types all of which possess features in common that are not found in other genera of the family)
Species......*Felis concolor* (The cougar or mountain lion. Distinguished from other members of genus by size, color, and other minor features)

## Classification of Fossil and Living Invertebrate Animals

The *classification of invertebrate animals* which follows emphasizes those types which have left a significant fossil record in Michigan. Only the major categories (phyla, classes, orders) are listed. Detailed descriptions of families, genera, and species are beyond the scope of this book but may be sought in the standard references to paleontology. The generalized drawings used as illustrations supplement the brief descriptions and illustrate general form or structures of groups; the drawings are modified from those in a laboratory manual for historical geology by Dorr and Briggs (1953). The photographs of actual specimens illustrate, without description, some common genera and species likely to be found in Michigan by amateur collectors. The photographs are so arranged that all representatives of a particular group appear together regardless of their geologic age. The descriptions of the figures give the geologic age and formation in which each type occurs. A list of fossil collecting localities in Michigan follows the classification.

Note on the drawings that when both hard shell or skeleton parts and soft parts are illustrated the hard parts are shown in black. Normally, the soft parts will not be preserved in fossils.

The more rare but especially interesting fossil vertebrates (animals with backbones) found in Michigan are discussed in Chapter XIV; Chapter XV deals with the fossil plants.

The description of each animal group includes statements concerning its habitat and way of life (useful in historical interpretation of rocks), and the stratigraphic range (distribution through geologic time) of the group. An example of a living member of each group is given if the group is not extinct. An asterisk (*) in front of a group name signifies that the group occurs in the fossil state, in contrast to some groups having no fossil record and mentioned simply for the sake of completeness. A double asterisk (**) indicates the group is extinct with no close living representatives; the space sign (#) indicates a group common as fossils in Michigan. The stratigraphic time range given is for the occurrence of a group everywhere on earth, not just in Michigan; it is possible that the group does not actually occur in Michigan through all of the time span shown.

*Phylum PROTOZOA. Unicellular ("one-celled") or acellular organisms. Similar cells may group together as a colony but, without cell differentiation, the cells remain alike. Small sized. Principally aquatic. If more than one nucleus is present in the cell, no one nucleus has control of a special part of the cell material. In all but the most complex colonial types, each cell carries on all life functions for itself (digestion, excretion, reproduction, respiration, etc.). In some modern classifications, the protozoan "animals" and some unicellular "plants" (such as bacteria) are removed from the animal and plant kingdoms and placed together in a third separate kingdom named Protista. In this book, Protozoa are considered animals because they are mobile and lack photosynthetic substances such as chlorophyll.

*Subphylum SARCODINA. Single-celled, amoeboid forms (Fig. XIII-2). No flagella in main

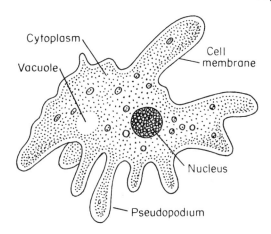

Figure XIII-2. *Amoeba*, a living protozoan without preservable hard parts, but closely related to the calcareous test producing Foraminiferida shown in Figure XIII-3.

life stage. Pseudopodia for locomotion, attachment, and food gathering. With or without either an internal or external skeleton. ? Precambrian, Cambrian-Recent.

##Order FORAMINIFERIDA (or Foraminifera) of the Class Rhizopoda and Subclass Lobosia. The "forams." With a hard, chambered skeleton, usually composed of $CaCO_3$, but in some of chitin or cemented particles (Figs. XIII-3 and 23).

*Habitat:* Chiefly marine. Early types were bottom crawlers. Surface floaters appeared in Cretaceous.

*Example:* The living *Globigerina*.

*Stratigraphic range:* Cambrian to Recent (still living).

*Notes:* Usually tiny to microscopic; rarely visible to naked eye, but some as large as wheat grains or even larger than a thumbnail. Best found by washing and screening shales. Important in correlation of rock strata found in oil wells. Also important sedimentary rock formers, as in case of chalk.

*Order (or subclass) RADIOLARIA of Class Actinopoda (Fig. XIII-4). A perforated, usually siliceous (opaline), nonchambered skeleton with a perforated internal capsule, and often with long, slim, radiating exten-

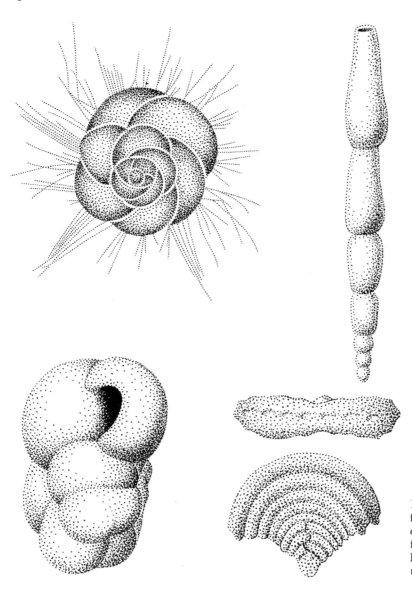

Figure XIII-3. Foraminiferida. A few, representative, calcareous tests of foraminiferans common in the fossil record. Type in lower right has been found in Michigan. All tiny (pinhead to thumbtack) in size.

sions called rays. Fine pseudopodia radiate through perforations. Microscopic in size.

*Habitat:* Entirely marine. Mainly surface floaters (plankton).

*Example:* The living *Trochodiscus*.

*Stratigraphic range:* Cambrian to Recent but relatively rare. Locally abundant enough to form sediment (radiolarian ooze) which forms the rock radiolarite.

*Note:* Other groups in the phylum Protozoa with little or no fossil record include the "Thecamebians," of the Class Reticularea (Cambrian–Recent), the Subclass Heliozoa (Pleistocene–Recent) of the Class Actinopoda, the Subphylum Sporozoa (no fossil record, all parasitic), and the various members of the Subphylum Ciliophora of which only the Suborder Tintinnina (Late Jurassic to Recent) has a fossil record. Many of these are important and abundant living types with which zoologists deal at great length. See a standard invertebrate zoology text or the *Treatise on Invertebrate Paleontology* (R. C. Moore, Editor) for descriptions of these groups.

**Fossil Invertebrates**

291

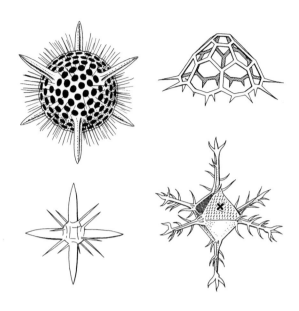

Figure XIII-4. A few types of fossil Radiolaria illustrating the great diversity of shapes found in this group. These microscopic protistans produce siliceous (silicon dioxide) hard parts. Not common in Michigan.

*Phylum PORIFERA. (Figs. XIII-5 and 23). The sponges. Many celled, with cells differentiated to perform special functions. A special cell type, the collared flagellate cell or choanocyte, is present. Body wall pierced by numerous pores. No tissue differentiation. No nerve cells. No mouth; the major opening (osculum) is a passage for water leaving interior of body.

*Habitat:* Sponges are chiefly, but not entirely marine, stationary bottom-dwellers. Fossil freshwater sponges are extremely rare compared with marine types.

*Hard parts:* The soft body is supported by numerous small spicules of calcium carbonate or silica, and/or by interwoven fibers of spongin (as in the bath sponge). The spicules may be separate, loosely intermeshed, or fused together. Usually, the soft body decomposes upon death so that the fossil record, except in rare cases, consists only of the loose spicules. Fossil sponges occur in Michigan but are very rare and are found in few localities. Paleontological classifications of sponges commonly base their distinctions upon the nature of the hard parts as follows:

*Class CALCAREA. Skeleton of calcareous spicules that are usually separate and with one, three, or four rays.

*Example:* the living *Ascon*.
*Stratigraphic range:* Cambrian to Recent.

*Class HEXACTINELLIDA. Skeleton of 6-rayed, siliceous spicules.

*Example: Euplectella*, the modern glass sponge, or the fossil type, *Hydnoceras* (extinct).

*Class DEMOSPONGIA. Spicules not 6-rayed. Skeleton of spicules of silica, spicules and spongin fibers, or of spongin alone.

*Example:* the common bath sponges.
*Stratigraphic range:* Cambrian to Recent.

**Phylum ARCHEOCYATHA (or PLEOSPONGIA). A calcareous, horn-shaped or cup-shaped or cylindrical skeleton whose hollow interior is surrounded by a perforated single or double wall. The walls are united by porous, septa-like, radial (vertical) partitions and by horizontal plates. The relationships of this group are uncertain because it is extinct and has no close living relatives. Resembles both corals and sponges. Not common in Michigan. Occurs only in the Early Cambrian and Middle Cambrian part of time.

##Phylum COELENTERATA (Fig. XIII-6). Many-celled animals with cells differentiated for various special functions. Well-developed tissues and nerve cell coordination. Two primary body layers, the ectoderm (outer) and endoderm (inner), separated by a jelly-like mesoglea containing widely separated, nontissue-forming cells. Adults with radial symmetry. Mouth surrounded by tentacles and opening into a single internal cavity, the gut. No anus, respiratory, nor circulatory systems. Two body forms may be present, sometimes in alternating generations. One is the sack-like, nonfreeswimming *polyp* form which in some coelenterates secretes a calcareous external support, the exoskeleton. The other body form is the jellyfish-like *medusa*. All coelenterates are aquatic. Most are marine although some live in fresh water.

# Geology of Michigan

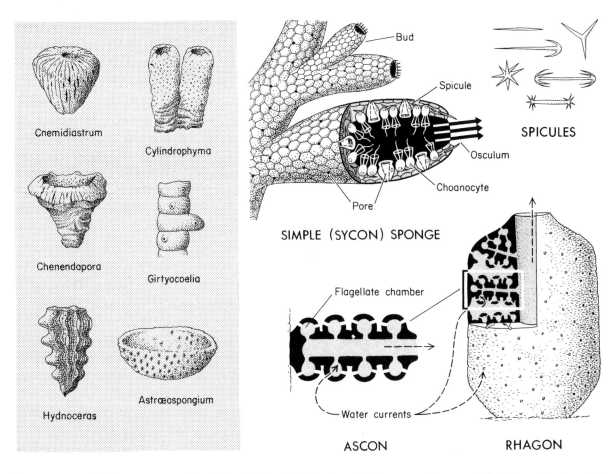

Figure XIII–5. Phylum PORIFERA. Fossil (at left) and living (at right) sponges. Increased folding of body wall found in progressively more complex "sycon" and "rhagon" types. Spicules are minute skeletal bodies of silica or calcite. (In part modified from Buchsbaum, 1948, *Animals Without Backbones,* permission University of Chicago Press. Copyright 1938, 1948, by University of Chicago.)

\*Class HYDROZOA. Polyp and medusa stages both present. Gut simple and unfolded internally. Margin of the umbrella-shaped medusa not notched.

*Hard parts:* Except for extinct stromatoporoids, most have no hard part or only a simple cup.

*Living examples: Hydra, Obelia.*

*Stratigraphic range:* Cambrian to Recent times.

\#\#\*Subclass STROMATOPOROIDEA (Fig. XIII–23). Extinct colonial organisms that formed a calcareous skeleton of highly variable size and shape depending upon the growth conditions in the environment. Skeleton may consist of either closely spaced, concentric layers joined by radial pillars, or of many tubular structures in a "spongy" matrix. The nature of the soft body is unknown, so the relationships of this group are uncertain. Have been classified as coelenterates, sponges, and even as algae (simple plants). They were important contributors to calcareous reefs in the marine waters of the Paleozoic. Especially common in the Silurian and Devonian rocks of Michigan.

*Example: Stromatopora.*

*Stratigraphic range:* Cambrian to Cretaceous. Most common in Ordovician to Devonian.

Fossil Invertebrates

293

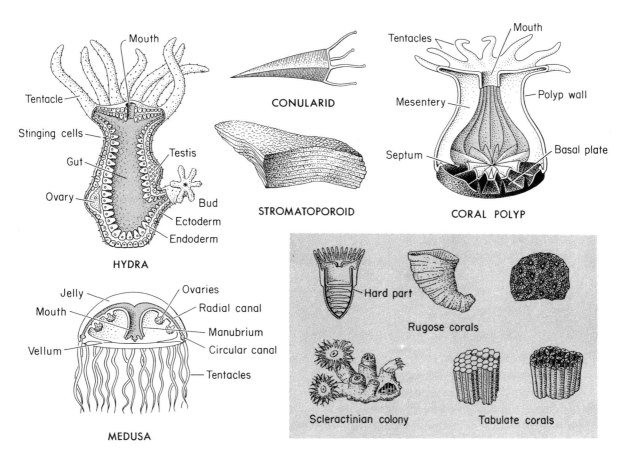

Figure XIII–6. Phylum COELENTERATA. Cross sections through living polyps and a medusa, and some representative hard parts of fossil types. (See Figs. XIII–23, 24, 25.)

*Class SCYPHOZOA. The medusa form is dominant; polyp stage limited or absent. Gut is folded. Margin is notched. Four-part radial symmetry. Rare as fossils except for the extinct conularids.

*Habitat:* Free-swimming marine.

*Hard parts:* Lacking, hence jellyfish rarely fossilized and then only as impressions or as sand or mud fillings of interior cavities.

*Examples:* The many living true jellyfishes. The extinct conularids.

*Stratigraphic range:* Cambrian to Recent.

##Class ANTHOZOA (or ACTINOZOA) (Figs. XIII–6, 24, and 25). Polyp form dominant. Colonial and solitary types. Gut wall and floor folded into mesenteries and basal ridges which increase digestive surface.

*Hard parts:* Base of polyp secretes calcium carbonate support (exoskeleton) called the *corallum*. Polyp occupies the upper, cup-shaped surface called the *calyx*. Ridges, deposited in the upfolded portions of the polyp base, radiate outward from the center of the calyx and are called *septa*. Periodically during its growth the polyp moves upward, in some groups depositing a new and higher "floor" of calcium carbonate called a *tabulum* beneath itself. The septa, which are extended upward, and the successive tabula formed during growth of the polyp may be seen in cross sections of the corallum. The septa, if present, also show inside the calyx. Solitary or "horn" corals build coralla which are separate for each individual. Colonial or "reef-building" corals build coralla which merge and are joined or fused together. The shapes of the coralla and the patterns of colonial growth are more or less constant for each different species of coral. Colonial corals were important in Michigan in the past, particularly during the Silurian and Devonian, as builders of thick and extensive limestone reef deposits. Commonly, the coralline limestone of the reefs will contain fossil remains of other organisms that lived in or on the reef; the associates included abundant stromatoporoids, brachiopods, and bryozoans, which also contributed to the growth of the reef. Corals also are common in smaller groups or individually in many normal marine limestones.

*Habitat:* Corals, particularly the colonial reef builders, are especially useful indices to the environment of deposition of rocks containing them because they were fixed, marine bottom dwellers that favored warm, shallow sea waters.

*Examples:* See Figures XIII-24 and 25.

*Stratigraphic range:* Ordovician to Recent. Common in the Paleozoic limestones of Michigan.

*Classification of Anthozoa:* This class includes the sea anemones as well as the corals. Five subclasses of corals are: *Rugosa* (or *Tetracorallia*), Ordovician to Permian time range, both solitary "horn corals" and colonial types, four-fold symmetry with four main ("primary") septa; *Scleractinia* (or *Hexacorallia* or *Zoantharia*), Triassic to Recent, with six-fold symmetry; *Tabulata*, mainly Ordovician to Permian, but a few continued to the Cretaceous; colonial forms with septa reduced or absent but tabulae prominent, *Alcyonaria* (or *Octocorallia*), and the *Schizocorallia*.

Phylum CTENOPHORA. Radially symmetrical. Free-swimming; locomotion by cilia (hairs) arranged on 8 longitudinal rows of comb-like plates. No distinct polyp or medusa stages. No hard skeleton. No known fossil record. Marine swimmers and floaters. Called "comb-jellies."

*Note:* In contrast to the phyla described above, the remaining phyla are anatomically much more complex. In addition to ectoderm and endoderm, a third tissue layer, the mesoderm, is present, at least in rudimentary form. Mesoderm forms the bulk of the adult soft body including the musculature and many important internal organs. Similar cells aggregate into distinctive types of tissue layers which, in turn, form discrete organ systems. Adults (except for most of the Echinodermata) are bilaterally rather than radially symmetrical.

Phylum "VERMES." Several wormlike phyla with elongate bodies are placed here for convenience sake. Actually, each of the phyla listed in this category is biologically important and interesting in its own right. In a more complete and detailed zoological classification each would be listed separately. However, they generally lack hard parts (except for some chitinous jaw parts) and have left little or no fossil record; hence the common paleontologic practice is to lump them together with little further notice. The phyla included here are: PLATYHELMINTHES (flatworms, tapeworms, liver flukes), no certain fossil record; NEMERTEA (ribbon worms), no certain fossil record; NEMATODA (roundworms and threadworms), no certain fossil record; TROCHELMINTHES (the rotifers and gastrotricha); GORDIACEA ("horse hair" worms); ACANTHOCEPHALA (hook worms); CHAETOGNATHA (Arrow worms), one unusual fossil specimen found in the Middle Cambrian Burgess shale of British Columbia, but otherwise not found as fossils; and the phylum KINORHYNCHA, tiny marine animals.

##Phylum BRYOZOA (or POLYZOA) (Figs. XIII-7 and 26). Sometimes called "moss animals," these tiny, sedentary animals superficially resemble the polyps of corals but are much more complexly formed and generally much smaller. The tubular or sack-like soft bodied polyp has a mouth and an anus. The mouth (and in some, the anus) lies within a horseshoe-shaped, tentacle-bearing ridge called the lophophore. The gut is U-shaped. Large numbers of individual polyps grow together in colonies which may be moss-like, net-like, lacy, branching, or hemispherical. Often the colony encrusts an object such as a brachiopod shell or coral fragment.

*Hard parts:* The polyp occupies a tiny, tubular exoskeleton most commonly composed of calcium carbonate, but more rarely of chitin. These tubes or cups are generally tightly packed into a wide variety of colony forms

# Fossil Invertebrates

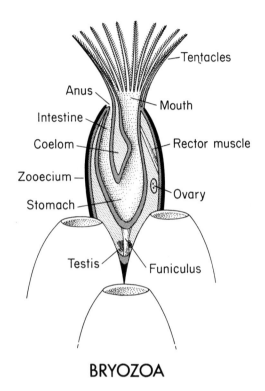
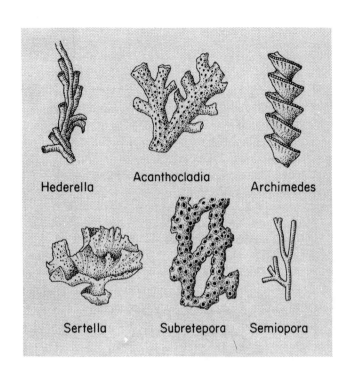

BRYOZOA

Figure XIII-7. Phylum BRYOZOA. Cross section of living type to show relation of tentacle bearing soft polyp to calcareous hard part in a colony, and some representative hard parts of fossil types. All highly magnified; tiny individual cup openings normally less than diameter of pinhead. (See Fig. XIII-26.)

mentioned above. The individual polyp chambers are barely visible to the naked eye and are better seen with the aid of a hand lens. Superficially, the colony may resemble the smaller varieties of tabulate colonial corals, but the bryozoan individuals are much smaller, and magnified cross sections show a lack of characteristic coral tabula and septa. Commonly, the colonies contributed substantially to coral reefs.

*Habitat:* Chiefly marine. Adults are fixed (attached) bottom dwellers.

*Stratigraphic range:* Middle Cambrian to Recent. Common in the Middle Paleozoic seas of Michigan.

Phylum PHORONIDA. Small, sedentary, colonial animals with a slender, tubular, unsegmented body lacking hard parts. Mouth lies within a scroll-shaped lophophore that bears tentacles. Occupy tubular burrows or construct upright tubes of cemented particles collected from the sea floor. Marine. No certain fossil record, although some so-called "fossil worm burrows" may have been formed by this group.

##Phylum BRACHIOPODA (Figs. XIII-8, 27, and 28). Bilaterally symmetrical, bivalved marine shellfish with an unsegmented soft body which has excretory and circulatory systems but no separate respiratory system. A complicated muscle arrangement opens the shell along the hinge line against the action of a resilient pad which closes the shell. A coiled, ciliated organ (the lophophore) about the mouth circulates water into the mantle cavity for respiration and feeding. A thin, soft *mantle* encloses the main body. The outer edge of the mantle secretes the shell. A fleshy stalk, the pedicle, extends through a hole, usually in one half of the shell near the hinge. That valve (or half of the bivalved shell) is called the pedicle valve. The other half internally bears arm-like extensions, called brachia, for support of the lophophore and hence that valve is the brachial valve. The pedicle serves to attach the animal to the bottom or other objects. In some brachiopods the pedicle passes through both valves at the hinge line.

*Hard parts:* A calcareous (or, more rarely, chitinous), bivalved (two-part) shell. The calcium carbonate of the shell is in the form of the mineral calcite. The valves of the shell are located on the upper and lower sides of the body (not, as in clams, on the right and left). The plane of bilateral symmetry passes *through* (rather than between) each of the two valves, through the beak and through the hingeline. Thus each half (valve) of the shell is bilaterally symmetrical and could itself be cut into two mirror image halves, but the one valve is not a mirror image of the other. (In contrast, the two valves of a clam or pelecypod shell usually are similarly shaped mirror images of one another). The two valves of some brachiopods interlock (articulate) by means of tooth-like structures along the hinge line; these are the "articulate" brachiopods. In "inarticulate" types the valves lack such structures. Radiating folds add strength to the shell. Growth lines show as roughly concentric lines paralleling the shell margin. A wide variety of shell forms occur. In some, both valves are convex, enclosing much space within. In others, one valve may be flat or

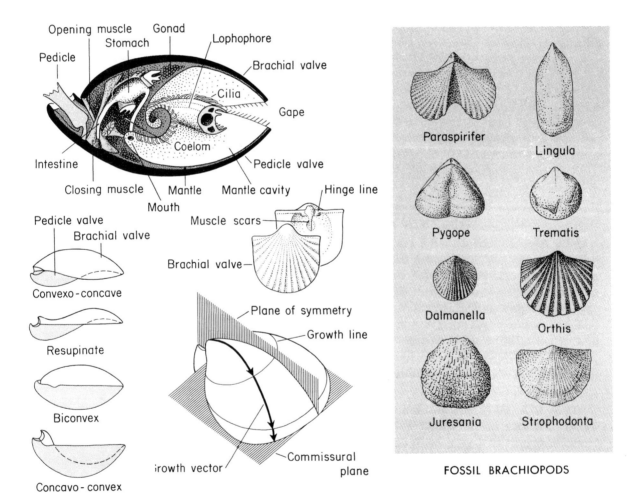

Figure XIII–8. Phylum BRACHIOPODA. Cross section of a living type to show relationship of soft body to calcareous shell, and representative shells of fossil types. (See Figs. XIII–27, 28.) (In part modified from *Treatise on Invertebrate Paleontology,* courtesy of the Geological Society of America and the University of Kansas.)

# Fossil Invertebrates

even concave, thus reducing the space between the valves. Muscle scars may appear on the interior surfaces of the valves.

*Habitat:* Marine bottom-dwellers. Good indices for marine origin of enclosing sedimentary rock.

*Examples:* See Figures XIII–8, 27, and 28.

*Stratigraphic range:* Early Cambrian to Recent. Climax in Middle to Late Paleozoic. Common in Paleozoic rocks of Michigan. Relatively rare (compared with past) in modern seas. May have occurred in sufficient abundance to form layers of shelly limestone rock.

*Phylum ANNELIDA (Fig. XIII–9). Elongate bodies divided from front to rear into a series of essentially similar segments each provided with a pair of short, bristle-like, nonjointed appendages. Internally, the annelids are complex, with well-developed excretory and circulatory systems, and, in some, a respiratory system. A biologically important group with little fossil record. Probably responsible for much mixing of ancient soils and of sediment in past seas.

*Hard parts:* Annelids lack external or internal hard parts except for chitinous jaw parts; hence have left little fossil record. Tiny, microscopic or nearly microscopic objects called

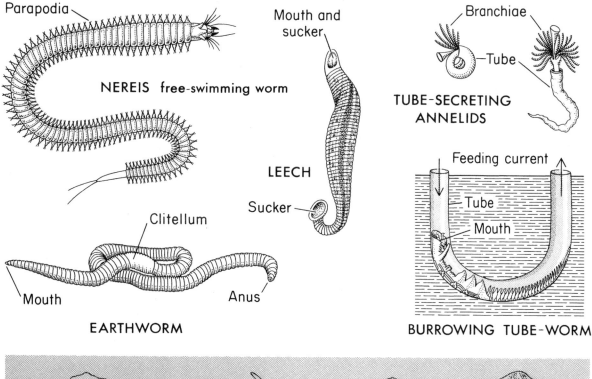

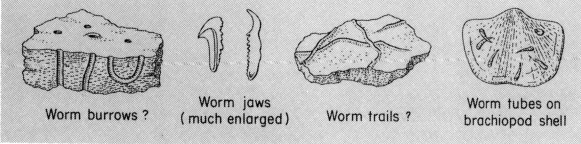

Figure XIII–9. Phylum ANNELIDA.

*scolecodonts* may in some cases be the fossilized jaw parts of annelid worms although many scolecodonts may also be the jaw parts of other worm-like animals (see the phyla collectively called "VERMES" described earlier). Certain markings or structures in sedimentary rocks resemble excretionary castings, trails, tubes, burrows, or borings constructed by modern worm-like animals in soft sediment today. Some of these fossils may have been the work of annelids, but in any case tell little about the creatures that formed them.

*Habitat:* Annelids include marine, freshwater, and terrestrial types.

*Examples:* Earthworms, leeches, marine sand worms (*Nereis*).

*Stratigraphic range:* Cambrian to recent.

*Phylum ONYCHOPHORA. Rare, "caterpillar-like" animals which combine some features of both the Phylum Annelida and the Phylum Arthropoda (described later), but also with peculiar features of their own. Some classifications place members of this group within the Phylum Annelida. Some zoologists believe the onychophorans may be closely related to some ancient and yet undiscovered group of animals intermediate in evolution between annelids and arthropods. Living onychophorans fall into a very few species all belonging to the single genus, *Peripatus*. Fossils of this group have not been found in Michigan, but are of paleontological interest because a single fossil locality in Middle Cambrian Burgess Shale in British Columbia has yielded the only known fossil record of this group, which consists of compressed carbon films of a soft bodied onychophoran called *Aysheaia*. Knowledge of this group evidently is very incomplete, which serves to demonstrate the generally incomplete nature of the fossil record of soft-bodied animals of all types that lack hard parts. In this phylum excretory, circulatory, and respiratory systems are present. Segmentation of the body is well developed internally but poorly developed externally; however, pairs of stubby, clawed, *jointed* legs correspond to the internal segments. The head consists of three fused body segments.

##Phylum MOLLUSCA (Figs. XIII–10-14, and 29 and 30). Animals with a short, unsegmented body (not divided into a series of similar sections). A muscular organ called the "foot" on the lower (ventral) surface. The soft body lies within a thin, fleshy envelope called the mantle. A wide variety of variations on this basic plan are shown by the classes described below.

*Hard parts:* The edges of the mantle commonly secrete calcium carbonate (in the mineral form of both calcite and aragonite) and chitin to form a calcareous shell of one or more parts, but a shell may not be present.

*Class AMPHINEURA (Fig. XIII–10). Elliptical body with a large, ventral foot. Head present but without eyes or antennae.

*Hard parts:* The mantle usually secretes a chitinous or calcareous shell consisting of 8 overlapping plates.

*Habitat:* Marine bottom-crawlers.

*Examples:* The chitons or sea-mice. *Chiton* is a common genus.

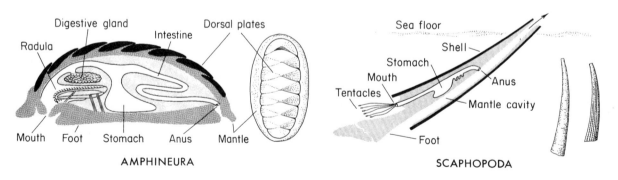

Figure XIII–10. Phylum MOLLUSCA, Amphineura and Scaphopoda. (See Fig. XIII–29.)

# Fossil Invertebrates

*Stratigraphic range:* Cambrian to Recent but not common. Rare in Michigan.

**Class SCAPHOPODA** (Figs. XIII–10 and 29). Elongate soft body. Mouth surrounded by tentacle-like organs.

*Hard parts:* Body enclosed in a simple (undivided), tapering, slightly curved, calcareous shell which is open at both ends.

*Habitat:* Stationary, marine bottom dwellers.

*Examples:* The tooth-shells or tusk-shells. *Dentalium.*

*Stratigraphic range:* Ordovician to Recent.

**Class GASTROPODA** (Figs. XIII–11 and 29). A distinct head with eyes and antennae. A flattened, muscular, ventral "foot," and a dorsal "hump." The hump commonly twists or coils and is enclosed in a soft mantle.

*Hard parts:* The mantle commonly secretes a one-piece, spirally coiled, nonchambered (internally undivided), calcareous shell. In some, the shell opening may be capable of being closed by a flat plate, the operculum. The calcium carbonate is in the mineral form of aragonite which dissolves more readily in water than does calcite, hence the actual shell material in fossil snails is often removed by solution, leaving only a sediment filling (mold) of the interior.

*Habitats:* Bottom crawlers and vegetation climbers found in both freshwaters (streams, lakes, ponds) and in the sea. Some also are land dwellers.

*Examples:* Snails, slugs, conches, limpets.

*Stratigraphic range:* Early Cambrian to Recent.

**Class PELECYPODA** (Figs. XIII–12 and 29). Head rudimentary. Ventral foot enlarged for crawling. Laterally compressed. Bilaterally symmetrical unless secondarily adapted and modified for fixed or attached bottom-dwelling. Well-developed gills. A pair of tubes (siphons) circulate water in and out of the mantle cavity, carrying in food and oxygen and expelling wastes. Large anterior (front) and posterior muscles close the two halves of the shell. An elastic ligament, stretched when the shell is closed, pulls the two halves (valves) of the shell open when the muscles relax; thus, upon death the tendency is for the shell to open or even for the two valves to separate.

*Hard parts:* The free outer edge of the mantle secretes calcium carbonate to form the bivalved (two-halved) shell. The shell

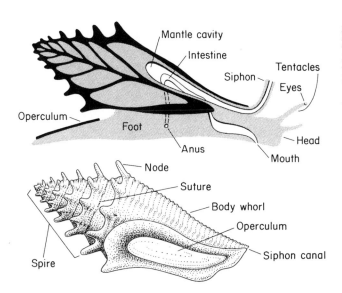
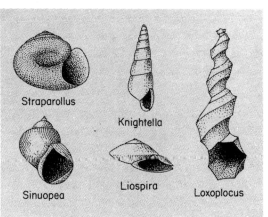

FOSSIL GASTROPODS

Figure XIII–11. Phylum MOLLUSCA, Gastropoda. (See Fig. XIII–29.)

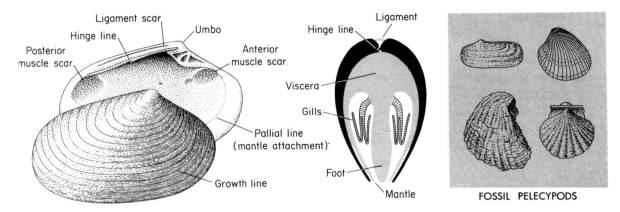

Figure XIII–12. Phylum MOLLUSCA, Pelecypoda. (See Fig. XIII–29.)

consists of hinged right and left halves. This contrasts with the shell of brachiopods in which the halves of the bivalved shell are top and bottom. In most pelecypods, the two valves are mirror images of one another and a plane of bilateral symmetry passes between the two halves. The exceptions are in forms, such as the oysters, wherein the shell is modified for attachment to the bottom or to some object, so that one half becomes much enlarged, the other reduced, and the basic symmetry distorted. The shell is three-layered. An outer, chitinous layer protects the calcareous shell layers from solution by water. The inner or "mother of pearl" layer is calcium carbonate in the mineral form of aragonite. The middle layer is calcium carbonate in the form of calcite. Ordinarily the outer, chitinous layer decays after death and is not preserved in the fossil state.

*Habitats:* Pelecypods may be either marine or freshwater types. They are chiefly bottom-crawlers, though some may be fixed (attached) forms, such as oysters, or boring types.

*Examples:* Figures XIII–12 and 29. The many different clams and oysters.

*Stratigraphic range:* Early Ordovician to Recent.

#*Class CEPHALOPODA (Figs. XIII–13, 14, and 30). A well-developed head with eyes. Mouth ringed with tentacles which bear sucking disks or hooks.

*Hard parts:* A wide variety. See Subclasses below. Some lack hard parts.

*Habitat:* Marine bottom-crawlers, swimmers, and floaters.

*Stratigraphic range:* Late Cambrian to Recent.

#*Subclass NAUTILOIDEA. Two pairs of gills (tetrabranchiate) in the living types and presumably also in extinct forms. Many tentacles around mouth.

*Hard parts:* A straight or coiled, calcareous, external shell with centrally placed tube (the siphuncle) in the interior behind the soft body. Sutures, the lines formed where internal partitions join the inner surface of the outer shell, are simple, nearly straight lines. Suture lines, however, are seen only if the outer shell is removed. Surface ornamentation is simple or absent. The interior of the shell is divided by partitions, called septa, laid down by the mantle behind the soft body as the animal grew. At any stage in its life, the soft body occupied only the front or living chamber.

*Habitat:* Marine bottom-crawlers and surface swimmers.

*Example:* Figures XIII–13 and 30. The sole living genus is *Nautilus,* the chambered or "pearly" nautilus.

*Stratigraphic range:* Late Cambrian to Recent.

# Fossil Invertebrates

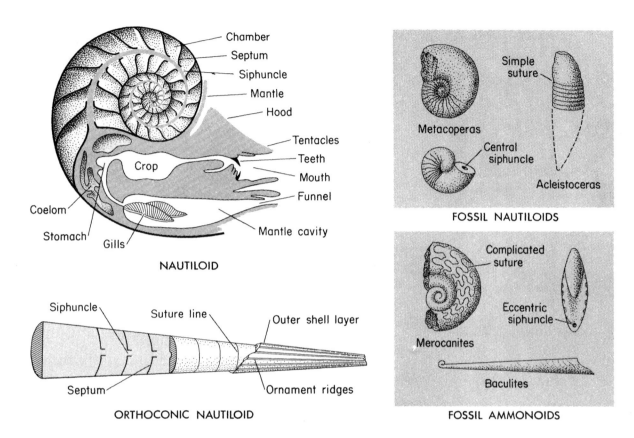

Figure XIII-13. Phylum MOLLUSCA, Nautiloidea and Ammonoidea. (See Fig. XIII-30.)

#***Subclass AMMONOIDEA. All members extinct; hence the number of gills and the nature of the other soft parts are poorly known. Probably 2 pairs of gills (tetrabranchiate). Assumed to have been much like the better-known nautiloids in most respects.

*Hard parts:* Shell either straight or coiled, calcareous, and more or less highly ornamented compared with that of nautiloids. The margins of the septa (partitions) of ammonoids are folded and wrinkled. Therefore, the suture lines (septal edges) are wrinkled. The degree of wrinkling of the sutures varies, from gently looped to intricate patterns that resemble a complicated root system. The internal siphuncle or tube running from the living chamber back to the earliest chambers is off center, usually lying along the lower edge of straight coned shells, or along the outer edge of each turn in coiled forms.

*Habitat:* Probably both marine swimmers and bottom-crawlers, if habits resembled those of still-living nautiloids.

*Stratigraphic range:* Late Silurian to end of the Cretaceous. Very abundant during the Mesozoic Era.

#**Subclass COLEOIDEA (Fig. XIII-14). Two gills (1 pair, dibranchiate). Eight to 10 tentacles around the mouth. Soft body cigar-shaped (as in squids) or sack like (as in the octopi). Fleshy lateral fins often present.

*Hard parts:* The hard part, if present at all, is mainly or entirely "internal," that is, enclosed within the soft body. Members of this group are thought to have evolved from an ancestral stock similar to nautiloids in such a fashion that the soft body grew out of and back around

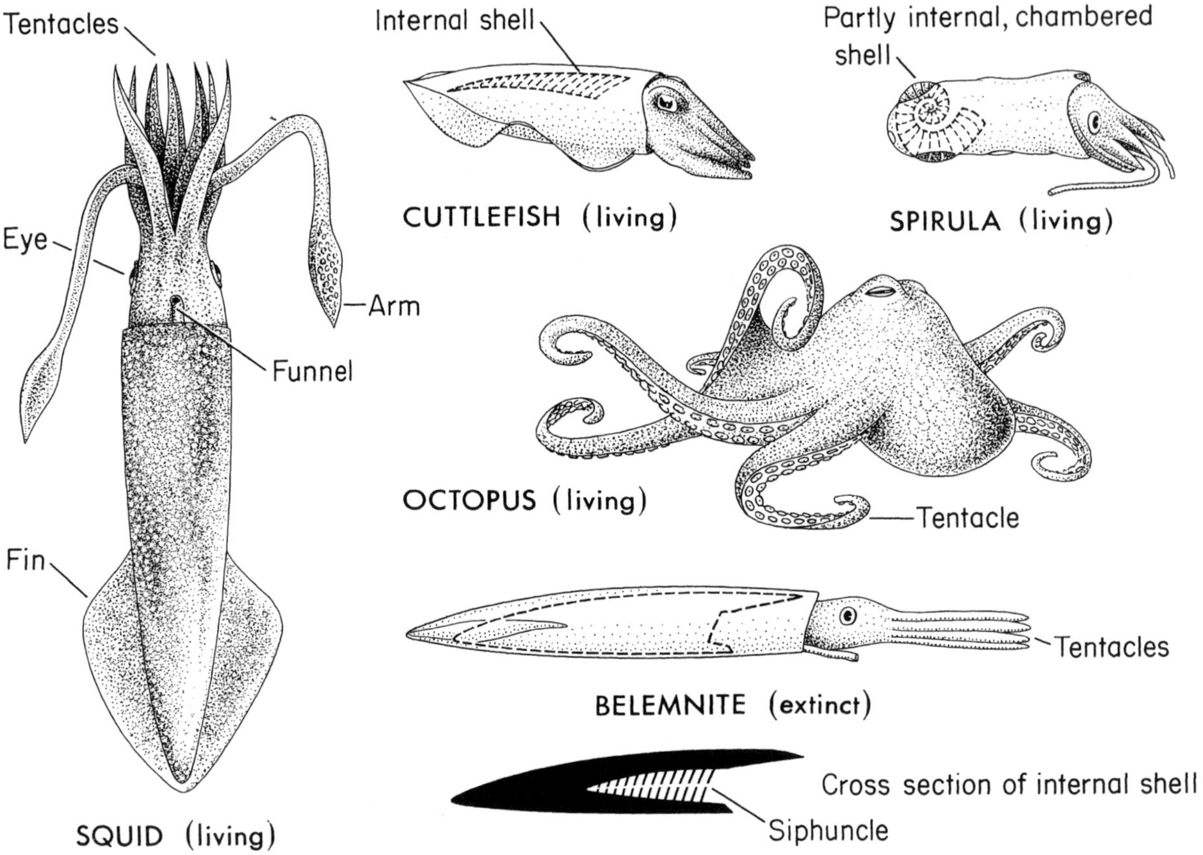

**Geology of Michigan**

302

Figure XIII-14. Phylum MOLLUSCA, Coleoidea.

the shell; the shell then became reduced to various degrees, or wholly lost as in the octopi. The types showing the least amount of reduction or modification of the hard part are the now-extinct belemnites. The belemnites resembled modern squids in their body form but the internal hard part was a cigar-shaped object which resembles a straight-coned type of nautiloid shell with greatly thickened walls.

*Habitat:* Marine swimmers; bottom crawlers.

*Examples:* Squid, cuttlefish, octopus, and the extinct belemnites.

*Stratigraphic range:* Late Mississippian to Recent.

#\*Phylum ARTHROPODA (Figs. XIII-15, 16, 17, 31, 32). Body segmented but with various modifications resulting from fusion and/or specialization of segments. Phylum takes its name from the presence of paired *jointed* legs (or other appendages) variously modified for the functions of walking, swimming, and food-handling. Excretory, circulatory, and respiratory systems present. Many different types. The most highly diversified of all the invertebrate phyla.

*Hard parts:* A strong, chitinous exoskeleton (outside the body) whose exact nature varies. See classes described below.

*Habitat:* Very widely distributed and diverse (see classes that follow). Marine, freshwater, terrestrial, and flying types.

*Stratigraphic range:* Cambrian to Recent for Phylum as a whole.

*Special note:* All of the classes of arthropods described below have left some fossil record, but only the trilobites, eurypterids (see Class Arachnoidea), insects, and ostracods (see

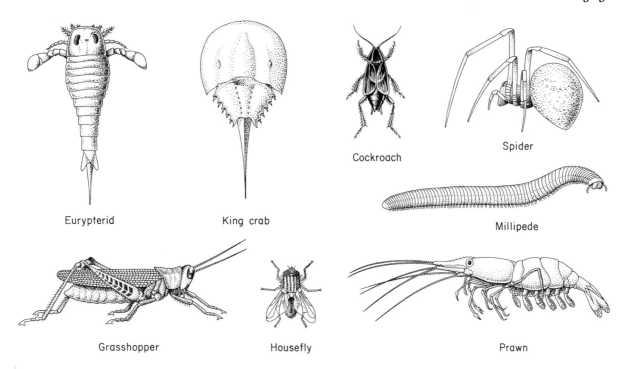

Figure XIII–15. Phylum ARTHROPODA.

Class Crustacea) are common as fossils. Of these, only trilobites and ostracods are common in Michigan. The ostracods are such tiny creatures that special training and techniques are required to find any but the largest of them.

#***Class TRILOBITA. The trilobites. Extinct, so soft body features known only from study of probable relations to hard parts.

*Hard parts:* A chitinous exoskeleton, called the carapace, covers top of body. Carapace divided into three parallel "lobes"; hence the name tri-lobite. The axial lobe runs the length of the body from head to tail down the center. The two pleural lobes lie one on each side of the axial lobe. Body and carapace also divided, front to rear, into a series of segments, each of which bears a pair of double-branched (biramous) appendages. A single pair of antennae. Several anterior (front) segments are fused together into a head region, the glabella, on which are found a pair of many-faceted eyes similar to those of a fly or other insect. Segments in the thorax (main body) region are not fused but flexibly hinged allowing the animal to roll up much like a pill bug. Several segments at the rear are fused into an inflexible structure, the pygidium (or "tail"). Trilobites, like many other arthropods, periodically molted or shed the carapace during growth. Moreover, the head, tail, and body segments tended to come apart due to decay after death. Hence, parts and pieces of trilobites are much more common as fossils than are whole individuals. The jointed legs must have been weakly joined to their respective segments because few specimens have been found with the legs in place.

*Habitat:* Marine, because other fossils found with them are always marine forms. Probably both swimmers and bottom crawling types existed.

*Examples:* Figures XIII–16 and 31.

*Stratigraphic range:* Cambrian through Permian. Long extinct.

#**Class CRUSTACEA. Jawlike appendages ("pincers") on the fourth segment. Two pairs of antennae, on the second and third segments.

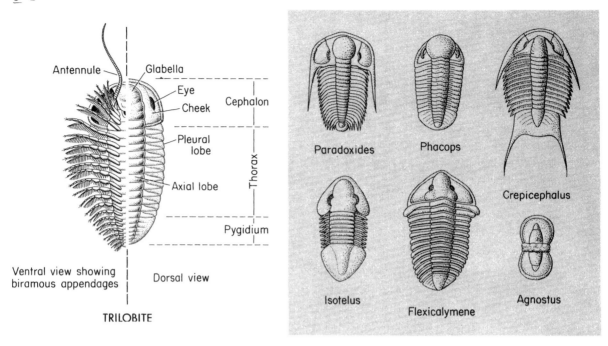

Figure XIII-16. Phylum ARTHROPODA, Trilobita. (See Fig. XIII-31.)

Variety of body forms.

*Hard parts:* A partially segmented, chitinous exoskeleton.

*Habitats:* Marine, freshwater, and terrestrial. Swimmers, bottom-crawlers, and forms attached to bottom and objects. Very diverse.

*Examples:* Lobsters, crayfish, shrimp, barnacles, and ostracods. Of all these, only the ostracods are common fossils in Michigan.

The ostracods, abundant in some Paleozoic age strata in the state, are, nevertheless, so tiny (about pin-head size or less) that they are unlikely to be found by one without special training and equipment. See Figures XIII-17 and 32.

*Note:* Several other living arthropod classes are important today and in some rocks are important fossils, but they have left little or no known fossil record here in Michigan, so we merely list them with examples.

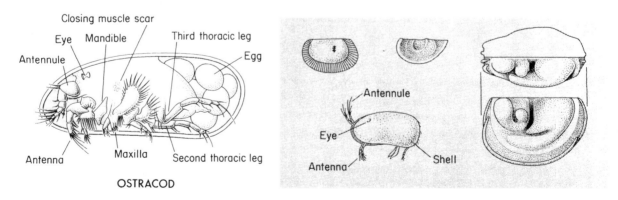

Figure XIII-17. Phylum ARTHROPODA, Ostracoda. (See Fig. XIII-32.) In part modified from Kesling, after Sars.)

*Class MYRIAPODA. Centipedes and millipedes ("hundred- and thousand-leggers").

*Class INSECTA. The insects.

*Class ARACHNOIDEA. Spiders, scorpions, crabs, mites, fleas, lice, ticks, and the long extinct *eurypterids* (late Cambrian to Permian).

##Phylum ECHINODERMATA (or ECHINODERMA) (Figs. XIII–18 and 19). This group is very different in many respects from all other animal phyla except certain simple or "primitive" members of the Phylum Chordata (see below), to which vertebrates and we ourselves also belong. This has led many students of evolution to consider that the echinoderms and chordates had a distant common ancestor quite apart from the other phyla. Adults of many echinoderm groups, including all living types, develop a radial type of symmetry that commonly is 5-rayed although it may be more than 5-rayed. However, the larvae of living echinoderms and the adults of several ancient and now extinct types possess or possessed bilateral symmetry as do the chordates. Circulatory and respiratory systems are present but an excretory system is lacking. A rather complicated internal water vascular system is present. This may connect to the tube feet which, in turn, are involved in locomotion, grasping, and food catching, as are also the "arms" when present. Radially arranged "food grooves," often numbering 5 (but more or less in some) lead to the mouth. The arms, if present, bear the food grooves. In some (sea urchins and sand dollars) locomotion is by means of movable spines.

*Hard parts:* The skeletons of echinoderms are of many forms, but all are *internally* formed and located, close to the surface, within the body of the animal. In life they are covered by a thin covering of soft tissue. This is in contrast with the major skeletal hard parts of most other invertebrate animal phyla which instead produce external exoskeletons. The

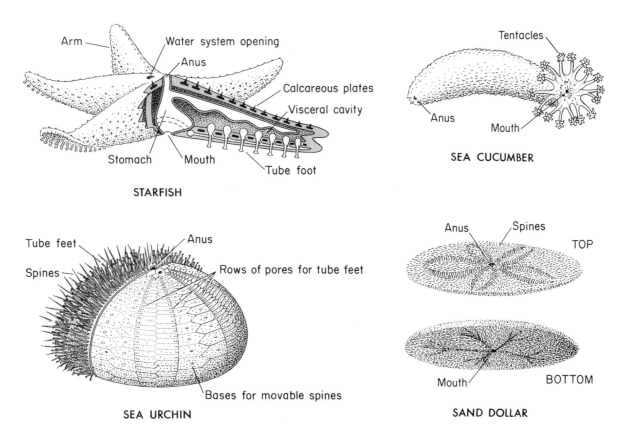

Fig. XIII–18. Phylum ECHINODERMA, Eleutherozoa. (See Fig. XIII–33.)

echinoderm hard parts consist of a variable number of flexibly hinged or interlocking, calcareous plates. Each plate is formed of a single unit of crystalline calcite. Thus, when broken out of rock, fragments of echinoderm skeletons commonly split along shiny, flat faces which are the mineral cleavage planes of the calcite.

*Habitat:* So far as can be told echinoderms are and always have been exclusively marine creatures. Thus, their presence in rocks is a good index to the marine origin of the sediment in which they were fossilized. Some are bottom-crawlers. Others were attached to objects or to the sea floor. The attached forms (crinoids and blastoids) commonly were affixed to the bottom or to other objects by means of a flexible stem-like organ, the column or stalk. The column, with the root-like extensions of its base, gave those forms a somewhat plant-like appearance.

*Stratigraphic range:* The phylum as a whole (including all groups) ranges from Cambrian to Recent. The ranges of individual groups are shorter, however, as shown below.

*Subphylum ELEUTHEROZOA (Fig. XIII–18). Unattached, freely moving types, including starfishes (Stelleroidea), sea urchins, and sand dollars (Echinoidea), and sea cucumbers (Holothuroidea). None of these is a common fossil in Michigan; Cambrian to Recent in time in areas outside Michigan.

#**Subphylum PELMATOZOA (Figs. XIII–19 and 33). Highly evolved forms with complex skeletons. Adults attached to objects or to the sea floor during major part of life. Several classes are known as listed below. Those common as fossils in Michigan are so indicated.

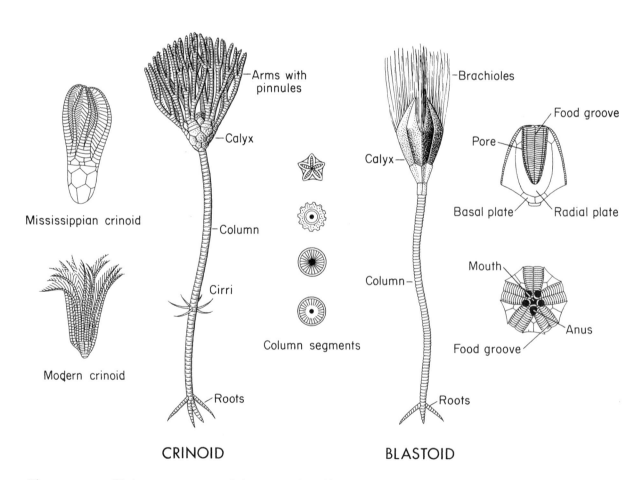

Figure XIII–19. Phylum ECHINODERMA, Pelmatozoa. (See Fig. XIII–33.)

# Fossil Invertebrates

###Class CYSTOIDEA. Cystoids and blastoids. Ordovician to Permian. Now extinct. Blastoids common in Michigan.

##Class CRINOIDEA. The crinoids or sea lilies. Ordovician to Recent but reduced in number of kinds today in comparison with past abundance. Common in the Paleozoic rocks of Michigan. Complete skeletons are rather rare, but portions and isolated ring-like columnals (column sections) of the stalks are extremely common.

**Class EOCRINOIDEA. Cambrian and Ordovician.

**Class PARACRINOIDEA. Ordovician.

###Class EDRIOASTEROIDEA. Cambrian to Pennsylvanian.

**Subphylum HOMALOZOA. Simple, bilaterally symmetrical, with skeleton of relatively few and irregular plates. May have been attached (nonfreeswimming) in some cases. Not common in Michigan or, for that matter, anywhere. Includes the classes CARPOIDEA (Cambrian to Devonian) and MACHAERIDIA (Ordovician to Devonian).

**Subphylum HAPLOZOA. Simple, unattached forms with relatively few plates. Rare. Includes radially symmetrical Class CYAMOIDEA from the Cambrian, and the bilaterally symmetrical Class CYCLOIDEA, also from the Cambrian.

##Phylum CHORDATA (Fig. XIII–20). This important phylum includes the VERTEBRATES (animals with backbones, such as fish, amphibia, reptiles, birds, mammals, and some extinct types) as well as some simpler and perhaps less well-known forms that lack vertebrae. The fossil vertebrates of Michigan are described separately in the next chapter (XIV). In all chordates, a *notochord,* consisting of unsegmented tissue runs part or all of the length of the body, forming a dorsal (upper) support at least in early stages of growth. In vertebrates the unsegmented notochord appears in the embryo, but is later partly or completely replaced by the segmented vertebrae. In the "lower" chordates described below, the notochord persists into the adult stages. A *dorsal nervous system* lies above the notochord in chordates. *Branchial arches* (becoming gill arches in some) and *branchial clefts* (becoming gill clefts in some) are present in all chordates, even in man, at least in embryonic stages. Chordates are complex, bilaterally symmetrical animals.

*Geologic occurrence of chordates:* Some types of chordates, most of them vertebrates, occur as fossils, but they rarely are as common as the many invertebrate types of animals described earlier. It also is unfortunate that few chordates are common in Michigan and most of those are from the most recent part of the geologic past, that is, the Pleistocene. The reasons for this are described in Chapters VI and XIV.

*Subphylum HEMICHORDA. A short, rod-like structure above and in front of the mouth resembles a partial notochord. Short dorsal and ventral nerve cords. Paired gill slits. The larva strongly resembles that of certain echinoderms.

Class ENTEROPTNEUSTA. Free-living, solitary (noncolonial) worm-like. An elongate proboscis (digging organ) and "collar" at front end.

*Hard parts:* None, hence *no* known *fossil* record.

*Habitat:* Marine bottom burrowers.

*Examples:* The "acorn" worms including *Balanoglossus.*

*Class PTEROBRANCHIA. Colonial types. Individuals occupy separate chambers joined into branching colonies by tubular structures of chitin.

*Habitat:* Marine, attached to bottom or other objects.

*Examples: Cephalodiscus.*

*Stratigraphic range:* ? Ordovician. Cretaceous-Recent.

**Class GRAPTOZOA (Figs. XIII–21 and 32). Extinct colonial organisms. Soft parts unknown.

*Hard parts:* Individual animals occupied chitinous cup-like or tube-like chambers (theca) serially arranged along one or more thread-like structures. Details of skeleton resemble those found in

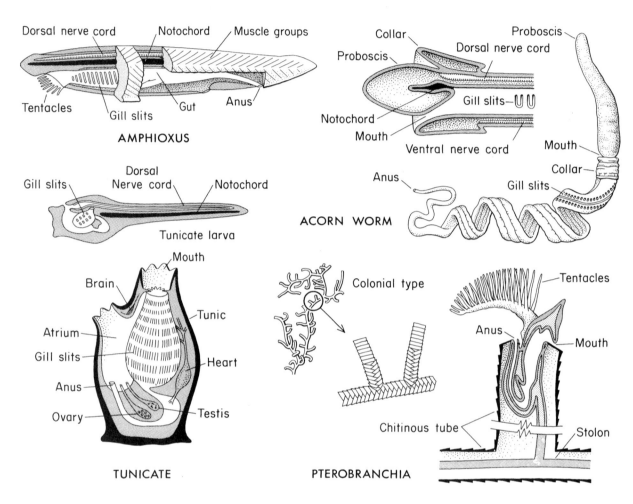

Figure XIII–20. Phylum CHORDATA. (See Chapter XIV.)

Pterobranchia described above. Carbonized impressions and films of the chitinous exoskeleton are rather common, widespread and distinctive in rocks of Paleozoic age and hence are valuable index fossils.

*Habitat:* Surface floaters, both free and attached to floating objects. Exclusively marine.

*Examples:* The graptoloid and dendroid graptolites.

*Stratigraphic range:* Cambrian to Mississippian. If these are truly chordates (there is some doubt), then they are the oldest known Chordates in the fossil record.

Subphylum UROCHORDA. Small, marine animals with no certain fossil record. Adults are solitary, sessile, motionless, animals covered by a leathery "tunic," hence the name "tunicates" is often applied to them. Notochord and nerve cord are vestigial in adults. The internal gills are expanded and highly modified into elaborate, basket-like respiratory and feeding organs. The *larva*, however, is a free-swimming, bilaterally symmetrical, tadpole-shaped creature with well-developed notochord, dorsal nerve cord, and gill slits.

*Hard parts:* Lacking, hence *no fossil record.*

*Habitat:* Marine, attached (fixed) bottom dwellers.

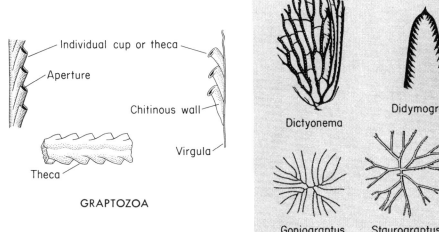

Figure XIII–21. Phylum CHORDATA, Graptozoa. (See Fig. XIII–32.)

*Examples:* The sea squirts or tunicates or ascidians.

Subphylum CEPHALOCHORDA. Small, simple, "minnow"-shaped chordates. No fins, limbs, jaws, or teeth. Small, simple brain with rudiments (or vestiges) of sense organs. The notochord and nerve cord run the full length of the body.

*Hard parts:* Lacking, hence *no certain fossil record*. However, some paleontologists believe that two peculiar fossil specimens of a genus called *Jamoytius* may have been closely related. The cephalochorda are important because they most closely approach what is believed to have been the structural condition of the bilaterally symmetrical, free-swimming ancestors of vertebrates.

*Habitat:* Marine.

*Example:* The lancelets, including *Amphioxus*.

Subphylum POGONOPHORA. A newly discovered and as yet poorly known group of living chordates somewhat resembling the acornworms (Hemichorda) but lacking a digestive tract. Food catching *and* digestion both carried on externally by means of tentacles around the mouth. No hard parts. No known fossil record.

##Subphylum VERTEBRATA. Chordates possessing a segmented axial support, the spinal column or "backbone" (although it may be cartilage, not bone), which largely or completely replaces the notochord in adults. These are the Vertebrates, including some extinct fossil classes (*Ostracoderms* and *Placoderms* of aquatic form), as well as the many fossil and living types of *fish, amphibians, reptiles, birds,* and *mammals*. The fossil record of vertebrates is excellent elsewhere, but they are relatively rare in Michigan except for the Pleistocene mammals. *See Chapter XIV that follows for a description of the fossil vertebrates of Michigan.*

## The Significance of Fossils

The study of fossils serves several very important biological and geological purposes. For the biologist, fossils provide documentary evidence in support of the concept of organic evolution. Research on living animals and plants yields a host of clues which lead to the *hypothesis* that all organisms have ancestor-descendant relationships and have evolved through time, but only paleontology provides the evidence to prove that evolutionary processes really did operate in the geologic past.

## Geology of Michigan

Figure XIII-22. Index map to some Michigan fossil collecting localities. See detailed list in text; this chapter.

Thus, evidence from fossils serves to elevate the evolutionary concept from the status of hypothesis to *theory*. Beyond this, some ancient animals now are extinct, and only fossils can reveal their nature and the proof of their former existence. We shall not dwell on the biological significance of fossils because this is a book about geology, but the subject of evolution is a fascinating study.

### Geological Uses of Fossils

The geologist uses fossils to aid him in the reconstruction of earth history. That history, of course, includes the development of life itself, but fossils also assist in the dating of rocks and in the determination of the environment in which sedimentary deposits originated. We briefly discussed the use of fossils for telling time in Chapter II and will say more about that subject later. Then we will examine the subject of paleoecology, the name given to studies of the environments occupied by ancient animals and plants. One of the most fascinating aspects of the fossil record is the opportunity it affords to reconstruct rock history and to visualize the ancient environments that prevailed in Michigan, or elsewhere, in the past. Fortunately, too, even an amateur can interpret the paleoecological significance of the fossil record in surprising detail if he gives some thought to the fossils which he collects.

Few fossils serve equally well as both relative time indicators (index or guide fossils) and as environmental indicators (facies fossils). Let us examine the reasons and, at the same time, discover more about the nature of the fossil record itself.

To serve well as *time indicators* or *guide fossils* for dating rocks, ancient animals must have had, at the time they lived, the following general characteristics: (1) Geographically widespread distribution, (2) abundance, (3) capability of having been preserved in a wide variety of sediments in a large number of different environments, (4) membership in a rapidly evolving group, (5) capability

# Fossil Invertebrates

Figure XIII-23. Fragment of a stromatoporoid (Phylum COELENTERATA) colony from the Petoskey Limestone in the Middle Devonian age Traverse Group at Petoskey.

*Top*—A surface of growth cracked out approximately along 1 lamination.

*Bottom*—Same specimen in side view showing undulatory laminations of successive growth stages. About natural size.

of rapid dispersal over long distances, and (6) possession of *complex* hard parts.

Let us consider each of these characteristics.

If a fossil is to tell us that sedimentary rocks deposited in two widely separated areas actually are equivalent in age (time correlative), the same animal must have lived in widely separated places, i.e., Europe and North America, or Canada and eastern Asia. Modern examples are certain genera and species of marine snails and clams which occur along the coasts of both California and Japan on opposite sides of the Pacific Ocean Basin. Upon death, their shells accumulate in the sediments currently being deposited in both areas, and in the future, they could be found in rocks of both places. Polar bears, to mention an extreme contrast, live only in the Arctic. Their fossil remains would not occur in sediments deposited elsewhere in the world.

Unless the individual members of an animal or plant species are abundant, their chances of preservation anywhere are poor, and the likelihood of their fossil remains being found are even slighter. Invertebrate animals generally are more abundant than vertebrates, so the former are usually more useful as guide fossils.

The presence of a hard shell or skeleton which can resist destruction by the elements and decay makes it more likely that some fossil record of the animal that possessed it will be preserved. Fleshy soft parts alone rarely leave even an impression in the rocks.

If the shell or skeleton is complex, a small evolutionary change in some part of it is more easily detected and thus successive members of an evolving lineage can be distinguished from one another.

Sediments of any given age often accumulate in a wide variety of different environments. If a fossil is to show that rocks in many separate places are the same age, the animal represented must either have been able to live in a wide variety of the ancient environments involved, or it must have been a type which could have been transported into those environments shortly after death. Surface floaters or swimmers, for example, might easily fall into many different kinds of bottom sediments in streams, lakes, or the sea. Sedentary bottom dwellers, on the other hand, would tend to be fossilized only in or near those sedimentary environments which they actually occupied at the time of life. A fish or a floating foraminiferan might fall into near-shore beach sands, lagoonal deposits, reef beds, or even deep water oozes. Corals, in contrast, especially the colonial reef builders, would be restricted in their potential occurrence largely to the reef environments themselves. Colonial corals would be even further restricted in their occurrence by the fact that they can live only in relatively warm, clear, shallow seas. Their geographic distribution in the fossil record would be "patchy." They might be found in the fossil state in reefal limestones

**Geology of Michigan**

312

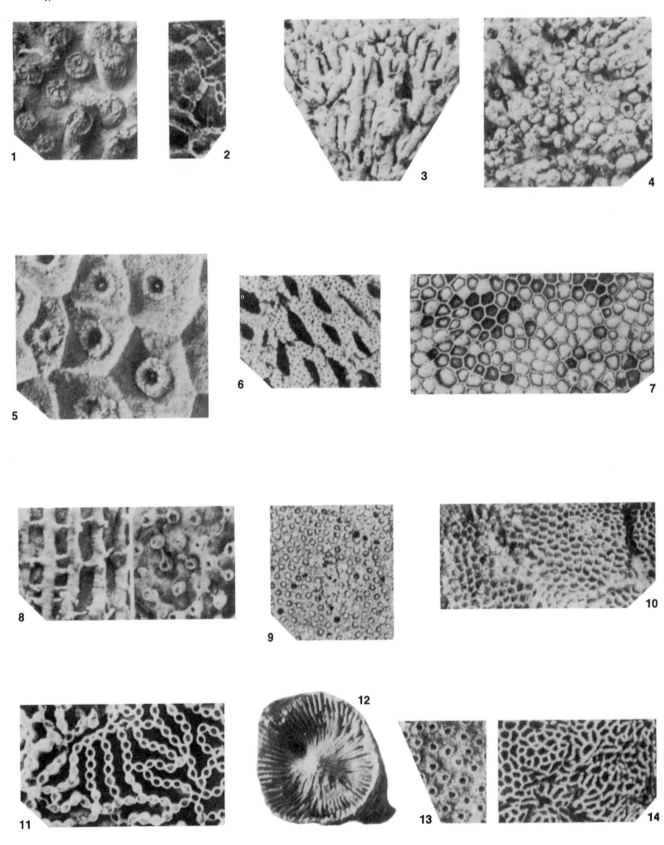

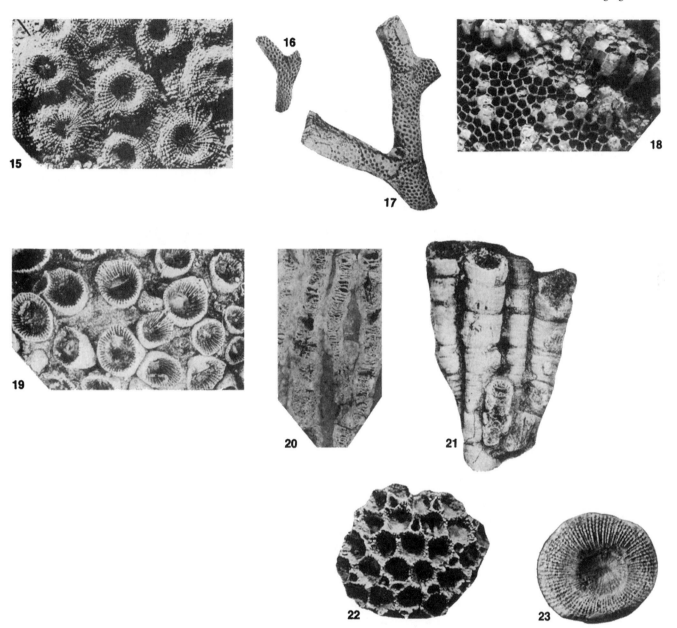

Figure XIII-24. Some representative genera of fossil corals (Phylum COELENTERATA) that occur in Michigan. All approximately natural size.

1—*Paleophyllum,* Late Ordovician.
2—*Halysites,* Late Ordovician.
3, 4—*Paleophyllum,* side and top views, Early Silurian.
5—*Arachnophyllum,* Middle Silurian.
6—*Coenites* (or *Cladopora*), Middle Silurian.
7—*Favosites favosus,* Middle Silurian.
8—*Syringopora,* Middle Silurian, side and top views.
9—*Propora,* Middle Silurian, small species.
10—*Favosites obliquus,* Middle Silurian.
11—*Halysites,* Middle Silurian.
12—*Dinophyllum?,* Middle Silurian.
13—*Propora,* Middle Silurian, large species.
14—*Catenipora,* Middle Silurian.
15—*Billingsastraea,* Middle Devonian.
16—*Coenites* (or *Cladopora*), Middle Devonian.
17—*Thamnopora,* Middle Devonian.
18—*Favosites biloculi,* Middle Devonian.
19, 20, and 21—*Cylindrophyllum,* top, side in cross section, and exterior side views, Middle Devonian.
22—*Pleurodictyum?* (*Procteria*), Middle Devonian.
23—*Heliophyllum,* Middle Devonian.

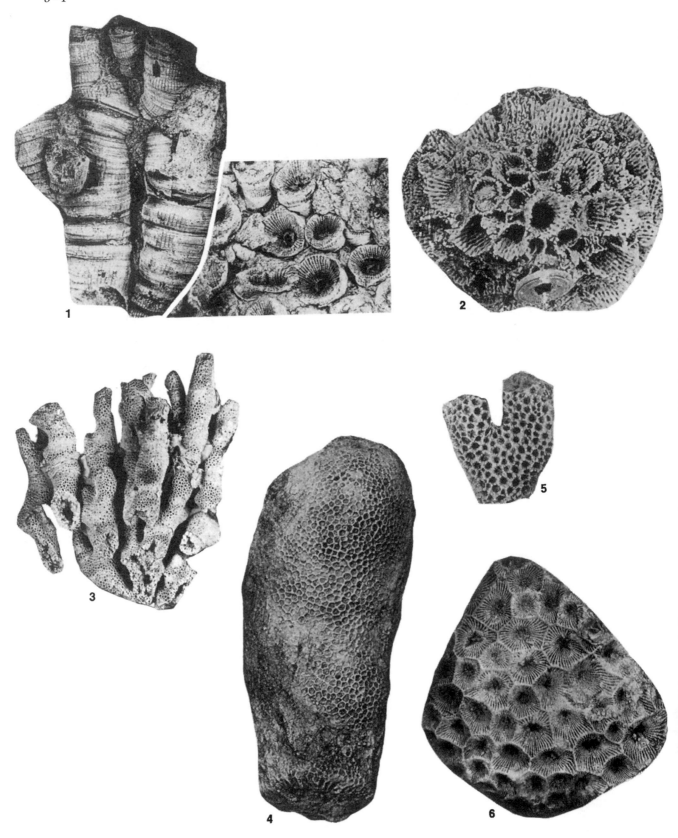

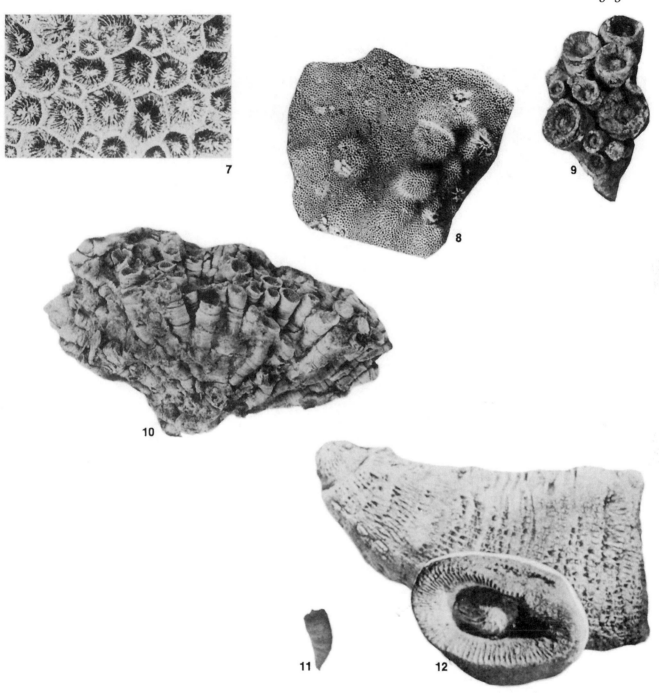

Figure XIII–25. Additional representative genera of fossil corals (Phylum COELENTERATA) that occur in Michigan. Natural size.

1—*Eridophyllum*, side and top views, Middle Devonian.
2—*Antholites*, Middle Devonian.
3—*Favosites clausus*, Middle Devonian.
4—*Favosites alpenensis*, Middle Devonian.
5—*Striatopora*, Middle Devonian.
6—*Hexagonaria anna*, Middle Devonian.
7—*Spongophyllum*, Middle Devonian.
8—*Favosites mammilatus*, Middle Devonian.
9—*Lithostrotion*, Late Mississippian.
10—*Cylindrophyllum hindshawi*, Middle Devonian.
11—*Lophophyllum*, Early Pennsylvanian.
12—*Acrophyllum?*, Middle Devonian, side and top (inset) views.

# Geology of Michigan

316

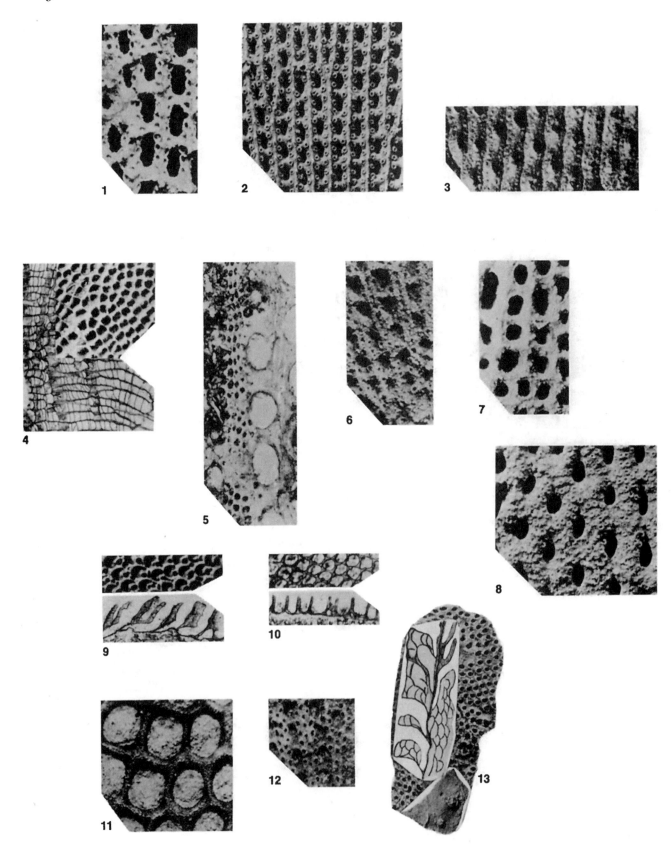

# Fossil Invertebrates

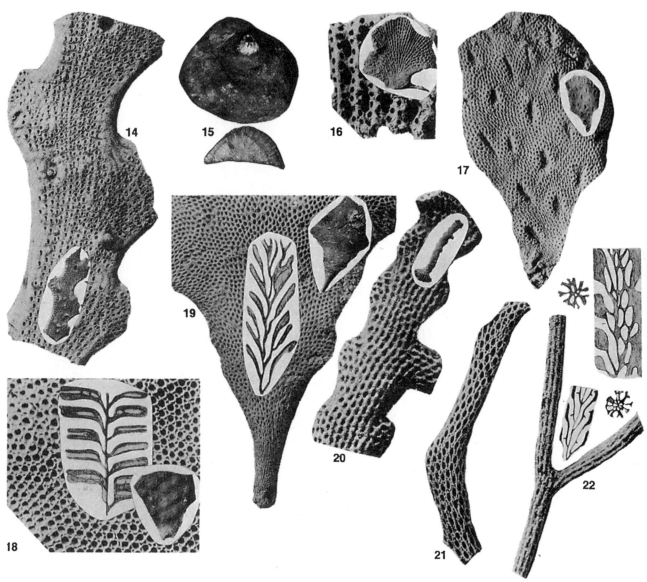

Figure XIII-26. Some representative genera of fossil bryozoans (Phylum BRYOZOA) that occur in Michigan. Magnifications as shown. All Devonian in age except 15.

1—*Fenestella*, ×12 magnification.
2—*Fenestrellina?*, ×10.
3—*Semiscocinium*, ×12.
4—*Atactotoechus*, ×10, cross section and surface views.
5—*Polypora?* (or *Protoretopora*), ×12.
6—*Polypora exemplaria*, ×12.
7—*Isotrypa*, ×12.
8—*Anastomopora*, ×10.
9—*Cyclopora? lunata*, ×20, cross section and surface.
10—*Calacanthopora*, ×20, cross section and surface.
11—*Polypora magnifica*, ×12.
12—*Polypora modesta*, ×12.
13—*Phractopora*, bottom inset natural size, cross section inset ×15, magnified surface ×5.
14—*Sulcoretopora*, inset natural size, magnified surface ×5.
15—*Prasopora* colony one half natural size, surface (top) and cross section, Ordovician in age.
16—*Lyropora?*, ×10 and inset natural size.
17—*Ceramella*, ×5 and inset natural size.
18—*Intrapora traversensis*, ×10, inset surface view natural size, and cross section ×15.
19—*Stictoporina*, ×5, inset surface view natural size, and cross section ×15.
20—*Euspilopora*, ×10 and inset natural size.
21—*Streblotrypa*, fragment of branching colony, ×10 with transverse cross section ×10 and longitudinal cross section ×15.
22—*Acanthoclema*, ×10 with transverse and longitudinal cross sections.

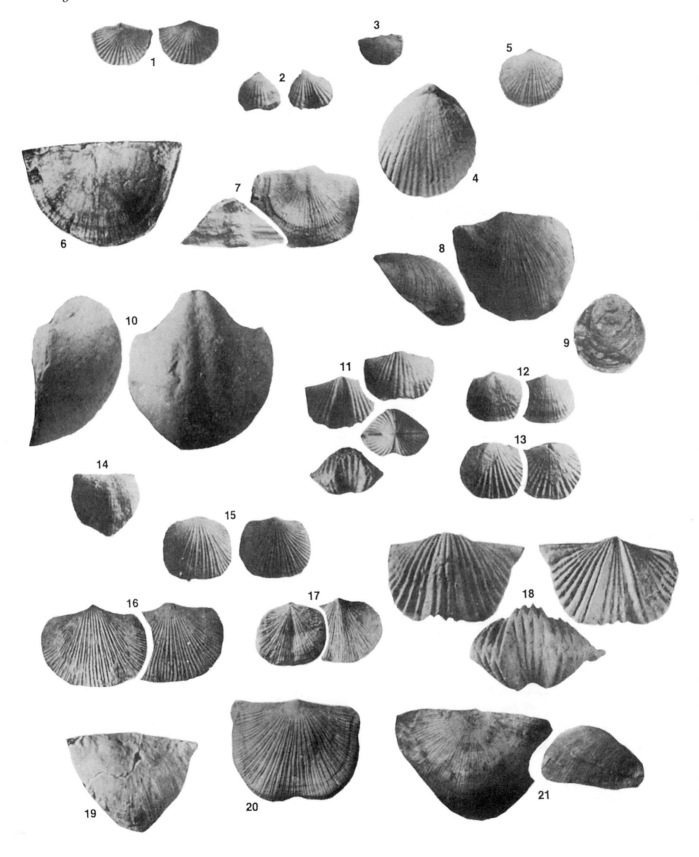

# Fossil Invertebrates

Figure XIII-27. Some representative genera of fossil brachiopods (Phylum BRACHIOPODA) that occur in Michigan. Natural size unless otherwise indicated. Different views of same specimen closely spaced.

1—*Hesperorthis*, 2 views, Middle Ordovician.
2—*Rhynchotrema minnesotense*, 2 views, Middle Ordovician.
3—*Sowerbyella*, Middle Ordovician.
4—*Zygospira*, ×4, Middle Ordovician.
5—*Resserella?*, ×2, Middle Ordovician.
6—*Strophomena filitexta*, Middle Ordovician.
7—*Vellamo*, 2 views, Middle Ordovician.
8—*Rafinesquina deltoidea*, 2 views, Middle Ordovician.
9—*Lingula changi*, Middle and Late Ordovician.
10—*Cyclospira*, ×5, 2 views, Middle Ordovician. Mold.
11—*Platystrophia trentonensis*, 4 views, Middle Ordovician.
12—*Glyptorthis*, 2 views, Middle Ordovician.
13—*Plectorthis*, 2 views, Middle Ordovician.
14—*Trigrammaria*, Middle Ordovician.
15—*Plaesiomys*, 2 views, Middle Ordovician.
16—*Hebertella occidentalis*, 2 views, Late Ordovician.
17—*Dalmanella*, 2 views, Late Ordovician.
18—*Platystrophia*, large species, 3 views, Late Ordovician.
19—*Strophomena huronensis*, Late Ordovician.
20—*Hebertella alveata*, Late Ordovician.
21—*Strophomena neglecta*, 2 views, Late Ordovician.
22—*Plaesiomys (Dinorthis)*, large species, 2 views, Late Ordovician.
23—*Rafinesquina alternata*, Late Ordovician.
24—*Zygospira recurvirostris*, 4 views, Late Ordovician.
25—*Virgiana*, 2 views, Lower or Middle Silurian.
26—*Pentamerus*, 2 views, Middle Silurian. Mold.
27—*Pentameroides*, 2 views, Middle Silurian. Mold.
28—*Camarotoechia?* (or *Leiorhynchus*), 4 views, Middle Silurian.
29—*Whitfieldella*, Late Silurian.
30—*Delthyris*, ×3, 2 views, Late Silurian.
31—*Orbiculoidea*, ×3, Late Silurian.
32—*Coelospira*, ×2, 2 views, Middle Silurian.
33—*Leptatrypa*, ×2, 3 views, Middle Silurian.

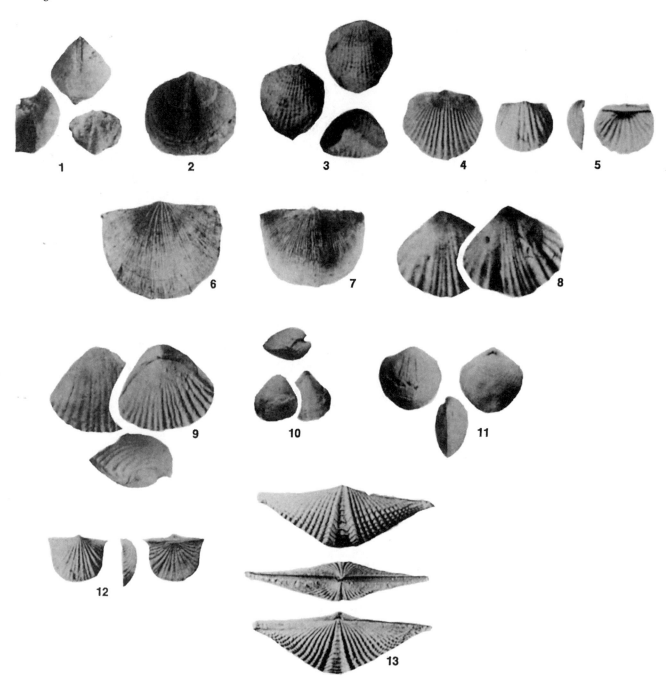

**Fossil Invertebrates**

321

Figure XIII-28. Additional genera of fossil brachiopods (Phylum BRACHIOPODA) that occur in Michigan. Natural size unless otherwise indicated. Different views of same specimen closely spaced.

1—*Prosserella*, 3 views, Middle Devonian.
2—*Rhipidomella*, Middle Devonian.
3—*Atrypa costata*, 3 views, Middle Devonian.
4—*Tropidoleptus*, Middle Devonian.
5—*Stropheodonta erratica*, 3 views, Middle Devonian.
6—*Stropheodonta demissa*, Middle Devonian.
7—*Chonetes coronatus*, Middle Devonian.
8—*Leiorhynchus*, 2 views, Middle Devonian.
9—*Gypidula*, 3 views, Middle Devonian.
10—*Pentamerella*, 3 views, Middle Devonian.
11—*Athyris*, 3 views, Middle Devonian.
12—*Chonetes emmetensis*, ×2, 3 views, Middle Devonian
13—*Mucrospirifer*, 3 views, Middle Devonian.
14—*Paraspirifer*, Middle Devonian.
15—*Rensselaeria?*, Middle Devonian.
16—*Amphigenia*, Middle Devonian.
17—*Plethorhyncha?*, Middle Devonian.
18—*Leptaena*, Middle Devonian.
19—*Lingula carbonaria*, parts of 4 specimens in black shale, Early Pennsylvanian.
20—*Juresania*, Early Pennsylvanian.

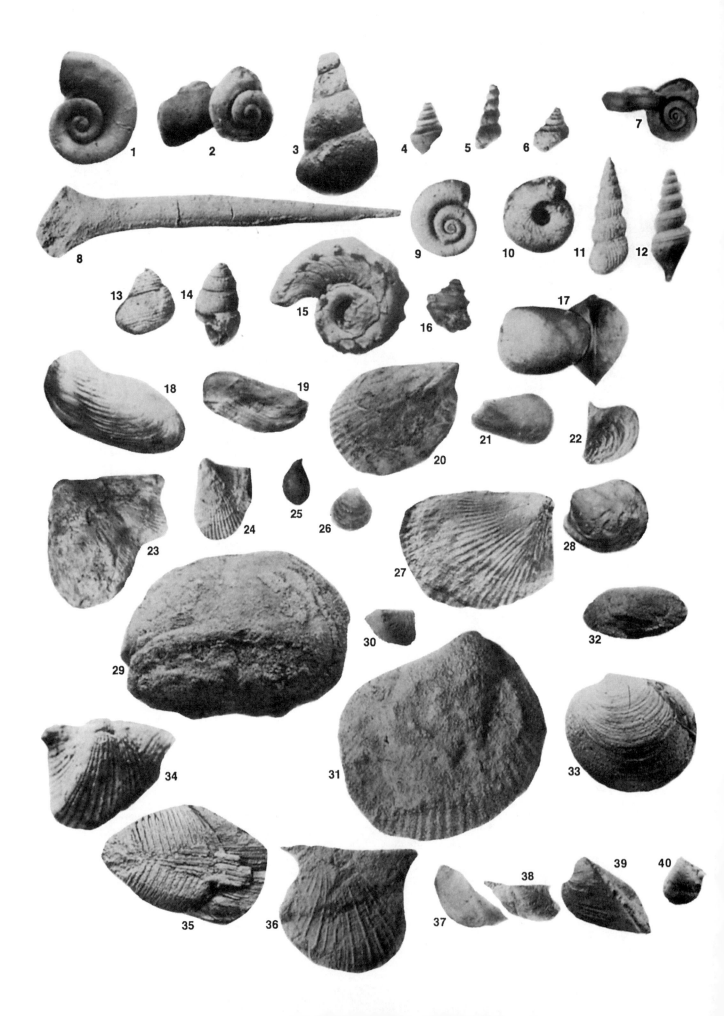

Figure XIII-29. Some representative genera of fossil molluscs (Phylum MOLLUSCA) that occur in Michigan. All about half natural size unless otherwise indicated. Numbers 1–16 are snails (Class GASTROPODA), 17–61 are clams (Class (PELECYPODA).

1—*Maclurites*, Middle Ordovician.
2—*Holopea*, 2 views, Middle Ordovician.
3—*Fusispira*, Middle Ordovician.
4—*Lophospira*, Late Ordovician.
5—*Hormotoma*, Late Ordovician.
6—*Trochonema*, Late Ordovician.
7—*Helicotoma?*, 2 views, Late Ordovician.
8—*Mastigospira*, a noncoiled type, Middle Devonian.
9—*Platyoron?*, Middle Devonian.
10—*Tropidodiscus*, Middle Devonian.
11—*Loxonema*, Middle Devonian.
12—*Murchisonia*, Middle Devonian.
13—*Rhineoderma?*, Middle Devonian.
14—*Acanthonema?*, Middle Devonian.
15—*Omphalocirrus*, ×1/4, Middle Devonian.
16—*Worthenia*, Early Pennsylvanian.
17—*Vanuxemia*, Middle Ordovician.
18—*Cuneamya*, Middle Ordovician.
19—*Psiloconcha*, Late Ordovician.
20—*Byssonychia*, Late Ordovician.
21—*Modiolopsis*, Late Ordovician.
22—*Pterinea*, Late Ordovician.
23—*Opistholoba*, Late Ordovician.
24—*Byssonychia*, small species, Late Ordovician.
25—*Clionychia*, Late Ordovician.
26—*Trematis*, Late Ordovician.
27—*Anomalodonta*, Late Ordovician.
28—*Cyrtodonta*, Late Ordovician.
29—*Megalomus*, Middle Silurian.
30—*Pterinea*, small species, Late Silurian.
31—*Panenka*, Middle Devonian.
32—*Cypricardinia*, Middle Devonian.
33—*Paracyclas*, Middle Devonian.
34—*Conocardium*, Middle Devonian.
35—*Conocardium*, Middle Devonian.
36—*Follmannella*, Middle Devonian.
37—*Diodontopteria*, Middle Devonian.
38—*Leptodesma*, Middle Devonian.
39—*Goniophora*, Middle Devonian.
40—*Limoptera*, Middle Devonian.
41—*Actinopterella*, Middle Devonian.
42—*Nuculoidea*, Middle Devonian.
43—*Solenomorpha*, Middle Devonian.
44—*Gosselettia*, Middle Devonian.
45—*Lophonychia*, Middle Devonian.
46—*Phenacocyclas*, Middle Devonian.
47—*Liromytilus*, Middle Devonian.
48—*Palaeosolen*, Mississippian.
49—*Nuculopsis*, Mississippian.
50—*Schizodus*, Mississippian.
51—*Parallelodon*, Mississippian.
52—*Grammsyia*, Mississippian.
53—*Solenomorpha*, Mississippian.
54—*Prothyris*, Mississippian.
55—*Polidevcia*, Mississippian.
56—*Ctenodonta*, Mississippian.
57—*Allorisma*, Mississippian.
58—*Sphenotus*, Mississippian.
59—*Sanguinolites*, Mississippian.
60—*Palaeoneilo*, Mississippian.
61—*Naiadites*, Early Pennsylvanian.

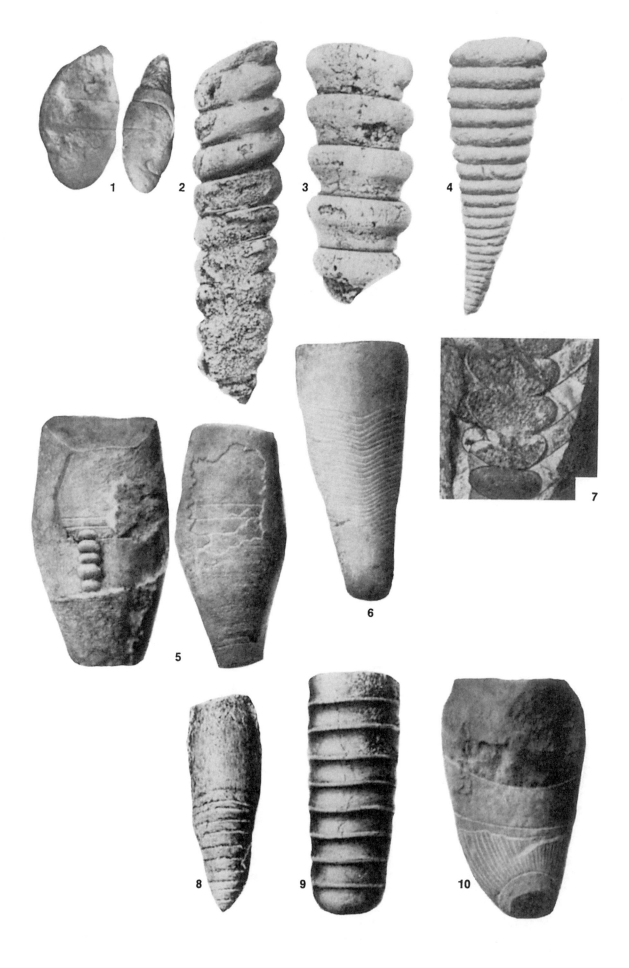

Figure XIII-30. Some representative genera of fossil cephalopods (Phylum MOLLUSCA, Class CEPHALOPODA) that occur in Michigan. All illustrations are about one-half natural size unless otherwise indicated. Numbers 1–15 are nautiloids (NAUTILOIDEA), 16–20 are ammonoids (AMMONOIDEA).

1—*Billingsites,* 2 views, Late Ordovician.
2—*Armenoceras,* a siphuncle from interior of shell, Middle Silurian.
3—*Huronia,* a siphuncle from shell interior, Middle Silurian.
4—*Stokesoceras,* siphuncle from shell interior, Middle Silurian.
5—*Acleistoceras,* 2 views, suture lines show where exterior shell has been removed, interior siphuncle seen on left. Note straight, uncomplicated, nautiloid sutures. Middle Devonian.
6—*Alpenoceras,* Middle Devonian.
7—*Armenoceras,* cross section of part of shell to show relationship of outer conch wall to interior septa and siphuncle. Silurian.

8—*Poterioceras,* Early Mississippian.
9—*Cycloceras,* Early Mississippian.
10—*Nephriticerina,* Middle Devonian.
11—*Endolobus,* Early Mississippian.
12—*Vestinautilus?,* 2 views, Late Mississippian.
13—*Mooreoceras,* Early Mississippian.
14—*Kionoceras,* 1.5 times natural size, Early Mississippian.
15—*Pseudorthoceras,* Early Pennsylvanian.
16—*Imitoceras,* 3 views, note complexly folded ammonoid type of suture lines. Early Mississippian.
17—*Munsteroceras,* 3 views, note complex suture lines, Early Mississippian.
18—*Merocanites,* Early Mississippian.
19—*Beyrichoceras,* 2 views, Early Mississippian.
20—*Gattendorfia,* natural size, Early Mississippian.

Figure XIII-31. Some representative genera of fossil trilobites (Phylum ARTHROPODA, Class TRILOBITA) that occur in Michigan. All illustrations natural size.

1, 2—*Prosaukia,* a head and tail, Late Cambrian.
3—*Ogygites,* Middle Ordovician.
4—*Ceraurus,* Middle Ordovician.
5—*Illaenus,* Middle Ordovician.
6—*Isotelus,* Middle Ordovician.
7—*Flexicalymene,* Middle Ordovician.
8—*Bumastus,* Middle Ordovician.
9—*Anchiopella,* a head and tail, Middle Devonian.
10—*Dechenella,* a nearly complete specimen and a head, Middle Devonian.
11—*Greenops,* 3 nearly complete individuals, Middle Devonian.
12—*Trimerus (Dipleura),* Middle Devonian.
13—*Calymene,* Middle Devonian.
14—*Crassiproteus,* Middle Devonian.
15—*Phacops,* top view, Middle Devonian.
16—*Phacops,* side view of specimen tightly enrolled so that tail is beneath head. Note facets on eye. Middle Devonian.
17—*Kaskia,* top and side views of an extended specimen, Late Mississippian.

**Geology of Michigan**

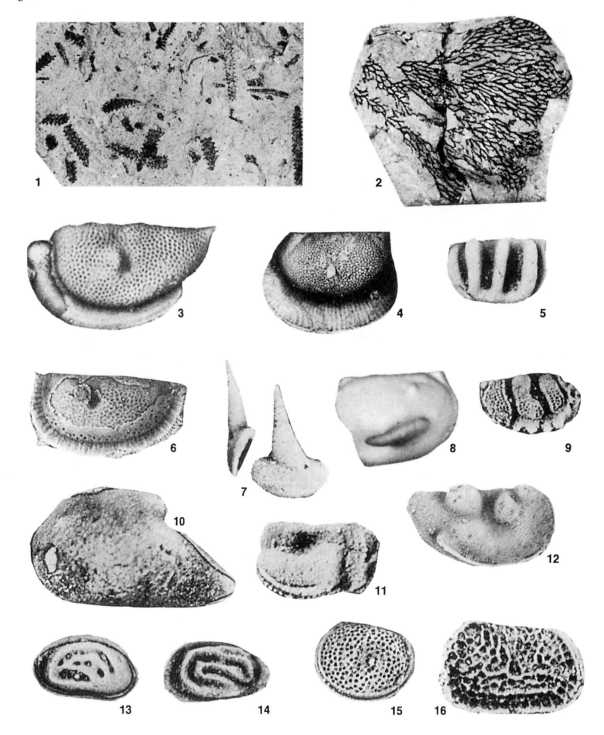

# Fossil Invertebrates

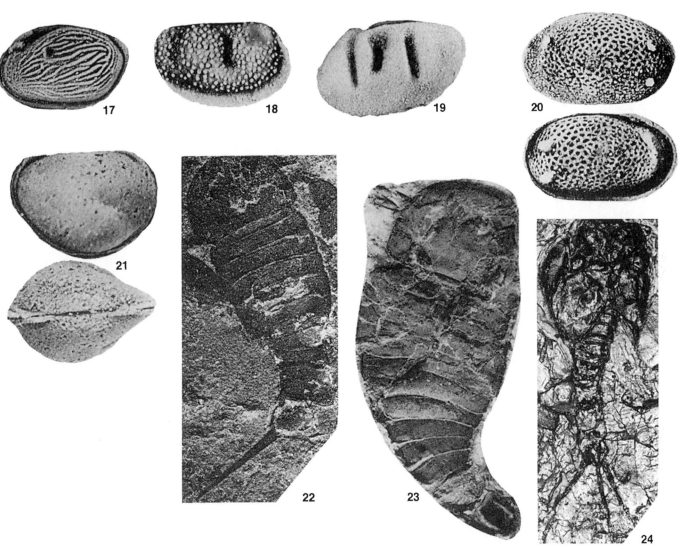

Figure XIII-32. Graptozoans (Phylum CHORDATA, Subphylum HEMICHORDA, Class GRAPTOZOA) in illustrations 1 and 2. Additional arthropods (Phylum ARTHROPODA) in 3-24. Illustrations 3-21 are ostracods (OSTRACODA), 22 and 23 eurypterids (EURYPTERIDA), 24 a crustacean. All fossil types occurring in Michigan. Magnifications as indicated.

1—*Glossograptus,* about twice natural size, Late Ordovician.
2—*Dictyonema,* natural size, Middle Ordovician.
3—*Levisulculus,* ×50, Middle Ordovician.
4—*Platybolbina,* ×25, Middle Ordovician.
5—*Quadrijugator,* ×30, Late Ordovician.
6—*Eurychilina,* ×20, Middle Ordovician.
7—*Aechmina,* ×40, front and left sides of one half of shell. Note large lateral spine. Middle Devonian.
8—*Herrmannina,* ×5, Middle Silurian.
9—*Ctenoloculina,* ×28, Middle Devonian.
10—*Bairdia,* ×30, Middle Devonian.
11—*Sulcicuneus,* ×40, Middle Devonian.
12—*Hollinella,* ×30, Middle Devonian.
13—*Octonaria,* ×31, Middle Devonian.
14—*Euglyphella,* ×31, Middle Devonian.
15—*Macronotella,* ×30, Middle Devonian.
16—*Arcyzona,* ×40, Middle Devonian.
17—*Barychilina,* ×31, Middle Devonian.
18—*Ctenobolbina,* ×28, Middle Devonian.
19—*Dizygopleura,* ×30, Middle Devonian.
20—*Ponderodictya,* ×30, right (bottom) and left (top) halves of shell. Middle Devonian.
21—*Phlyctiscapha,* ×30, dorsal and right side view, showing bivalved nature of an ostracod shell. Devonian.
22—*Erieopterus,* a eurypterid, natural size. Nearly complete specimen from Late Silurian of New York State.
23—*Erieopterus,* ×2, less complete specimen from Late Silurian of Michigan.
24—A newly discovered rhinocarid crustacean, natural size, from the Middle Devonian age Silica Formation, found just across state boundary in Ohio.

**Geology of Michigan**

330

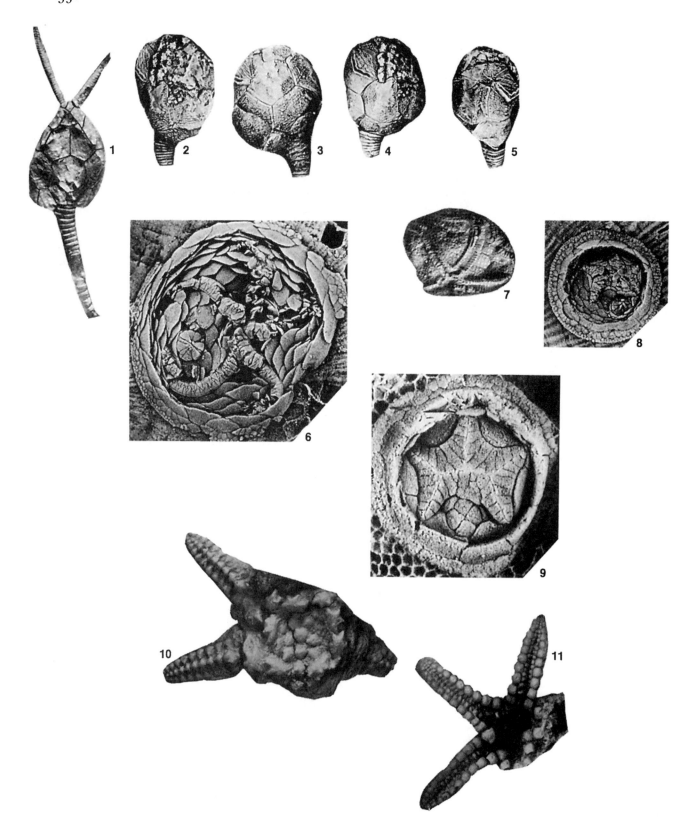

# Fossil Invertebrates

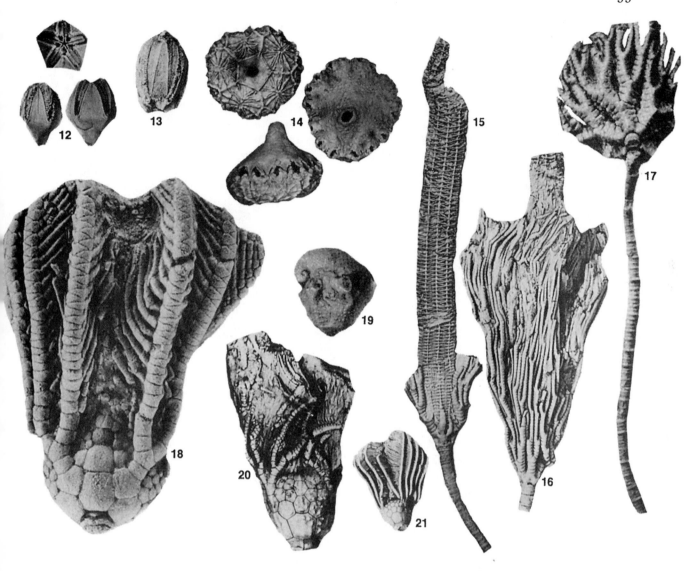

Figure XIII-33. Some representative genera of fossil echinoderms (Phylum ECHINODERMA) that occur in Michigan. Magnifications as indicated. Illustrations 1-5 are cystoids (CYSTOIDEA), 6-9 edrioasteroids (EDRIOASTEROIDEA), 10-11 starfish (ASTEROIDEA), 12-13 blastoids (BLASTOIDEA), and 14-21 crinoids (CRINOIDEA).

1—*Pleurocystites*, 3/4 natural size, Middle Ordovician.

2-5—*Lipsanocystis*, about natural size; anal, right, left, and antanal views respectively. Middle Devonian. Note tendency toward bilateral rather than radial symmetry.

6—*Agelacrinites*, about ×4, an edrioasteroid attached to a brachiopod shell. Middle Devonian.

7—*Edrioaster*, ×3/4, Middle Ordovician.

8—*Hemicystites*, ×6, attached to a brachiopod shell, Middle Devonian.

9—*Timeischytes*, about ×4, attached to a bryozoan colony, Middle Devonian.

10, 11—*Protopalaeaster*, about ×4, Middle Ordovician.

12—*Hyperoblastus*, ×1 1/2, 3 views, Middle Devonian.

13—*Nucleocrinus*, ×1 1/2, Middle Devonian.

14—*Dolatocrinus*, ×3/4, 3 views, Middle Devonian. Calyx only. Arms and stem missing.

15—*Proctothylacocrinus*, ×3/4, part of stem, and calyx with arms largely removed, Middle Devonian.

16—*Proctothylacocrinus*, ×3/4, calyx with branched arms largely intact, Middle Devonian.

17—*Euryocrinus*, ×3/4, stem nearly complete and calyx with arms nearly complete, Middle Devonian.

18—*Opsiocrinus*, ×7 1/2, calyx magnified to show arrangement of plates and the pinnules on arms, Middle Devonian.

19—*Arthroacantha*, ×3/4, Middle Devonian.

20—*Corocrinus*, ×2, calyx with plates outlined in ink, Middle Devonian.

21—*Corocrinus*, ×3/4, Middle Devonian.

# Geology of Michigan

but not in the shales or sandstones laid down close around the reef.

Rapid evolution is important for a guide fossil group because it produces many unique but short-lived new animal and plant forms over a relatively brief span of geologic time. Thus, rapidly evolving forms of one time, differing from those of the next, could serve to distinguish different periods of time. If a genus were so well adapted that it lived on, virtually unchanged, from Devonian to Recent times, as the ostracod genus *Candona* has done, one could not use it to tell a Devonian rock from any rock deposited during the hundreds of millions of years thereafter. Amongst the brachiopods, the genus *Lingula*, a common inhabitant of muddy bottoms in shallow marine waters, is a classical example of a long-lived genus that remained evolutionarily stationary, a so-called "living fossil." It has changed so little that Cambrian types are essentially like those living today. Nevertheless, *Lingula* is a good brachish marine facies indicator. In the Pennsylvanian rocks of the Saginaw Formation at Grand Ledge, Michigan (Chapter V), the presence of this brachiopod in a thin zone between coaly deposits and stream channel sandstones indicates a brief, shallow-water marine invasion of the area between times of swamp and river sediment deposition. On the other hand, some foraminifera and ammonites, common invertebrate fossil groups, evolved so rapidly that many new and different species, genera, and even families appeared during the course of a single geologic epoch.

Rapid dispersal is important if a fossil group is to be a good index to time, because this would make it possible for newly evolved forms to appear nearly simultaneously in widely separated places. Even though a few thousand or even tens of thousands of years might be involved in such a spread, this would still be a mere tick of the slowly running geologic clock. Certain types of organisms are less able to disperse rapidly than are others. Brachiopods spend their adult lives more or less fixed on the sea floor, moving slowly, if at all, over short distances. The brachiopod larva is free floating and could be carried by currents, but the larval stage lasts only about 24 hours. Thus, newly evolved brachiopods could not disperse from their places of origin as rapidly as could an animal with a free swimming or floating adult stage. Brachiopods, too, would be more or less restricted to those areas where they could be carried by currents. Their spread into other areas would be very slow. Careful comparative studies of Paleozoic brachiopod and graptolite faunas from Europe and North America have shown that, whereas newly evolved graptolites appeared almost simultaneously in both of those widely separated regions, the newly evolved brachiopod types appeared first in one area and then much later in the other, crossing over the more nearly time-parallel zones of the successive graptolite faunas. Rocks with the same brachiopods in them might be much older (or younger) in one place than in another. Graptolites, being floaters with rapid dispersal capability, make better time index fossils. Unfortunately, they are rare in the Paleozoic rocks of Michigan, whereas brachiopods are abundant.

Free floating or free swimming animals, in general, provide the best time indices, not only because of their rapid dispersal rates, but also because they live in waters above the bottom. There they are more or less independent of the bottom environments. Upon death they may settle into whatever type of bottom sediment happens to lie below, making it possible for their remains to be preserved in sediments of a wide variety of types. Among the marine invertebrates, we have mentioned the graptolites, ammonites, and foraminifera as being especially good time indicators for rock age correlation. To these we should add the trilobites. All these were abundant, evolved rapidly, dispersed quickly and widely, and had intricate hard parts.

Sluggish or attached, bottom-dwelling animals rarely make good time indices, but often are very sensitive indicators of the nature of the bottom environment on or in which they live. This is because most of these animals are adapted to and tolerant of a rather restricted range of bottom conditions. We have already noted that the reef building corals are good *facies* indicators. Brachiopods, clams, snails, bryozoans, and the attached types of echinoderms (crinoids, blastoids) should be added to the list. This brings us to the next important use of fossils: their significance in studies of ancient environments of sedimentary deposition.

# Fossil Invertebrates

## Fossils and Ancient Environments

A preceding section in this chapter, dealing with the classification of animals, includes brief statements concerning the habitat preferences of each of the major invertebrate animal groups. Figure XII–10 in Chapter XII in the section concerning sedimentary rocks, provides additional information of paleoecological significance. That information is pertinent to the discussion that follows.

Paleoecological studies lean heavily on the Principle of Uniformity, a concept which, as we have already seen (Chapter II), is also very important in studies of physical geology. Application of that principle to paleoecology involves the assumption that fossil remains represent once living organisms which obeyed the same general biological laws as do their living relatives. Animals of the past must have been adapted to their environments just as are their modern counterparts. There must have been limits to their ranges of tolerance to conditions such as temperature, light, water chemistry, bottom-sediment composition, water turbulence, climatic change, and a host of other factors in their physical environment. Ancient animals must also have lived alongside the other members of their community, in balance with predators, food supply, and available space. According to the uniformitarian principle, animals of the past can be identified as having been marine, freshwater, or terrestrial types if they can be classified biologically into groups which today are restricted to one or another of those environments. It is often possible to recognize that certain ancient animals had even more restricted habitat preferences, such as shallow water marine or freshwater lake shore niches in the environment. In certain cases this line of reasoning can be extended to include extinct animal types and even fossil forms with no close living relatives. The trilobites have been extinct since the end of the Paleozoic Era, over 200 million years ago, and they are not closely akin to any other living arthropods. Nevertheless, they are considered marine animals because, when found in association with other fossil animals, the latter always are marine types.

Although the variety of ecologic niches (each niche with its own special set of environmental factors) available for animal occupancy may seem endless, the many possibilities commonly are classified into a few major subdivisions. The nonmarine or continental environment may be occupied by animals (and plants) of aquatic types which inhabited freshwater lakes, streams, and swamps. These would include fishes, most amphibians, certain turtles, alligators, and crocodiles, semiaquatic or aquatic mammals, and semiaquatic birds. Certain of the invertebrates also are freshwater types, but one ordinarily must be a specialist to recognize those types. For instance, some groups of snails, clams, and bryozoans live in freshwater, while others inhabit the sea. One would have to identify the particular group (order, family, genus, or species) to which such a fossil animal belonged before he could tell what kind of an ancient environment was indicated. Figure XIII–34 summarizes the major environmental situations occupied by both living and ancient animals and we may use the terms shown on that chart in the discussion that follows.

Normally, each of these environments is occupied by a wide variety of animals (and plants) suited for life therein, but when one considers the great length of geologic time, the vast expanse of land and sea, and the many geologic and geographic changes the earth has undergone, one realizes that there have always been limits as to where certain organisms could live. For some, a once suitable environment to which they had become well adapted disappeared, in which case they either evolved rapidly enough to readapt to the new conditions or else they became extinct. The mastodons of Michigan (Chapter XIV) are an example of a group which failed to adapt to changing conditions and became extinct. Some organisms lived in a region until conditions there changed. Then they moved elsewhere or survived only in a limited portion of their former range. Musk oxen in the Pleistocene of Michigan (Chapter XIV) illustrate this kind of adaptation, as do the linguloid brachiopods of more ancient deposits in the Great Lakes Region. Some organisms may have been in existence for a long time before environmental changes in a certain region provided habitats suitable for their occupancy. Amphibians first appear in the Late Devonian, but could not occupy Michigan then because this region, at that time, was covered by the sea and amphibians cannot tolerate saline water. Reef building corals, stromatoporoids, and algae could not thrive in the Paleozoic seas of Michigan un-

## Figure XIII-34

MAJOR TYPES OF ENVIRONMENTS AVAILABLE FOR ANIMAL OCCUPANCY

| | | | |
|---|---|---|---|
| **NONMARINE OR CONTINENTAL** | Aerial | | As in terrestrial environments, a wide variety of adaptive types. Semiaquatic types included |
| | Terrestrial | | Dry land surface dwellers. Wide variety of climates, elevations, vegetation zones, and other factors lead to many different adaptive types of animals including semiaquatic types |
| | Freshwater | | *Fluviatile*—Rivers and streams. Aquatic and semiaquatic types |
| | | | *Lacustrine*—Lakes. Aquatic and semiaquatic types |
| | | | *Palludal*—Swamps. Aquatic and semiaquatic types |
| | Underground | | *Cave* dwellers. Terrestrial and aquatic types<br>*Soil* dwellers. Terrestrial types |
| **MARINE** Saline or brackish waters and related environment | *Pelagic* types, including floaters (free or on drifting material) and free swimmers | | *Neritic*—Low tide level out to edge of continental shelf. Waters less than about 600 feet. Aquatic types |
| | | | *Oceanic*—Deep waters (over 600 feet) beyond the edge of the continental shelf. All aquatic |
| | *Benthonic* types, including free-moving bottom crawlers (vagrant benthos), sedentary or fixed types (sessile benthos), and burrowing types that live within the bottom or beach sediments | | *Upper Beach*—Shore line just above high tide level. Terrestrial types |
| | | | *Littoral*—Beach zone between high and low tide levels. Aquatic and semiaquatic types |
| | | | *Sublittoral*—Low-tide level to continental shelf edge (water depths of about 600 feet). All aquatic types |
| | | | *Bathyal*—Continental shelf edge (about 600 feet deep, down to average ocean depths, 12,000 feet. All aquatic types |
| | | | *Abyssal*—Oceanic deeps, about 12,000 feet down to over 30,000 feet below sea level. All aquatic types |

less the waters shallowed to less than about 200 feet. In any case, once a suitable environment became available, it must have taken some time for it to become occupied by animals and plants, the rate of colonization and sequence of colonizers being dependent on the dispersal rates of the organisms involved.

The Paleozoic rocks of Michigan are rich in fossil animal types which individually provide significant paleoecological insights. They furnish the paleontologist with fascinating glimpses of conditions in the geologic past. The exclusively marine types of animals disclose that the rocks containing them were deposited in the sea. These silent historians of the past include most foraminifera, archeocyathids, corals, stromatoporoids, brachiopods, most bryozoans, trilobites, all cephalopods including nautiloids and ammonoids, and all echinoderms and graptolites.

Modern corals inhabit normal marine environments. They cannot tolerate brackish water that has been diluted by inflowing streams near shore, nor can they thrive in excessively saline waters of evaporite basins. Except for these restrictions, the solitary corals are broadly distributed and apparently can tolerate a relatively wide range of water temperatures and depths. Colonial reef building corals, in contrast, require warm water, thriving only where the mean annual water temperature does not fall below 68°F. They require clear water because mud tends to smother them. These corals propagate in abundance only in relatively shallow water (200 feet or less), because they are dependent upon, and grow in close association with, algae. Algae, like almost all other plant life, require sunlight for photosynthesis, and there is insufficient light penetration for algal growth in deeper waters. The Paleozoic corals (Figs. XIII-6, 24, 25) belong to extinct groups, Tabulata and Rugosa, which, nevertheless, are thought to have required conditions similar to those described for modern corals.

# Fossil Invertebrates

Trilobites, another common marine animal group in Paleozoic rocks of Michigan (Figs. XIII-16, 31), appear to have been mobile bottom dwellers (vagrant benthos) for the most part, although some may have swum in the surface waters (pelagic types) and others, those without eyes, are thought to have been bottom burrowers.

The graptolites were either free floating or drifted along attached to floating objects such as seaweed. Some had buoyant disk-shaped or balloon-shaped float structures. Thus, they were carried widely across the face of the sea and at death fell into a wide variety of bottom sediments. Preservation of graptolites is best, however, in black shales. Perhaps this is because the organic hard parts, composed of chitin rather than calcium carbonate, tended to decay in oxygen-rich sediments but did not do so in the oxygen-poor black shales.

Crinoids, blastoids, and other pelmatozoan echinoderms (Figs. XIII-19, 33) grew attached to the sea floor (sessile benthos). Attachment was by means of long, stem-like columns. Often great numbers of individuals grew close to one another in crinoid "gardens" on the sea floor. In these areas the disarticulated skeletal remains, particularly the disk-shaped columnals (column segments), accumulated in such great quantities that they formed thick layers of crinoidal limestone rock. Other thick accumulations of crinoid remains are essentially coquinas, consisting of skeletal fragments of thousands of individuals that were torn loose from the bottom by excessive storm wave turbulence and swept together by the waves and currents. Perhaps some were thrown up on the beach as are the clam shell coquinas along the Florida coast today.

Nautiloid and ammonoid cephalopods (Figs. XIII-13, 30) included pelagic floaters and swimmers of both the neritic and oceanic types (Fig. XIII-34), as well as bottom crawlers (vagrant benthos). Their remains, beautiful and often highly ornate shells, were widely distributed by oceanic currents, just as are the shells of the modern chambered nautilus. By these means, these marine creatures were deposited in a wide variety of bottom sediments which do not necessarily represent the environment in which the animal itself lived. As mentioned, cephalopods make good time index fossils (guide fossils) because of their rapid evolution, abundance, rapid and random dispersal, and complex shell morphology.

Certain fossil types are very commonly associated (Figs. V-21, XIII-35) in the Paleozoic rocks of Michigan. One such group consists of brachiopods (Figs. XIII-8, 27, 28), solitary corals, and bryozoans (Figs. XIII-7, 26). These animals lived a sedentary existence (sessile benthos) on the sea floor. Clams, snails, and certain cephalopods crawled amongst them, and the flower-like crinoids and blastoids waved above them. In the waters overhead swam other cephalopods, sharks, and a variety of primitive fishes of the placoderm and bony fish groups (Chapter XIV).

One should be aware that certain fossil assemblages may be collections of individuals brought together after death by waves or currents (Fig. XIII-36). Such a grouping, called a *thanatocoenose*, does not really represent any one of the environments occupied by members of the group at the time they lived. Highly fragmented, badly worn shell or bone fragments generally indicate extensive transportation and are a clue to the possibility that one is dealing with a thanatocoenose. Conversely, if shell or skeletal remains of very fragile sorts are found nearly intact in the rock, this indicates that the fossil remains were not moved far from the place where the animals themselves lived. One could then conclude that this assemblage of animals actually lived near one another (a *biocoenose*). Thick accumulations of shell fragments, nearly all of one size, result from the sorting action of strong currents or waves and are usually death assemblages. Coquinas, mentioned above, exemplify this type of deposit.

Land animals and plants often are subject to long distance transport before their remains are deposited in sediment. The skeletal remains of a horse that dwelt beside a river might have been picked up by flood waters, carried along by the current, and then deposited in the sands of the stream channel itself, on the adjoining flood plain, in the delta where that stream entered a lake or the sea, or may even have been swept some distance out to sea. One may judge that a fossil assemblage is truly indicative of a continental (as opposed to marine) origin for the enclosing sediments, if the fossils themselves show little evidence of fragmentation or wear due to long distance transport, and if the assemblage itself consists solely of terrestrial organisms with no marine forms included.

Fossil vertebrates may be especially sensitive in-

# Geology of Michigan

336

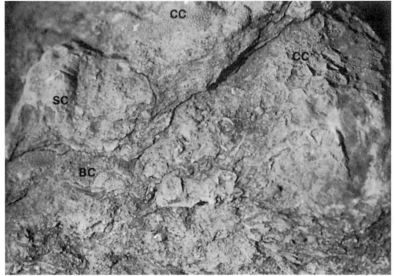

Figure XIII-35. Typical marine fossil assemblages from rocks of the Michigan Basin. (Scales are in inches.) A few individuals of each type present are labeled as follows: BC–bryozoan colony; BR–brachiopods; CC–colonial coral; CR–crinoid stem fragment; NC–nautiloid cephalopod (note internal septa where outershell is weathered off); SC–solitary coral; TR–trilobite.

*Top* and *center*—rocks of Devonian age.

*Bottom*—Mississippian age Coldwater Shale.

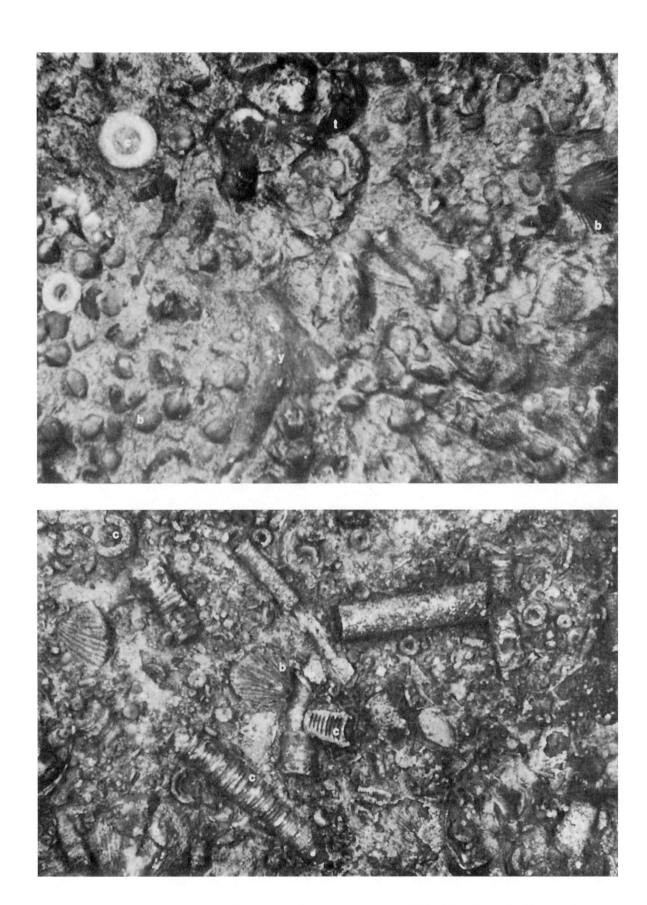

Figure XIII–36. Fossiliferous rocks of marine origin. Symbols are: b–brachiopod, c–crinoid stem fragment, t–trilobite head, y–branching bryozoan colony.

Figure XIII–37. Coral-Stromatoporoid association in Devonian rocks of Michigan. CC–Colonial coral; SC–Solitary coral; ST–Stromatoporoid.

*Right*—Distant and close-up views of limestone in roadcut, west side of "old" US 27 just south of Mackinac City.

*Left*—Close-up of stromatoporoid. Cross section at top shows successive growth layers. Below is surface of a growth layer split from above. Scale is 1 inch long.

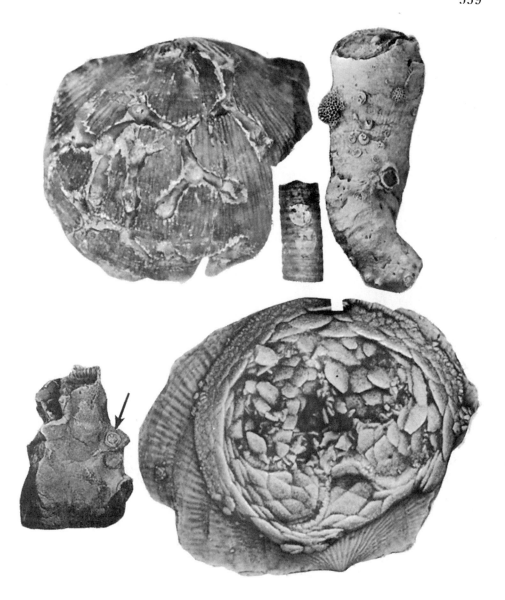

Figure XIII-38. *Top left*—Small, branching corals of the genus *Aulopora* commensal on a shell of the brachiopod genus *Stropheodonta* of Middle Devonian age; magnified 2.5 times.

*Top center*—Tiny, oval shells of the brachiopod genus *Crania* attached to the side of a Pennsylvanian age crinoid stem; natural size.

*Top right*—Clusters of honeycomb-like colonial corals, and spiral-shaped tubes of the polychaete annelid worm *Spirorbis* attached to the side of a Middle Devonian solitary coral of the genus *Tabulophyllum;* natural size.

*Bottom left*—*Timeischytes,* an edrioasteroid echinoderm (at tip of arrow), attached to a bryozoan colony which in turn is encrusting a horn coral of Devonian age. Natural size.

*Bottom right*—*Agelacrinites,* another edrioasteroid, attached to a brachiopod shell of Devonian age, magnified ×5.

dicators of different types of continental environments because the variety of land vertebrates is great and many of them were specifically adapted to a rather restricted set of environmental conditions. Arboreal primates, open plain grazers, such as horses, tundra dwelling wooly mammoths, and freshwater fishes are good examples. Unfortunately, the Paleozoic rocks of Michigan so far have yielded only marine or brackish water fishes, and there are no Mesozoic vertebrates of any type known from the state. The fossil record of terrestrial vertebrates in Michigan is restricted, so far, to deposits laid down shortly after the retreat of the last (Wisconsinan) Pleistocene continental ice sheet (Chapters VII and XIV). These remains are all less than 14,000 years old. Nevertheless, they indicate that since that time Michigan has been inhabited by a host of animals, such as the mammoths, mastodons, musk oxen, peccaries, caribou, giant moose, giant beaver, even walruses and whales (Chapter XIV), which are no longer in existence here. Some of these were cold climate types which left the state (some to become extinct) shortly after the withdrawal of the ice sheet. The fossil plants, especially pollen, found in association with these vertebrate animals, tell an interesting story of changing climate and vegetation in postglacial Michigan (Chapters XIV and XV).

Some types of Paleozoic fossils typically occur in particularly close association. The relationship implies that one or both of the pair of types enjoyed some advantage in this relationship. Such a relationship is called *symbiosis* (living together). Stromatoporoids and corals are one example (Fig. XIII–37). It is uncertain what advantages were enjoyed in this case, but possibly the open coral framework offered space and protection from wave action within which the stromatoporoids could thrive, while the calcareous deposits of the stromatoporoids helped to cement and strengthen the coral framework.

Commensalism, a special type of symbiosis, is a relationship in which one organism benefits, whereas the host is neither helped nor harmed. The tiny tabulate coral *Aulopora,* growing on the shells of brachiopods and bryozoans (Fig. XIII–38) appears to illustrate this kind of relationship. The shell of the brachiopod provided support for the coral up above the muddy bottom. There is no evidence that the coral injured the host. The coral polyp often was oriented toward the shell opening of the host, suggesting that the corals were aided in obtaining food by the water currents set up by the brachiopod filtering system. If, as suggested by Agar (1963, p. 264), the stinging cells (nematocysts) of the coral helped to protect the brachiopod, then this would be a full symbiotic relationship, mutually advantageous to both animals and not merely commensalism.

The growth forms of some Paleozoic fossils found in Michigan often indicate something about the manner in which the animals lived. Individual members of coral colonies tended to "fan out" or diverge upward as they increased in diameter and individually required more space and food. Corals and bryozoans often grew upward in delicately branching or fan-shaped forms in deep water below wave base, but took on rounder, more compact shapes when they extended upward into the turbulent zone of strong wave action. In some cases the shape of a whole coral reef was elongated in the direction of wind or current movement (see coral reefs, Chapter V). Shell and skeletal fragments from dead animals (and plant remains, too) tended to be moved and then deposited by waves and currents in rows in such a way that the long axes of the fragments were oriented more or less consistently either parallel with, or perpendicular to, the current direction (Fig. XIII–39). This orientation provides the geologist with a means to determine the directions taken by ancient currents. Echinoderm skeletal fragments, especially the stems of crinoids and blastoids, were particularly susceptible to trans-

Figure XIII–39. Paleozoic age, marine limestone showing current oriented fragments of branching bryozoan colonies tending to parallel one another. Natural size.

portation, size sorting, and oriented deposition because of their composition. The hard parts of members of that phylum are composed of a very porous type of calcium carbonate which is of low density and tends to float more readily than other types of shell material. This porous quality favored the transportation of even large echinoderm fragments and may help to explain why these fragments are often found mixed in with smaller sized shell debris of other animals. This tendency to float also helps to explain the origin of crinoidal coquinas and the commonly observed lineation of crinoid stems.

In certain areas, where the sedimentary rock sequence and the fossil record are relatively complete, a succession of faunas (and floras) are revealed to have occupied that region as environments changed. Agar (1963, p. 268–69) describes a fossil sequence distributed vertically in the rocks which was interpreted as showing a gradual change in the composition of marine invertebrate faunas in certain Ordovician and Silurian rocks of Great Britain. The environmental change through time was from deep water conditions, in which graptolites were dominant, to progressively shallower water. The fossil record revealed that a graptolite fauna gave way to a mixed assemblage of graptolites and trilobites, which, in turn, were replaced by a dominantly trilobite fauna, followed by trilobites and small brachiopods, and, finally, that the environment was occupied by a shallow-water fauna of large brachiopods. Assuming this example and its interpretation are valid and that the conclusions may be applied to other regions, the fossil faunas in some of the marine Paleozoic rocks of Michigan reveal significant information of environments of the past. The Middle Devonian Silica Formation of southeastern Michigan and northern Ohio, for instance (Chapter V), includes zones of fine clastic deposits (shales for the most part), rich in associated trilobites and brachiopods. As in the example analyzed above, this assemblage, as well as the fine clastic composition of the formation itself, may indicate deposition in an intermediate sublittoral environment (Fig. XIII–34) well out from the beach but still within relatively shallow water.

Major changes in the environment of sedimentary deposition may also be reflected by significant changes in the fossil record. These changes may sometimes be repetitive or even cyclic, which indicates frequent recurrence of similar sets of conditions. The Pennsylvanian age floras and faunas of Michigan provide an exceptionally interesting picture of such changes. The record is discussed at some length in Chapters V and XV. The rocks involved compose the Saginaw Formation in the central part of the Michigan Basin. Especially good exposures in quarries near Grand Ledge record several influxes of shallow, brackish marine waters followed by withdrawal of the sea and the reestablishment of continental types of environments. One such marine advance and retreat, recorded in the compositions of both the sediments themselves and their fossil content, is shown in Figure XIII–40.

The Paleozoic coral reef faunas of the Great Lakes Region, unlike faunas that reflect radical alterations in the physical environment, reveal more gradual environmental changes brought about by the evolution of the biological community itself. Figure XIII–41 shows the correlative changes in water depth, wave turbulence, availability of food and oxygen, and faunal composition that took place with the upward growth of a typical reef.

Such a reef was pioneered in water depths below wave base by reef building corals, the plant-like "Stromatactis" and stromatoporoids. Initial development began in relatively deep water, but not in excess of 200 feet. In the beginning the reef fauna was sparse and showed little diversity. When upward growth brought the reef top above storm wave base into the zone of occasional turbulent water, the destructive force of the waves was offset by the more abundant food and oxygen supply, so that the reef not only continued to thrive but the richness of its fauna increased as well. This richness is reflected in both the greater diversity of animal types and the increased abundance of individual animals of each type. Conditions in the sea itself remained stable, but the reef community modified those conditions, shaping its own biological and physical environment as it rose. Surprisingly enough, as shown in Figure XIII–41, the reef enjoyed its greatest abundance and diversity of life when it reached up into the zone of strongest surface water turbulence above normal wave base, where, at the same time, the destructive force of the waves was at its maximum.

Additional discussions and examples of paleoecological interpretation and analyses of other

# Geology of Michigan

### Figure XIII-40

GEOLOGIC HISTORY INTERPRETED FROM THE SEQUENCE OF ROCKS AND FOSSILS IN THE PENNSYLVANIAN AGE SAGINAW FORMATION IN THE VICINITY OF GRAND LEDGE, MICHIGAN

(See Figure V-30)

| | Observations | | | Interpretation |
|---|---|---|---|---|
| Younger | Rock Types and Features | Fossil Content | Significance of Observations | of the Ancient Environment |
| *TIME* ↑ | 9–Limestone with little or no clastic sediment | Abundant solitary corals, brachiopods of several types, crinoid stems, snails, nautiloid and ammonoid cephalopods | Abrupt change from unit 8 to 9 indicates rapid environmental change. Limestone is a normal marine deposit in this case. Absence of clay, silt, and sand indicates land either too far away or too low-lying to provide clastic sediment. The fossils listed are marine types. The cephalopods are normal marine animals indicating area of deposition connected with open sea | Relatively abrupt reentry of marine waters and submergence of the land. Open marine conditions. Land now either far away, low-lying, or both |
| | 8–Thick, cross-bedded lens-shaped sandstone with interbedded coal lenses interfingering laterally in short distance into blocky, poorly bedded claystone | Many logs and stem fragments of *Calamites, Sigillaria, Neuropteris* and *Cordaites* in sandstone. Adjacent clays rich in leaves of same plants | Sandstone lens is a river channel deposit. Adjacent clays were laid down on floodplain adjoining channel. Land vegetation was lush and included same varieties as found in units 6 and 7 below | Meandering streams with sand-filled channels periodically overflowed onto broad, flat, heavily vegetated flood plains. Swampy ox-bow lakes in abandoned river meander bends became small coal-forming swamps |
| | 7–1 to 2-foot thick coal seam | Coal formed from common Pennsylvanian plants (see 8 above) | Coal forms in stagnant, shallow, swampy conditions, but not in tropics. Pennsylvanian plants were warm-temperate to subtropical types. Purity of coal indicates little influx of clastic sediment. Several hundred years required to produce sufficient vegetation to compact into 1 foot of coal | Region occupied by coal forming swamp of several hundred years' duration in humid, warm-temperate or subtropic climate |
| | 6–Top of zone 5 below locally scoured by channels; bedding cut off. Channels filled with sandstone of variable thickness | *Stigmaria* (roots of *Sigillaria*) common; main axis upright (vertical); lateral rootlets horizontal. Large, sand-filled log fragments. Nearly complete leaves of *Calamites, Sigillaria, Neuropteris,* and *Cordaites* common | Scoured channels due to stream erosion. Lensing sandstone is a stream deposit. Plant fragments of local origin. *Stigmaria* are oriented in position of growth | Emergence of land complete. Sea has withdrawn. Streams locally erode underlying beds and fill channels with sand. Abundant plants grow along stream courses, some even in sand banks in streams |

Figure XIII-40—*Cont.*

GEOLOGIC HISTORY INTERPRETED FROM THE SEQUENCE OF ROCKS AND FOSSILS IN THE PENNSYLVANIAN AGE SAGINAW FORMATION IN THE VICINITY OF GRAND LEDGE, MICHIGAN

(See Figure V–30)

| | | | |
|---|---|---|---|
| 5–Alternating light-colored siltstone and fine sandstone in ½ to 2 inch layers. Wave and current ripple marks common | Nearly entire leaves and log fragments of *Calamites, Sigillaria, Neuropteris,* and *Cordaites* abundant | Sandstone is coarser grained than siltstone in unit 4 below, indicating shallower water. Ripple marks are caused by waves and currents. More complete nature of plant fragments than in 4 below means shorter transport from land | Shallowing of sea continues. Heavy vegetation on land. Streams carry plants short distance to sea. Beach littered with drift wood and leaves. Strong waves and currents |
| 4–Alternating layers of dark, carbonaceous shale and thin lighter colored siltstone. Current and wave ripple marks | Small fragments of land plants abundant | Siltstone consists of coarser clastic sediment than shale in unit 3 below, requiring greater energy in currents or waves for transport. Abundant land plant fragments indicate transport from nearby heavily-vegetated land | Sea begins to withdraw, waters become more shallow, and waves and longshore currents are stronger. Nearby land still heavily vegetated |
| 3–Dark shale and blocky claystone, with mud cracks and symmetrical ripple marks | Leaf and stem fragments of land plants. The brachiopod *Lingula* common | Small plant fragments indicate distant transport from land. *Lingula* is a marine and brackish water tidal flat animal (littoral and shallow sublittoral zone) | Maximum submergence of land. Marine waters deepest. Tidal flats and shallow offshore waters alternate. Streams from nearby heavily vegetated land carry plant remains into the sea |
| 2–Thinly bedded, dark, carbonaceous shales with plant fragments, and with thin interbedded coal seams and lenses | Coal and land plant fragments | Dark colors originate in an acidic, reducing environment; stagnant water indicated. Water shallow. Coal is from swamp-dwelling land plants | Beginning of land submergence. Swampy conditions. Lush growth of land plants in stagnant shallow swamp water. Periodic influx of mud, carried into swamps by streams |
| 1–Red zone with concretions overlies gray zone which lacks stratification. Gray zone grades downward into stratified shale | Scattered root cavities in red and gray zones | Red color is due to oxidation in contact with atmosphere. Gray is due to chemical leaching by plant acids. Lack of stratification due to action of roots and burrowing soil organisms | Period of land emergence. Oxidation, leaching, produce soil. Warm humid conditions. Good drainage. Heavy vegetative cover |

TIME ↑

Older

## Geology of Michigan

### Figure XIII-41

NIAGARAN REEFS IN THE GREAT LAKES REGION—PHYSICAL DEVELOPMENT AND FAUNAL CHANGES

(From work of Lowenstam as summarized in Agar, 1963, pp. 282-84, and Fig. 16.4)

| | Physical Environment | Availability of Food and Oxygen | Diversity of Animal Types and Abundance of Individuals | Dominant Reef Builders | Associated Dwellers on and Within Reef in Order of Abundance |
|---|---|---|---|---|---|
| | | | Surface of the Sea | | |
| ↑ Time, and Upward Direction of Reef Growth ↑ | Rough water. Shallow depth. Constant wave turbulence | Abundant | Both great | Stromatoporoids and corals | Crinoids (stout types), brachiopods, bryozoans, solitary corals, trilobites, cystoids, nautiloid cephalopods (coiled and curved), gastropods, pelecypods, sponges, blastoids, conularids |
| | | | Normal Depth of Wave Turbulence | | |
| | Semirough water. Intermediate depths. Storm waves regularly reach this depth. Periodic strong turbulence | Intermediate | Intermediate | "Stromatactis" and, in increasing abundance, stromatoporoids, form open lacy network. Tabulate corals, such as *Favosites*, *Syringopora*, *Halysites*, and *Heliolites* in minor numbers | Greater variety than below, including crinoids (more types), brachiopods (especially on reef flanks), bryozoans, trilobites, and sponges. Also rare nautiloid cephalopods (straight uncoiled types), gastropods, and pelecypods |
| | | | Deepest Storm Wave Penetration | | |
| | Relatively deep, quiet waters, below reach of even greatest storm waves. Low turbulence | Scarce | Both small | Pioneer reef builders are *Favosites* and *Syringopora* (80 percent), both tabulate corals. These are followed by "Stromatactis" (an alga or stromatoporoid) which build an open, lacy reef network | Crinoids dominant, followed in decreasing abundance by bryozoans, trilobites, brachiopods, and sponges |
| | | | Original Sea Floor Base of Soft Sediment Upon Which Reef Growth Begins | | |

types of environments of sedimentary deposition in the Michigan Basin in the geologic past are to be found elsewhere in this book. We have mentioned the sections on coral reefs and coal swamps in Chapter V. The sections dealing with Paleozoic fishes and Pleistocene (Ice Age) vertebrates and plants in Chapter XIV and dealing with paleobotany in Chapter XV also include paleoecological interpretations of Michigan's geologic past. Reference to Agar (1963) and Imbrie and Newell (1964) in the Bibliography will provide added details on the general subject of paleoecology.

### Michigan Fossil-Collecting Localities

The following is a list of the best invertebrate fossil collecting localities in Michigan (and adjacent parts of Ontario and Ohio) for rocks of various ages. See Figure XIII-22 for map. It does not include every outcrop in the state, but only those which are known to have been productive. Serious collectors are urged to search the literature for further information. The references listed will assist in search and identification. Sections, townships, and ranges may be located most readily on

Michigan Conservation Department county road maps. The age relationships of the various rock formations listed below are illustrated on Figure v-2 in Chapter V.

WARNING: This list does *not* constitute permission to collect fossils on private lands nor does it imply that permission will be granted by the owners! Obtain permission personally before entering private property!

### CAMBRIAN

The Cambrian rocks in Michigan are for the most part unfossiliferous, but a few fossil remains have been found as follows:

*Munising Formation.* The upper 5 feet produce the only known fossils.

1. Ledges along the north side of U.S. Highway 2, 0.1 mile east of the junction with Foster City Road, 0.5 mile north of Waucedah, Dickinson County, Michigan. Trilobites and phosphatic brachiopods.
2. Ledges at top of abandoned Breen Mine, Sec. 22, T39N, R28W, Dickinson County, Michigan. Trilobites, mainly fragmental.
3. Base of the north side of abandoned Pewabic Mine, Iron Mountain, Dickinson County, Michigan. Trilobites and phosphatic brachiopods.

### References

Stumm, E. C. "Upper Cambrian Trilobites from Michigan." *Contr. Mus. Paleontology, Univ. Michigan,* v. XIII, no. 4, 95–102, 1956.

Hamblin, W. K. "The Cambrian Sandstones of Northern Michigan." PhD thesis, Univ. Michigan (also published as a *Michigan Dept. Conservation, Geol. Surv. Div., Bull.*) 1958.

### ORDOVICIAN

*Au Train Formation* (Middle Ordovician, Black River Group). Poor collecting.

4. Miner's Castle, Alger County, Michigan. 1 cephalopod and 1 gastropod species.
5. Ledges along top of abandoned Austin Mine #2, NE¼, NE¼, SW¼, Sec. 20, T48N, R25W, Marquette County, Michigan. 1 cephalopod species.
6. Sault Point on south shore of Whitefish Bay, Lake Superior, Chippewa County, Michigan. 1 gastropod genus.
7. Au Train Falls, Alger County, Michigan.
8. Top of the Pictured Rocks Cliffs.

*Bony Falls Formation* (of Black River Group). Good collecting.

9. Bony Falls on the Escanaba River, Delta County, Sec. 1, T41N, R24W, 26 feet of Black River Formation, 13 feet of Trenton Formation. Species: 4 gastropods, 2 corals, 5 brachiopods, 3 trilobites, 1 stromatoporoid, 1 pelecypod, 2 cephalopods, and 1 ostracod.
10. Exposures just south of Trenary, Alger County, Michigan, in a railroad cut, in a quarry west of the road, and along U.S. Highway 41, south of the quarry. Mainly brachiopods and ostracods, with some trilobites and molluscs.
11. 0.3 mile northwest of Spalding on U.S. Highway 2, Menominee County, Michigan. A few brachiopods.
12. 0.7 mile northwest of Spalding, near Spalding Fire Tower, Menominee County, Michigan. 4 species of brachiopods and 1 of cephalopods.
13. Exposure at Cornell, southwestern Delta County, Michigan. Both Black River and Trenton Formations. Numerous brachiopods, gastropods, trilobites, pelecypods, and bryozoa.

*Trenton Group*

14. Groos Quarry Formation, at the abandoned Bichler Quarry at Groos, 5 miles north of Escanaba, Sec. 1, T39N, R23W, Delta County, Michigan. Species: 1 bryozoan, 1 pelecypod, 3 brachiopods, 1 gastropod, 2 trilobites, and 1 crinoid.
15. Chandler Falls Formation, at Chandler Falls on the Escanaba River, 3 miles north of Escanaba, Delta County, Michigan. T39N, R22W. Species: 1 plant, 1 sponge, 1 coral, 3 echinoderms, 22 brachiopods, 9 gastropods, 2 cephalopods, 9 trilobites.
16. Along the Escanaba River 1.5 miles east of Cornell, Delta County, Michigan. Several species of brachiopods and gastropods.
17. Abandoned quarry 7.6 miles southwest of Cornell, Delta County, Michigan, with the same fauna as that at Groos Quarry.

18. Bed of the Rapid River, Delta County, Michigan, just east of the town of Rapid River. An extensive fauna of brachiopods, molluscs, trilobites, and bryozoa.
19. Island in the middle of the Escanaba River 0.25 mile northwest of Groos Quarry, Delta County, Michigan. Groos Quarry Formation. Pelecypods and brachiopods.
20. Haymeadow Creek Member, along Haymeadow Creek, NW¼, Sec. 19, T42N, R20W.
21. Roadside exposure 0.5 mile northwest of Shaffer, Delta County, Michigan. Groos Quarry Formation, with fauna similar to that at Groos Quarry.
22. Poor roadside exposure 0.5 mile east of Harris, Sec. 12, T38N, R25W. A fauna of pelecypods.
23. Abandoned quarry 0.25 mile west of Perkins, Sec. 5, T41N, R22W. Groos Quarry Formation. Typical fauna (see 14 above); crinoid plates especially common.
24. Bed of Ford River, Sec. 5, T38N, R23W, Delta County, Michigan. Lower Groos Quarry Formation. A few brachiopods and a scyphozoan.
25. Roadside exposure 2 miles east, 1 mile south of Wilson, Menominee County, Michigan. Upper Chandler Falls Formation, with a few brachiopods and gastropods.
26. Road cut 2 miles north of Bark River, Delta County, Michigan. Chandler Falls Formation. Mainly gastropods.
27. Several other smaller exposures, some fossil-bearing, are listed by Hussey in his various publications, and the interested collector is referred to them (see Ordovician references following locality 38).

*Richmond Group* (Upper Ordovician)

28. Bill's Creek Shale, eastern part of Sec. 12, T41N, R21W, and western part of Sec. 7, T41N, R20W, Delta County, Michigan. Species that may be found: 1 graptolite, 7 pelecypods, 1 echinoderm, 3 bryozoans, 5 brachiopods, 3 trilobites, 3 ostracods.
29. Bill's Creek Shale, along banks of Haymeadow Creek in northern part of Sec. 19, T42N, R20W. Fauna similar to 28.
30. Collingwood Formation, in loose surface blocks in gravel deposits near the town of Newberry, Michigan. Species: 3 trilobites, 3 graptolites, 2 articulate brachiopods, 2 inarticulate brachiopods, 1 cephalopod.
31. Stonington Formation (Bay de Noc Limestone Member), and Bill's Creek Shale, exposed at end of road between Secs. 14 and 23, T39N, R22W, Delta County, Michigan, along the shore of Little Bay de Noc. Good collecting. Numerous trilobites, bryozoans, brachiopods, ostracods, molluscs, and rare graptolites, worms, echinoderms.
32. Bay de Noc Limestone Member of the Stonington Formation, in road-cut in NW¼, Sec. 24, T39N, R22W. 15 feet exposed. Good collecting: 3 corals, 7 bryozoans, 24 brachiopods, 9 pelecypods, 5 gastropods, 1 cephalopod, 1 ostracod.
33. Ogontz Limestone Member of the Stonington Formation, south shore of Little Bay de Noc, SE¼, Sec. 26, T39N, R22W. (Bay de Noc Member also exposed here, with same fauna as above.) Good collecting: 1 coral, 2 worms, 1 echinoderm, 6 bryozoans, 20 brachiopods, 18 pelecypods, 15 gastropods, 3 cephalopods, 4 trilobites.
34. Big Hill Formation, on south slope of Big Hill (Hinkins Hill), Sec. 11, T39N, R22W. Numerous species: 1 alga, 1 sponge, 9 anthozoan corals, 3 hydrozoans, 2 bryozoans, 10 brachiopods, 1 pelecypod, 1 gastropod, 2 cephalopods.

*Other Ordovician Localities*

35. Drummond Island. Upper Ordovician, Richmond Group. Excellent collecting, particularly near Raynolds and Chippewa Points.
36. The west Neebish Channel, located at the eastern end of the Upper Peninsula of Michigan, south of Sault Sainte Marie, Chippewa County, connects Hay and Munuscong Lakes west of Neebish Island. Material blasted from the channel to improve passage is now piled up in a long row along the side of the channel, and is Early Trenton in age. Good collecting: 5 trilobites, 1 cephalopod, 1 scyphozoan, 11 brachiopods, 6 bryozoans, 2 corals.
37. St. Joseph Island, Ontario, at old Quarry Point. Black River Group. Good collecting: 2 corals, 2 bryozoa, 5 brachiopods, 4 pelecypods, 5 gastropods, 2 cephalopods, 4 trilobites.
38. Sulphur Island, Ontario. North channel of Lake Huron, 2.75 miles north and 1 mile east of Poe Point, Drummond Island, Michigan.

Black River Group. Good collecting: 4 corals, 2 crinoids, 13 bryozoans, 14 brachiopods, 2 trilobites.

*References*

Hamblin, W. K. See reference under CAMBRIAN.

Hussey, R. C. "The Middle and Upper Ordovician Rocks of Michigan." *Michigan Dept. Conservation, Geol. Surv. Div.*, Pub. 46, Geol. ser. 39, 89 p., 1952.

——— "The Ordovician Rocks of the Escanaba-Stonington Area." *Guidebook, Michigan Geol. Soc.*, 1950. (Out of print.)

——— "The Richmond Formation of Michigan." *Contr. Mus. Geol., Univ. Michigan,* v. II, no. 8, p. 113–188, 1926. (Out of print.)

——— "The Trenton and Black River Rocks of Michigan." Occ. Paper on Geol. Mich., Part III. *Michigan Dept. Conservation, Geol. Surv. Div.*, Pub. 40, Geol. ser. 34, 1936. (Out of print.)

SILURIAN

Collecting in general is only fair, but some places are very good, especially where the fossils are silicified.

39. Hendricks Quarry, Mackinac County, NW¼, Sec. 6, T44N, R8W, and NE¼, Sec. 1, T44N, R9W. Hendricks Dolomite of the Burnt Bluff Group. Collecting fair: ostracods, corals, brachiopods, trilobites.

40. The abandoned quarry of the Scott Quarry Company, about 1 mile southeast of Cordell, Chippewa County, SW¼, Sec. 29, T44N, R4W. Cordell Dolomite. Very good coral collecting, also brachiopods.

41. 9 feet of Manitoulin Dolomite exposed in small abandoned quarry 0.1 mile north of Highway 98 and about 1 mile north of Manistique Lake, Luce County, Michigan (SW¼, Sec. 20, T45N, R12W). Scarce collecting: brachiopods, corals, molluscs.

42. Top of Cataract Group above Cabot Head Shale, 14 feet of the "Moss Lake Formation" (of Ehlers and Kesling, 1957), exposed in ditch on north side of road just west of the southeast corner of Sec. 35, T41N, R19W about 1.5 miles west of Isabella, Delta County, Michigan. A few poorly preserved brachiopods and crinoid columnals.

43. 6 feet of the Lime Island Dolomite (Burnt Bluff Group) lying below 28 feet of the Byron Dolomite, exposed along the shore of St. Mary's River and in the small abandoned quarry adjacent to the old lime kiln on the west side of Lime Island, about 1300 feet north-northeast of the power house of the Northwestern-Hanna Fuel Company. Numerous brachiopod molds and a few corals.

44. The Schoolcraft Dolomite of the Manistique Group, and the Hendricks and Byron dolomites (Burnt Bluff Group). About 250 feet of section exposed in Burnt Bluff on the east side of Big Bay de Noc about 9.5 miles southwest of Garden Village, Michigan. Occasional brachiopods, corals, stromatoporoids, and ostracods.

45. Hendricks Dolomite (Burnt Bluff Group) exposed in rock-cut along Highway 94 and in abandoned Sawheidle Quarry on west side of this highway, about 1 mile west and 4.75 miles north of Manistique. Corals, brachiopods, stromatoporoids, and crinoid columnals common.

46. Cordell Dolomite and Schoolcraft Dolomite (Manistique Group) exposed in abandoned White Marble Lime Co. Quarry, Manistique, Michigan. Numerous corals, brachiopods, ostracods, bryozoans, and a few molluscs.

47. Middle Silurian (Manistique Group) Schoolcraft Dolomite overlain by Cordell Dolomite, top of east-facing escarpment about 1.5 miles south and 1 mile east of Raber, along an old logging road in Chippewa County, Michigan. Very good collecting: numerous corals, stromatoporoids, brachiopods, cephalopods.

48. Numerous localities in the Niagara Escarpment region of Peninsular Ontario. See reference to Mozola below.

*References*

Ehlers, G. M. *et al.* "Pleistocene and Early Paleozoic of the Eastern Part of the Northern Peninsula of Michigan." *Guidebook, Michigan Geol. Soc.*, 1948. (Out of print.)

——— and Kesling, R. V. "Silurian Rocks of the Northern Peninsula of Michigan. *Guidebook, Michigan Geol. Soc.*, 1957. (Out of print.)

Mozola, A. J. *et al.* "The Niagara Escarpment of Peninsular Ontario, Canada." *Guidebook, Michigan Geol. Soc.*, 1955.

## DEVONIAN

Excellent fossil collecting in many localities.

49. A series of quarries in the vicinity of Silica and Sylvania, Lucas County, Ohio, just south of the Michigan-Ohio line. Especially good are the North and South Quarries of the Medusa Portland Cement Company. Many brachiopods, gastropods, pelecypods, bryozoa, and a few crinoids, corals, and a trilobite. Excellent collecting in the Dundee Limestone and Silica Formation.
50. Brunner, Mond Canada, Ltd. Quarry, 1.25 miles northeast of Amherstburg, Ontario. Dundee Limestone, Anderdon Formation, and Lucas Formation. Numerous brachiopods and corals. A few other types are less common.
51. Solvay Process Company quarry at Sibley, about 2 miles north of Trenton, Wayne County, Michigan. Dundee Limestone, Anderdon Formation, and Lucas Formation. Mainly brachiopods and corals.
52. Near Arkona, Ontario. Hungry Hollow, along north bank of Au Sable River about 2 miles east and 0.25 mile north of Arkona in West Williams Township, Middlesex County, Ontario. Middle Devonian Traverse (Hamilton) Group. Not in Michigan, but close—and collecting is very good. Mainly brachiopods, but also trilobites, corals, molluscs, bryozoans, and others.

The following are all excellent localities for Devonian fossils, though not so well-known as the first four. Look especially for corals, brachiopods, crinoids, cephalopods, and trilobites.

53. Penn-Dixie Cement Co. quarry, about 1.5 miles west of Petoskey, Emmet County, SW¼, Sec. 2 and SE¼, Sec. 3, T34N, R6W. Gravel Point Formation and lowermost Charlevoix Limestone.
54. Northern Lime Co. quarry, bordering Little Traverse Bay near east end of Petoskey, Emmet County, Sec. 32, T35N, R5W. Charlevoix Limestone and Petoskey Limestone.
55. Kegomic Quarry, south shore of Mud Lake just east of Harbor Springs Rd., Mich. 131, and about 0.25 mile north of its termination on U.S. 31, 1 mile east of Bay View, Emmet County, SE¼, SW¼, Sec. 27, T35N, R5W. The Petoskey Limestone (*Gypidula petoskeyensis* zone) with Potter Farm fauna.
56. Bluffs on the northeast shore of Partridge Point, 4 miles south of Alpena, Alpena County, SE¼, Sec. 11, T30N, R8E. Thunder Bay Limestone.
57. Abandoned quarry (Kelley's Island Lime and Transport Co.) at Rockport, Alpena County, Michigan, Sec. 6, T32N, R9E. Upper Bell Shale, Rockport Quarry Limestone, and lower Ferron Point Formation. Bryozoans especially common.
58. Michigan Alkali Co. quarry, east edge of Alpena, Alpena County, Michigan, Sec. 13, T31N, R8E. Upper Genshaw Formation, Newton Creek Limestone, and Alpena Limestone (type section).
59. Abandoned shale pit of Alpena Portland Cement Co., about 1 mile east and one-eighth mile north of Genshaw School and 8 miles northeast of Alpena, Alpena County, Michigan, SE¼, Sec. 18, T32N, R9E. Upper Ferron Point Formation and lower Genshaw Formation (type section). (Private property; permission to enter usually not granted.)
60. Thunder Bay Quarries Co. quarry, east edge of Alpena, Alpena County, Michigan, SE¼, Sec. 14, T31N, R8E. Alpena Limestone and basal Dock Street Clay member of the 4-Mile Dam Formation.
61. Small shale pit at northwestern corner of Alpena cemetery (Evergreen Cemetery), Alpena County, Michigan, SW¼, Sec. 21, T31N, R8E. Potter Farm Formation.

### References

Ehlers, G. M., E. C. Stumm, and R. V. Kesling. "Devonian Rocks of Southeastern Michigan and Northwestern Ohio." *Guidebook, Geol. Soc. America*, 1951. (Out of print.)

Kelly, W. A., 1940. "Guidebook, 10th Annual Excursion to Afton, Onaway District." Michigan Academy of Science, Arts, and Letters, Section of Geology and Mineralogy, 19 pp. (Mimeographed, out of print).

## MISSISSIPPIAN

Very good collecting in places.

62. Lower Marshall Sandstone of Marshall Formation. Boulders in the Blue Ridge glacial esker approximately 5 miles south of Jackson,

Michigan, along U.S. 127. The best localities on the esker for fossils are 0.75 mile southeast along Blue Ridge Road from U.S. 127 in a large gravel pit and in a road cut and gravel pit just west of Meyers Road on Wetherby Road. Very abundant pelecypods, cephalopods, ostracods, gastropods, brachiopods, and a rare trilobite.

63. Lower Marshall Sandstone. Stoney Point quarry about 2.5 miles by road southwest of Hanover, Michigan (15 miles west of Jackson). NE¼, Sec. 31, T4S, R2W, Jackson County. Clams, crinoids, nautiloids, ammonoids, and ostracods are common.

64. Grindstone quarries, Pointe Aux Barques and Grindstone City. Thumb region of Michigan. "Marshall Sandstone" (probably age equivalent to Coldwater Shale, not true Marshall Formation).

65. Around Marshall, Michigan. Lower Marshall Sandstone. Same fauna as 62. See reference to Winchell, below, for old localities, now difficult to find.

66. In Hillsdale County, near Moscow. NW¼, NW¼, Sec. 4, Jefferson Township, and SW¼, SW¼, Sec. 26, Allen Township. Lower Marshall Sandstone.

67. At Union, Branch County. Exposures of the Coldwater Shale.

68. Near Holland, Ottawa County. Lower Marshall Sandstone.

69. Battle Creek region, Calhoun County, and Columbia, Jackson County, and in a railroad cut 3 miles north of Napoleon, Jackson County. Lower Marshall Sandstone.

70. Numerous other small scattered localities, all given by Winchell.

*References*

Winchell, A. "Description of Fossils from the Marshall and Huron Groups of Michigan." *Phila. Acad. Sci.,* v. 33, 1862.

PENNSYLVANIAN

Rare outcrops, sporadic collecting.

71. Basal Saginaw Formation ("Parma Sandstone"), northeast of the town of Parma, Michigan, in Old Titus Quarry, NW¼, Sec. 29, T2S, R2W; farther east in Sec. 27, T2S, R2W, along the road; in Sandstone Creek, 0.25 mile east in Sec. 34; main highway between Jackson and Lansing, 0.25 mile south of Bentley Corners—all in Jackson County. Also in the Fisk Quarry north of Albion, Calhoun County, and in road cuts nearby and south of Devereaux on the New York Central Railroad, Sec. 6, T2S, R3W. Rare fossils, mainly plants, especially *Calamites.*

72. Saginaw Formation. Chief outcrops are near Grand Ledge, Eaton County, Michigan. Also in clay pits, northwestern part of T4N, R5W (type section), especially along the bluffs of the Grand River and in the abandoned shale pits of the American Vitrified Co., the Grand Ledge Clay Products Co., and the Grand Ledge Face Brick Co. Very fossiliferous in places. Four major plant groups common. Some brachiopods, foraminiferans, bryozoans, pelecypods, gastropods, and cephalopods. Rare trilobites, ostracods, and fish remains.

73. Saginaw Formation, in the Saginaw Valley (type locality). No outcrops but once exposed in numerous coal mines in this region; some fossil material remaining on the dumps. A large fauna and flora similar to that of Grand Ledge but collecting now poor, old mines difficult to find and dumps overgrown with vegetation.

74. Scattered localities in Genessee County, Tuscola County, Shiawassee County, Clinton County, Ingham County, and Jackson County, many of them fossiliferous, are reported by Kelly; most are small. Refer to Kelly's localities and maps.

*References*

Kelly, W. A. "The Pennsylvanian system of Michigan." Occasional Paper on Geol. Mich., Part III, *Michigan Dept. Conservation, Geol. Surv. Div., Pub.* 40, Geol. ser. 34, 1936.

# XIV

# Fossil Vertebrates in Michigan

## Introduction

Some of the most spectacular fossil animals found in Michigan, such as whales, walruses, musk oxen, giant beaver, ancient fishes, and sharks, are vertebrates. Man himself is a vertebrate. A prehistoric Indian skeleton, for example, is a vertebrate fossil, just as is a mastodon or mammoth. Vertebrate animals were classified in the preceding chapter as a subphylum of the Phylum Chordata. There we also defined the vertebrates more specifically as being those animals with a segmented axial support, the spinal column. Although often called the "backbone," the spinal column may be composed of cartilage rather than of bone. The segments are the vertebrae (singular vertebra). Fish, amphibians, reptiles, birds, and mammals all are vertebrates.

## Relative Abundance of Fossil Vertebrates

Vertebrates occupy the small upper tip of what biologists call the "food pyramid," which is a way of saying that they depend for food upon the far larger numbers of "lower" (invertebrate) animals and plants. This means, too, that at any time there are relatively few vertebrates compared with the numbers of most other living things. The truth of this as it applies to living animals can be proven by observation today. The Principle of Uniformity (Chapter I) implies that the same was true in the geological past. Vertebrate fossils are relatively rare even though they possessed skeletal hard parts which preserve well. The reasons for this are several. Not all individuals that lived in the past ever were preserved as fossils. The chances of a given type of animal (or plant) being preserved in the historical record depends in part on how many individual animals lived. Furthermore, the record is not equally good for any or all vertebrates everywhere. Relatively speaking, fossil vertebrates are rather rare in Michigan compared with many other areas. A few types of Paleozoic fishes have been found here, but Ice Age animals that lived here during the latest Pleistocene (Late Wisconsin) and post-Pleistocene time are the most common. Although a serious amateur fossil collector stands a good chance of finding fossil invertebrates in this state, he might never be fortunate enough to find even a scrap of fossil vertebrate material.

# Fossil Vertebrates

Figure XIV-1. Michigan at the close of the Pleistocene. American mastodons (*Mammut americanum*, extinct) interrupt their browsing to trumpet a warning at approaching woodland musk oxen (*Symbos cavifrons*, extinct). Heavy horns indicate these are male musk oxen; lighter-horned females, sometimes called *Boötherium*, are illustrated in Figure XIV-17. Restorations such as this are based on actual specimens such as shown in Figures XIV-43 and 47. Artist showed all tusks of mastodons as being intact, but often one of pair is broken and worn from heavy use. (Drawing by the late Carleton W. Angell, courtesy C. W. Hibbard, The University of Michigan Museum of Paleontology.)

Nevertheless the fossil vertebrate record in Michigan is more extensive than one might think. It is brought completely together in one place for the first time in this chapter.

## Problems of Identification and Restoration

Mention of "scraps" brings us to another point. Often in science fiction, or in cartoons, vertebrate paleontologists are mistakenly attributed the ability to "reconstruct" an entire animal from a single bone. This is true *only* in a very limited sense. It can be done if the bone or bone fragment is an identifiable part of some previously well-known fossil or living animal, or if the part is clearly similar to well-known parts of closely related animals. If relatives of the animal represented by a single fragment of bone or incomplete skeleton are still living, then the task of reconstruction is made simpler. By analogy, if one found identifiable pieces of a broken piston and crankshaft fragments from a Cord automobile in a town dump, one might safely say that a Cord or an automobile much like one, with an internal combustion engine, once existed in or near that town. If no one nearby had a thorough knowledge or recollection of Cords, the fragments might be compared with parts of old automobiles in museums until similar structures were found. If no such automobile was in collections, there might still be pictures or drawings of its parts. Even if no knowledge of Cords existed the parts could be identified as having come from an internal combustion engine, and so on. The paleontologist works the same way. Depending on the degree of completeness of the specimens he has found and the state of knowledge of related forms, he might, for example, identify fossil remains very specifically as being those of a certain species of extinct mammoth or, less specifically, from the proboscidean order, or at least from a mammal, or at the very least as being from a vertebrate animal. Fossil animals with a good geologic record and many close living relatives are easiest to identify. Conversely, poorly known, completely extinct types with no living relatives are very difficult to study. This matter should be appreciated and well understood because statements will be made later in this chapter to the effect that certain types of vertebrate animals lived in Michigan in the past. In all cases there will be some sort of specimen in a museum collection somewhere to support this, but that specimen may not be very spectacular or even worth putting on exhibit. The documentary historical proof might be a mere fragment of a tooth or bone, recognizable and meaningful only to a specialist. For example, the proof, to be cited again later, that lung fishes lived in Michigan 275 million to 350 million years ago during the Mis-

### Geology of Michigan

sissippian and Pennsylvanian periods, at present consists solely of some mud-filled lung fish burrows from Pennsylvanian rocks near Grand Ledge in the central part of the state. These objects are unique enough to be clearly identified as lung fish burrows, but only because similar burrows, with lung fish in them, have been found elsewhere. The Michigan specimens do not tell what genus or species of lung fish is represented. Therefore, in the lists at the end of the chapter, the lung fish material is identified only as belonging to the Dipnoi, the collective name applied to all lung fishes. A sort of "reverse twist" to this problem of dealing with incomplete material stems from the fact that isolated parts of poorly known extinct animals, if never found together in one complete specimen, may be impossible to relate to one another. Thus, a tooth and fin spine of one extinct Paleozoic fish, if found separately, might each be classified as a separate and new species until such time as these were found together in a more complete specimen. Some of the Michigan Paleozoic fish teeth and spines of the *"Cladodus"* and *Ctenacanthus* type probably fall in this category.

Another important point concerns "in-the-flesh" restorations, those attempts by artists and scientists to depict the appearance of an animal as it was in life. Again, the task is easiest if the same or very closely related animals still live today. His knowledge of comparative anatomy may be applied by the paleontologist to study extinct animals with no living relatives, in an effort to picture their form in life with some accuracy. The skeletons may show the sizes and locations of places of attachment for former muscles which give some hint of body shape. But the extent to which restorations can be made is limited. It usually is impossible to tell exactly what types of hair or skin coloring and covering were present because those characters are rarely preserved in the fossil record and bear no close relationship to internal skeletal structures. It might be a fairly safe conclusion that an animal had hair simply because its skeleton showed it was a mammal and all mammals have some hair at some stage in their life, but how much hair, in what locations, what length and color, usually are unanswerable questions. Let us take an example to illustrate the problem of restoration. There are at least 2 species and 3 subspecies of the white-footed deer mouse (Genus *Peromyscus*) living in Michigan today. To an amateur these look much alike. Many of the differences which expert mammalogists use to tell them apart involve hair color and pattern. Their external body forms are much alike. Although to a specialist their teeth are different too, one could not predict from the teeth what the hair color would be. Suppose the fossilized teeth of a deer mouse are found in a Pleistocene deposit in Michigan. The specialist might be able to tell that he had found a species different from any living here today, possibly an extinct species. But what kind of a picture could he draw of the animal in life? The tooth differences would not show on the outside and would tell nothing about the hair. He could draw a generalized picture of a deer mouse without specifying hair color or pattern, but it would look like any other deer mouse even to other specialists. In this chapter, fossil species of many sorts of vertebrates are listed for Michigan, but the restorations, if any, often will be generalized ones for the larger groups (genus, family, or order) to which those extinct species clearly belong. So much for the general principles and limitations of the study of fossil vertebrates; let us now move on to their occurrence.

### Finding Fossil Vertebrates in Michigan

How and where does one find fossil vertebrates in Michigan? Many discoveries simply are accidental. Workmen may happen upon fossil remains in the course of quarrying sand, gravel, or rock, excavating building foundations, making highway and road cuts, digging ditches to drain swamps, or plowing fields. Unfortunately, few such finds are recognized for what they are, still fewer are saved, and very few are brought to the attention of scientists who would know how to collect them properly and preserve them in museums. To conduct an intentional search successfully, one must not only know what to look for and where to look, but must have almost infinite patience and interest. Knowing what to look for requires training and experience. One must first become familiar with the appearance of skeletons and bone fragments. A good way to begin is to examine skeletons of living animals. Then study specimens on display in museums. Consult the lists of fossil localities that come at the end of this chapter because they give some idea of where fossils have been found in the past. There is no guarantee of success, but hours spent examining and cracking rocks of appropriate age in quarries, or carefully walking over new road cuts, old soil zones in sand dunes, drainage ditch spoil ridges, and newly plowed fields especially in boggy areas, may produce a "find." Talk with experts at museums and with local amateurs to get their advice. The first indication of a buried fossil vertebrate specimen may be a single bone scrap, often found partway down a slope at some distance below the level of the remainder of the specimen. Most serious collectors are constantly alert for fresh new excavations in their areas, and they periodically recheck old sites after weathering has had a chance to expose specimens on the surface. Offer your help on expeditions or "digs." Get down on your hands and knees to look for small objects as well as scanning broad areas for large material. Although at any time, present or past, there are far more small than large vertebrate animals living, most fossil remains found are those of large animals. At present, there are more mastodon teeth than mouse teeth in fossil collections from Michigan. Obviously, it is easier to see a mastodon tooth than a mouse tooth, but in the geologic past mice were biologically more important than elephants, just as they are today; though not so spectacular small animals may be scientifically as important and interesting. Remember, too, that the fossil remains of a certain animal will occur mainly or only in deposits of appropriate age and type. One does not find nearly modern, land dwelling Pleistocene mastodons in extremely ancient Paleozoic marine rocks. Mastodons did not live in oceans, nor had they evolved as early as the Paleozoic period. In the last analysis, finding fossils is as much an art as a science, and few people are really good at it. You can't expect to just jump out of your car anywhere and "strike it rich."

## Special Problems

Sometimes fossil vertebrates are found in what may seem to be anomalous situations. For example, fossil whales and walrus got into this state in Pleistocene times when the geography was quite different. The great Wisconsin glacial ice sheet had just recently retreated, the land was still low because it had been depressed by the weight of the ice, water connections between the Great Lakes and the Gulf of Mexico and the St. Lawrence were extensive, and lake waters lay upon parts of the state from which they have since retreated (Chapters VII and VIII). The far more ancient Paleozoic fishes, another example, lived here when extensive arms of the sea covered much or all of the state (Chapter V).

The nature of animal life in Michigan changed extensively through time. Some animals evolved early, others much later. Some became extinct long before others appeared. The physical environment, including the geography and climate, often changed, hence inhabitants of warm ocean waters were succeeded by land-dwelling forms or vice versa. As climates changed, some existing forms of life emigrated as others immigrated. For example, musk oxen and caribou that occupied the state just after the glaciers had receded, eventually moved northward; later on the opossum spread northward into Michigan from the south. Even the restricted fossil record of Michigan vertebrates reflects extensive change through time, often in recurring cycles.

Another problem that hinders studies of past vertebrate life of Michigan arises from "gaps" in the geologic record. That record is preserved in ancient sediments, now for the most part turned into sedimentary rock. Not all environments are suitable for deposition of sediment, and not all deposits survive subsequent erosion. For long periods of time no sediment accumulated in Michigan. At other times much of what had accumulated was eroded away. With the exception of the unexposed and poorly known "red beds" in the central part of the Lower Peninsula (Chapter VI) there is no known rock record in the state for a long period of time from about 280 million years ago to less than 15 thousand years ago (Late Pennsylvanian–Late Pleistocene). This historical gap includes the Permian period of the late Paleozoic Era, practically all of the Mesozoic Era, and all but the very last part of the Cenozoic Era (Chapters II and VI). What this means in terms of the history of fossil vertebrates in Michigan is that there are no records here of the rise of modern fishes, nor of the last blossoming of Paleozoic and Mesozoic amphibia. The rise and heyday of reptiles, including dinosaurs, flying reptiles, and

# Geology of Michigan

marine reptiles is unrecorded here. No records of the evolution of birds, of the early origin and flowering of mammals, nor of the rise of primates and man have been found here. The historical gap continues to the last part of the Pleistocene. Many of the animals mentioned above must have lived here during that long "time gap," but they left no record, probably because Michigan was then an upland in which relatively little sediment accumulated and on which much erosion occurred.

Some gaps in the record more likely are due to lack of discovery rather than absence of suitable rocks. For example, the most primitive known forms of vertebrate life are the ancient ostracoderms, a group of early Paleozoic, fish-like aquatic animals. These have been found in other areas in sedimentary rocks as old as Late Ordovician, and they were relatively abundant in the following Silurian and Devonian periods as well. Rocks of all 3 of those periods occur in Michigan. Although as yet no ostracoderms are known from this state, more extensive search is probably all that is required to turn them up. The same may be true for amphibians and reptiles which should be found sooner or later in rocks of Pennsylvanian age and appropriate type in the central part of the Lower Peninsula near Grand Ledge. Amphibian and reptile remains have been found in similar Coal-Age swamp deposits elsewhere in North America. The recent discovery of lungfish burrows in those rocks illustrates how a gap in the fossil record may eventually be filled.

## Documentation of the Michigan Record

The fact that every fossil species, genus, or larger taxonomic unit of vertebrates mentioned in this chapter actually lived in Michigan at one time or another can be documented by reference to one or more actual specimens in museum collections. At the end of this chapter is a list of fossil vertebrates that have been found in Michigan and that are represented in collections. The list is intended mainly for illustration and reference, but it also is the foundation for the story told next in this chapter. No group is mentioned unless it is represented by at least a specimen whose geological age and site of discovery are recorded with reasonable accuracy in museum records. In most cases the exact rock formation, date of collection, and collector are known. The majority of specimens are in the collections of The University of Michigan Museum of Paleontology in Ann Arbor. This can be assumed to be the case for all specimens unless otherwise stated. The letters UMMP preceding catalog numbers of specimens in that collection will identify them. Specimens in other collections are given special mention. All reputable public and private museums keep careful records of specimens in their care, and the specimens are available for examination or study by professionally qualified persons upon request to the curators of those collections. Some of the better, more complete, or most significant specimens are on public display. Thus, if one has sufficient knowledge and training, he can check the statements that follow for himself. Specimens in the hands of private collectors, or those once reported but lost before they were subject to scientific study, are not mentioned. The reason for this is explained in the remarks that follow.

## The Value of Fossil Vertebrates

What is a fossil worth? Paradoxically, in most cases the answer is both "a great deal" and "nothing." Most fossils are rare, some more so than others, and fossil vertebrates especially so. Some are literally one of a kind! If lost or destroyed another may never be found. Collected and preserved properly, they form the material basis for scholarship. Their acquisition results from a combination of hard looking, luck, knowledge, and training. They are a *nonrenewable natural resource* in the fullest sense of the word. We believe that they are of great value and should be considered part of the "natural estate" of all men; an inheritance that should be cared for, profited from in the fullest intellectual sense, and passed on to later generations for their use, enjoyment, and instruction. Though fossils may be found by some private individual or group, perhaps even on private land, their greatest value will be realized only if they find their way into qualified hands, preferably public or large, well-staffed private museums, where they can be adequately recorded, identified, studied, displayed, and preserved. The days of the private collection should

## Fossil Vertebrates

have passed long ago for all but the most common and replaceable of such natural objects. Private individuals who turn specimens over to museums are cited for their generous acts and are given credit in scientific journals if the specimen warrants publication. Donors can see their gifts in the research collections or perhaps even on public display whenever they wish, and they have the satisfaction of knowing that their finds are being put to the best possible use. Qualified donors often study and report on their materials in scientific journals with the help of professional museum staff members. But it is an unfortunate fact that most museums operate literally on a financial "shoestring." Competitive bidding and high prices for specimens simply are out of the question as far as such institutions are concerned. Only rarely can they purchase fossil material and then only to fill significant gaps in their collections. Usually, such purchases repay only a fraction of the original cost to the collector. Many good specimens have disappeared into attics and basements, have been lost or destroyed, or have become separated from significant and necessary data on their occurrence so as to become scientifically useless simply because some amateur collector had the mistaken idea that a museum could pay a high price for his find. If this were the way the world had to run there would be far fewer specimens in natural history museums and all men would be the poorer for it; certainly this chapter would be much shorter.

### General References

Scientific publications on the occurrence of specific fossil vertebrates in Michigan are cited individually later on. However, the following are included in the reference list at the back of the book because they are useful general works covering the broad field of vertebrate paleontology; E. H. Colbert, 1955; A. S. Romer, 1941 and 1966; W. B. Scott, 1913; and R. A. Stirton, 1959.

### General Geological History and Evolution of Vertebrates

Historical data of any kind are best appreciated when seen as part of the broader picture of the

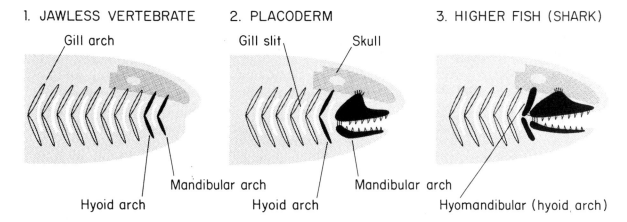

Figure XIV-2. Origin and development of jaws in vertebrates

1) Jawless condition found in modern lampreys and in extinct ostracoderms. Mandibular and hyoid gill arches and the gill slit between, are unmodified, as are those behind.

2) First appearance of jaws in early, primitive, now extinct fishes called placoderms. Mandibular arch modified to form jaws, but gill slit and hyoid arch behind retain primitive character of a gill arch and slit.

3) Fully developed jaws found in higher fishes receive additional support from the enlarged hyoid arch, now called the hyomandibular, which is attached by ligaments to the jaw joint and abuts the cranium at its upper end. The first gill slit becomes compressed to a small spiracle and eventually eliminated. (Modified from Simpson, Pittendrigh, and Tiffany, 1957, *Life—An Introduction to Biology*, permission of Harcourt, Brace and World.)

## Figure XIV-3
### Vertebrate Animal Relationships and Distribution Through Geologic Time

Those groups found in Michigan, and their time span, shown by heavy bar. Dotted lines mean gaps in record. Dashed lines indicate probable ancestor-descendant relationships.

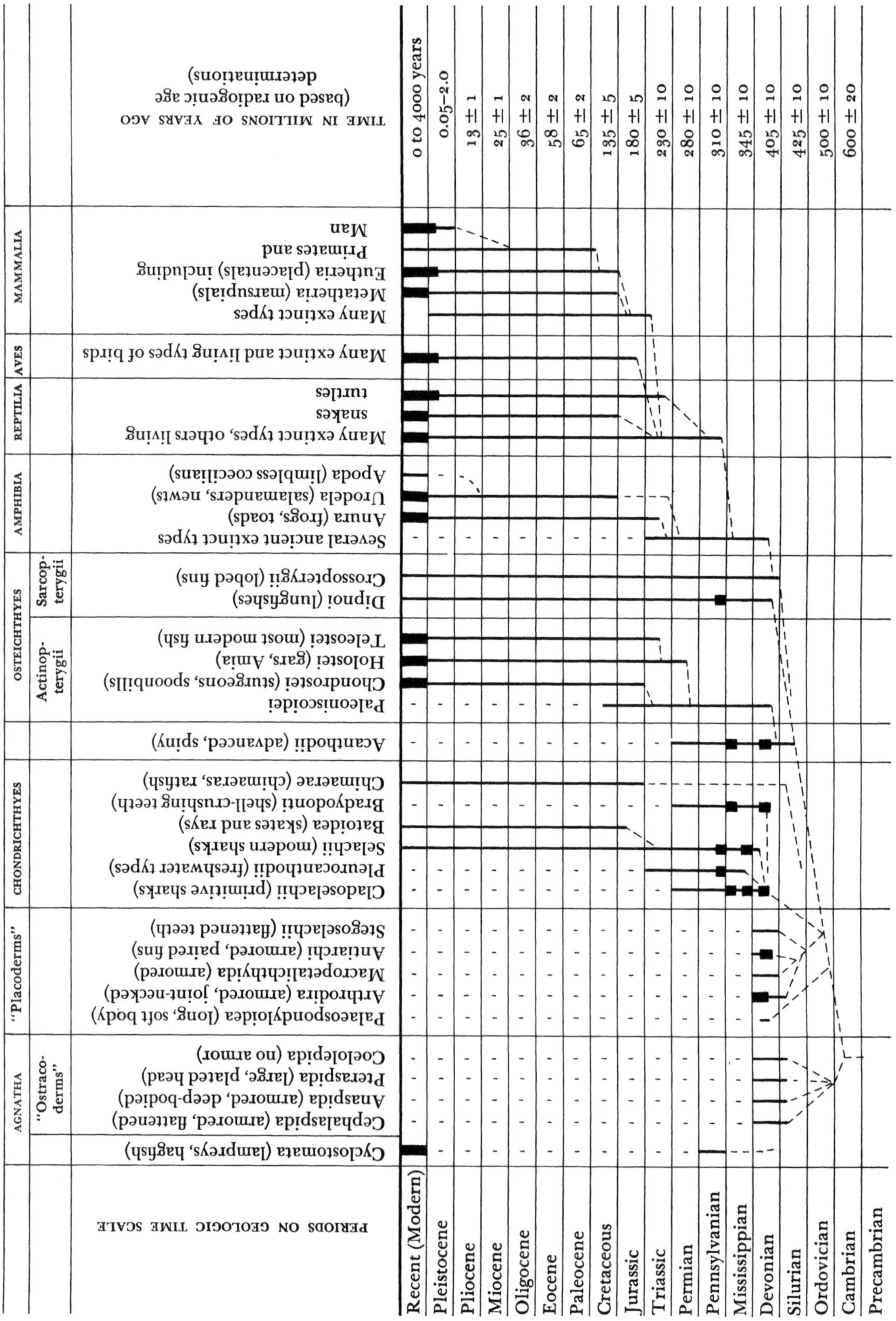

**Fossil Vertebrates**

357

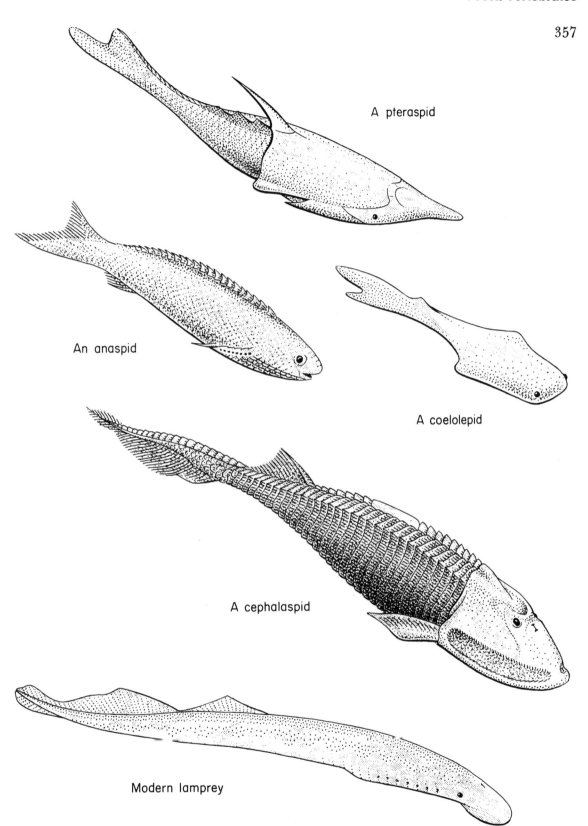

Figure xiv-4. Some typical agnathan (jawless) fishes. All but the lamprey are long extinct. (In part modified from Colbert, 1955, *Evolution of the Vertebrates*, permission of John Wiley and Sons.)

past, some aspects of which they reflect. Therefore, this section reviews the evolution and occurrence of all the major groups of vertebrates through geologic time, indicating which are found as fossils in Michigan. A review of the vertebrate groups found in association with one another here in Michigan as a series of successive faunas through time then follows. And at the end of the chapter is a detailed list of all the genera and species that have been found here, including their geologic ages, geographic locations, and the rock formations they were discovered in.

The most primitive vertebrates, and also the first to appear in the fossil record, are the Agnatha or "jawless" forms (Figs. xiv–3, 4). Fossil members of this group sometimes are called "ostracoderms." *A* means without and *gnatha* means jaws. Fossil fragments of the bony external armor of these fish-like animals have been found in the Late Ordovician-Harding Sandstone near Canon City, Colorado, and elsewhere. Primitive features of the Agnatha include the lack of paired pectoral and pelvic fins and the absense of jaws in the mouth. This last characteristic is well illustrated by the modern lampreys and hagfishes which are the only living agnathans. Directly behind the mouth lay the first of many pairs of unmodified gill arches, no doubt used mainly for breathing or for filter-feeding (Fig. xiv–2). The "ostracoderms" had a cartilaginous internal skeleton but had an armor of bony external plates. Several distinct groups of "ostracoderms" are well represented in rocks of Silurian and Devonian age but so far no "ostracoderm" material has been identified with certainty from Michigan. R. C. Hussey (1947, pp. 130–31) described and illustrated an object, once thought to be an "ostracoderm" scale, found in the Middle Ordovician-Black River rocks 3 miles south of Bony Falls along the Escanaba River in the Upper Peninsula. Unfortunately, the actual specimen has been lost and only an impression of it remains in the rock. That impression is in the University of Michigan Museum of Paleontology collection. Some experts believe that the specimen is a plate from some extinct invertebrate. The "ostracoderms" became extinct at the close of the Devonian period of the Paleozoic Era and the only Agnatha living today are the lampreys and hagfishes. So far the only fossil lamprey known is a single specimen from Pennsylvanian age rocks of Illinois (Bardack and Zangerl, 1968).

The placoderms (Class *Placodermi*) are the next group to appear in the geologic history of vertebrates. They are illustrated in Figure xiv–5. Placoderms first appeared in the Late Silurian and were fish-like forms whose remains are found in sedimentary rocks that also contain many marine invertebrates. All were extinct by the end of the Permian period. Six major groups (orders) are known. Three of these, the *Arthrodira, Antiarchi,* and *Acanthodii* are found in Michigan as well as elsewhere. The remaining three, *Macropetalichthyida, Stegoselachii,* and *Paleospondyloidea,* have not yet been found here. All were so distinct and uniquely evolved as to suggest that they had originated earlier than the time of their first appearance. Some experts even believe that these were so different from one another that they should not even be classified together; in other words the term "placoderm" may not be valid except as a handy, informal name. The *Acanthodii* or acanthodians, in particular, seem to be a much

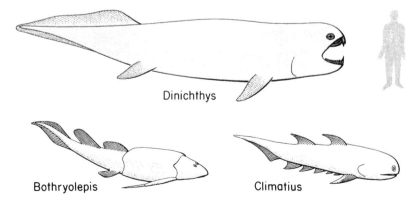

Figure xiv–5. Some typical placoderms (extinct) found as fossils in Michigan. (In part modified from Romer, 1966, *Vertebrate Paleontology,* after Stensiö, and Traquair and Watson, permission of University of Chicago Press.)

more advanced group than the other "placoderms," perhaps directly ancestral to higher fishes. In any case, all so-called "placoderms" had advanced evolutionarily by developing crude jaws. Paleontological, anatomical, and embryological evidence indicates that this change was accomplished by a modification of an anterior gill arch (Fig. XIV–2). Tooth-like structures were present on the jaws. Two or more pairs of side fins had appeared as well. Hence placoderms were better fitted for life in water. It may be that in competition with the more primitive jawless fishes the latter became extinct. In spite of their worldwide success during Devonian times, placoderms later dwindled. All but the acanthodians became extinct at the end of the Devonian Period.

Even the acanthodians disappeared at the end of the Permian. Long before, however, in the Devonian, 2 new major fish groups had appeared, and representatives of these are still living today. These are the Class *Chondrichthyes* (*chondri* meaning cartilage, *ichthyes* meaning fishes; hence fishes with cartilaginous internal skeletons) and the Class *Osteichthyes* (*oste* meaning bone, hence fishes with bony internal skeletons). The *Chondrichthyes* include the Elasmobranchii (sharks, skates, rays), and the ratfish (chimaeras) as well as several extinct orders (Fig. XIV–6). The *Osteichthyes* include many extinct groups and all other fishes of the past and today (Figs. XIV–7, 8). Recent fossil evidence strongly suggests that the ancestors of bony fish were the acanthodians. The ancestry of the Chondrichthyes is less certain. They first appear in sedimentary rocks of marine origin and have special physiological adaptations for life in salt water, hence appear to have originated in the sea. Most Chondrichthyes continued to live in the sea to the present day, although the pleuracanths were a minor, long extinct freshwater group and some few modern sharks do find their way, mainly by accident, into freshwater rivers and lakes today. The Osteichthyes, in contrast, have anatomical and physiological characters, such as kidney construction, which suggest they originated as freshwater forms. The fossil record is not yet clear on this point. However, the bony fishes did enter the sea, virtually overwhelming the Chondrichthyes. Fossil and living forms of both the Chondrichthyes and Osteichthyes exhibit certain similar characters which indicate both were evolutionarily advanced beyond placoderms. These characters again primarily involve the jaw and gill region. In both groups the first gill arch pair (one on each side) moved forward (Fig. XIV–2), attached itself to the brain case at its upper end, became connected to the jaws at their joint, and thus was transformed from a breathing (gill) structure into a support for the jaws called the hyomandibular. The first gill slit behind the jaws in sharks became compressed and reduced to a small opening, a spiracle. In most bony fishes it was entirely eliminated.

The Class Chondrichthyes includes at least 5 major orders. Of these, 3 are extinct (*Cladoselachii, Pleuracanthodii, Bradyodonti*), and 2 are

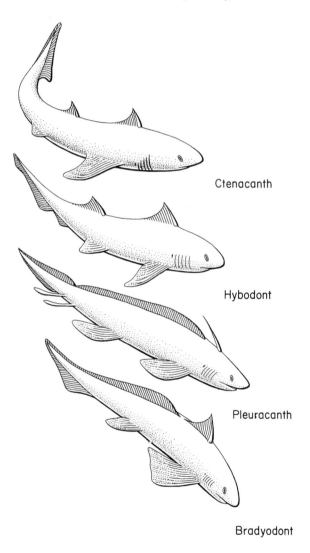

Figure XIV–6. Some typical chondrichthyan fishes found as fossils in Michigan.

still living (*Selachii,* or the modern sharks, and *Batoidea,* or the skates and rays). Another group, the *Chaemaerae* or ratfish has long been classified as a sixth chondrichthyan order, but is so unique that some authorities today believe it may belong in a separate class of its own. The first 4 orders of chondrichthyans all occur in the Michigan fossil record (Fig. XIV-3). The class as a whole first appeared in the geologic record in rocks of Middle Devonian age. The presently known Michigan record is restricted to rocks of the Devonian, Mississippian, and Pennsylvanian periods. There are no sharks or their allies living in Michigan today, and their presence in ancient rocks implies that portions of this region once were covered by marine waters. This conclusion is supported by the associated fossil record of invertebrate animals in the state (Chapter XIII).

The Osteichthyes represent the most successful group of advanced fishes. Throughout most of their history (Early Devonian to the present, Fig. XIV-3) there have been a multitude of both freshwater and marine forms. Almost any fish that you can think of, except the sharks and their allies, is an osteichthyan or bony fish. But now we must distinguish several major subdivisions of the Class Osteichthyes in order to appreciate their history.

The Subclass *Actinopterygii* (*actino* meaning ray, *pterygium* meaning fin; hence "rayfinned" fishes) is the group of bony fishes which includes most familiar fossil and living fishes (Fig. XIV-7). The structurally most primitive rayfins and also the first to appear in the geologic record, is the Order *Paleoniscoidea*. The oldest of these are found in rocks of Middle Devonian age (Fig. XIV-3). The *Chondrostei,* a primitive offshoot, includes some specialized types, for example the sturgeons, which still live today and occur in freshwaters of Michigan, but most chondrosteans are now extinct. No fossil paleoniscids or chondrosteans have been found yet in this state. The *Holostei,* more advanced forms of rayfinned fishes, descended from unspecialized paleoniscids (Nikol'skii, 1954). The holostean fishes (Fig. XIV-7) first occur in rocks of Middle Permian age (Fig. XIV-3). Many fossil forms are known, but most of this group are now extinct. No fossil holostean fishes have yet been found in Michigan but two of the most common, still-living, types do occur in the state today. These are the gars or gar-"pike" and *Amia* (often called the Bowfin or Freshwater Dogfish). Examine the hard, shiny, parallelogram-shaped overlapping scales of a gar sometime, and you will see that they are of the primitive and very ancient enamel covered or ganoid type, a common form of scale in ancient fishes. The *Teleostei* or teleosts (Fig. XIV-7) are the most advanced rayfinned fishes. Teleosts arose from holosteans and their earliest known fossil record is in rocks of Late Triassic age (Fig. XIV-3). This is the most widespread and successful bony fish group throughout both salt and fresh waters of the world today. Many familiar teleosts live in Michigan today. Perch, bluegills, bass, and pike are good examples. We could review the scientific classification of fishes, as we have discussed it so far, by saying that if you have ever caught one of these common types of fish you have caught an osteichthyan (bony fish)-actinopterygian (rayfinned bony fish)-teleostean (bony, rayfinned, advanced fish). Quite a mouthful, you might say. No very ancient teleosts have yet been found in Michigan rocks, but fossils of this group have been found in the Late Pleistocene and Post-Pleistocene deposits of the state (Fig. XIV-3). A list of these fossils follows later in this chapter.

Now that we have looked briefly at the actinopterygian or rayfinned bony fish, let us go back

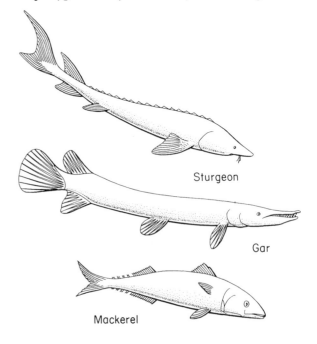

Figure XIV-7. Some typical osteichthyan ("bony") fishes found in the fossil record.

to the other major group, which is the *Subclass Sarcopterygii* (*sarco* meaning fleshy, *pterygium* meaning fin; "fleshy-finned" or "lobe-finned" fishes). Another name for these, used in some books, is *Choanichthyes* (*choana* meaning nostril, hence the nostril fishes). The term fleshy-finned or lobe-finned refers to the fact that some of the fin muscles extend out along the pectoral and pelvic fins beyond the margin of the body. Internally, the fin is supported mainly by a rather few bones very similar to those in the feet of the first amphibian, rather than by fin rays. The alternative name, Choanichthyes, refers to the additional fact that an air passage connected the external nostril openings with the throat. Well-developed lungs also were present to receive air after it was brought into the throat through the internal nostril passage. The combination of strong, muscular fins, nostrils connected to the throat, lungs, and a strong, bony spinal column, made it possible for the Sarcopterygii not only to breathe air by lungs in addition to getting oxygen through the gills, but also to crawl around on the bottom, on logs and even out on mud flats. Most sarcopterygians lived in fresh water. These abilities preadapted some of these fishes to be ancestors of those first true land vertebrates, the amphibia. The first group of Sarcopterygii to appear were the *crossopterygians* (Figs. xiv-3, 8). These occur in rocks as old as the Early Devonian. This group has since disappeared almost entirely. Most crossopterygians were extinct by the end of the Jurassic, and the only known living genus is the coelacanth (*Latimeria*), a recently discovered marine form. Shortly after they appeared, the crossopterygians gave rise to the first amphibia in the Late Devonian. No crossopterygians live in Michigan today, and evidence that they occurred here in the geologic past still is lacking.

The other group of Sarcopterygii is known as the *Dipnoi* (Fig. xiv-8). This group includes many fossil and a few living "lung fishes." Most of these have flattened, fan-shaped, highly specialized teeth for crushing shellfish. This group is not on the main evolutionary line to higher land dwelling vertebrates, but is interesting, nonetheless, because the living lung fishes give us a clue to how their relatives lived in the past. Modern lung fishes, such as the Australian Genus *Neoceratodus*, the African *Protopterus* and the South American *Lepidosiren*, actually breathe air into their lungs which helps them to obtain oxygen during times of low or foul water. The African lung fish will, in fact, drown if not allowed to come to the surface to breathe air. In aquaria, as for example in Detroit on Belle Isle, they can be observed using their weak fins as legs to crawl on objects on the bottom, much as the ancient crossopterygian ancestors of amphibians must have used their more muscular lobed fins to flop across mud flats from a drying pool back into water. *Protopterus* and *Lepidosiren* both burrow into the mud of their drying streams and pools during times of drought and in this fashion "aestivate" until the waters rise again. Thus, in their structure and habits the Dipnoi also anticipate characters to be expected in the ancestors of land-dwelling vertebrates. Many fossilized lung fish burrows have been found in sedimentary rocks of Late Paleozoic and Early Mesozoic age in North America. Apparently, these once hollow burrows often became filled with sediment. Sometimes lung fishes became entrapped in their burrows and thus fossilized. Recently, Dr. R. L. Carroll found similar lung fish burrows in Pennsylvanian age sandstones of the Saginaw Formation near Grand Ledge, Michigan. This discovery is discussed in more detail later, but serves to illustrate how paleontologic discoveries may continue to be made in Michigan and how, little by little, historical gaps are filled in.

Many evolutionary steps remain to be taken

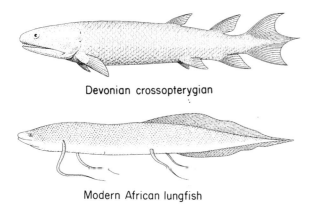

Devonian crossopterygian

Modern African lungfish

Figure xiv-8. Lungfishes (sarcopterygians), fossil and living. (Modified in part from Colbert, 1955, *Evolution of the Vertebrates*, permission John Wiley and Sons, and from Romer, 1941, *Man and the Vertebrates*, permission of University of Chicago Press. Copyright 1933, 1937, 1941 by University of Chicago.)

from fish to higher land dwelling vertebrates such as mammals. Unfortunately, little of this remaining history is recorded in Michigan except the very last stages of Late Pleistocene and near Recent time.

In the Late Devonian, primitive amphibia evolved from crossopterygian fish (Fig. xiv–3). Many now extinct types in a wide variety of shapes and sizes are known. Some giants were over 6 feet long, and heavily armored, weighing more than a man. Amphibia enjoyed their evolutionary climax of abundance and diversity in the Permian and Triassic, and declined thereafter. Most are now extinct. Paleozoic fossil amphibia have yet to be found in Michigan. This is partly due to the fact that the only rocks of the right age and of nonmarine origin presently known in Michigan are those of the Pennsylvanian-Saginaw Formation and these are very poorly exposed, over a small area, mainly in old quarries. Careful search over the years may eventually produce some fossils, however, because rocks of the same type and age in neighboring Ohio, Indiana, and Illinois have yielded fossil amphibia. Some eventually should be found in the Pleistocene deposits too. More looking is needed. The amphibia living in Michigan today represent 2 of the 3 surviving amphibian groups of the modern world, the *Anura* (frogs and toads) which first appeared in Late Triassic times, and the *Urodela* (salamanders and newts) whose first appearance is in Early Cretaceous rocks. The third living group, *Apoda* or caecilians is a peculiar lot of limbless, snake-like, burrowers; all are southern forms, not appearing in North America.

Although the amphibia were the first land vertebrates, they still had not fully solved the problem of life on land. The absence of a shell and protective membranes around the egg required that amphibians return to lay their eggs in water (like fish) or at least in moist places, and the lack of a large yolk for food required early hatching into an aquatic larval or tadpole stage followed by metamorphosis. However, the reptiles *(Class Reptilia)* which arose from amphibia in the Middle Pennsylvanian period, solved most of these remaining problems for living on land by evolving a shell-covered large-yolked egg. They thus became the dominant land-dwelling vertebrates of their time and remained so for millions of years thereafter. Nearly everyone knows about the vast number and endless variety of reptiles that inhabited the earth in past ages. Their adaptive forms included giants and tiny types, fleet footed and plodding forms, meat eaters, fish eaters, and plant eaters, heavily horned or armored monsters, upland, swamp and marine forms, and even some that could fly. Nevertheless, the reptiles eventually passed their evolutionary peak and near the end of the Cretaceous Period began a swift decline. Still many are left—the snakes, lizards, alligators, crocodiles, and turtles. In spite of their peculiarly specialized shell, turtles are the most primitive. Most of this vast reptilian procession also passed unrecorded in Michigan. The reasons are the same as those given for the absence of amphibia, and again it is possible that some ancient reptiles may eventually be found in some part of the Pennsylvanian Saginaw Formation or in the Grand River Group in this state. The only known rocks in Michigan that date back to the Mesozoic "Age of Reptiles" are the "red beds" located in the central part of the Lower Peninsula. These recently were recognized by studies of their pollen (Chapter XV) as being of Late Jurassic or Early Cretaceous age. They are about the same age as the Morrison Formation which, in the Rocky Mountain region, has produced so many fossil reptiles, including dinosaurs. However, the Michigan "red beds" are nowhere exposed at the surface and the chances of accidentally finding large vertebrate fossils in oil or water well drill holes are slim indeed. So far the only record of fossil reptiles in Michigan consists of turtles from the Late Pleistocene and Post-Pleistocene deposits, and these are the same types of turtles as live in Michigan or elsewhere nearby today.

Birds *(Class Aves)* arose from reptiles sometime in the Late Jurassic, but their historical record is relatively poor everywhere. In Michigan, again for the reasons given earlier, fossil birds are unknown except for a few Pleistocene and Post-Pleistocene specimens, also the same as or closely related to modern types.

Mammals *(Class Mammalia)*, those warm-blooded animals, with hair and mammary glands, that suckle their young, arose from reptiles in the Late Triassic or Early Jurassic (Fig. xiv–3). Mammals evolved in such a way as to enjoy an advantage over reptiles which probably was a principle factor leading to their eventual dominance over all other land vertebrates. They can control their body temperature at a more or less constant even level. Efficient hearts and circulatory systems, and insulating hair, make this possible. Mammalian intelligence eventually increased also. Thus, mammals could and did adapt more successfully to a much wider range of geographic and climatic conditions than could other vertebrates. Today they range from mountain tops to oceans and to the air (bats), and from the arctic to the equator. Many extinct types are known from rocks elsewhere in North America and the world, but again the record in Michigan is restricted to Late Pleistocene and Post-Pleistocene types (Fig. XIV–11). What there is of this record is extremely interesting, however. The list that follows later includes such types as peccaries, musk oxen, mammoths, mastodons, walruses, whales, and a host of other types no longer found in this state. Unfortunately, the lower primates are completely missing even from the Pleistocene deposits here, and paleo-Indians did not appear in Michigan until a relatively late date long after early man had appeared in the Old World. Many major groups of mammals are known from the fossil record, but only 3 main groups live today. The rare, egg-laying *Monotremata* (duck-billed platypus and spiny echidna) are not known from the fossil record and do not occur here. The *Metatheria* or *Marsupialia* (Cretaceous to Recent) are the pouched mammals including the primitive opossums, kangaroos, and a host of others geographically isolated, mainly by geologic accidents of continental separation, in South America and Australia. The *Eutheria* (placental mammals) occur over most of the rest of the world. Of these 3 living groups, only the marsupials and placentals are of any real importance in the world today. Pleistocene and Post-Pleistocene fossil mammals of Michigan are all placental (Eutherian) types. The only marsupial in the state, even today, is the primitive opossum, a Recent migrant from the warmer southlands who spread rapidly northward into the state only within the last 50 years or so.

This brief review of the major features of vertebrate origin and evolution points out the fact that the historical record of those events in Michigan suffers from great gaps. As mentioned, this is mainly due to the absence of rocks of the proper type and age during the time span from the Late Pennsylvanian to the Late Pleistocene. Of course this does not mean that this area of the continent was completely uninhabited by vertebrate animals during that time. On the contrary, it is probable that all of the major vertebrate classes, most of the orders, and a large sampling of the tremendous variety of other smaller groups represented in the fossil record in neighboring regions, lived here. To take an extreme example from the modern world, consider the Tibetan Plateau. A host of vertebrate animals inhabits that geographically and climatically hostile environment today. Yet it is an upland region now undergoing erosion. Little in the way of a sedimentary rock record is accumulating there, and most of that will be lost again to later erosion, hence the record of present life there will be lost to the future, although it may be represented to some extent by fossils present in deposits laid down around the fringes of that region. Chapter VI, The Lost Interval, discusses in greater detail the nature of Michigan during that great time gap, and the biologic, geologic, and geographic conditions that prevailed here then.

## Fossil Vertebrate Faunas of Michigan and Their Ancient Environments

All of the vertebrate animals living together in a given region at a given time constitute the vertebrate fauna of that region. For instance, we could list all the vertebrates (fish, amphibians, reptiles, birds, and mammals including man) known to have lived in Michigan during historical time and that could be called the "Recent Vertebrate Fauna of Michigan." "Recent," spelled with a capital R, stands for the geologic present or near-present. We could also speak only of the mammalian fauna, or, restricting ourselves to a particular habitat or environment, speak only of the aquatic fauna of the Great Lakes, or the land fauna of southern Michigan. A discussion of an ancient fossil fauna, vertebrate or otherwise, must begin with a definition of the limits in time and space of what is to be called "the fauna." These limits are arbitrary and may be as broad or narrow as we wish, but they must be understood and agreed upon in advance so we are certain to appreciate the significance of what we are saying and the uncertainties involved. An analysis of ancient faunas involves special geologic problems,

# Geology of Michigan

too. The animals represented in the fossil "fauna" may have been found at only one or a few scattered localities; hence we cannot be certain that they occurred over the whole of the region we are studying. The fossils actually found may represent only a small fraction of all the animal types that actually lived in the region at the time being considered, in which case we can only make an educated guess concerning their probable but as yet undocumented associates. An even more important problem stems from the uncertainties of geologic dating. For example, a geologist or paleontologist might establish that 2 animals both lived at Michigan at some time during the Devonian Period of geologic time, but the best of modern knowledge based on radiogenic dating shows that the Devonian Period was about 60 million years long (Fig. xiv-3). One animal might have lived at the beginning of that period and the other at the end, yet they might never have been actual contemporaries. With more exact geologic dates we might be able to say that both lived in the Middle Devonian, but still that was a long time. To be absolutely certain that 2 or more fossil animals actually were temporal associates, the fossil specimens representing them would have to have been found close together at the same level in the same rock formation. Even so they might not actually have lived together; perhaps they were merely preserved together. For example, a fish that died in the sea and was washed up onto the beach might have been buried in beach sand near a land animal that had fallen into a stream and was floated down to the sea. Thus, they were contemporaries, but they lived in separate places and in different habitats. Realizing all these limitations that must restrict our statements when we attempt to reconstruct the fossil record, let us agree that a fossil vertebrate fauna, as we shall use the term, will include those animals actually known from the fossil record to have occupied at least some part of Michigan during some part or all of what may have been a rather long but definable period of time. Thus we

Figure xiv-9. Late Devonian seascape. Primitive shark *(Cladoselache)* prowls past tall, solitary corals (at right) and over a garden of sponges. At left, a cephalopod mollusk, tentacles extended in search of food, approaches 2 starfish. (Courtesy Rochester Museum of Arts and Sciences, New York.)

# Fossil Vertebrates

can speak, for example, of the Middle Devonian fauna of Michigan and still recognize the possibility that not all the animals mentioned lived exactly at the same instant nor in all the same places or environments over the whole region. The actual evidence for the statements that follow is presented in the fossil lists at the end of the chapter and elsewhere in this book (Chapters II, III, V, VI, VII, XIII, and XV).

## Middle Devonian Faunas

The oldest fossil vertebrates yet found in Michigan are Middle Devonian in age. The geological record from other regions tells us that vertebrates actually had appeared long before that. Some types more ancient than Middle Devonian probably did live in Michigan, but we cannot yet prove it directly with actual specimens. During the Middle Devonian large areas of the interior of North America, including most of what is now Michigan, were under water. Great extensions of sea had spread inland over the continent. Sediments that slowly accumulated on the sea floor eventually formed the sequence of Middle Devonian rock formations known in Michigan as the Dundee limestone, the Rogers City limestone, and the several formations of the Traverse Group (Chapter V, Figure v–2). Figures xiv–9 and xiv–10 illustrate the vertebrates that might have been seen in the Middle Devonian marine environment of the interior of North America. The warm waters of the Middle Devonian epeiric seas also favored a host of invertebrates such as corals, bryozoans, brachiopods, snails, clams, cephalopods, and crinoids (Chapter XIII). Along with these lived the now extinct, fish-like vertebrates called placoderms. Especially common among these were the giant, heavily armored, and joint-necked arthrodires such as *Arctolepis, Titanichthys, Protitanichthys, Dinichthys, Holonema, Mylostoma, Dinomylostoma, Ptyctodus,* and *Eczematolepis*. Most specimens of arthrodires found in

Figure xiv–10. Devonian seascape. *Coccosteus,* a large arthrodiran placoderm pursues one of a school of small acanthodian placoderms into a bed of several kinds of sponges. (Courtesy New York State Museum and Science Service, Albany, New York.)

Michigan occur in normal marine limestones. *Holonema farrowi*, described by Stevens (1964), occurred in a deposit near Onaway, the Koehler limestone, which appears to have been laid down in a semi-isolated lagoon. Judging from their formidable jaws and teeth, arthrodires probably were voracious predators that fed upon other fish and invertebrate animals. A less common group of placoderms, the acanthodians, more closely resembled modern fish and appear to have been fast and agile swimmers, probably favoring the warm, sunlit surface waters. The acanthodian fins were supported by long spines which most commonly were preserved as fossils. The spines may have been protective or else served as cutwaters. At least 3 genera of acanthodians, *Onychodus*, *Gyracanthus*, and *Machaeracanthus*, lived then. Less common placoderms were the grotesque antiarchs, represented solely by *Bothryolepis*, a flattened, heavily armored, ungainly looking creature who may have been a slow-moving, bottom-dwelling scavenger.

Several chondrichthyan fishes related to sharks also inhabited the Middle Devonian seas of Michigan. One of these was the primitive cladoselachan called *Ctenacanthus*. Cladoselachans were members of the ancestral shark group. Other chondrichthyans, of the bradyodont group (Fig. XIV–6), may be represented by spines identified as belonging to the genera *Oracanthus* and *Acondylacanthus*. The bradyodonts, well known from better specimens found outside Michigan, had peculiar, flattened teeth which are thought to have served as shell crushers, possibly indicating that these fish fed on such bottom-dwelling invertebrates as clams, snails, and brachiopods.

## Late Devonian Vertebrate Faunas

Late Devonian marine vertebrate life is poorly recorded in Michigan. This could mean that fishes were less abundant and diverse here then. However, most of the Late Devonian rocks are dark shales which represent an accumulation of sediment in a muddy environment which may have been less favorable for fish life than the clearer seas in which the Middle Devonian limestones had formerly accumulated. Another possible explanation, leading to a somewhat different conclusion, is that fossil fish remains are simply harder to find in the Late Devonian shales. There are fewer Late Devonian shale quarries in the state than there are Middle Devonian limestone quarries, and soft shales tend to break up and fall apart. Although these explanations may be inconclusive, the fact remains that only a few fragments of 3 arthrodiran placoderms have been found in the Late Devonian rocks of the state. *Trachosteus* (?) and *Aspidichthys* are slightly different from the Middle Devonian types. The third genus, *Dinichthys*, is a carryover from the Middle Devonian.

## Early Mississippian Vertebrate Faunas

The early Mississippian rocks of Michigan, like those of the preceding Devonian Period, were deposited in the sea. The Coldwater Shale is a moderately deep water, offshore marine deposit, whereas the overlying Marshall and Napoleon sandstones are near-shore marine and marine beach deposits (Figs. v–3, 29). Evidently, the seas had begun to recede and the water was shallower than in the Devonian. The sandstones indicate that land must have stood close by, but no actual land-laid deposits of early Mississippian age have been found in the state. This is unfortunate because the Amphibia, first of the land-dwelling vertebrates, had come into existence just before then, in the Late Devonian, and the rock record elsewhere shows that by Early Mississippian time the amphibians were widespread and abundant. Some may have lived in Michigan, and it would be interesting to find proof in the rocks. The only clue to the nature of life on land at this time in Michigan comes from the occurrence of driftwood fragments found in these rocks (Dorr and Moser, 1964). The fragments indicate that there was vegetation on land, including trees of the *Calamites* and *Lepidodendron* or *Sigillaria* type (Chapter XV); so we know there was plant food available to support land animal life.

Chondrichthyan fishes still roamed the Mississippian marine waters of the state. Their sparse fossil record is a poor reflection of the actual abundance and diversity of marine vertebrate life that existed then, but we can reason that since they were predators there must have been other animals present upon which they fed. Primitive cladoselachan sharks were one of the types pres-

# Fossil Vertebrates

ent. These were little-changed survivors from Middle Devonian time. *Cladodus* and *Ctenacanthus* are the only genera that have been found. But a more modern type of shark, a selachian, also occurred here; this is the genus *Orodus* (Fig. XIV-34).

There is some evidence that during the Middle Mississippian the seas receded briefly from at least some parts of the state, producing a local unconformity or break in the rock record and a time gap in history.

By the Late Mississippian the seas had returned, but the nature of the marine environment was changed. Warm, dry conditions and periodic semi-isolation of the Michigan basin produced a strongly evaporating environment. The frequent layers of evaporitic gypsum interbedded in the marine shales of the Late Mississippian Michigan Formation testify to this. The paleogeography of this time is described in detail in Chapter V. The Bayport Limestone which overlies the Michigan Formation suggests that the evaporitic environment may have given way finally to more normal marine conditions toward the very close of the Mississippian Period. Primitive cladoselachan sharks (*Cladodus* and *Ctenacanthus*) still occupied these marine waters; their remains occur in the Michigan Formation. In addition, the bradyodont shark group was represented slightly later in time by fossil remains of the genera *Helodus*, and *Psephodus* and by some unidentified "cochliodont" teeth, all from the Bayport limestone. Recently, a fin spine of *Ctenacanthus* was found in a thin shale layer within what petroleum geologists call the "triple gyp," a name given to 3 distinct but closely spaced gypsum layers which are interbedded with shales in the Michigan Formation. The discovery was made underground in the walls of the Bestwall Gypsum Company mine beneath Grand Rapids. Associated with the spine were peculiar, bean-shaped coprolites which in this case are fossilized pellets of fish excrement. Judging from the associated spine, it appears that sharks lived and fed in the area during the brief freshwater interval of time represented by the thin shale layer, sandwiched between gypsum layers. No fossils occur in the gypsum itself, probably because the highly saline evaporation environment represented by those deposits was unfavorable for life. Dorr and Moser (1964), speculating on the historical significance of this discovery and on other evidence from Moser's more extensive studies of the sedimentary rocks themselves, conclude that, "the presence of fossil ctenacanth shark remains in association with marine invertebrates and driftwood in the near shore marine sandstones of the Marshall Formation in southern Michigan, and the presence of ctenacanth shark remains in association with coprolites in the thin, sandy shale parting within what may be the near age equivalent portion of the evaporite sequence in the Michigan Formation at Grand Rapids, is significant. We suggest that the sharks and associated organisms lived in the more normal saline waters freshened by entering streams from the land along the margins of the Mississippian aged Michigan Sea at the same time that evaporites were accumulating in off shore areas of deposition. The salt-saturated off shore waters ordinarily would have been intolerable for such creatures except during such times, as represented by the sandy shale partings in the Michigan Formation at Grand Rapids, when current changes or increased influx of fresh water from the land freshened the sea for a greater than normal distance from the shore. The sharks of that time may have skirted around the margins of the sedimentary basin, avoiding the central portions."

This analysis of the historical meaning of a minor fossil discovery may seem of purely academic interest, but remember that such scraps of historical analysis, pieced together with the results of other and often more extensive studies, may assist economic geologists to reconstruct the paleogeography of the past and thus to predict the location of new mineral deposits. In this case the "triple gyp" is a key correlation horizon in subsurface studies of petroleum geologists, and gypsum itself is mined in the Grand Rapids area.

A fish scale recently discovered in rocks near Pigeon at the tip of the thumb of the Lower Peninsula reveals the presence in Michigan, perhaps for the first time in the Late Mississippian of Osteichthyes, the bony fishes. Studies of this specimen and the details of the geology at the discovery site still are incomplete.

## Pennsylvanian Vertebrate Faunas of Michigan

By the Pennsylvanian Period the Michigan scene

had changed still further. Very shallow seas, often mere tidal shallows, fluctuated in and out of the state at least a half dozen different times. Low swamp lands and sluggish streams occupied the area in times between the seas. Thus, the Saginaw Formation, a deposit of late Early Pennsylvanian age was deposited as a series of cyclic layers of alternating marine and freshwater origin. Included are swamp deposited coals, stream channel sands, river flood plain clays and silts, and tidal flat muds. Marine and freshwater fossil animals and swamp-dwelling plants alternate, too. The details of this history are given in Chapters V and XV. As for the vertebrates, the peculiar and now extinct fresh water shark *Pleuracanthus* (Figs. xiv–6, 33) lived in the fresh waters of the swamps and streams. The recent discovery of the lung fish burrows (R. L. Carroll, 1965) in basal sandstones of the Saginaw Formation near Grand Ledge proves that fish of the dipnoan group also occupied the ponds and sluggish streams that flowed through those Pennsylvanian swamps. So far none of the burrows found actually contain fossil fish remains, but the burrows are similar to those made by the fossil lung fish *Gnathorhiza*, which is common in rocks of somewhat younger Permian age in such widely scattered places as Texas, New Mexico, and Prince Edward Island. Bones of *Gnathorhiza* have also been found in Arkansas and Oklahoma in rocks of the same age as those in Michigan. The Michigan burrows are significant because they provide the earliest evidence yet found that some lungfish had evolved the habit of aestivating in mud or sand during times when waters became foul or temporarily dried up. This habit must have served them in good stead during those Pennsylvanian times in Michigan when the waters of their swampland pools were excessively fouled by mud or decaying vegetation, and when the streams temporarily dried up or shifted course. The lung fish's ability to breathe air also would have been a distinct ad-

Figure xiv–11. Southern Michigan shortly after retreat of the Late Pleistocene (Wisconsinan) ice sheet. A herd of American mastodons *(Mammut americanum)* browses its way into a clearing in a swamp. At right, a giant beaver *(Castoroides ohioensis),* large as a black bear, clambers up a huge, ice-carried erratic boulder. Distant morainic ridges and lake-filled depressions, left behind by the ice, are mostly clothed by a "pioneer" spruce-fir forest. (Diorama, The University of Michigan Exhibits Museum.)

## Figure XIV–12

### Late Wisconsin (Latest Pleistocene) and Post-Wisconsin (Post-Pleistocene) Fossil Vertebrates of Michigan For Which $C^{14}$ (Radiocarbon) Dates and/or Associated Pollen Analyses Are Available

| Common Name and $C^{14}$ Date Reference Number (M stands for Michigan Memorial Phoenix Project Laboratory where $C^{14}$ analysis was made) | | County or Locality in Which Specimen Was Found | $Carbon^{14}$ Date in Years Before Present (± shows number of years of uncertainty in date) | Percentages of Dominant Tree Pollen Out of Total Tree Pollen (not percent of all tree pollen nor of all pollen) |
|---|---|---|---|---|
| Prairie vole | M208 | Leelanau (Sleeping Bear Dune Locality) | 730 ± 250 | |
| Plant assemblage | M1149 | Old Saline River Terrace near Milan | 4080 ± 200 | Oak, 28; basswood, 16.7; walnut, 11; elm, 10.3; sycamore, butternut, hickory, beech and ash, combined, 29 |
| Caribou | M294 | Genesee | 5870 ± 400 | |
| Mastodon | M347 | Lapeer | 5950 ± 300 | |
| Mastodon | M281 | Lenawee | 7820 ± 450 | |
| Mammoth | M1400 | Berrien | 8200 ± 300 | |
| Mastodon | M1254 | Gratiot | 10,700 ± 400 | Spruce, 40; pine, 35 |
| Musk ox | M1402 | Kalamazoo | 11,100 ± 400 | Pine, 68.6; spruce, 5.2; balsam fir, 5.6; larch, 4.8; birch, 3.0 |
| Mastodon, Texas Bionuclear Lab. | | Oakland | 11,900 ± 350 | Spruce, 82.8; pine, 7.7; oak, 4.3; poplar, 2.6 |
| Mammoth | M507 | Jackson | 12,200 ± 700 | |
| Musk ox | M639 | St. Joseph | 13,200 ± 600 | Spruce, 86.9; balsam fir, 1.4; birch, 2.0; larch, 1.0; pine, 0.5 |

vantage in foul swamp waters, where the oxygen content must frequently have been depleted by decaying vegetation.

Now, at a very exciting point in the drama of vertebrate evolution, just as the reptiles of the Middle Pennsylvanian were about to appear on the scene, the Michigan fossil record goes blank! Abundant fossil evidence from elsewhere proves that in the Late Paleozoic reptiles appeared, and shortly thereafter the Mesozoic Age of Reptiles dawned. Giant amphibians flowered, passed their peak, and became extinct. Dinosaurs rose, then disappeared again along with a multitude of other reptilian forms. Birds appeared and mammals, too. During the 60 million years or so of the Cenozoic Era, from the Paleocene Epoch to the Pleistocene, the so-called Age of Mammals, primates and many other mammalian orders evolved. Toward the end of the Age of Mammals, man himself appeared. Many of these important animal groups must have lived in parts of Michigan, but no records remain in the rocks. Michigan probably was an emergent upland during much of this time. If any sedimentary deposits with fossil remains were laid down here, the forces of erosion destroyed them again. Several times toward the close of the Cenozoic Era the great continental ice sheets of the Pleistocene advanced and retreated across the state, grinding away what little record may have remained before. The record of fossil vertebrates in Michigan does not begin again until after the last major Pleistocene ice sheet, the Wisconsin stage, finally began to withdraw. But that may have been as little as 13 thousand to 14 thousand years ago and by that time vertebrate life was much changed from what it had been in the Pennsylvanian Period. So many evolutionary advances and animal extinctions had

## Figure XIV–13

### GROUPS OF TWO OR MORE FOSSIL VERTEBRATES FOUND TOGETHER IN POST-PLEISTOCENE DEPOSITS OF MICHIGAN AND PRESUMABLY CLOSE TEMPORAL ASSOCIATES

(Modified from R. L. Wilson, 1965, and with additions from other sources)

| General Location | Fishes | Turtles | Birds | Mammals |
|---|---|---|---|---|
| East side of Fenton Lake, Genesee County | Lake trout, whitefish, muskellunge, silver redhorse, white sucker, long-nosed sucker, quillback or carpsucker, walleye | Soft-shelled turtle | Small duck (scaup or ringneck), bald eagle | |
| Sleeping Bear Dune, Leelanau County | Channel catfish | | | Short-tailed shrew, woodchuck, eastern chipmunk, red squirrel, gray squirrel, southern flying squirrel, Canada beaver, white-footed deer mouse, muskrat, northern red-backed vole, prairie vole, meadow vole, pine vole, raccoon, marten, red fox, wolf |
| Farmington Township, Oakland County | | Snapping turtle, painted turtle | | Muskrat |
| Sec. 36, Ann Arbor Township, Washtenaw County | | | | Muskrat, moose |
| Washtenaw County | | | | American mastodon, Jefferson mammoth |
| Thread Lake near Flint Genesee County | | | | American elk (wapiti), Virginia white-tailed deer |
| Moorland, Kent County | | | | Mastodon, musk ox (*Boötherium*) |
| White Pigeon, St. Joseph County | | | | White-tailed deer, musk ox |
| Adrian, Lenawee County | | | | White-tailed deer, mastodon |
| Millington Township, Tuscola County | | | | Giant beaver, wolf |

occurred that the Late Pleistocene and Post-Pleistocene vertebrate faunas had an entirely different aspect from those of the warm, Late Paleozoic seas and swamps.

## Pleistocene and Post-Pleistocene Vertebrate Faunas of Michigan

The Pleistocene and Post-Pleistocene vertebrates of Michigan are especially fascinating because they occupied such a new and radically different environment. The margin of the glacial ice melted backward slowly and unevenly toward the north, with many fluctuations along an irregularly receding front. The central part of the Lower Peninsula was freed first from the ice, while the Great Lakes basins remained clutched in the icy fingers of the glacier long after. The story of the physical changes that followed on the land and in the Great Lakes is told in Chapters VII and VIII, and the succession of plants that followed the ice northward is discussed in Chapter XV.

Referring to Figures XIV-12, XIV-13, and XIV-14 for the basic data, let us now look at the vertebrate faunas that appeared here after the ice withdrew.

The woodland musk ox, *Symbos cavifrons*, a tall, slender, long-legged, extinct genus and species was one of the first mammals to enter the state. Apparently, these musk oxen followed closely behind the gradually retreating edges of the Wisconsin glacial ice. Musk oxen arrived in southern Michigan about 13,200 years ago, when a spruce forest blanketed what little of the Lower Peninsula had been exposed from beneath the ice. Somewhat later, perhaps, but at least by 12,200 years ago, came those giant, hairy elephants, the Jefferson mammoths (Figs. XIV-15, 44).

As early as 11,900 years ago (Fig. XIV-14) musk oxen, mammoths, and mastodons roamed the primeval spruce forests of the southern Lower Peninsula in large numbers, no doubt accompanied by a host of other as yet unrecorded mammals of that time. The dominance of spruces at that time has been established by a recent study by Stoutamire and Benninghoff (1964). The youngest musk ox for which a radiocarbon date is available at this time is 11,100 years old. So far as the data go then, musk ox may have disappeared from the state around that time, but mammoths and mastodons persisted here. However, the forests had begun to change. Tree pollen associated with the radiocarbon dated 11,100 year old musk ox found near Kalamazoo indicates that pines had come to dominate the forest at least in that area and spruces had become far less abundant, while other trees and plants filled in the remaining niches available to plants (Semken, Miller, and Stevens, 1964).

About 10,700 years ago musk oxen apparently had left the state; at least none of that age are yet known, but mammoths and mastodons still were present. This date is based on a radiocarbon age determination made on a mastodon jaw from Gratiot County. Oltz and Kapp (1963) called this the "Smith" mastodon because it was found on the Albert Smith farm 6 miles southeast of Alma in 1909. Oltz and Kapp were able to obtain some sediment from cavities in the cranium of the skull of that mastodon and from this obtained samples of pollen. Presumably, the sediment had filled the skull cavities shortly after the animal died. Of all the tree pollen varieties present, spruce constituted 40 percent and pine 35 percent. Since pine trees produce more pollen than spruces, pine pollen may be overrepresented in the sample. Anyway spruce pollen clearly exceeded pine pollen suggesting that the spruce forest still dominated the central part of the Lower Peninsula at that time. The climate must still have been cool and moist compared with that of today. This would seem to indicate little change in forest cover from 13,200 to about 10,700 years ago in the area near Alma. Why then does pine dominate spruce in the case of the 11,100 year date at Kalamazoo? One answer may simply be that there were local differences in the forest. A second and more interesting possibility is that by 11,100 years ago pine had taken over near Kalamazoo while farther north in Gratiot County spruces still prevailed. Climatic change in Post-Pleistocene time must have been progressive from south to north and vegetational changes probably progressed northward as well.

Radiocarbon dating indicates that Jefferson mammoths lived in the state as recently as 8200 years ago. They may have become extinct here and elsewhere shortly thereafter. Mastodons, however, are known to have lived here as recently as

# Geology of Michigan

Figure XIV-14

TIME DISTRIBUTION IN MICHIGAN OF SOME LATE PLEISTOCENE AND POST-PLEISTOCENE VERTEBRATES, EARLY MAN, AND DOMINANT TREE TYPES, BASED ON CARBON[14] DATING, OR ON POLLEN ANALYSES

References in parentheses provide details on vertebrates and Early Man.
(See Chapter XV, Fig. XV-24, for details and references on forest succession.)

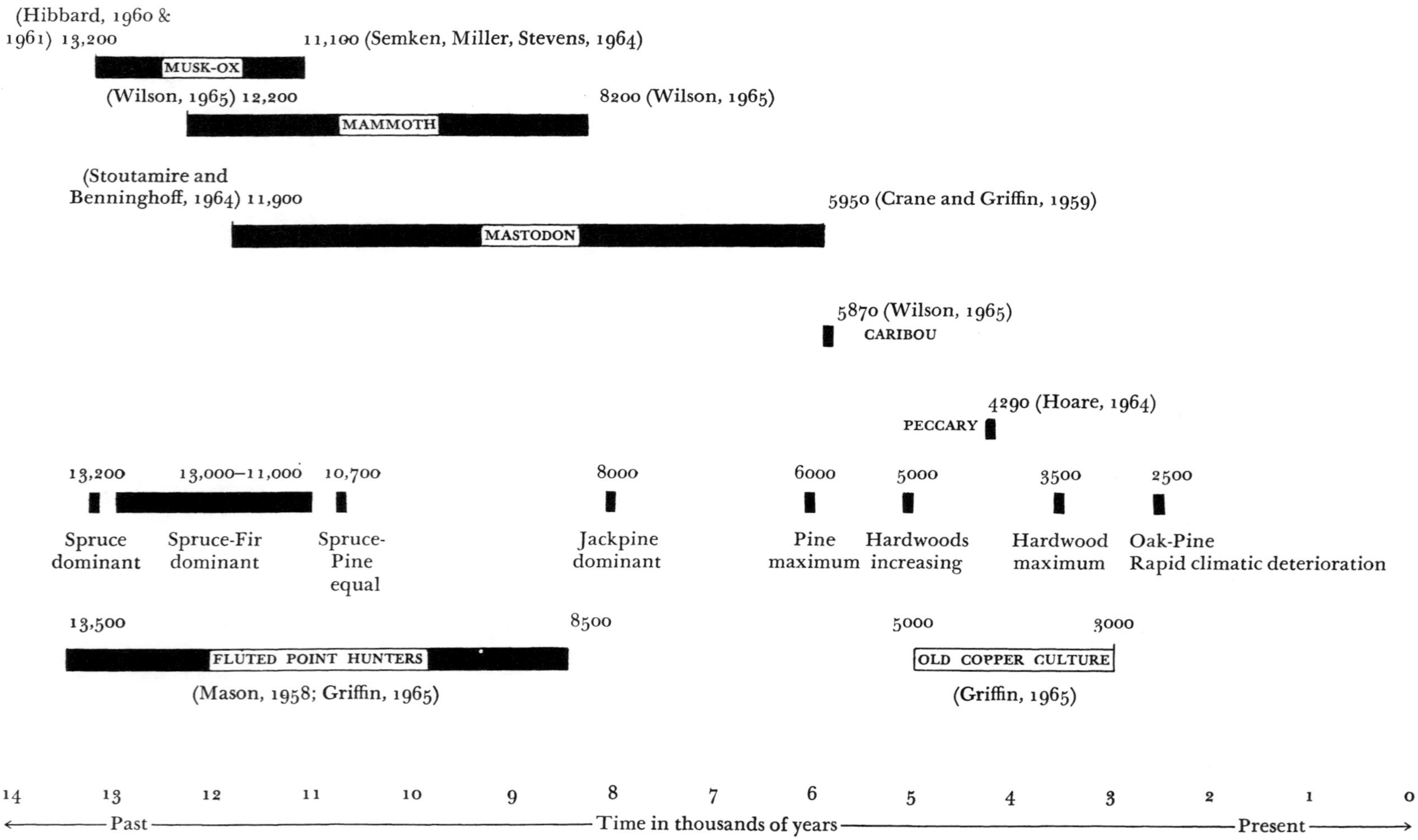

(Hibbard, 1960 & 1961) 13,200

11,100 (Semken, Miller, Stevens, 1964)

MUSK-OX

(Wilson, 1965) 12,200

8200 (Wilson, 1965)

MAMMOTH

(Stoutamire and Benninghoff, 1964) 11,900

5950 (Crane and Griffin, 1959)

MASTODON

5870 (Wilson, 1965)
CARIBOU

4290 (Hoare, 1964)
PECCARY

| 13,200 | 13,000–11,000 | 10,700 | | 8000 | | 6000 | 5000 | | 3500 | 2500 |
|---|---|---|---|---|---|---|---|---|---|---|
| Spruce dominant | Spruce-Fir dominant | Spruce-Pine equal | | Jackpine dominant | | Pine maximum | Hardwoods increasing | | Hardwood maximum | Oak-Pine Rapid climatic deterioration |

13,500

FLUTED POINT HUNTERS

(Mason, 1958; Griffin, 1965)

8500

5000 3000

OLD COPPER CULTURE

(Griffin, 1965)

14  13  12  11  10  9  8  7  6  5  4  3  2  1  0
←——Past————————————Time in thousands of years————————————Present——→

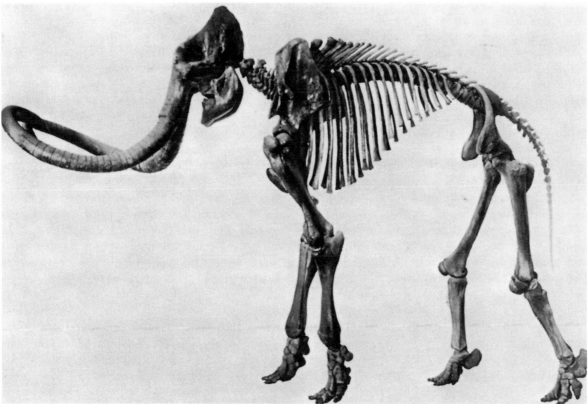

Figure XIV-15. Big bull leads herd of Late Pleistocene mammoths across a clearing (drawing by C. R. Knight). Full skeleton of *Mammuthus jeffersoni*, below, is from Indiana. Many similar but less complete specimens have been found in Michigan. (See Fig. XIV-44.) (Courtesy American Museum of Natural History, New York.)

5950 years ago, at about the same time as recorded human history had begun in Egypt!

Fossil remains of several caribou (Fig. XIV–46) have been found in the state. One from Genesee County, well south of Saginaw Bay, has been dated by radiocarbon as being 5870 years old.

Fossil peccaries (Figs. XIV–17, 45) have been found in Michigan, but so far none has been dated by radiocarbon. Fortunately, a specimen from Sandusky County, Ohio, not far south of the southeastern corner of Michigan has yielded a radiocarbon date of 4290 years Before Present (Hoare, 1964; and Michigan Memorial Phoenix Laboratory date M-1516). These pig-like animals certainly lived in Michigan about the same time as they did in Ohio.

The next to the last significant radiocarbon date shown on Figure XIV–14 is 4080 years Before Present. Although no fossil vertebrates are involved, Kapp and Kneller (1962) have shown by pollen analysis of sediments in an old terrace of the Saline River, near Milan, that by then the vegetation covering that part of southeastern Michigan was dominated by hardwoods rather than by conifers such as pine or spruce.

The last radiocarbon date, 730 years ± 250 Before Present, is little more than yesterday as geologic time is reckoned. It is the age of a fauna from Sleeping Bear Dune in Leelanau County in the northwest part of the Lower Peninsula. This fauna, described by Pruitt (1954), is essentially modern. The fossil remains that occurred in association there, although they may have been brought together by owls and did not necessarily all live in that exact spot, must have lived within a relatively short radius of the dune. Included in that fauna are the following: channel catfish, shorttailed shrew, woodchuck, eastern chipmunk, red squirrel, gray squirrel, southern flying squirrel, Canada beaver, white-footed deer mouse, muskrat, northern red-backed vole (much like a field mouse), prairie vole, pine vole, meadow vole (common field mouse), raccoon, marten, red fox, and possibly wolf.

It is unfortunate that as yet there are no reliable radiocarbon dates for many of the other interesting Post-Wisconsin animals that lived here in Michigan at one time, no doubt in some cases in company with the musk oxen, mammoths, mastodons, caribou, and peccaries. Without radiocarbon dates it is difficult to date these others exactly unless their remains are found in datable Pleistocene Great Lakes beaches or other deposits. Rather than trying to make intricate and elaborate "educated guesses" let it simply be noted that fossils of the following have been found together, hence represent animals that probably lived together (Fig. XIV–13): muskrat with moose, mammoth with mastodon, American elk or Wapiti with deer, mastodon with musk ox, deer with mastodon, and wolf with giant beaver (Figs. XIV–16, 40). Other interesting types not yet dated nor found associated with others include the black bear, modern moose, bison, and the giant extinct Scott's moose, *Cervalces* (Figs. XIV–16, 46).

Using geological and palynological evidence to arrange all these in their probable order of occupancy in the state, the first to appear clearly were the musk oxen, mammoths, and mastodons. Caribou, the modern moose and the extinct Scott's moose, deer, American elk (or Wapiti), mastodons, wolves, black bears, giant beavers, and muskrats probably lived together here at one time or another. The peccaries and bison probably were open hardwood forest and grasslands types that arrived here relatively recently.

Two very nearly Recent fossil faunas, that from Sleeping Bear Dune listed above and one from the Fenton Lake locality are of special interest for several reasons. The Fenton Lake locality was discovered and the material collected mainly through the efforts of Mr. Robert A. Hard of Fenton, Michigan. The specimens were found in material thrown up in the course of dredging operations along the east side of Fenton Lake in Genesee County. Both faunas reveal, in part, the nature of the Michigan fauna just a few hundred years before the advent of European explorers. The Sleeping Bear Dune fauna is unique in its own right because the majority of animals represented are small mammals. Relatives of at least some of these must have lived in Michigan much earlier, along with musk oxen, mammoths, and mastodons, but their remains have not been found, probably in large part because, being small, they have been overlooked whereas even a single tooth of a mammoth or mastodon can hardly be missed. The Fenton Lake locality is unusual because, in contrast with the mammalian types previously mentioned, the Fenton Lake fossils are fishes, reptiles, and birds. So far the Fenton Lake reptiles and birds are the oldest of those two major verte-

Figure XIV-16. Michigan at the close of the Pleistocene. (Drawings by the late Carleton W. Angell, courtesy C. W. Hibbard, The University of Michigan Museum of Paleontology.)

*Top*—A Scott's Moose (*Cervalces scotti*, extinct) and a heron stand in a morainic lake, startled by the crack of an aspen being felled by a giant beaver (*Castoroides ohioensis*, extinct). There are strong reasons to doubt that the beaver actually fed in this manner (see text).

*Bottom*—Wolves (*Canis lupus*) cut out a winter-weakened old bull from a small group of woodland caribou (*Caribou caribou*).

Figure XIV-17. Michigan at the close of the Pleistocene. (Drawings by the late Carleton W. Angell, courtesy of C. W. Hibbard, The University of Michigan Museum of Paleontology.)

*Top*—A band of Peccaries (*Platygonus compressus*, extinct) rests by a birch grove, unconcerned at the close approach of a herd of familiar Jefferson mammoths (*Mammuthus jeffersoni*, extinct).

*Bottom*—A herd of musk oxen seeking water from one of the Glacial Great Lakes, flushes a flock of vultures from the stranded carcass of a whale. Small horns of the musk oxen suggest that these are females of the species *Symbos cavifrons* (extinct), although some paleontologists consider them a separate genus, *Boötherium* (see Fig. XIV-1).

brate classes (Reptilia and Aves) so far found in the state. The Fenton Lake forms include the following: lake trout, whitefish, muskellunge, silver redhorse, white sucker, long-nosed sucker, quillback or carpsucker, walleye, soft-shelled turtle, a small duck (scaup or ringneck), and bald eagle. That notorious fish eater, the bald eagle, certainly must have found abundant food in the area.

Relatively recent fossils found elsewhere but of about the same age as those above include black bear, wild turkey, painted turtle, and freshwater drum fish.

Thus, we see that during the last 12,000 years in Michigan improvement of the postglacial climate was accompanied by a number of other striking environmental and faunal changes. Many of the unusual animals found here raise questions of special interest about which we may speculate beyond the limits of the fossil evidence alone.

Most of the musk oxen found here are the Woodland type *(Symbos cavifrons)* not the Barren Ground or tundra musk ox *(Ovibos moschatus)*. But the latter has been reported from Post-Pleistocene deposits in Ohio, Indiana, Pennsylvania, New York, Minnesota, Iowa, and Ontario (Semken, Miller, and Stevens, 1964, p. 823). Thus, we may expect sooner or later to find fossils of the Barren Ground or tundra musk ox here in Michigan as well. Semken, Miller, and Stevens (1964, p. 835) conclude that the Woodland musk ox, in contrast with the familiar, short-legged Barren Ground musk ox of today, was taller than a bison, though more slender. Uncertainty and disagreement exist concerning the relationships of another musk ox specimen found in Muskegon County and named *Boötherium sargenti* by Gidley (1908). A recent restudy of that specimen by Semken, Miller, and Stevens (1964, p. 824) concludes that the specimen may not be a distinct species, but simply a female of the Woodland musk ox type. Only more and better fossil finds will resolve this question. While we are thinking about relationships, anatomical comparisons and studies of blood serum indicate that although called musk "oxen," those animals probably are more closely related to sheep than to cattle! In Chapter XV, in the section on paleopalynology (fossil pollen study) the case is discussed of a Woodland musk ox from near Climax, in Kalamazoo County in southern Michigan, which was dated by radiocarbon as being 13,200 ± 600 years old. Pollen associated with that specimen indicates that a spruce forest dominated southern Michigan then. More recently (Semken, Miller, and Stevens, 1964), a Woodland musk ox, found one half mile south of Scotts, in Kalamazoo County, was dated by radiocarbon as 11,100 ± 400 years old. Fossil pollen found in the cranial cavity of that animal gives a picture of the vegetative cover of the time. About 81 percent was tree pollen, indicating that a heavy forest cover still was present, but of all the trees represented by pollen, grains of pine pollen were most abundant. The next most abundant tree pollen was oak, followed in decreasing amounts by very small percentages of balsam fir, larch, birches, elm, and maple. Not only was the musk ox from Scotts about 2000 years younger than the one found near Climax, but over the course of those 2000 years the dominant forest cover had changed from spruce to pine near Kalamazoo.

The few fossil caribou *(Rangifer)* specimens from Michigan also leave some unanswered questions. The 2 types of modern caribou are the more northern Barren Ground caribou and the more southern Woodland caribou. Neither occur in Michigan today, but Burt (1948, p. 262) records that the last living Woodland caribou died on Isle Royale about 1900. Formerly, the 2 types were classified as separate species, but they now are considered simply as separate subspecies or races of one species. The fossil specimens from Michigan consist mainly of antler fragments (Fig. XIV–46), and the antlers of the 2 living types are similar in many respects. Thus, there still is doubt which caribou lived in Michigan in the Pleistocene, although C. W. Hibbard (1951) has identified some specimens as being the Barren Ground type and Mikula (1964) identified another as the Woodland type. It seems probable that both would have been present at one time or another here as tundra gave way to forest. Both were cold-adapted animals, which indicates that the climate of Michigan formerly was different.

The giant beavers are an enigma! Hay (1914, p. 457) stated that the total length of a nearly complete skeleton from Indiana, now at Earlham College in Richmond, was 7 feet, 2 inches, measured along the curve of the back. Hay reasoned that since this specimen was not fully grown, mature adults might have reached 8 or 9 feet in length. Stirton (1965, p. 273) concluded that this would give a large *Castoroides ohioensis* adult about 8 times the bulk of an average, 60-pound modern beaver *(Castor canadensis)*, or a weight of about 480 pounds, making *Castoroides* the largest known beaver and very nearly the largest known rodent that ever existed. Its size would have been comparable with that of the largest modern Michigan black bears (Burt, 1946). Hay's (1914) studies of the feet and limbs of the Indiana specimen suggested that *Castoroides* was even shorter legged than the modern beaver and an excellent swimmer. Yet the short transverse processes on the tail vertebrae, little larger than those of the relatively small modern beaver, indicate that the tail was rather narrow. The great incisor teeth of *Castoroides* extend far out beyond the bony margins of the tooth sockets in the jaws, thus appearing to have been rather weakly supported. This, coupled with the fact that those gnawing teeth terminated in blunt, convex, gouge-shaped tips, unlike the ever-sharp, chisel-like incisors of modern beavers, led Stirton (p. 274) and others before him to the conclusion that the giant beaver was ill fitted to gnaw and fell large, hard, woody trees, hence probably had food habits more similar to those of modern muskrats. The latter feed on softer, aquatic vegetation. No fossil remains of *Castoroides* have yet been found in association with material that could be interpreted as once having been a dam. One specimen from New Knoxville, Ohio, occurred in a peaty layer in association with willow poles about 3 inches in diameter. The arrangement of these seemed to suggest that they had been part of a house 3 to 4 feet high and perhaps 8 feet square, but this interpretation is open to doubt, and the original evidence is no longer available for confirmational study. Stirton (1965, p. 274) concludes: "*Castoroides* evidently preferred lakes and ponds bordered by swamps which were common from Illinois to Ohio and farther north after retreat of the glaciers. Farther west and elsewhere they occupied similar environments as well as large streams where similar conditions prevailed. Therefore there was no necessity of their constructing dams to create suitable bodies of water for swimming and protection. The eventual reduction and disappearance of these environments probably was influential in the extinction of the giant beavers near the end of the Pleistocene." Pilleri (1961), however, made a careful study of the shape of *Castoroides*' brain. This was made possible by taking a cast of the interior of a skull from Michigan, now in The University of Michigan Museum of Paleontology collections in Ann Arbor. Pilleri concluded that the brain of the giant beaver was considerably more primitive than that of modern *Castor*, suggesting that other factors as well may have contributed to the extinction of the giant beaver.

Perhaps the most familiar of all the Post-Wisconsin mammals of Michigan were the American mastodons and Jefferson mammoths. R. L. Wilson (1965) lists the references in which are reported 174 separate discoveries of mastodons alone in Michigan. The distribution of these is shown in Figure XIV–19. The number grows each year. Wilson also lists 36 discoveries of Jefferson Mammoths (Fig. XIV–18). Many but not all of these finds are now in museums. The mastodons were essentially browsers whose teeth bore blunt cusps which worked up and down against one another in the jaw when those animals fed on twigs and leaves. The mammoths were primarily grazers whose teeth slid forward and backward past one another as they ground grassy vegetation. Figure XIV–20 shows the nature of these mechanisms. Margaret Skeels Stevens (M. A. Skeels, 1962) has recently written a comprehensive summary of the occurrences of mastodons and mammoths in Michigan with remarks on the nature of those animals. She notes that most of the fossil remains of those animals found in Michigan have occurred in swamps and bogs which filled depressions left behind by melting ice. As these depressions or kettle holes filled with sand, silt, and clay, they began to support a growth of aquatic and semiaquatic vegetation. The bog which formed as a result included plant material growing on the surface of the water and some extending down to the bottom thus forming what is known as a "quaking bog." Heavy animals venturing out on such unstable material, particularly in the winter when the surface was semifrozen, commonly broke through, became mired, and died. Mastodons and mammoths apparently sought out such places for protection in the winter time and frequently fell through, sinking into the ooze and becoming preserved. The lure of protection from the wind and abundant food served to draw them into natural

**Fossil Vertebrates**

379

Figure XIV–18. Discovery sites of the extinct Jefferson mammoth (*Mammuthus jeffersoni*) in Michigan. Solid black dots represent specimens in museum collections. (From M. A. Skeels, 1962.)

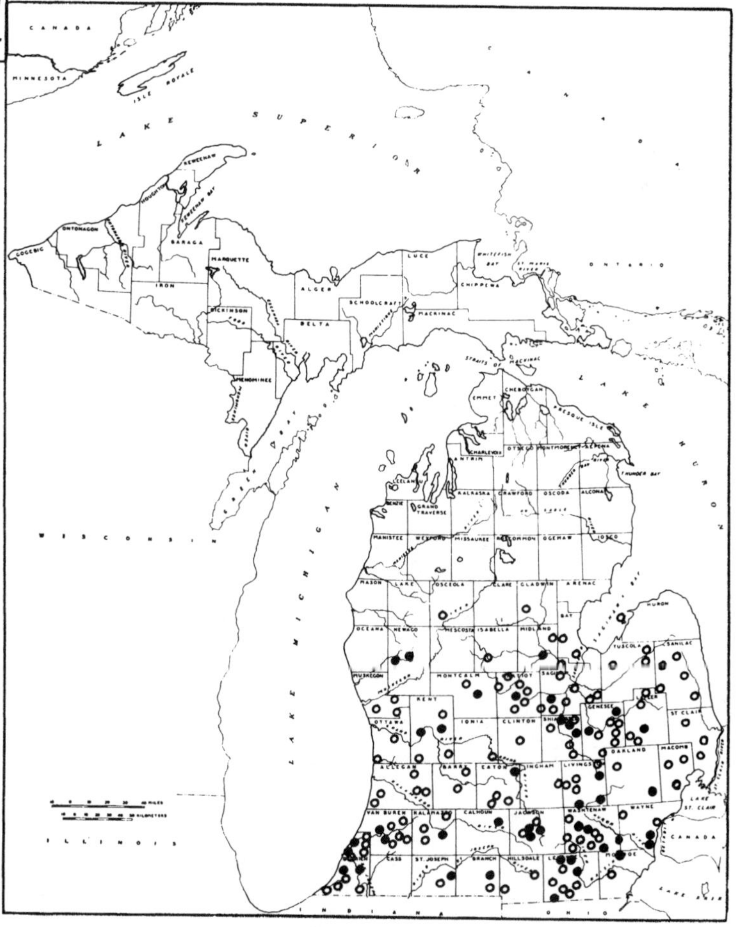

Figure XIV–19. Discovery sites of the extinct American mastodon *(Mammut americanum)* in Michigan. Solid black dots represent specimens in museum collections. (From M. A. Skeels, 1962.)

# Geology of Michigan

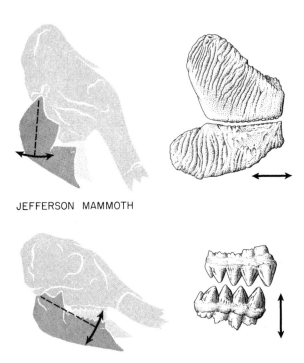

JEFFERSON MAMMOTH

AMERICAN MASTODON

Figure XIV–20. Contrast in mechanics of mammoth and mastodon jaws and teeth. Arrows show that the *grinding* action of grazing mammoth involves a fore-and-aft motion, whereas the *crushing* motion of the browsing mastodon is accomplished by an up-and-down motion of the jaw.

traps. There they stood a good chance of being preserved as fossils because they were buried immediately, before predators and scavengers could scatter their bones, and before decay could destroy them. Their fossil remains are often colored dark brown or black due to the acids produced by decaying plant matter in the bogs.

The ancestors of the American mastodon *(Mammut americanum)* reached North America from Eurasia by crossing the Bering Land Bridge approximately 18 million years ago, during the Miocene Epoch. These early ancestors were warm climate types. During the Pleistocene, the American mastodon evolved to adjust to cooler conditions and near the end of the Ice Age occupied the cold northern forests. Skeels notes that mastodons and mammoths, although often popularly called either mastodons or elephants, represent 2 very different types of animals. They were similar only in being large, with pillar-like limbs, a trunk and large tusks. They occupied different habitats, however. In specimens of the American mastodon the upper tusks are large and round while the lower tusks, if present, are small and straight. Mature individuals usually lost the lower tusks. Tusks in mastodons are enlarged upper and lower second incisor teeth. Tusks consist of ivory, a complex type of dentine, which does not lend itself to fossilization. Well-preserved tusks of mastodons or mammoths have not been found in Michigan. Those found tend to break up and the layers of ivory split apart into small pieces.

Skeels also notes that one of the differences between the American mastodon and the Jefferson mammoth *(Mammuthus jeffersoni)* is the way the tusks leave the skull. In mastodons they project horizontally from the skull, curving first outward and then inward. In mammoths the tusks leave the skull nearly vertically, curve downward and outward, and then inward to bring their hooks toward one another. Tusk size in both differs between sexes. Those of males usually are heavier, larger, and longer. Mastodons apparently used their tusks to pry off and break up branches into small pieces much as living African elephants rip limbs from trees, stand on them with both feet, break them up, and feed upon them. Skeels points out that only one tusk is used in this operation. Thus, elephants and mastodons were, as she says, either "right" or "left-handed." When both tusks are found together, one is usually shorter, often broken, and polished at the tip. The molar or cheek teeth of the American mastodon numbered 6 in each jaw top and bottom, or 24 in all. The first 3 in each jaw were milk teeth called premolars and the last or back 3 were permanent molars. Figure XIV–20 shows the construction of these teeth. The American mastodon tooth typically has transverse ridges or lophs each formed by 2 cusps. The name, mastodon, comes from the Greek and translates "nipple tooth." Mastodon teeth increase in size from the first premolar backward to the last molar.

Skeels says concerning mastodon teeth, "Tooth replacement in mastodons continued throughout much of their life. In the majority of mammals milk teeth are pushed out by the upward or downward growth of the permanent teeth that lie at their roots. In the proboscideans . . . the permanent teeth are formed in the back part of the jaw behind the milk teeth. As the milk teeth are worn down, their replacements have to move forward and upward in the lower jaws and forward and downward in the upper jaws." As the anterior teeth wore down they fell out.

Mastodons were smaller and shorter than mammoths but had larger skulls. The skull of the mastodon lacked the high crest and point which is prominent at the top and back of mammoth skulls. Mastodons presumably were forest animals that browsed on vegetation along bogs and streams. Skeels quotes a number of descriptions of material, presumably stomach contents, preserved with mastodons that have been found in other states. This material resembled crushed branches of trees, twigs of trees broken into 2-inch lengths, finely divided leaves, and various other vegetable material.

Skeels notes that the Jefferson mammoth reached North America from Eurasia later than the mastodon, apparently in Middle Pleistocene time. The ancestral stock gave rise to the Imperial mammoth and the Columbian mammoth (both southern kinds) and also to the Jefferson mammoth. The well-known wooly mammoth, a tundra dweller, arrived in northwestern North America later in the Pleistocene, but did not reach Michigan. Mammoths are most closely related to living elephants, and their teeth are much alike. Mammoth teeth consist of alternating ridges of cement, dentine, and enamel ridge plates. When these teeth slide backward and forward across or past one another they do a very efficient job of grinding, thus mammoths and their elephant relatives are grazers, feeding in part on grass. The premolar teeth are smaller than the molars and have fewer plates. The third molar is the largest tooth with the most plates. The Jefferson mammoth usually has 25 plates in the upper molar and 24 in the third lower molar.

The most complete specimen of a mammoth found in Michigan is a partial skeleton (number 8298) in the Michigan State University Museum in East Lansing. Unfortunately, no skull or pelvis was found with that individual.

Mammoths, like the living elephants, fed on a variety of vegetation. They probably lived mainly on the open grasslands rather than in deeply forested areas. The Jefferson mammoth apparently was less abundant than the mastodon in Michigan, perhaps because meadow areas were limited. If they preferred grassy upland habitats they may seldom have ventured into bogs where they could have become entombed. Undecayed stomach contents taken from the frozen carcasses of wooly mammoths found in Siberia show that these closely related animals were primarily grass eaters. They may have grazed in the summer and supplemented this diet with leaf browse and twigs in the wintertime.

There has been much speculation concerning the reasons why mastodons and mammoths became extinct, the latter disappearing from Michigan about 6000 years ago and the mammoths somewhat earlier. The reasons really are not known and may have involved a number of factors rather than a single cause. It has been suggested that large predators or Indians killed off the giant proboscideans. This seems unlikely as the sole cause, however, because biological and anthropological studies indicate that it is very rare or never that predators or primitive peoples kill off their entire food supply. Usually, the number of predators drops as the food supply diminishes and then picks up again with an increase in the animals used for prey. Predation may have been a contributing factor, however. Another suggestion is that as climate changed and the cold temperature zones shifted northward the mastodons and mammoths were unable to follow this climate northward because of barriers formed by the Great Lakes and other glacial lakes to the north. This explanation fails to recognize the probability that both types of animals lived on either side of the Great Lakes in New York and Ontario or in Wisconsin and Minnesota and hence could well have moved northward from there. Another suggestion is that as climate changed, and vegetation along with it, the food supply of the browsing and grazing mastodons and mammoths disappeared. This seems like the most reasonable explanation although some combination of those given may be the real reason. It rarely is possible to pinpoint the exact cause for the disappearance of any ancient and now extinct animal group.

C. W. Hibbard (1951), in a very popular article entitled "Animal Life in Michigan During the Ice Age," drew an illuminating picture of the Pleistocene and Post-Pleistocene conditions here in the state. With his permission we quote extensively from that article as follows: "As the moun-

### Geology of Michigan

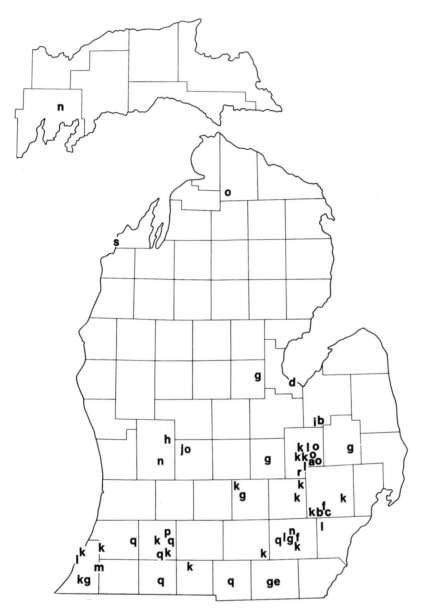

Figure XIV–21. Discovery sites in Michigan of Pleistocene age vertebrate fossils *other than* mastodons, mammoths, and marine mammals. (Modified from R. L. Wilson, 1966.)

a—Largemouth bass
b—Snapping turtle
c—Painted turtle
d—False map turtle
e—Wild turkey
g—Muskrat
f—Extinct giant beaver
h—Black bear
i—Wolf (or dog)
j—Extinct peccary
k—American elk
l—Whitetail deer
m—Extinct Scott's moose
n—Moose
o—Caribou
p—Bison (or cow)
q—Extinct woodland muskox
r—Fenton Lake locality, see Figure XIV–13 for list of animals found there
s—Sleeping Bear Dune locality, see Figure XIV–13 for list of animals found there

tain and continental glaciers began to develop and areas became covered with snow and ice, those animals that had lived in these regions shifted their ranges away from the ice fields by population spread. The rate of development and advance of a continental glacier is slow enough to permit the plants and snails to reproduce, grow to maturity, and reproduce again and again so that there is a very gradual population shift into new areas in front of the glacier." Thus he shows the slow rate at which glacial and associated biological changes occur.

Continuing, he says:

It is doubtful whether any mammals lived on the ice fields that covered the state during any of the four major glaciations, at the time the glaciers reached their maximum southern limits. The recession of the ice was probably just as slow as its advance, since the plants and animals were able to follow its slow retreat by an extension of their ranges northward by population spread.

Only a few of the vertebrates (fishes, amphibians, reptiles, birds, and mammals) that are known to live in Michigan at the present time have been found in Pleistocene deposits, though these forms possess a long geologic history. The reason that so little is known of most of them is that they are small and their remains go unnoticed in the deposits. As far as is known the Pleistocene fishes, amphibians, and reptiles differed but little from the present day forms. . . .

The ancestor of the American Mastodon . . . reached North America from Eurasia by the Alaskan-Siberian land bridge . . . approximately 10 million years before the Ice Age. The American mastodon browsed along the edges of the wooded streams, especially in eastern North America. . . .

Bands of peccaries fed along the edge of the forests and meadows on roots, tender plants, fruits, nuts, and smaller animal life. . . .

New forms, such as the elephants (mammoths), musk ox, bison, etc., arrived from Eurasia via the Alaskan-Siberian land bridge. . . .

Herds of the Barren Ground Caribou which fed on the tundra, were indistinguishable from those living today. . . .

Adjoining the tundra on its southern border were the forested and intermittent grassland areas with their successive faunas working their way northward. The hairy American Mastodon lived in large numbers along the wooded areas and browsed on the shrubs and trees. It is safe to say that there is at least one mastodon skeleton, or part of one, in every fair-sized bog in southern Michigan. . . .

The giant moose-like *Cervalces,* a late arrival from Asia or the northwest part of our continent, fed along the edges of the lakes with the American mastodon. . . . Its bones have probably been confused many times with those of the moose, which it closely resembles. . . .

Foraging along the edge of the lakes and forests were the Woodland Musk Ox *(Symbos)* and the Woodland Caribou. . . .

The Jefferson Mammoth, a large hairy elephant, stood about eight feet at the shoulder, being slightly larger than the American Mastodon. It was a distant cousin of the Wooly Mammoth that lived to the northwest in the tundra, and it grazed in the meadows around the lakes and between the timbered areas. Because of the absence of extensive grasslands, it was not as abundant as the American mastodon. . . .

The peccaries, a distant relative of the domestic pig, roved in small bands along the edge of the hardwood timbers and the intermittent grasslands. They stood about 26 inches high at the shoulder and weighed about 100 pounds. They were larger than their living relatives, the Collared Peccary of Texas, New Mexico, and Arizona, or the White-Lipped Peccary of Central and South America.

The occurrences of fossil whales and walruses in Michigan were reviewed by Handley (1953). These remains throw a strange and fascinating sidelight on the history of Post-Pleistocene life here in the state. As Hibbard (1951) comments, it is highly improbable that Indians would have carried such large objects as whale ribs and vertebrae long distances into Michigan and then have had sufficient knowledge of the Great Lakes geology to have buried those remains precisely in glacial lake beach deposits. The best explanation for their presence is that the whales and walruses themselves found their way into the Great Lakes region during postglacial times when the major rivers flowing from those lakes to the sea were swollen by glacial meltwater and hence sufficiently deep to float even a whale. Details of the Great Lakes history may be studied in Chapter VIII. Let us simply review those aspects of glacial Great Lakes history pertinent to this problem.

Handley (1953, p. 253), states that specimen UMMP 14101, in the collections of The University of Michigan Museum of Paleontology, is the single rib of a finbacked whale *(Balaenoptera)* from a cellar excavation ten miles northeast of Mt. Morris, in Genesee County. The rib was standing vertically in loose sand of a beach formed by Glacial Lake Arkona. Figure XIV–22 shows the position of Glacial Lake Arkona and the location of the discovery site relative to the approximate shore of that lake. The finbacked whale could have gained entry by coming up the Mississippi and Illinois River drainages into Glacial Lake Chicago and thence up the Grand River Valley from Lake Chicago into Lake Arkona.

Handley (1953, p. 253) says with regard to the sperm whale *(Physeter),* that it is represented by a lumbar vertebra and two ribs, UMMP 14102. The specimen was found in a swamp in the northeast corner of Lenawee County near a beach deposit formed by Glacial Lake Whittlesey. Figure XIV–20 shows that essentially the same possible route as described above was available into Glacial Lake Whittlesey at a later stage.

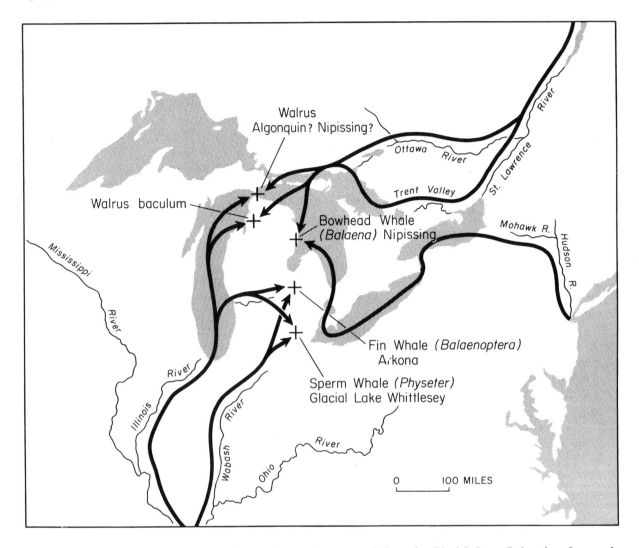

Figure XIV-22. Some possible routes of entry for marine mammals into the Glacial Great Lakes, based on estimated age and lake stage of beach deposits at or near fossil discovery sites. See Figures VIII-3-12 for maps of Glacial Great Lakes shore lines, outlets, and relationships of the latter to modern drainage features. Map does *not* imply that the large mammals swam up the modern rivers as they exist today, but rather that during the earlier stages of the Great Lakes certain connections were complete and the glacial lake water outlets deeper and broader. There is some evidence, for example, of a direct sea level connection from the Atlantic, down the St. Lawrence Valley into the lakes at one stage, which would have given access much the same as into Hudson Bay today.

Handley notes that the bowhead whale *(Balaena)*, represented by specimen UMMP 11008, a single rib, came from a schoolhouse excavation at Oscoda, in Iosco County. The rib was found 5 feet below the surface in sands of a Glacial Lake Nipissing beach. Figure XIV-20 shows that 3 routes were possible through which the Bowhead Whale might have entered the Great Lakes region. One was via the Mississippi River as before. Another was via the St. Lawrence, through lakes Ontario, Erie, St. Clair, and Huron. A third and more northern route was across Canada via the St. Lawrence River and then up what are now valleys of the Ottawa, Matawa, and French rivers across Canada into Georgian Bay.

Two walrus specimens from Michigan also are reported by Handley. One is a baculum (penis bone) found in a gravel pit about 7 miles northwest of Gaylord, Michigan. The specimen, number 490, is in the Museum of Anthropology and

Archaeology of The University of Michigan. The second walrus specimen is the front part of a skull, now in The University of Michigan Museum of Paleontology, catalog number 32453. The locality data for this specimen are also somewhat in doubt, but it is reported to have come from a glacial lake beach deposit on Mackinac Island. The lake beach may have been either Algonquin or Nipissing in age. Unfortunately, the original data were lost in a fire. The tusks are missing, but appear to have been removed after discovery of the skull. Some short, parallel grooves on the snout region may be human carvings, but this is uncertain.

The geological data on the walruses being incomplete, it is still an open question how they entered the state, but obviously they may have followed one of the same routes available to the whales.

This brings us nearly to the close of the story of Michigan fossil vertebrates. We have come a long way from the time of Middle Devonian seas, nearly 400 million years ago, when giant placoderms and primitive sharks cruised the warm seas covering the state. We have passed through the critical times of Pleistocene glaciation. When does man enter the picture?

Figure XIV-23. Some typical examples of fluted points left by Early Man in Michigan. (Courtesy James B. Griffin, The University of Michigan Museum of Anthropology.)

# Geology of Michigan

### Prehistoric Methods of Obtaining Copper Were Simple.

The hard rock enclosing copper nuggets was shattered by means of fire and crushed with heavy hammerstones. Tens of thousands of these hammers, some weighing over 50 pounds, have been seen in and near the ancient mines.

Loose chunks known as float copper were picked up; they were torn from their matrix by Pleistocene glaciers and transported south as far as northern Indiana, Illinois, and Iowa

Nuggets projecting from outcrops of Keeweenawan lava were "quarried."

On Isle Royale and on the southern shores of Lake Superior, thousands of open pits averaging 3 to 4 feet in depth were scooped out of the soil-covered copper-bearing rock. Little tunneling was attempted, for the simple stone tools of the miners were inadequate for penetrating the dense, massive lava beds.

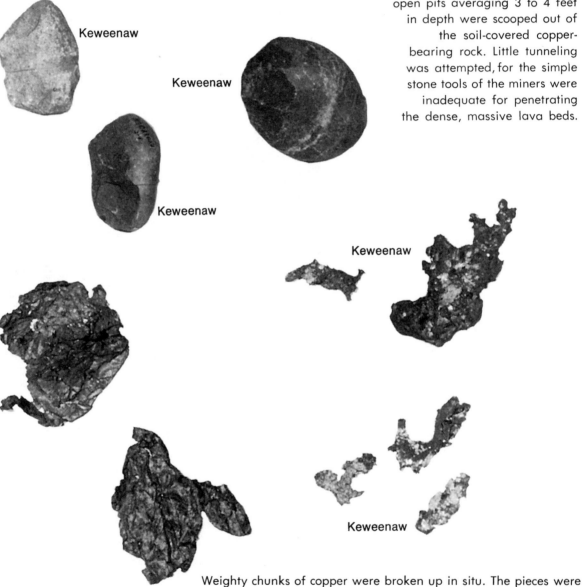

Weighty chunks of copper were broken up in situ. The pieces were flattened out but not shaped into objects near the mines.

Figure XIV-24. Michigan Paleo-Indian "Old Copper" Culture. (Courtesy The University of Michigan Exhibits Museum and the Museum of Anthropology.)

**Fossil Vertebrates**

387

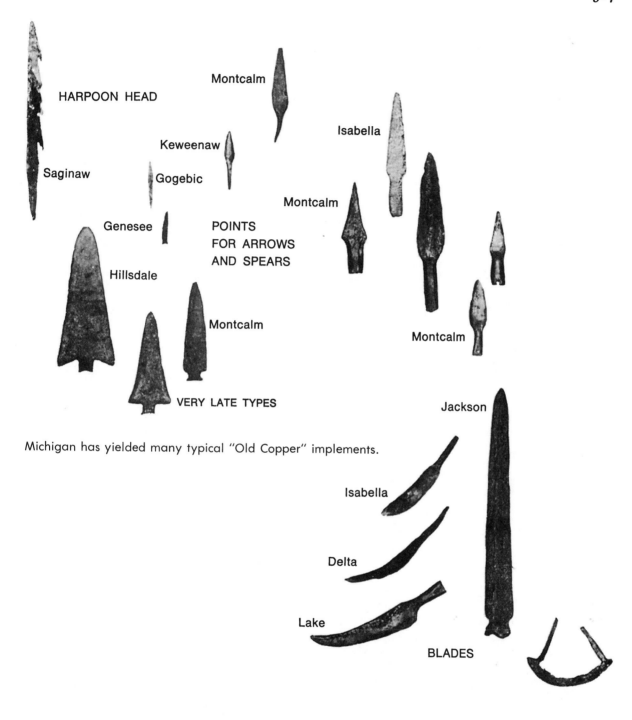

Michigan has yielded many typical "Old Copper" implements.

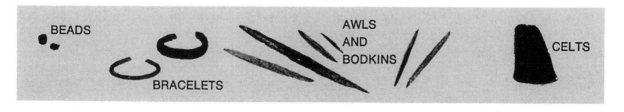

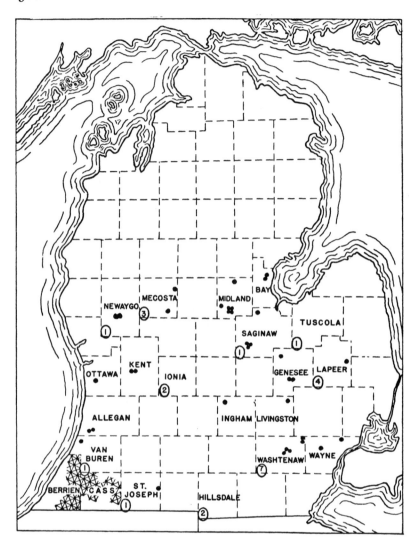

Figure XIV-25. Some "fluted point" finds in Michigan. Cross-hatched area in Berrien and Cass counties indicates region where exceptionally large number of points have been found. Circled numbers indicate how many additional points for which source data are lacking have been found in those counties. (From R. J. Mason, 1958.)

## Early Man in Michigan

Man arrived in North America long after his first appearance in Eurasia and Africa. He reached this continent via the Bering Strait land bridge, and traveled down through Alaska and Canada just as did some of his mammalian contemporaries. Griffin (1965, p. 656) states that the technology and general way of life of these prehistoric people were derived, with some cultural modifications and losses, from the Advanced Paleolithic hunters of central Siberia. Eventually, these paleo-Indians spread entirely over the New World.

Mason (1958) and Griffin (1965) have reviewed the problem of man's first entry into Michigan. Let us close this chapter and bring man into the picture by summarizing some of their conclusions.

Evidence for the existence of Early Man in the state is based on finds of what are called *fluted points* (Figs. XIV-23, 24, and 25). Fluted points of the Clovis and Folsom types have been found at Early Man sites elsewhere in North America as well. Griffin believes that the earliest primitive men in unglaciated southeastern United States were the Fluted Point Hunters, who may have been there as early as 14,000 years ago. Fluted points are common there and appear to be older than in the northeast. He suggests that these people, along with the plants and animals of their time, moved into northeastern United States close behind the retreating margin of the last Pleistocene (Wisconsin) ice sheet.

# Fossil Vertebrates

When Mason wrote (1958), 115 fluted points with reliable data had been found on the surface in Michigan. He believed that although some of the fluted points are well made, they are inferior in workmanship to Folsom points and more like the Clovis type.

Mason states (p. 17): "Because the Wisconsin ice sheet in Michigan or, more exactly, the ice lobes of the adjacent Great Lakes basins, completely covered the state during the Cary Maximum [mid-Wisconsin, see Chapter VIII] and extended southward into Ohio, Indiana, and Illinois, it is clear that the maximum age of all fluted blades found in Michigan is limited by the presence of the early Cary ice. The areas in which these projectile points have been discovered were not ice-free and, hence, were not open to human penetration until after the Cary recession began." The Cary recession began prior to 14,000 years ago.

Discussing finds in Berrien and Cass counties in extreme southwestern Michigan, Mason says: "Estimates of the earliest possible entry of man into Michigan must be largely based upon finds made in this area because of the pattern of glacial retreat in the state. . . . The fluted points in this area, then, cannot antedate Middle Cary features but may be as old as Late Cary. A maximum date for penetration of Early Man into Berrien County . . . circa 13,500 B.P. seems most probable."

According to Griffin, no fluted points have been found in areas occupied by Glacial Lake Algonquin, but they have been found close to its margin and to the margins of earlier and higher lakes such as Lundy and Warren (Chapter VIII). This seems to indicate that the Fluted Point Hunters occupied Michigan before and until Glacial Lake Algonquin time, but not thereafter, that is, during the retreat of the glacial ice from the position of the Port Huron Moraine, but before the lake waters dropped to the Two Creeks Low Water Stage (Chapter VIII). Thus, Griffin (1965, p. 660, 665) dates the time of man's first arrival in Michigan at 11,000 B.C. (about 13,000 Before Present), which is close to Mason's estimate.

The date for the end of occupation of the northeast by the Fluted Point Hunters, and their replacement by people of the Early Archaic Culture, is not yet well established, but Griffin (1965, p. 660) tentatively places it at about 9000 B.C. (about 11,000 B.P.) because the Early Archaic sites clearly are associated with Glacial Lake Algonquin and post-Algonquin shorelines.

Thus, it appears that prehistoric man, manufacturing fluted points, arrived in Michigan somewhere between about 13,500 and 13,000 years ago and remained here until about 11,000 years ago. All this is based upon artifacts because no skeletal remains of Fluted Point Hunters have been found here. The oldest Indian skeletal remain yet found in Michigan is the Early Archaic "Union Lake Skull" from a peat bog in Oakland County (Griffin, 1965, p. 662). Pollen from the ear and nasal cavities of that specimen was compared with the radiocarbon dated pollen spectrum from sediments in nearby Sodon Lake and compares most closely with a pollen level there dating 5050 ± 200 B.C. (about 7000 B.P.).

According to Mason, the manufacturers of the fluted points were in an Upper Paleolithic cultural stage and lived primarily by hunting. It is uncertain what they hunted because none of the artifacts was found associated with any fossil mammals of that time. Our review of Pleistocene vertebrates from Michigan established that musk oxen, mammoths, and mastodons were present here then. Probably also present were moose, elk, deer, caribou, and giant beaver. However, we do not yet know which of these animals the Fluted Point Hunters pursued or were able to kill.

The remainder of man's history in Michigan must be left to anthropologists and archeologists. Bibliographies will be found in the articles by Mason and Griffin referred to above.

## Detailed List of Fossil Vertebrates from Michigan

The review of the geological history of vertebrates in Michigan presented in the preceding section is based upon documentary proof in the form of specimens of vertebrate fossils that have been found in the state. The following list of those specimens is set up in the standard form of the formal biological classification of animals. The nature of such a classification and its subdivisions was explained near the beginning of Chapter XIII. Every fossil species, genus, or larger taxonomic group listed as once having occurred in Michigan is supported by evidence in the form of a specimen in some museum. Future studies of those specimens might show that some have been misidentified, thus altering the list slightly but not destroying the evidence itself. New discoveries in the future may add to the list. In broad outline, however, it is correct though still incomplete. Most of the fossil specimens are in The University of Michigan Museum of Paleontology in Ann Arbor. The letters UMMP will signify that fact. If a specimen is preserved in another museum this will be specially noted. This is the first comprehensive summary of all fossil vertebrates from Michigan to be published in recent years, so catalog numbers are included for all pre-Pleistocene aged specimens. These mostly are Middle and Late Paleozoic fishes. Catalog numbers are not given for the Late Pleistocene and Post-Pleistocene specimens because a recent paper (1967) by Richard L. Wilson, entitled, "The Pleistocene Vertebrates of Michigan," contains all the catalog numbers of such "Ice Age" and "post-Ice Age" specimens and, in fact, served as one of the important sources of information for this chapter. Anyone interested in verifying the occurrence of a fossil vertebrate in Michigan could, then, consult further with the curators of the museums in which all the specimens are housed. The list also will provide the interested collector with information about the kinds of rocks, rock ages, and locations in which fossil vertebrates have been found in the past.

Specimens in private hands are not listed because there is no guarantee that they will be adequately preserved for future study, nor that important data on the place of their discovery is or will be properly recorded. Such specimens do, in fact, often become lost or separated from their data.

The classification of vertebrates only shows those found as fossils in Michigan. There are many other types living here now but not found as fossils, and many fossil types not found here.

Phylum CHORDATA. Animals with a notochord, pharyngeal (gill) slits and bars, and a dorsally located nerve cord at some stage in their development.

    Subphylum VERTEBRATA. Chordates with a segmented, longitudinal axial support, the spinal column, composed of either bone or cartilage or both, the segments being called vertebrae (singular, vertebra).

Class AGNATHA (Fig. XIV-4). Vertebrates without jaws. Includes the living lampreys and hagfishes and the extinct ostracoderms. UMMP 43936, possibly an ostracoderm scale impression in rock, scale itself lost. Pictured in Hussey (1947, Fig. 80). From Ordovician–Black River Formation 3 miles south of Bony Falls along Escanaba River, Delta County.

Class PLACODERMI. Fish-like forms with true jaws but with unmodified first gill slit and arch. Paired fins. Often with bony dermal armor (Fig. XIV-5).

    Order ARTHRODIRA. Joint-necked, heavily armored predators.

*Arctolepis,* species undetermined.
Middle Devonian Dundee Limestone near Trenton: UMMP 14320.

*Titanichthys?,* sp. undet.
Middle Devonian Dundee Limestone, Sibley Quarry, Wyandotte: UMMP 26114.

*Trachosteus? clarkii.* Upper Devonian Antrim Shale, 1 mile north of Norwood: UMMP 18206.

*Protitanichthys rockportensis.* (Fig. XIV-26). Middle Devonian (Traverse Group) Rockport Quarry Limestone, Kelly Island Limestone Co. Quarry, Rockport, Alpena County: UMMP 3842, 4065, 12975–12986, 13044, 13046, 13049, 13052–13054, 13536.

*Dinichthys,* sp. undet.
Late Devonian, Antrim Shale, Squaw Bay 4 miles south of Alpena on U.S. 23: UMMP 15432.

Middle Devonian Traverse Group.
    1—Alpena Limestone, Alkali Quarry, Alpena: UMMP 16152.
    2—Rockport Quarry Limestone, Rockport Quarry, Rockport: UMMP 16156.

*Holonema rugosum?* Middle Devonian Devonian (Traverse Group) Koehler Limestone, Onaway Stone Quarry, north edge of Onaway, Presque Isle County: UMMP 46647, 46648.

*Holonema rugosum?* Middle Devonian Traverse Group.

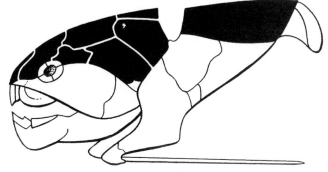

Figure XIV-26. A marine, Middle Devonian age, joint-necked arthrodiran placoderm, *Protitanichthys rockportensis,* about the size of a large sturgeon. Cephalic shield (left) and median dorsal plate (right), from Alpena County. Specimens v–12980 and v–13049 in The University of Michigan Museum of Paleontology. The two plates are from different individuals. Five centimeter long scale for size. Drawing (after Heintz) shows head region of related form, *Dinichthys,* for location of parts. See Figures XIV–9 and 10 for full restorations of related forms.

1—Rockport Quarry Limestone, Rockport Quarry, Alpena County: UMMP 3898.

2—Killians member of Genshaw Formation, French Road near Long Lake, near Rockport, Alpena County: UMMP 3899.

*Holonema,* sp. undet. Middle Devonian Traverse Group, Rockport Quarry Limestone, Rockport Quarry, Alpena County: UMMP 3843, 12987–12989, 12991, 12992, 13040, 13043, 13048, 13050, 13051, 15439.

Holonemiid, plate of unidentified genus, Middle Devonian, old Crawfords Quarry (now U.S. Steel Co. Rogers City Quarry, much expanded): UMMP 3130.

Holonemiid, plate of unidentified genus, Middle Devonian Traverse Group, Gravel Point Formation, exposures along Lake Michigan shore at South Point (also called Gravel Point and Pine River Point), Little Traverse Bay, Charlevoix County: UMMP 3129.

*Aspidichthys clavatus.* Late Devonian Antrim Shale, shore of Grand Traverse Bay near Norwood, Charlevoix County: UMMP 3127.

*Mylostoma,* species undet. Middle Devonian Traverse Group, Rockport Quarry Limestone, Rockport, Alpena County: UMMP 13612.

*Dinomylostoma,* species undet. (Fig. XIV–28). Middle Devonian Traverse Group, Rockport Quarry Limestone, Rockport,

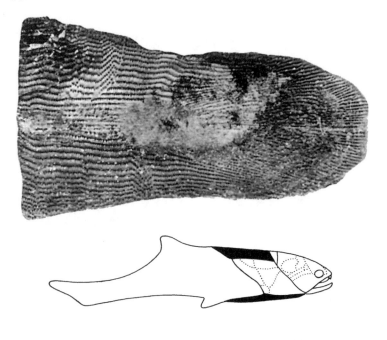

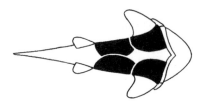

Figure xiv-27. Plates from the dorsal and ventral shields of *Holonema farrowi*, an arthrodiran placoderm from a quarry in the Middle Devonian age Koehler Limestone near Onaway, Michigan. Darkened areas on sketches show position of plates. Note scale in inches for size. Specimens in The University of Michigan Museum of Paleontology.

Figure xiv-28. Jaw (gnathal) elements of arthrodiran placoderms. Sketch shows location of parts. All specimens in The University of Michigan Museum of Paleontology.

*Top*—Upper gnathal element of *Dinomylostoma*. Middle Devonian age Traverse Group near Rockport, Alpena County.

*Bottom left*—Same as above. Left lower gnathal. Note worn, semipolished grinding surface surrounded by toothlike tubercles. (5-cm. scale is reduced.)

*Bottom right*—Tuberculated lower gnathal of *Trachosteus,* a much smaller arthrodire. Much enlarged scale is in millimeters. Late Devonian age Antrim Shale, 1 mile north of Norwood, Charlevoix County. Specimen still embedded in rock matrix.

Alpena County: UMMP 3046, 12974, 13041, 13042, 13056, 13148, 16158.

*Ptyctodus,* species undet. (Fig. xiv-29). Middle Devonian.
1—Dundee Limestone?, Trenton, Wayne County: UMMP 14321.
2—Traverse Group, "quarry floor just west of Petoskey," no other data, Emmet County: UMMP 14712.
3—Potter Farm Formation?, Traverse Group, "old Wamer's Brickyard southwest of Alpena," Alpena County: UMMP 21718.
4—Traverse Group, loose block in Alkali Quarry, Alpena, Alpena County: UMMP 16157.
5—Traverse Group, Potter Farm Formation, west edge of Alpena Cemetery, Alpena, Alpena County: UMMP 21817.
6—Traverse Group, Rockport Quarry Limestone, Rockport, Alpena County: UMMP 13045.
7—Traverse Group, Thunder Bay Limestone, Partridge Point, Alpena County: UMMP 3023.

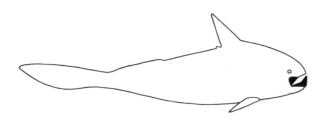

Figure xiv-29. Ptyctodont arthrodiran placoderms from Devonian rocks of Michigan. Specimens, still partially embedded in limestone, are in The University of Michigan Museum of Paleontology. Graduated scales are in millimeters.

*Top*—Flattened and polished upper, crushing, gnathal element. Sibley Quarry, Trenton, Michigan.

*Bottom left*—Sharp-edged left lower gnathal. Partridge Point, Alpena County.

*Bottom right*—Left lower gnathal. South side Cheboygan Lake, Cheboygan County. Sketch showing location of jaw parts is modified from Watson.

8—Traverse Group, Bell Shale, dump of Bell Shale, Rogers City, Presque Isle County: UMMP 14460.

*Eczematolepis,* species undet. Middle Devonian Traverse Group, Locality 650 of Winchell Survey, Alpena, Alpena County: UMMP 14374.

Arthrodire, genus and species undet. Middle Devonian Petoskey Limestone of Traverse Group, abandoned Kegomic (Mud Lake) Quarry 1 mile east of Bay View, Emmet County: UMMP 44646.

Order ANTIARCHI. Dorsoventrally flattened, armored, with anterior pair of movable fins.

*Bothryolepis,* species undet. (Fig. xiv-30). Middle Devonian Traverse Group, Genshaw Formation, near Posen, Presque Isle County: UMMP 4169.

Order ACANTHODII. Streamlined, elongate forms with 2 or more pairs of lateral fins, commonly with fin spines.

*Onychodus sigmoides.* Middle Devonian Dundee Limestone, Monroe County: UMMP 22006.

*Onychodus,* species undet. (Fig. xiv-31). Middle Devonian Dundee Limestone, Sibley Quarry, Wyandotte, Wayne County: UMMP 26113.

*Onychodus?,* species undet. Middle Devonian Traverse Group, South Point (Gravel Point), shore of Little Traverse Bay, Charlevoix County: UMMP 14370.

*Gyracanthus,* species undet. Late Mississippian, Bayport Limestone, west side of Saginaw Bay, Arenac County: UMMP 13663.

**Fossil Vertebrates**

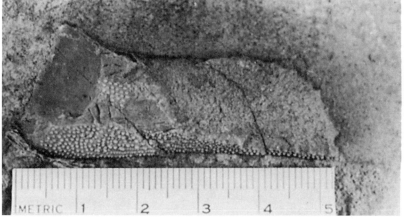

Figure XIV-30. *Bothriolepis?*, an antiarchan placoderm from rocks of Devonian age. Centimeter scales for size.

*Top*—Nearly complete specimen in a nodule from Quebec, Canada, to show nature of the whole animal.

*Center*—Lateral plate from near Posen, Michigan (University of Michigan Museum of Paleontology specimen v-4169.)

*Bottom*—Restoration, after Patten and Goodrich, to show location of the plate in a whole animal.

# Geology of Michigan

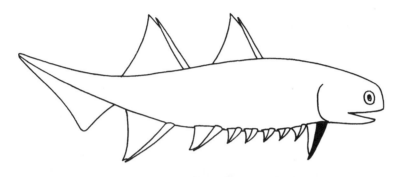

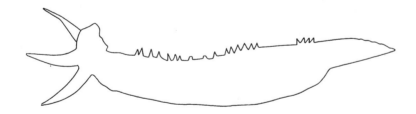

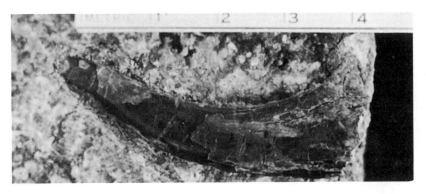

Figure XIV–31. Fossil remains of acanthodian fishes from Devonian age rocks of Michigan. Specimens in The University of Michigan Museum of Paleontology. Restorations after Traquair and Watson, and after Newberry.

*Top*—Fin spine of *Machaeracanthus*, Monroe County.

*Center*—Lower jaw with teeth of *Onychodus*, from Sibley Quarry near Wyandotte.

*Bottom*—Symphysial tooth from front of jaw of *Onychodus sigmoides*, Monroe County.

**Fossil Vertebrates**

397

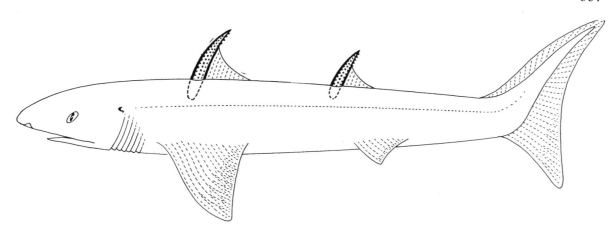

Figure XIV-32. Ctenacanth ("comb-spined") shark remains from Middle Mississippian age rocks of Michigan. The genus *Ctenacanthus*. Specimens in The University of Michigan Museum of Paleontology. All specimens reduced to one-half natural size.

*Left side*—An anterodorsal spine fragment, and some coprolites (fecal pellets) containing bone fragments, found in association in the Michigan Formation in the Bestwall Gypsum Company mine in Grand Rapids.

*Right center*—A mold (impression) of spine from Marshall Sandstone near Marshall.

*Right*—A mold of a spine associated with fossil marine clams. (*Palaeoneilo sulcatina*) from Marshall Sandstone in old Stony Point Quarry near Hanover.

Middle Devonian Traverse Group
  1—Rogers City Limestone, Rogers City Quarry, Rogers City, Presque Isle County: UMMP 3130.
  2—Gravel Point Formation, South Point (Gravel Point), Little Traverse Bay, Charlevoix County: UMMP 1329.

*Machaeracanthus,* species undet. Middle Devonian Dundee Limestone, Monroe County: UMMP 3521.

Placoderm? fragments of undetermined order, genus or species.
  1—Middle Devonian Dundee Limestone, Sibley Quarry, Wyandotte, Wayne County: UMMP 26111, 26112.
  2—Middle Devonian Traverse Group, Koehler Limestone, Onaway Stone Quarry, north edge of Onaway, Presque Isle County: UMMP 47691, 47692.

Class CHONDRICHTHYES. Fishes with cartilaginous internal skeletons. Includes sharks, skates, rays, and some extinct groups. Figure XIV–6.

Order CLADOSELACHII. Early, primitive, unspecialized, ancestral sharks.

*Cladodus romingeri.* Early Mississippian, lower part of "Marshall Formation," near Grindstone City, Huron County: UMMP 13613.

*Ctenacanthus,* resembling *C. varians,* associated with abundant coprolites (Fig. XIV–32). Late Mississippian, Michigan Formation, "triple gyp" horizon, Bestwall Gypsum Company mine, Grand Rapids, Kent County: UMMP 45738.

*Ctenacanthus,* sp. undet. Early Mississippian, lower member (Marshall Sandstone) of Marshall Formation.
  1—Stony Point Quarry near Hanover southwest of Jackson, Jackson County: UMMP 45739.
  2—Abandoned quarry, south side of Marshall, Calhoun County: UMMP 23895.
Middle Devonian Traverse Group, Rockport Quarry Limestone, Rockport Quarry, Alpena County: UMMP 13147.

Order PLEURACANTHODII. Extinct, freshwater sharks.

*Pleuracanthus,* resembling *P. arcuatus* (Fig. XIV–33). Later part of Early Pennsylvanian, Saginaw Formation, old St. Charles-Garfield Coal Mine at Eastwood (Garfield), Saginaw County: UMMP 33463.

Order SELACHII. Advanced sharks including most modern living forms (Fig. XIV–34).

*Orodus,* resembling *O. ramosus.* Early Mississippian, "Marshall" Sandstone, Grindstone City, Huron County: UMMP 8296.

*Edestus minor.* Latter part of Early Pennsylvanian, Saginaw Formation. Found by Dr. W. A. Kelly of MSU in 1932 along "Six Mile Creek"; no other data: UMMP 14411.

Order BRADYODONTI. Extinct sharks with teeth flattened for shell crushing.

*Helodus modestus?* (Fig. XIV–35). Late Mississippian, Bayport Limestone, Grand Rapids, Kent County: UMMP 21719.

*Psephodus,* species undet. Late Mississippian, Bayport Limestone (Point au Gres Limestone), near Grand Rapids, Kent County: UMMP 23741, 23742.

*Oracanthus?,* species undet. Middle Devonian Traverse Group, Norway Point Formation, Four Mile Dam about 3.5 miles northwest of Alpena, Alpena County: UMMP 23495.

"A cochliodont tooth," Family Cochliodontidae. Late Mississippian Bayport (Point au Gres) Limestone, Charity Island, Arenac County: UMMP 23743.

*Psammodus,* sp. undet. Late Mississippian, Bayport (Point au Gres) Limestone, Grand Rapids, Kent County: UMMP 23739, 23740, 21816.

Ichthyodorulites. Paleozoic fish spines of uncertain relationships, probably placoderms.

*Acondylacanthus gracillimus.* Middle Devonian Traverse Group, Rockport

# Fossil Vertebrates

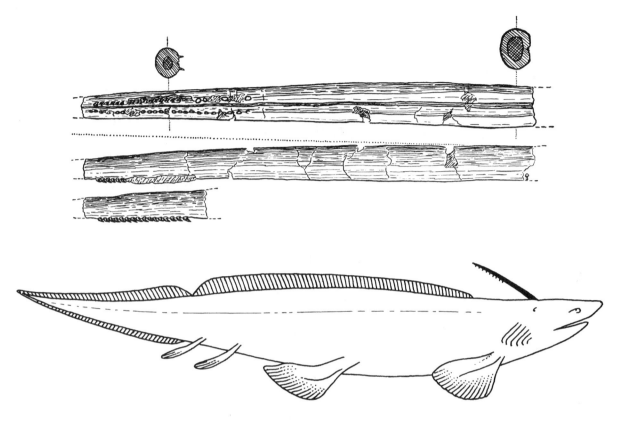

Figure XIV-33. *Pleuracanthus*, a freshwater shark. Head spine from Pennsylvanian age Saginaw Formation near Saginaw, Michigan. Side, back, and cross section views of spine, about one half natural size. Note double row of strongly curved denticles on back edge. Spine may have served as a cut-water. Specimen in The University of Michigan Museum of Paleontology.

Figure XIV-34. Single tooth of an extinct, edestid shark of Pennsylvanian age, still embedded in rock matrix. Shiny enamel preserved near tip shows serrated tooth edges. Specimen in The University of Michigan Museum of Paleontology.

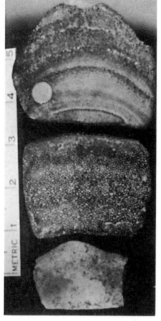
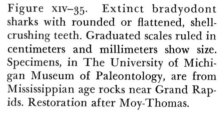

Figure XIV-35. Extinct bradyodont sharks with rounded or flattened, shell-crushing teeth. Graduated scales ruled in centimeters and millimeters show size. Specimens, in The University of Michigan Museum of Paleontology, are from Mississippian age rocks near Grand Rapids. Restoration after Moy-Thomas.

*Top left*—PSEPHODUS.

*Bottom left*—HELODUS.

*Right*—PSAMMODUS, upper and lower surfaces.

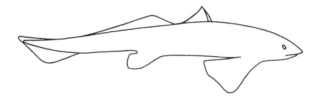

Quarry Limestone, Rockport, Alpena County: UMMP 13047.
Spine. Middle Devonian Dundee Limestone, Sibley Quarry, Wyandotte, Wayne County: UMMP 26523.

Class OSTEICHTHYES. Fishes with partially or entirely bony skeleton.

Subclass SARCOPTERYGII (Fig. XIV-8). Fleshy finned fishes with internal nostrils and usually with functional lungs.

Order DIPNOI (Fig. XIV-36). Lungfishes with teeth modified for shell crushing. Burrow fillings collected by Dr. Robert L. Carroll (McGill University, Montreal, Canada) may represent this group. Specimens, now in Redpath Museum, McGill University (Carroll, 1965) are from bottom of Grand Ledge Clay Products Company Quarry, just west of Grand River in Section 34, Eagle Township, Clinton County. Rocks containing burrows are Early Pennsylvanian Saginaw Formation, probably "Cycle A" of W. A. Kelly (1936). Burrows resemble those found elsewhere formed by *Gnathorhiza*.

(*Note*: From this point on in the classification, specimens are from Late Pleistocene or Post-Pleistocene deposits in Michigan; see R. L. Wilson, 1967, for complete data including location of specimens and catalog numbers.)

Subclass ACTINOPTERYGII (Fig. XIV-7). Bony fishes in which the portion of the fins external to the body margin is supported by numerous slim, flexible, closely spaced fin rays. The "rayfinned fishes."

Order TELEOSTEI (Fig. XIV-37). Evolutionarily advanced ray-finned fishes. The teleosts. (Modern distribution data from Hubbs and Lagler, 1958).

Family SALMONIDAE (Salmon, Trout, Char).

*Salvelinus namaycush*—Common Lake Trout. Fenton Lake Locality, East side of Fenton Lake, NE¼, Sec. 14, T. 5 N.—R. 6 E., Genesee County. Found by Robert A. Hard of Fenton and donated to the University of Michigan Museum of Paleontology (UMMP). Found today (Hubbs and Lagler, 1958) in main Great Lakes and a few cold northern inland lakes.

Family COREGONIDAE (Whitefishes).

*Coregonus clupeaformis*—Lake Whitefish. Fenton Lake Locality. Found and donated to UMMP by R. A. Hard. Occurs in Michigan today.

Family ESOCIDAE (Pickerel, Pike, Muskellunge).

*Esox masquinongy*—Muskellunge. Fenton Lake Locality. Found and donated to UMMP by R. A. Hard. Occurs in Michigan today but becoming rare, especially in southern Michigan.

Family CATOSTOMIDAE (Suckers).

*Moxostoma*, comparing closely with *M. anisurum*—Silver Redhorse. Fenton Lake Locality. Found and donated to UMMP by R. A. Hard. Lives in Great Lakes Region today (Hubbs and Lagler, 1958).

*Catostomus*, comparing closely with *C. commersonnii*—White Sucker. Fenton Lake Locality. Found and donated to UMMP by R. A. Hard. Widespread in Great Lakes Region today (Hubbs and Lagler, 1958).

*Catostomus catostomus*—Longnose Sucker. Fenton Lake Locality. Found and donated to UMMP by R. A. Hard. Found in main Great Lakes, and in inland lakes in the Lake Superior drainage today; spawns in streams (Hubbs and Lagler, 1958).

*Carpoides*, comparing closely with *C. cyprinus*—Quillback or Carpsucker. Fenton Lake Locality. Lives in Great Lake Region, except for Lake Superior, today (Hubbs and Lagler, 1958).

Family ICTALURIDAE (North American freshwater Catfish).

*Ictalurus punctatus*—Channel Catfish. Sleeping Bear Dune Locality, Leelanau County (Pruitt, 1954). Specimen in UMMP.

Family PERCIDAE (Perches, Darters).

*Stizostedion vitreum*—Walleye. Fenton Lake Locality. Found and donated to UMMP by R. A. Hard. Common in Great Lakes and many inland lakes and

Figure XIV-36. Sand filling of former lungfish burrow in Early Pennsylvanian age Saginaw Formation, Grand Ledge Clay Products Company north quarry, near Grand Ledge, Michigan. (Discovery and photo by Dr. R. L. Carroll, Redpath Museum, McGill University, Montreal, Quebec, Canada.)

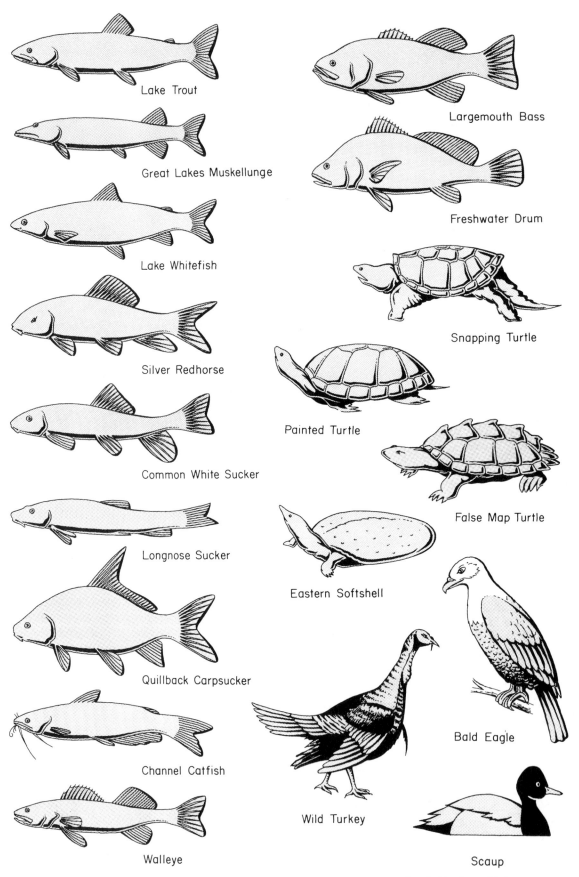

Figure xiv–37. Fish, reptiles, and birds so far found represented by fossil remains in the Late Pleistocene or post-glacial deposits of Michigan.

rivers of region today (Hubbs and Lagler, 1958).

Family CENTRARCHIDAE (Bass, Sunfishes).

*Micropterus salmoides* — Largemouth Bass. From old lake deposit 60 yards south of Cook Road, east side of U.S. 10 Expressway off northbound lane, Genesee County. Specimen in UMMP. Still common throughout Great Lakes Region today.

Family SCIAENIDAE (Drums).

*Aplodinotus grunniens* — Freshwater Drum. Reported by Hubbs (1940) from Cheboygan County. Specimen in UMMP. Occurs in southern Lake Michigan, Saginaw Bay, and Lake Erie today.

Class AMPHIBIA (Frogs, toads, salamanders, many extinct types). None yet found in Michigan fossil record.

Class REPTILIA (Reptiles) (Fig. XIV-37).

Family CHELYDRIDAE (Snapping Turtles). *Chelydra serpentina*—Snapping Turtle. One specimen, in UMMP, found and donated by R. A. Hard. From beach sand below peat in drainage ditch, Millington Twp., Tuscola County. Another specimen, now in the UMMP, found by Robert J. Lutz in peat at a depth of 15 feet, half way between 13 and 14 Mile Roads in Farmington Township, Oakland County. Snapping turtles are common in Michigan today.

Family TESTUDINIDAE (Aquatic, semiterrestrial, and terrestrial Turtles).

*Chrysemys,* species undet.—Painted Turtle. Specimen in UMMP. Same locality as for snapping turtle above. Common in Michigan today.

*Graptemys pseudogeographica*—False Map Turtle. Specimen in UMMP. Found and donated by R. Hard in Section 2, T. 14 N.—R. 5 E., Bay County. Not found in Michigan today but still occurs in Ohio, Indiana, and several states to the west.

Family TRIONYCHIDAE (Softshell Turtles).

*Trionyx,* comparing closely with *T. spinifer,* the eastern spiney softshell turtle. Found and donated to UMMP by R. A. Hard. Fenton Lake Locality. Occurs over the southern half of Michigan today.

Class AVES. Birds (Fig. XIV-37).
Family ANATIDAE (Ducks, Geese, Swans). *Aythya,* species undetermined. Some resemblance to both *A. affinis* (lesser scaup) and *A. collaris* (ringnecked duck). A small duck. Found and donated to UMMP by R. A. Hard. From Fenton Lake Locality. Lesser scaup common in Michigan today, ringnecked duck occurs here but less common.

Family MELEAGRIDAE (Turkeys).

*Meleagris gallopavo*—Wild Turkey. Found and donated to UMMP by Gerald Larson of Adrian. Found in material dredged from Wolf Creek Valley, Adrian Township, Lenawee County. Occurred in state in historic time. Recently replanted and thriving in many areas of Southern Michigan.

Family BUTEONIDAE Broadwinged Hawks and Eagles.

*Haliaeetus leucocephalus*—Bald Eagle. Found and donated to UMMP by R. A. Hard. From Fenton Lake Locality. Occurred in southern part of state in historic times.

Class MAMMALIA. Warm-blooded vertebrates with hair, that give birth to their young alive (except for the monotremes or egg-laying mammals) and suckle them by means of mammary glands. Many extinct and 3 living subclasses. The living subclasses are the Prototheria (or Monotremata, Platypuses, Spiny Echidnas), the Metatheria (or Marsupialia, Opossums, Kangaroos, etc.), and the Eutheria (or Placentalia, the placental mammals). There are no fossil or living monotremes in the state. Marsupials are represented in Michigan by the living Opossum who

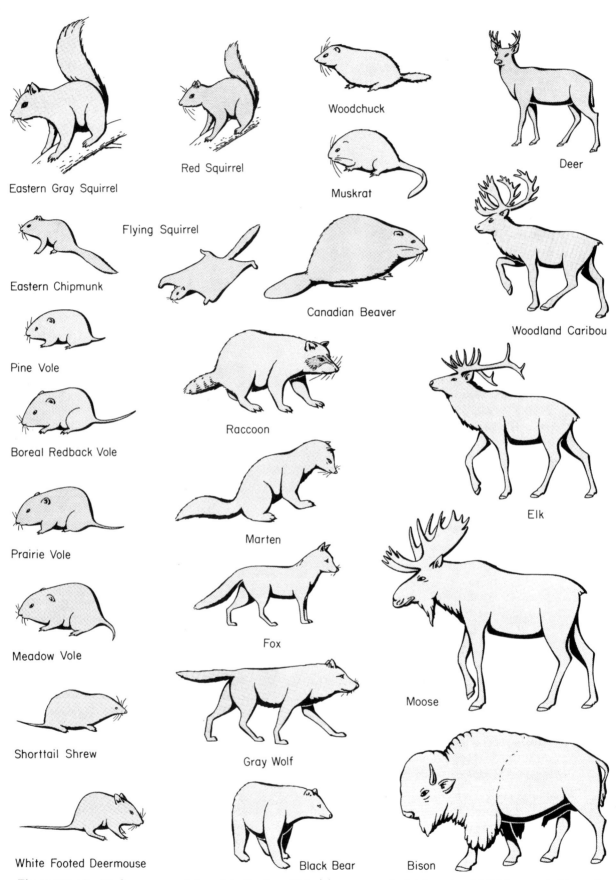

Figure XIV-38. Modern mammals found in the Late Pleistocene or postglacial deposits of Michigan and either still living in the state, or recorded as having been seen here within historic time. Small forms on the left, intermediate in center, and large types on right. (Modified from Burt and Grossenheider, 1952, *A Field Guide to the Mammals*, permission of Houghton Mifflin Company.)

**Fossil Vertebrates**

Figure xiv-39. Mammals found in the Late Pleistocene or postglacial deposits of Michigan and either now extinct or at least never recorded from the state within historic time. (In part modified from Burt and Grossenheider, 1952, *A Field Guide to the Mammals*, permission of Houghton Mifflin Company.)

reached here only within the last few decades; no fossil marsupial remains have been found here. Eutherians or placental mammals, both fossil and living, are common in the state and the fossil finds are listed below. Data on modern distributions are from Burt and Grossenheider (1952), and Burt (1948).

Subclass EUTHERIA (Placentalia; placental mammals) (Figs. XIV-38, 39).

Order INSECTIVORA (Shrews, Moles, and related forms).

Family SORICIDAE (Shrews).

*Blarina brevicauda*—Shorttail Shrew. Found and donated to UMMP by Jerome S. Miller. Reported and identified by Pruitt (1954). From Sleeping Bear Dune Locality (see text). Occurs in Michigan today.

Order RODENTIA (Rodents).

Family SCIURIDAE (Squirrels, Woodchucks, Chipmunks).

*Marmota monax*—Woodchuck. Specimen in UMMP. From Sleeping Bear Dune Locality (Pruitt, 1954). Occurs in Michigan today.

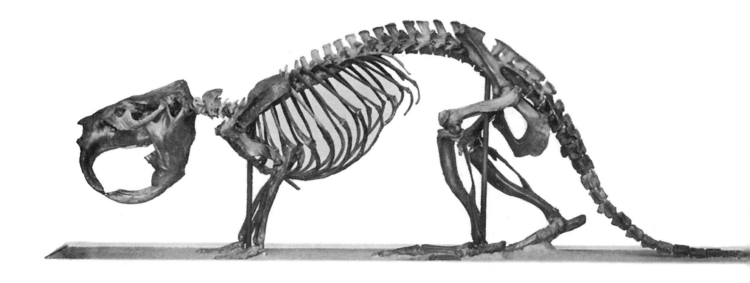

Figure XIV-40. Giant beaver, *Castoroides ohioensis*.

*Top*—Complete fossil skeleton, about 7 feet long. (Courtesy of the Field Museum of Natural History, Chicago.)

*Bottom*—Skull from Late Pleistocene of Michigan with modern beaver (*Castor*) skull to same scale for size comparison. Ruler is 5 inches long. Michigan specimen in the Museum of Paleontology. (Photo courtesy C. W. Hibbard.)

**Fossil Vertebrates**

*Tamias striatus*—Eastern Chipmunk. Specimen in UMMP. From Sleeping Bear Dune Locality (Pruitt, 1954). Occurs in Michigan today.

*Tamiasciurus hudsonicus*—Red Squirrel. Specimen in UMMP. From Sleeping Bear Dune Locality (Pruitt, 1954). Occurs in Michigan today.

*Sciurus carolinensis*—Eastern Gray Squirrel. Specimen in UMMP. From Sleeping Bear Dune Locality (Pruitt, 1954). Occurs in Michigan today.

*Glaucomys volans*—Southern Flying Squirrel. Specimen in UMMP. From Sleeping Bear Dune Locality (Pruitt, 1954). Occurs in Michigan today.

Family CASTORIDAE (Beavers).

*Castor canadensis*—Beaver. Specimen in UMMP. From Sleeping Bear Dune Locality (Pruitt, 1954). Occurs in Michigan today.

*Castoroides ohioensis* (Fig. XIV–40). Extinct Giant beaver. Pilleri (1961) reports that the brain of the Giant beaver was much more primitive than that of the modern Canadian beaver. Seven specimens from Michigan are as shown below:
- 1 skull, in The University of Michigan Museum of Paleontology; found at Ann Arbor, Washtenaw County (Wood, 1914).
- 1 lower jaw, in the UMMP; found at Owosso, Shiawassee County (Wood, 1914).
- 3 specimens, from Berrien, Lenawee, and Lapeer counties (Hay, 1923).
- 1 front part of skull, now in Michigan State University Museum at East Lansing; found in Ingham County in 1892 but not donated to MSU until 1962 (R. L. Wilson, 1967).
- 1 right upper incisor tooth, in the UMMP; collected near Midland by George F. Cristin in 1951 (R. L. Wilson, 1967).

Family CRICETIDAE (Mice and Rats).

*Peromyscus*, comparing closely with *P. leucopus*, the White-footed Deermouse. Two specimens, in the UMMP, collected and donated by Miller and Pruitt (Pruitt, 1954). Deermice are common in Michigan today.

*Ondatra zibethica*—Muskrat. Four specimens from Michigan as follows (all in the UMMP):
- 1 from Sleeping Bear Dune Locality (Pruitt, 1954).
- 1 collected by R. J. Lutz and donated by Richard Alde; from a depth of 15 feet in a peat deposit in Farmington Township, Oakland County. Found associated with remains of Snapping turtle. The 2 are common living associates in Michigan today.
- 1 collected and donated by Robert A. Hard of Fenton, Michigan; found at the Fenton Lake Locality in association with the many kinds of fossil fishes listed above for that locality.
- 1 collected and donated by Gerald Schultz; from a depth of 2 feet below the surface, 0.7 mile north of Ann Arbor, Salem Township, Washtenaw County.

Muskrats are common throughout the state today.

*Clethrionomys gapperi*—Boreal Redback Vole. Specimen in UMMP. Collected and donated by Jerome Miller; from Sleeping Bear Dune Locality, Leelanau County (Pruitt, 1954). Occurs in same area in Michigan today.

*Microtus ochrogaster*—Prairie Vole. Specimen in UMMP. From Sleeping Bear Dune Locality. This particular specimen was subjected to Carbon 14 analysis and gave a radiogenic age of about 730 years B.P. (Chapters II and XIV). This date establishes the approximate "absolute" geological age of the Sleeping Bear Dune Fauna (Chapter XIV). Occurs only in southwesternmost Michigan today.

*Microtus pennsylvanicus*—Meadow Vole (commonly called a field "mouse"). Voles (see those listed above as well) are mouse-like animals, commonly called "mice." Voles, however, have very short, almost hidden ears. Their teeth, too, differ from those of mice, being more closely similar to those of muskrats, a fact which a specialist can use to distinguish their fossil remains from those of mice. This specimen, in the UMMP, is from the Sleeping Bear Dune Locality in Leelanau County. Meadow voles are common in Michigan today.

*Pitymys pinetorum*—Pine Vole. Specimen in UMMP. Collected by Miller from Sleeping Bear Dune Locality (Pruitt, 1954). Common in Michigan today, but in deciduous forest, not pines.

Order CARNIVORA (Dogs, wolves, foxes, bears, cats, etc.)

Family URSIDAE (Bears).

*Ursus americanus*—Black Bear. Lower jaw, collected by Wayne Mol of Greenville, Michigan, and donated to the Grand Rapids Public Museum. From a marl layer in a drainage ditch excavation through a swamp in the northeast corner of Big Wabasis Lake, Kent County, and identified by Dr. Weldon D. Frankforter, Director, GRPM (Frankforter, 1966). Black bears are common in Michigan today.

Family PROCYONIDAE (Raccoons).
*Procyon lotor*—Raccoon. Specimen in UMMP. From Sleeping Bear Dune Locality (Pruitt, 1954). Common in Michigan today.

Family MUSTELIDAE (Weasels, marten, fisher, skunks, otter, wolverines, badger).

*Martes americana*—Marten. Specimens in UMMP. From Sleeping Bear Dune Locality (Pruitt, 1954). Lived in Michigan until late 1920's but none reported here recently (Burt, 1948).

Family CANIDAE (Dogs, wolves, foxes).

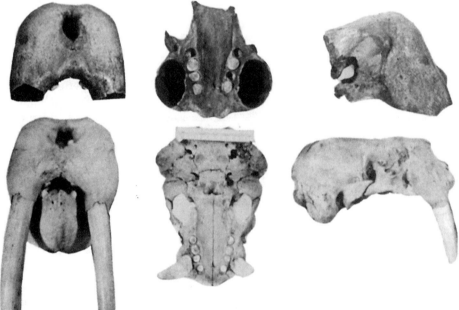

Figure XIV-41. *Top row*—Front, palatal and side views (left to right) of portion of fossil walrus skull, lacking tusks, found on Mackinac Island. Specimen in The University of Michigan Museum of Paleontology.
*Bottom row*—Similar views of modern walrus skulls for comparison. Large bull, with lower jaw included, in lower left. Younger animal center and right. Ruler on skull in center is 5 inches long.

# Fossil Vertebrates

*Vulpes fulva*—Red Fox. Specimen in UMMP. From Sleeping Bear Dune Locality (Pruitt, 1954). Common in Michigan today.

*Canis lupus*—Gray Wolf. Specimen in UMMP. Collected and donated by R. A. Hard from between a peat layer and beach sand of old lake in Tuscola County. Found in association with Snapping turtle remains. Found only in the Upper Peninsula of Michigan today.

*Canis,* species undet., wolf or dog. Specimen in UMMP. From Sleeping Bear Dune Locality (Pruitt, 1954).

Order PINNEPEDIA (Seals, Sea Lions, Walruses).

Family ODOBAENIDAE (Walruses).

*Odobenus,* species undet. (Fig. XIV–41).
1 specimen, a baculum, found in 1914 by Ezra Smith. Locality an old gravel pit about 7 miles northwest of Gaylord, Otsego County. Specimen now in The University of Michigan Museum of Archeology and Anthropology, Ann Arbor. Reported by Hinsdale, 1925.
1 specimen, the anterior part of a skull, reportedly found on Mackinac Island, probably in an elevated glacial Lake Nipissing or Algonquin beach. Specimen (No. 32453) now in the UMMP. Reported in detail and recently studied by Handley (1953).

Order CETACEA (Whales). For detailed reports see Hussey (1930) and Handley (1953).

Family PHYSETERIDAE (Sperm Whales).
*Physeter,* species undet. (Fig. XIV–42). Sperm Whale. One lumbar vertebra and 2 ribs, now in UMMP. Specimens dug from swamp in northeast corner of Lenawee County. Nearby beach sands are of Glacial Lake Whittlesey. Sperm whales widespread in Atlantic and Pacific oceans today.

Family BALAENOPTERIDAE (Finback Whales).

*Balaenoptera,* species undet. Finback

Figure XIV–42. Fossil whale remains from Late Pleistocene or postglacial shore line deposits of ancestral Great Lakes. All specimens in The University of Michigan Museum of Paleontology.

*Small rib*—V–11008, *Balaena?* (possibly a bowhead whale), found at Oscoda in Iosco County.

*Large rib*—V–14101, *Balaenoptera* (finback whale), found near Thedford Center northeast of Mt. Morris, Genesee County.

*Inset lower right*—A vertebra, V–14102, *Physter* (sperm whale), found in Lenawee County. Ruler is 6 inches long.

**Geology of Michigan**

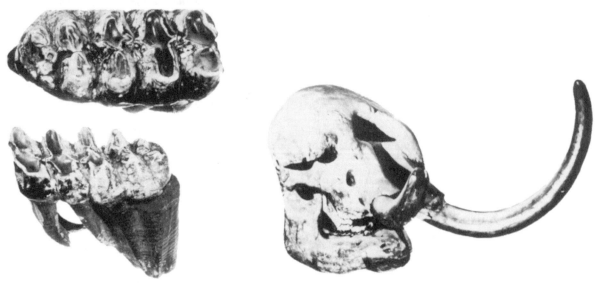

Figure xiv–43. American mastodon *(Mammut americanum)* skeleton found near Owosso, Michigan. On left, crown and side views of molar teeth, about 1 inches long. Same skull, seen from right front, shows how right tusk was broken and worn smooth again during life of animal. Also note enormous nasal pit and groove on forehead where trunk muscles attached. (Photos courtesy C. W. Hibbard, The University of Michigan Museum of Paleontology, the Exhibits Museum, and the University of Michigan News Service.)

Whale, or Rorqual. Single rib found in cellar excavation 10 miles northeast of Mt. Morris, Genesee County. Specimen now in UMMP, was found standing vertically, buried in beach sand of Glacial Lake Arkona. Genus occurs in Atlantic and Pacific oceans today.

Family BALAENIDAE (Baleen Whales).

?*Balaena*, generic identification uncertain. Possibly a Bowhead Whale. Specimen in UMMP. Discovered in 1928, 5 feet below surface, in Glacial Lake Nipissing beach sand, during course of excavation of a southwest corner of what was then the schoolhouse at Oscoda, Iosco County. Occur in polar and subpolar seas near ice edge today.

Order PROBOSCIDEA (Mastodons, Mammoths, Elephants)

Family MAMMUTHIDAE (Mastodons).

*Mammut americanum* (Figs. XIV-1, 19, 43). Extinct American Mastodon. These are the most commonly reported fossil mammals from the Late Pleistocene and Post-Pleistocene deposits of Michigan. A recent summary by R. L. Wilson (1967) of reported discoveries lists 174 separate occurrences of mastodons. These are too numerous to list here but are shown on the accompanying map and can be found in lists published by MacAlpine (1940), Skeels (1962), Hatt (1963), and Wilson (1967). The first was discovered here in 1839 and at least one is found nearly every year. Probably every bog and swamp in the state contains the remains of this giant animal and many will no doubt be discovered in the future. It is probable that not all finds have been reported and not all reported finds have been donated to museums. There are, nevertheless, many specimens in museums in the state. Good exhibits may be seen in The University of Michigan Museum of Paleontology in Ann Arbor, the Michigan State University Museum in East Lansing, the Grand Rapids Public Museum, and the Cranbrook Institute of Science in Bloomfield Hills.

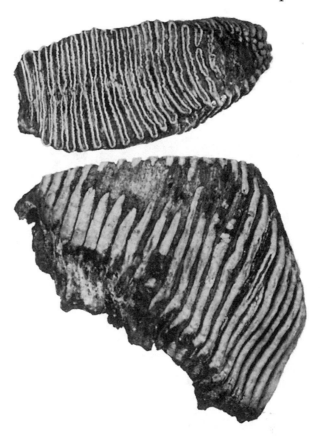

Figure XIV-44. Late Pleistocene age Jefferson mammoth (*Mammuthus jeffersoni*) molar tooth, about 7 inches long, from Michigan. Note closely spaced enamel plates (light colored) and cementum (dark colored) in side view at bottom. Crown view above shows how wear on the occlusal surface cuts through to interior of the enamel plates, exposing dark-colored dentine within, and thus forming alternating ridges of cement-enamel-dentine-enamel-cement. The sequence is repeated many times along the wearing surface. This is a grinding type of tooth. Compare it with the simpler, blunt-cusped, crushing type of tooth found in mastodons (Fig. XIV-43; also see Fig. XIV-15). (Photo courtesy C. W. Hibbard, The University of Michigan Museum of Paleontology.)

Family ELEPHANTIDAE (Mammoths, Elephants).

*Mammuthus jeffersoni* (Figs. XIV-15, 18, 44). Extinct Jefferson Mammoth. Less abundant than mastodons. Thirty-six discoveries are listed in Skeels (1962), Oltz and Kapp (1963), Hatt (1963), and Wilson (1967).

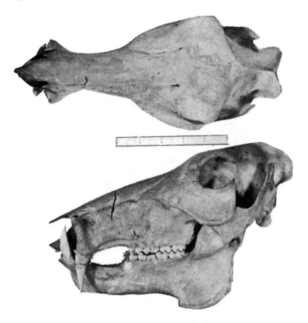

Figure XIV-45. Skull of *Platygonus compressus,* an extinct fossil peccary, 1 of 5 specimens collected by L. N. Tuttle in 1887 from Post-Pleistocene deposits in a peat bog near Belding, in Ionia County. Graduated scale is 5 inches long. Specimen v-7325 in The University of Michigan Museum of Paleontology.

Order ARTIODACTYLA (The cloven-hoofed mammals).

Family TAYASSUIDAE (Peccaries).

*Platygonus compressus* (Figs. XIV-17, 45). Extinct Peccary. Individual remains of 5 of these pig-like animals were discovered by L. N. Tuttle in 1877, in a peat bog near Belding, Ionia County. Some of these remains are still preserved in the UMMP. A skull of one was described by Wagner (1903). Hay (1923) also described their occurrence. Hoare (1964) reported a Carbon[14] radiogenic date of 4290± 150 years B.P. for a specimen found just south of Michigan in Sandusky County, Ohio. This specimen is significant because it gives the best available evidence of when peccaries may also have occurred in nearby Michigan.

Family CERVIDAE (Fig. XIV-46). (Deer, Elk, Moose, Caribou, etc.).

*Cervus canadensis*—American "Elk" or Wapiti. Some prefer more properly to call these animals wapiti because of a confusion of names. What we Americans later called an "elk" had long been known in Europe as a stag. The term elk, in Europe, is applied to what we call a moose. (This is why Norwegian elkhounds actually hunt "moose" by our usage.) Wilson (1967) lists 25 fossil specimens from Michigan. Examples may be found in the UMMP, The University of Michigan Museum of Zoology, and the Michigan State University Museum. Specimens have been found in Berrien, Genesee, Branch, Ingham, Jackson, Kalamazoo, Leelanau, Livingston, Oakland, VanBuren, and Washtenaw counties. Native elk were found in Michigan well into historic time. According to Burt (1948) the last native individual in the state died in 1870. Elk have since been reintroduced into the state and are common again in some areas, especially around Vanderbilt in Otsego County. Unfortunately none of the fossil specimens has yet been dated by Carbon[14]. Possibly few of the finds are of great antiquity, although one specimen found in association with the extinct giant beaver *(Castoroides)* must be quite ancient (Hay, 1923). Another specimen, found in association with remains of the whitetail deer *(Odocoileus virginianus),* a modern inhabitant of the state, may be more recent (Wilson, 1967).

*Odocoileus virginianus*—Whitetail Deer. Twelve specimens catalogued in UMMP, may not all be separate individuals. Fossils found in Berrien, Genesee, Leelanau, Washtenaw, and Wayne Counties. Common in Michigan today and fossil materials may not be very ancient.

*Cervalces scotti*—Extinct Scott's Moose. A prehistoric type of moose. Antler found in Berrien County and reported by E. A. Hibbard (1958).

*Alces alces*—Moose. Wilson (1967) lists 3 specimens as being rather clearly fossil.

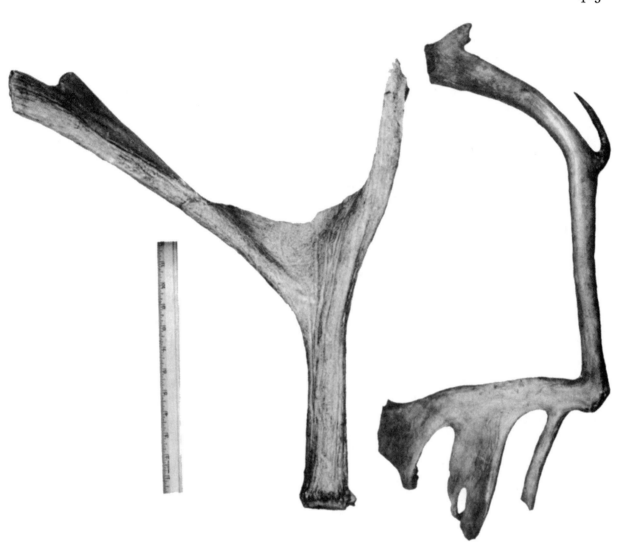

Figure xiv-46. *Left*—Antler fragment of extinct, giant Scott's moose *(Cervalces scotti)* found by E. W. Pudell in Late Pleistocene deposits on his farm in Berrien County. UMMP No. v-33666. One foot ruler for scale.

*Right*—Antler fragment of a caribou found in 6 feet of Late Pleistocene marl in Genesee County. UMMP No. v-44043. Note large brow tine at lower left. Radiocarbon determination run on part of specimen indicated age of 5870 ± 400 years B.P. Collected by Frank Trecha. (Photos courtesy C. W. Hibbard, The University of Michigan Museum of Paleontology.)

These come from near Grand Rapids in Kent County, from Washtenaw County, and from Delta County in the Upper Peninsula. Today, moose are restricted to the Upper Peninsula and to Isle Royale, but Burt (1948) states there are authentic records of moose as far south as Oakland County in the Lower Peninsula in historic time.

*Rangifer caribou*—Caribou. Burt (1942) reported an antler fragment from Sanilac County. That specimen, now in The University of Michigan Museum of Zoology, was identified by C. W. Hibbard (1951) as being the subspecies called the Barren Ground caribou, a northern form still living in Alaska and northern Canada today. Mikula (1964) recently reported the discovery on the Harold Reamer farm, Lapeer County, of what he identifies as an antler fragment (now in MSU Museum) of another subspecies, the Woodland caribou, a more southern form still living in Canada today. On the average the antlers of the Barren Ground caribou are smaller and less palmate, but there is considerable variation and overlap among individuals of both subspecies, so identifications based on antler fragments alone are debatable. The MSU specimen was found in a farm pond excavation in a peat bog, beneath 9 feet of peat on top of blue clay. This indicates considerable antiquity. R. W. Wilson (1967) lists other specimens from Genesee County (in UMMP), and from Cheboygan County south of Burt Lake (in UMMZ). An antler fragment (in UMMP) found by Frank Trecha in 1960 near Davison in Genesee County recently yielded a Carbon$^{14}$ radiogenic date of $5870 \pm 400$ years B.P. According to Burt (1948) there are authentic historic records of Woodland caribou occurring on Beaver and High islands in Lake Michigan and in the Upper Peninsula in recent times but none for the Lower Peninsula itself; the last native caribou, according to Burt, disappeared from Isle Royale "around the turn of the century."

Family BOVIDAE (Bison, Cattle, Sheep, Goats).

?*Bison bison*—(the Plains Bison) or *Bos* —(Domestic Cow). Skeletal remains of bovids are common in Michigan but usually turn out to be those of some domestic cow buried by a farmer. The skeletons of bison and cows are so similar that it is almost impossible to tell them apart unless one has either the skull and horns, or the vertebrae from the shoulder region. Unfortunately, no such parts, clearly identifiable as *Bison* remains, as yet have been found in Michigan. Wilson (1967) lists 1 specimen (in UMMP) found by Donald Haywood, near Scotts in Kalamazoo County, as possibly being a Plains Bison. According to Burt (1948) there are authentic historic records of Plains bison occurring in Michigan in the 1700's, but the last was recorded about the end of that century, shortly after the first European settlers arrived.

?*Boötherium sargenti*—A Pleistocene form of extinct Musk Ox. A skull, now in the Grand Rapids Public Museum, was described by Gidley (1908) as being a new species of this genus. The skull was found beneath the pelvis of an American mastodon, by E. R. Kalmbach, on the farm of Charles McKay, near Moorland, Kent County. Allen (1913) and Hibbard and Hinds (1960) believed this specimen actually was simply a female of *Symbos cavifrons*, the extinct woodland musk ox, described below. The genus *Boötherium* has other species which clearly are distinct from *Symbos*, but *B. sargenti* may not be one of them. Too few specimens of the latter are known to resolve this question.

*Symbos cavifrons*—Extinct Woodland Musk Ox (Fig. XIV-47). The following eight separate finds of this tall Musk Ox in Michigan have been reported and are preserved in museums:

1 Partial skull, UMMP, from 3 miles northeast of Manchester, Washtenaw County (Case 1915 and 1921; Hay, 1923).
3 vertebrae, UMMP, from near Kalamazoo (Hibbard and Hinds, 1960; Benninghoff and Hibbard, 1961); Carbon$^{14}$ radiogenic age 13,200 years B.P.
1 Skull, UMMP, Hillsdale County (Semken and others, 1964).
1 Skull, UMMP, Van Buren Co. (Semken and others, 1964).
Several bones, UMMP, near Scotts, Kalamazoo County, Carbon$^{14}$ age 11,100 B.P. (Semken and others, 1964).
Skull and tibia, UMMP, Marl Lake, St. Joseph County (Semken and others, 1964).
Part of first vertebra (atlas), UMMP, Berrien Springs, Berrien County (R. L. Wilson, 1967).

# Fossil Vertebrates

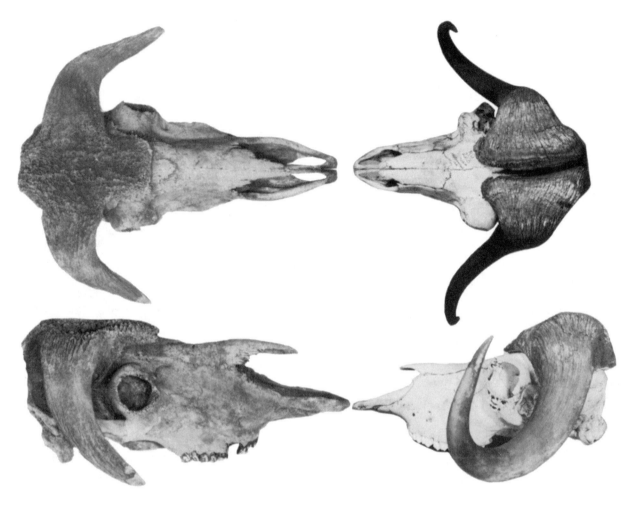

Figure XIV-47. *Left*—Extinct woodland musk ox *(Symbos cavifrons)*. A skull from near Scotts, in Kalamazoo County. Radiocarbon age is 11,100 ± 400 years Before Present. Specimen in Kalamazoo Public Museum.

*Right*—Modern tundra musk ox *(Ovibos moschatus)* from Greenland to same scale for size comparison. Horns in the extinct form appear smaller because only the bony cores remain, whereas the horny sheaths covering the cores are still present on the modern specimen. Note that the horn bosses merge completely across the forehead of the extinct form, but are separated by a narrow cleft on the modern animal. (Photos of fossil courtesy C. W. Hibbard, The University of Michigan Museum of Paleontology.)

1 Skull, Grand Rapids Public Museum, from Newago County (R. L. Wilson, 1967).

As noted above, *Boötherium sargenti* is believed by some authorities to be a female of *Symbos cavifrons*. The modern, tundra-dwelling Musk Ox of Europe, Asia, and North America, an arctic form, is named *Ovibos moschatus*.

# XV

## Fossil Plants in Michigan

### Introduction

The fossil record of plants in Michigan is limited to a few intervals of geologic time. Nevertheless, this record includes some very interesting localities and occurrences which offer brief glimpses of the long evolutionary history of the plant kingdom. The Michigan records fall into the following categories, listed in order of age from oldest (Precambrian) to youngest (near Recent) times: (1) Extremely ancient, Precambrian plant remains including algal limestones and cherts, graphite, anthracitic coal, microscopic thread-like and sack-like bodies, and hydrocarbon residues, (2) Paleozoic "fucoids," (3) Late Paleozoic (Mississippian and Pennsylvanian) coal-swamp floras, (4) Mesozoic (probably Late Jurassic) spores and pollen, and (5) latest Pleistocene and post-Pleistocene floras. Not all of these categories of fossil remains are equally well represented in this state. Fortunately, too, we need not rely on the Michigan record alone for our knowledge of plant history as a whole; the records are much better, although still by no means complete, elsewhere in North America and in parts of the rest of the world. In this chapter we shall first briefly review the main events of plant evolution to provide an historical setting, then we shall look more closely at the record in Michigan itself.

### General Review of Plant History

Plants first appear in the fossil record at least as far back in time as the Middle Precambrian (Chapter IV). Calcareous deposits in Africa, presumably formed by simple, lime-secreting algae, have been dated by a combination of geologic and radiometric means as being at least 2.7 billion years old (Rutten, 1962, pp. 76–83). Even older, carbonaceous, alga-like, spheroidal, cup-shaped and filamentous structures from the Onverwacht Series in South Africa recently have been dated as 3.2 billion years in age, occurring in what may be the oldest unaltered sedimentary rocks in the world (Engel, et al., 1968). Microscopic objects, also of Precambrian age, have been identified as the remains of blue-green algae, rod-shaped and coccoid bacteria, and simple fungi (Tyler and Barghoorn, 1954; Schopf, et al., 1965). These objects occur in rocks of the Gunflint For-

# Fossil Plants

mation of Ontario that are between 1.7 and 2.0 billion years old, according to radiometric dating methods (Tyler, Barghoorn, and Barrett, 1957, p. 1304; Schopf, *et al.*, 1965). Thus, it is certain that bacteria and simple plant life originated long ago and that the time of origin of life itself must have been even farther back in the past. Plants may have evolved some form of photosynthesis at least as early as 1 billion years ago (Barghoorn, Meinschein, and Schopf, 1965, pp. 465–67). Rutten (1959; 1962, p. 134) has suggested that this initiated the introduction of abundant free oxygen into the earth's atmosphere.

Aquatic plants more complex than the simple algae were in existence by Cambrian time, some 600 million years ago, but the evolution of terrestrial plants, probably from aquatic ancestors, was not accomplished until Late Silurian time. Prior to the Late Silurian the lands were barren of vegetation and must have presented a very bleak aspect. It also follows that erosion rates were more rapid before the appearance of land plants, which thereafter served to stabilize surface soils by retarding the erosive effects of water and wind. Even so, the Silurian land plants were few, relatively simple, and restricted mainly to environments around the margins of water bodies. By the Early Devonian, however, several complex and highly evolved groups of land plants had arisen and by Middle Devonian time these achieved sufficient abundance to produce forests, at least in swampy areas around the margins of inland seas. The fallen remains of plants of the mid-Devonian forests accumulated in such quantities as to produce thick layers of peat, later to be converted by heat and pressure into coal. Included among the Devonian land plants were "scale trees" (Lycopsida), scouring rushes (Sphenopsida), true ferns (Filicineae), "seed ferns" (Pteridospermales), and possibly even some very primitive conifers. None of those Devonian land plants is well represented in any known Devonian rocks in Michigan, no doubt because rocks of that age here are marine in origin, but those Devonian plants were the ancestors of the abundant coal-forming floras which did contribute extensively to the record of plants during Pennsylvanian time in Michigan. There are a few Mississippian records of those plants here in the state as well (Dorr and Moser, 1964; Chaloner, 1954).

True Flowering Plants (Angiosperms) of modern aspect make their first appearance in the fossil record elsewhere in 135 million year old rocks of Early Cretaceous age, but no rocks of that period have been positively identified here in the state. (Flowering Plants do not appear in the Michigan record until Post-Pleistocene time). During Late Cretaceous and Tertiary times the Flowering Plants truly "blossomed," evolving rapidly to become the dominant land plants of the modern world. This last phase, too, is missing from the Michigan rock record due to the absence of rocks of Late Cretaceous and Tertiary age here; after the Jurassic, the Michigan fossil plant record does not begin again until the last Pleistocene continental glacial ice sheet began its retreat from the Great Lakes Region, about 14,000 years ago.

A much simplified classification of plants appears in Figure xv–1 and a graphic resumé of their history in Figure xv–2.

Now let us take a detailed look at the plant groups that are represented in the Michigan rock record.

## Precambrian Plant Remains in Michigan

Beginning with the most ancient plant fossils in the state, algal deposits have been reported in calcareous rocks of Early Huronian (Early Middle Precambrian) age at several localities in the Upper Peninsula. In each case, the calcareous rocks originally were limestones or dolomites, but they have subsequently been metamorphosed into impure marble by the heat and pressure generated during crustal disturbances. The beautifully layered, "algal" structures bear a close resemblance to the types of deposits formed by some of the lime-secreting algae living today (Fig. xv–3). The rock formations in which these occur include the Kona "dolomite" near Marquette (Fig. xv–3), the Randville Dolomite in central Dickinson County (Fig. iv–11), and the Bad River Dolomite in the western Upper Peninsula. The chart of Precambrian rock formations (Fig. iv–3) in Chapter IV shows the position of those formations in geologic time. If these structures truly were algal in origin, then they represent an historical record of some of the very earliest forms of plant life on earth, a group of aquatic, probably marine, organisms over 2 billion years old, that lived at a time close to the dawn of life itself. The bands of calcareous

# Geology of Michigan

material illustrated in Figure xv-3, are *not* the structurally preserved remains of the actual plants themselves, but are the successive bands of limy material secreted by organisms during a number of stages of growth, the layer representing each stage being located above its predecessor as one of a series of either roughly concentric semicircles or subparallel bands.

Graphite, a form of relatively pure carbon, has long been known to occur in Precambrian rocks near the town of L'Anse at the southern end of Keweenaw Bay. These deposits were mined in years past as a source of pigment. The old Detroit Graphite Company of Detroit mixed this material into paints designed for use on exterior steel structures such as railroad bridges. Although graphite is known to originate, in some cases, as an inorganic deposit in igneous rocks, more commonly it appears to have been formed by the metamorphism of plant remains. In the latter case, which has been documented at many localities in younger rocks throughout the world, the graphite represents the end stage in the transformation of plant remains, by heat and pressure, first into peat, then to coal, and finally to graphite. If the organic mode of origin is accepted for the deposits near L'Anse, then we have additional evidence of the existence of abundant, although perhaps very primitive, plant life during Precambrian times in the state.

Moving forward through Precambrian time to the Middle Huronian, we discover that carbonaceous materials, derived from plants, are preserved in the Michigamme Slate of the Iron River District, in the south central part of the Upper Peninsula. The position of the Michigamme Slate in geologic time is shown in the center column of Figure iv-3 in Chapter IV. The absolute age of the deposit is not yet definitely established, but probably is less than that of the Gunflint Formation. A reasonable estimate would make it about 1.7 billion years old. The carbonaceous deposits in the formation were found in old test pits and more recent bulldozer excavations at the old Morrison Creek exploration site, about 6.5 miles due north and about 1 mile east of the city of Iron River, Michigan. Tyler, Barghoorn and Barrett (1957), who recently studied the material, described 2 types of fossil organic matter. The first type is an anthracitic coal which occurs in lenticular bodies interbedded in the Michigamme slates

## Figure xv-1

### Outline Classification of Plants

*Asterisk (\*) indicates groups known to occur in the fossil record in Michigan. Capitalized groups are common in the fossil record in the state (See Fig. xv-2 for general geologic history of plants.)*

*Nonvascular plants.* Lack food and water conduction systems, hence are structurally primitive and mainly aquatic.

* \*THALLOPHYTA. Bacteria, fungi, algae (including "sea weeds"). Algal remains occur in the Huronian (Middle Precambrian) and Keweenawan (Late Precambrian) rocks of the western Upper Peninsula. "Fucoids," which probably are mainly sea weed impressions, are common in Paleozoic marine sedimentary rocks of the state.

* \*BRYOPHYTA. Mosses and liverworts. Known from spores in Post-Pleistocene bog deposits in the state.

*Vascular plants.* With food and water conduction systems fitting them for life on land or at least in semiaquatic environments, although some have secondarily reverted to an aquatic way of life. An internal structure, the stele, including xylem and phloem, makes possible food and water conduction.

* Psilopsida. Among the earliest and most primitive vascular plants known from the fossil record. No fossils yet found in Michigan.

* \*LYCOPSIDA. The club-mosses and "scale trees." Living representatives are *Selaginella* and *Lycopodium*, the modern ground pine. *Lepidodendron* and *Sigillaria* are common types in Pennsylvanian rocks of Michigan. Spores of *Lycopodium* occur in Post-Pleistocene bog deposits here.

* \*SPHENOPSIDA. The "scouring rushes" or "horsetail rushes." *Equisetum* is a modern living representative found in the state. Rare in Mississippian rocks but common in Pennsylvanian rocks of central part of the Lower Peninsula. Spores of *Equisetum* occur in Post-Pleistocene bog deposits in the state. *Calamites, Annularia, Asterophyllites,* and *Sphenophyllum* are the common Pennsylvanian forms in state. *Calamites* has also been reported from Mississippian rocks in Michigan (Dorr and Moser, 1964, p. 111).

* \*Filicineae. The true ferns. Fossils known but not common in the Pennsylvanian rocks of the Lower Peninsula. Spores recognized in Jurassic red beds in central Lower Peninsula. Spores of modern types of ferns are common in Post-Pleistocene bog deposits here.

* \*GYMNOSPERMAE. Plants with true seeds which, nevertheless, lack protective and nutrient encasement of the ovary. Includes a number of orders, only some of which occur in the fossil record in Michigan.

Figure xv–1—*Cont.*

*Order PTERIDOSPERMALES. "Seed ferns." An extinct group common in Pennsylvanian rocks of the central part of the Lower Peninsula of Michigan. *Neuropteris*, *Alethopteris*, *Sphenopteris*, and *Rhacopteris* are the common genera.

Order Cycadeoidales. An extinct group not yet found in Michigan.

*Order Cycadales. Includes the living cycads. Represented in Michigan rocks in form of spores in the Late Jurassic "red beds" of the central Lower Peninsula.

*Order CORDAITALES. An extinct group, found in Pennsylvanian rocks of Michigan. *Cordaites* is the common Pennsylvanian genus here. Possibly the ancestral group that gave rise to conifers.

Order Ginkgoales. The modern ginkgos or maidenhair-trees are surviving remnants of this group. No known fossil record in Michigan.

*Order CONIFERALES. The conifers, including the evergreens. Pollen found in Late Jurassic "red beds" of central Lower Peninsula. Pollen also occurs in Post-Pleistocene bog deposits of the state. Pollen serves for fertilization of seed.

*ANGIOSPERMAE. The flowering plants. Seed is fertilized, protected, and provided with nutrient material, all within an ovary borne in a flower. Pollen serves for fertilization. Includes the deciduous or broad-leaved trees, weeds, flowers, grasses, etc. First fossil record in Michigan is in form of pollen from Post-Pleistocene bog deposits, although the group actually appears for the first time in rocks of Early Cretaceous age elsewhere.

(Fig. xv–4). The slates themselves originally were shales, that is sedimentary deposits, which were subsequently metamorphosed (Chapter IV). The coal lenses occur at several levels and at depths up to 1000 feet, and are scattered over an area of several miles, indicating that separate small but widely distributed accumulations of vegetation must have gone into their making (see Chapter V on the origin of coal). The largest lens of coal discovered was 5 feet long and 3 feet thick. The lenses lie parallel to the bedding planes of the slates within which they occur, showing that the carbonaceous material was of sedimentary origin, not introduced later by some igneous intrusive process. It also has been established that the coal is an anthracitic type that was not imported into the area by man. Microscopic examination of the coally material reveals textures and patterns best identified as the coalified remains of algal plant cells (Fig. xv–4). Biochemical analyses of the complex molecules of a pigment, extracted from the coal by means of an organic solvent, reveal that the pigment is of organic origin. This offers additional proof that the carbonaceous matter was a product of former life. Organic substances with similar properties have been identified in the Gunflint Formation of Ontario mentioned earlier (Tyler, Barghoorn, and Barrett, 1957, p. 1303).

The second type of carbonaceous material found in the Morrison Creek shales of the Michigamme Formation occurs in the form of circular, egg-shaped and elliptical bodies of graphite which are scattered profusely on bedding surfaces (Fig. xv–5). The graphitic material usually consists of a much flattened film, although some of the bodies still are 3-dimensional. They range from a fraction of a millimeter to a maximum of 12 millimeters in diameter. Tyler *et al.* (1957, p. 1297) concluded that the graphitized bodies are the remains of primitive, free-floating algae that grew in nodular colonies. They compared these with the modern blue-green alga called *Nostoc* (Fig. xv–5), although clearly stating that the ancient forms probably were *not* the same genus. The siliceous Morrison Creek shales in the Michigamme Formation are generally rich in both nitrogen and carbon, which offers further evidence that the carbon originated through some biologic process.

Continuing forward through Precambrian time we encounter the next fossil record of plants in Michigan in Keweenawan (Late Precambrian) rocks of the western Upper Peninsula. Algal structures occur in cherty deposits within the conglomerates of the Keweenawan Series. The banded, cherty structures have been observed in the vicinity of the "Devil's Washtub" along the Lake Superior shore of the Keweenaw Peninsula just west of Copper Harbor. A sign along the shore road identifies this site.

More significant, however, are the organic materials recently described from the Nonesuch Shale of the Keweenawan Series (Barghoorn, Meinschein, and Schopf, 1965). The Nonesuch Shale contains important copper ore deposits which are mined by the White Pine Copper Company at White Pine near Ontonagon. These copper deposits and their origin were discussed near the end of Chapter IV. The position of the None-

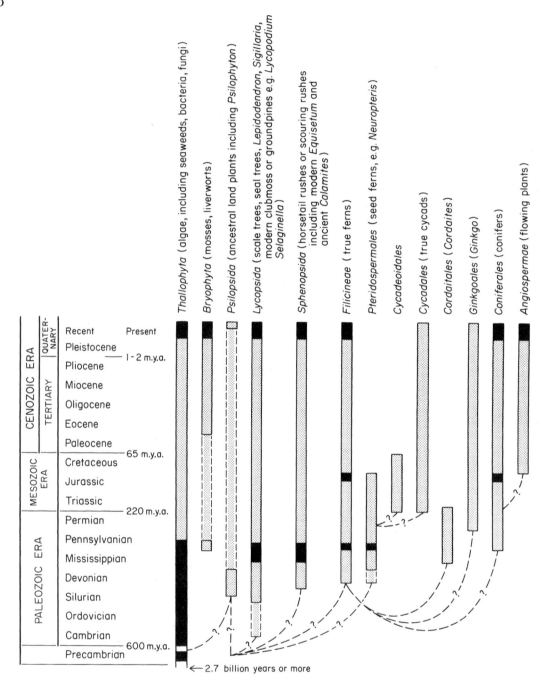

Figure xv-2. Distribution of major plant groups in geologic time. (Time spans *not* in true proportion to actual length.) Heavy, black bars indicate groups found in the Michigan fossil record and their time distribution in the rocks of this state; solid parallel lines the known record for the world as a whole; parallel dashed lines the inferred record based upon a fossil record known for times before and after the span shown, but for which no known actual specimens yet have been found; single dashed line indicates possible ancestral relationship.

such Shale in geologic time is shown on Figure iv–3 in that chapter. The radiogenic age of the Nonesuch Shale is 1046 ± 46 million (or 1.046 billion) years, the date being based on radioactive Strontium/Rubidium ratios (Barghoorn, Meinschein, and Schopf, 1965, p. 461). Detailed phys-

## Fossil Plants

ical and chemical analyses reveal that crude oil residues, a variety of complex organic compounds (alkanes, porphyrins, pristane, and phytane), and objects that appear to be the structural remnants of microscopic organisms, all occur in the Nonesuch Shale. Carbonaceous matter also is abundant in a widely disseminated state throughout the copper-bearing zone. The above mentioned authors state (p. 462) that the petroleum residues occur as devolatilized asphaltic material in layers and spheroids in the "Parting Shale" and "Upper Shale" members of the copper-bearing zone; the blobs of residue often enclose a minute particle of the copper mineral called chalcocite or a speck of native (metallic) copper. The fact that the percentage of chalcocite plus native copper varies in direct proportion to the percentage of organic matter present, leads to the conclusion that the organic matter itself played a role in the precipitation of the copper in the shale. As we discussed in Chapter IV, the copper appears to have entered the formation *after* deposition of the shale, *not* concurrently with deposition. If the organic matter facilitated precipitation of the copper and helped to control its distribution, then the organic matter must have been in the shale earlier, probably at the time of deposition of the muds that formed the shale. This simple, obvious conclusion is important because, if the organic matter is of biologic origin and is as ancient as the rock itself, then we have additional evidence for the early existence of life. The presence of vanadyl porphyrins is especially significant because porphyrins are complex organic pigments associated with metabolism and especially with respiration in living organisms. This evidence, and the presence of fragments of plant tissue in the shale, suggests that plants had evolved the important biologic process of photosynthesis at least as early as the time of deposition of the shale, about 1 billion years ago. As Barghoorn, Meinschein, and Schopf (1965, p. 465) state, this is the oldest presently known direct evidence of the photosynthetic mechanism. They further conclude (p. 467) that since porphyrins quickly break down chemically under high temperature, their presence in an undeteriorated state in the Nonesuch Shale indicates that the rock and its contents never were greatly heated. This, in turn, leads to the important conclusion that the copper-bearing ore solutions must have been introduced into the shale at low temperatures and that the copper deposits in the Nonesuch Shale cannot be of high temperature hydrothermal origin.

When polarized light is passed through certain carbon compounds called alkanes, extracted from the crude oil residues in the Nonesuch Shale, the light is "rotated" in a clockwise direction by optical refraction. This direction of rotation is typical of alkanes extracted from modern organisms, hence the crude oil residues in the Nonesuch Shale are probably of biologic origin. This is another source of evidence for the existence of life at the time of deposition of the shale. Barghoorn, Meinschein, and Schopf (1965, p. 469) conclude that since pristane, phytane, and porphyrins are complex organic compounds that may be derived from chlorophyll, the presence of those compounds in the shale suggests that even 1 billion years ago chlorophyll was an important photosynthetic substance in plants. If this is true, then one of the most important steps in the evolution of plant chemistry had already been taken. Those authors express this view (p. 470) as follows: "The presence in the Nonesuch Formation of porphyrins and alkanes that resemble porphyrins in geologically younger sediments and alkanes in living organisms suggests that life which existed in Precambrian time was metabolically similar to existing life."

In addition to the petroleum residues which fill pore spaces between mineral grains (Fig. xv–6), the copper-bearing zone of the Nonesuch Shale contains microscopic objects which are thought to be fragments of actual plant tissue. These shreds of tissue (Fig. xv–6) occur as interwoven filaments and sheets that often include what appear to be cell walls. Barghoorn, Meinschein, and Schopf (1965, p. 470) suggest that this plant tissue may be from algae or fungi, but warn that these identifications are very uncertain because the material is so fragmentary. Another line of evidence they cite to prove that the organic matter in the Nonesuch Shale is of biologic origin is the fact that the ratio of Carbon 12 to Carbon 13 is similar to the ratio of those two carbon isotopes found in modern plant tissue. There is an enrichment of $C_{12}$ in the shale, producing a *higher* $C_{12}/C_{13}$ ratio than is normal for modern *inorganic* matter.

In summary, the evidence indicates that organic matter in the 1 billion year old Nonesuch Shale

Figure xv-3. Modern and ancient algal deposits.

*Top* (including inset)—Algal layers over 2 billion years old in Kona Dolomite of Early Huronian (Early Middle Precambrian) age in old quarry just off US41 at north edge of Harvey, about 5 miles southeast of Marquette. Layering may be seasonal or may represent periodic flooding and renewed growth at undetermined intervals.

*Bottom left*—Vast, mud-cracked marine algal flats just above Gulf of Mexico water level near Moon Island in Laguna Madre, between Padre Island (an offshore barrier sand island) and the south Texas mainland, about halfway between Corpus Christi and Brownsville. (Courtesy Howard Gould, Stratigraphic Geology Division, Esso Production Research Company.)

*Bottom right*—Cross section dug in near-surface deposits on the algal flats of Laguna Madre. Note scale. White layers are gypsum, medium gray are clay, dark gray are blue-green algal mats. Average rate of sedimentation estimated at about 1 foot per 50 years. Layering here is result of periodic flooding, not seasonal or yearly varving. (Same source as above.)

was derived from living organisms of that time. The organisms probably were plants, and possibly were algae and fungi. Some plants then already had evolved the capacity for photosynthesis and the process probably was based on chlorophyll or some similar pigment. The organic matter influenced the deposition of the copper deposits in the Nonesuch Shale and the ore-bearing solutions were introduced into the shale at low temperature.

This concludes our review of the fossil evidence for Precambrian plant life in Michigan. Because of the fragmentary nature of the geologic record here in this state, we must now take a big jump forward in geologic time, to the Devonian Period of the Paleozoic Era, to continue with the state's plant history.

## Paleozoic Fucoids

Marine sedimentary rocks of Paleozoic age in Michigan (and elsewhere) commonly bear marks, impressions, or ridges whose mode of origin is not always certain. Collectively these are known as "fucoids." They may take a variety of forms. Some vaguely resemble the serpentine trails, burrows, or trackways that might have been left in soft sediment by a variety of bottom-dwelling organisms (clams, worms, and others) as they crawled across the sea floor or burrowed just below the surface of the bottom ooze. Other irregular, linear markings appear to be grooves formed by some object, organic or inorganic, that rolled or was dragged by currents along the sea floor. In many cases, however, the fucoids have a stem-like or root-like appearance. Many are branched in simple fashion. Often these occur in interlaced masses as though they had originally been parts of an intertwined mass of stems that later was compressed in the sea floor sediment. Probably these objects represent masses of marine vegetation, "sea weed" in the broad sense, but rarely is the preservation of the objects adequate for specific identification. Perhaps many were formed by brown algae because they are the common sea-weeds of modern times and are known to have existed very early in geologic time. One may picture these as having been parts of a "sargasso"-like mat of vegetation floating in the Paleozoic seas, or as "kelp" attached to the sea floor, later to be buried and compressed in the sediment and leaving only impressions and ridges in the sediment which eventually turned to stone. None of the actual plant material remains, only the impression or impression-filling. Often fucoids are oriented in roughly parallel fashion, as though aligned by some ancient current. Fucoids are most commonly found on the faces of sedimentary layers of limestone or shale after the layers have been split apart along bedding (stratification) planes. Some representative fucoids are illustrated in Figure xv-7.

## Devonian Plants

Chester A. Arnold, professor of botany and paleobotanist in The University of Michigan Museum of Paleontology, has found fragments of silicified wood in the Antrim Shale of Late Devonian age (Chapter V, Fig. v-2). The fragments are particularly abundant in the Paxton Quarry near Alpena. The woody structure has been identified (C. A. Arnold, written communication) as belonging to the genus *Callixylon*. Some fragments appear to have come from trees about 1.5 feet in diameter. On the basis of its advanced woody structure, *Callixylon* originally was classified in

# Geology of Michigan

424

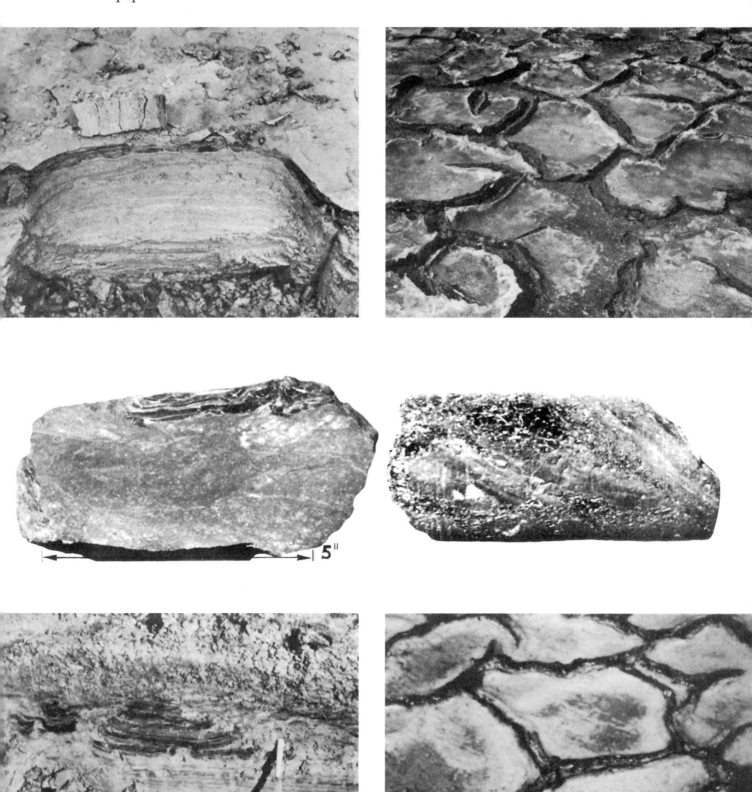

# Fossil Plants

Figure xv-4. Modern and ancient algal deposits. (Also see Fig. xv-3.)

*Top*—Modern mat of blue-green algae near Mesquite Rincon, along south Texas Gulf Coast between Padre Island and mainland. Polygonal surface cracks (at right) are due to drying. Cross section (at left) shows white gypsum and dark clay and algal layers. (Courtesy H. R. Gould, Esso Production Research Co.)

*Center*—Ancient algal layers, subsequently metamorphosed to dark-colored anthracitic coal, in 1.7 billion year old Michigamme Formation of Late Huronian (Late Middle Precambrian) age, about 6.5 miles north of Iron River, Michigan. (Courtesy E. S. Barghoorn, Harvard University.)

*Bottom*—Modern algal deposits in cross section and surface views. Dessication polygons about 1.5 feet in diameter. Southwestern margin of Persian Gulf. (Courtesy C. Kendall and Sir P. Skipworth.)

the Cordaitales (Fig. xv-1). More recently, however, Professor Charles B. Beck of The University of Michigan Department of Botany has found (in New York State) the spore-bearing, fern-like foliage of *Archaeopteris,* usually classified as a member of the fern group (Filicineae), attached to woody branches of *Callixylon.* Here is an example of a situation not at all rare in the fossil record —an organism that is still primitive in some respects (in this case foliage) while being advanced in other aspects (wood structure in the case of *Callixylon*). The association of *Callixylon* and *Archaeopteris* parts on a single plant also illustrates another common problem of paleobotany. Different parts of one plant, if originally found separately in the fossil record, may be given separate scientific names which only later are discovered to be synonyms. The best position for this plant, with its mixture of *Callixylon* and *Archaeopteris* characters, in the classification of the Plant Kingdom is yet to be settled. So far the two types of plant parts have not been found in association in the Michigan fossil record but this may simply be due to the incomplete record of Late Devonian plants here in this state. The Antrim Shale, in which the *Callixylon* occurs, is an offshore marine deposit and the plant fragments in it probably are fossilized pieces of ancient driftwood. Just as elsewhere in the world, the lands adjacent to the Late Devonian seas of Michigan must have been at least partially clothed with forests and the streams that flowed off the land carried plant material into the sea.

## Mississippian Plants

As with the Devonian, the known fossil record of land plants of Mississippian age also is scanty in Michigan. Again this is because the Mississippian rocks of the state are dominantly marine in origin. Dorr and Moser (1965, p. 111) reported water-worn woody fragments of *Calamites* (a sphenopsid, Figs. xv-11, 12) found in the mid-Mississippian Marshall Sandstone. The position of that formation in geologic time is shown on Figure v-2. The specimens were discovered in the old Stony Point Quarry, about 2.5 miles southwest of the town of Hanover, in Jackson County southwest of the city of Jackson. Dorr also has found woody stem and trunk fragments of *Lepidodendron* (a lycopsid, Figs. xv-9, 10) in the same quarry. *Lepidodendron* has not previously been reported from there, but its presence is significant in connection with the fact that fossil spores, probably also from *Lepidodendron,* have been reported by Chaloner (1954) from Mississippian rocks of nearly the same age near Grindstone City. Invertebrate animals of the nautiloid and ammonoid cephalopod groups (Chapter XIII) also are found in the sandstone in the Stony Point Quarry. The sandstone is a shallow-water deposit and the animals mentioned are marine types; thus it seems that the Marshall Sandstone there is a near-shore marine deposit and that the plant fragments again are pieces of driftwood that came to rest along the beach of an ancient sea.

As mentioned briefly above, Chaloner (1954) described and illustrated two species of Mississippian age spores from Michigan. Spores are the nearly microscopic reproductive bodies of primitive plants. The Michigan specimens came from a micaceous sandstone exposed on Willow Creek, about 1 mile south of Grindstone City, Huron County, Michigan. Chaloner used the conclusions of Monnett (1948, p. 677) who had decided that the rocks at this locality were Mississippian in age and belonged in either the upper part of the Coldwater Formation or the lower part of the Marshall Formation (Chapter V, Fig. v-2). There still is no doubt about the Mississippian age of the rocks there, but Dorr and Moser (1965, p. 112) recently concluded that the so-called "Marshall Sandstone" exposed in the Thumb Area of the Lower Peninsula is a deposit that is similar to but older than the true Marshall Formation which crops

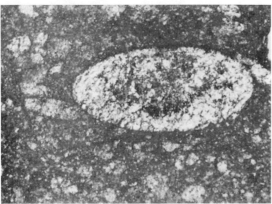

Figure xv-5. *Top*—Some globular living colonies of the modern blue-green alga, *Nostoc*, shown about one-quarter natural size.

*Center and bottom*—Oval, graphitic bodies, possibly fossil algae, found in 1.7 billion year old Michigamme Formation of Late Huronian (Late Middle Precambrian) age about 6.5 miles north of Iron River, Michigan. Possibly related to *Nostoc*, but probably not the same genus. Magnified about 3.5 times. Note smaller colonies adjacent to large ones. (All photos courtesy E. S. Barghoorn, Harvard University.)

out farther to the south and west. The sandstone from the Willow Creek locality was broken down by treatment with hydrofluoric acid, and the flattened spores were recovered from the resulting mud by sieving. The photographs (Fig. xv-8) which illustrate the spores were published by Chaloner who treated the specimens to make them transparent and then photographed them microscopically by transmitted light. The preparation method is described by Chaloner (1954, p. 25) and by Arnold (1950, p. 63). Both spore species found near Grindstone City belong to the genus *Triletes*, one being identified as *T. angulatus*, the other as *T. subpilosus* (forma *major*). Both types of spores were from plants that were widespread during Mississippian times, the first species having previously been reported from Pennsylvania and Poland, the second from Indiana, Scotland, and Turkey (Chaloner, 1954, p. 28-29). Chaloner (pp. 23, 32) concluded that the Michigan spores were produced by a Mississippian age flora of lycopods (Figs. xv-9, 10) consisting primarily of the genus *Lepidodendron*, possibly including *Lepidophloios*, but without any evidence of the presence of *Sigillaria* which in many other places was another common lycopod of the time. *Lepidodendron* wood fragments of about the same age were just mentioned above as having been found in the Stony Point Quarry to the southwest. Plants belonging to the genera just listed continued in existence into the following Pennsylvanian Period in Michigan (and elsewhere) as we shall see in the next section of this chapter. The Mississippian sandstones at Grindstone City also are of nearshore marine origin, like those in the Stony Point Quarry. Yet the plants themselves are terrestrial types. It seems, therefore, that the spores must have been transported, perhaps by wind, at least a short distance from land to the sea.

## Coal Swamp Floras of the Pennsylvanian Period

A wide variety of major plant groups had evolved by the Pennsylvanian Period. Many of these were vascular plants which possessed water and food conducting systems and thus were adapted for life in an erect position in freshwater swamps or even on dry land. The Mississippian and Pennsylvanian periods of geologic time often are referred

**Fossil Plants**

427

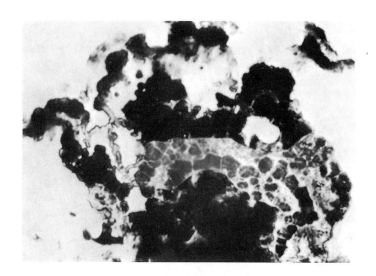

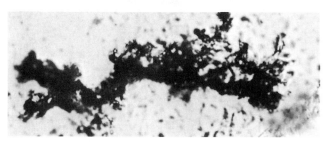

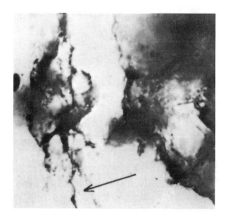

Figure xv–6. Fossil organic matter in the "Parting Shale" unit of the Nonesuch Shale of Late Keweenawan (Late Precambrian) age, about 1 billion years old, from White Pine, Michigan. Photographed by transmitted light through thin-sectioned rock. (All photos courtesy E. S. Barghoorn, Harvard University, from *Science,* Vol. 148, 1965. Copyright 1965 by the American Association for the Advancement of Science.)

*Top*—Concentrically layered, translucent, asphalt-like organic matter, probably devolatilized petroleum. Magnified about 1275 times.

*Center*—Sheet of filamentous plant tissue, probably cellular residues magnified about 1325 times.

*Bottom*—Interwoven, fragmented, plant filaments flattened on a stratification surface in siltstone. Arrow indicates cell wall of a filament. Magnified about 1275 times.

to, the world over, as the "Carboniferous" because of the abundance of coal deposits in rocks of those ages. The coal originated from vast accumulations of fallen plant remains that collected on the floors of widespread interior swamplands around the margins of Late Paleozoic epicontinental seas (Chapter V and Fig. v–31).

During the Pennsylvanian Period, shallow seas frequently alternated with swampy lowlands in Michigan. A half dozen sets or series of deposits in the Saginaw Formation of Michigan, each set including shallow water marine, swamp, and stream laid sediments, record the geographic fluctuations of the margins of those seas and swamplands. Each cycle of advance and retreat of the sea produced a cyclic deposit called a "cyclothem." Coal layers are common in the Pennsylvanian cyclothems of this state (Chapter V) and elsewhere. The coal de-

Figure xv–7. "Fucoids" on slabs of Devonian age marine limestone, from outcrops on the eastern and southern edges of the Michigan Basin. These may be sedimentary fillings (molds) of the hollow interiors of stems of a tangled mat of kelp-like seaweed. Pen, showing scale, is 6.75 inches long.

posits record the lush growth of vegetation that flourished in the Pennsylvanian swamps. Sluggish streams meandered through those swamps, locally depositing sand in their channels and silts and muds on the adjoining floodplains. Fossil plant remains are common in the channel and floodplain deposits as well as in the coals. The sands and muds provide the better specimens, however, because there the fossils occur as separated or isolated specimens, whereas in the coals so much plant material is compressed together that the individual plant remains are difficult to distinguish from one another.

Quarries in Early Pennsylvanian age rocks of the Saginaw Formation (Chapter V, and Fig. v–2) in the central part of the Lower Peninsula, in the vicinity of Grand Ledge west of Lansing, are good places to collect Pennsylvanian plant fossils (Chapter V, Fig. v–30). The list of fossil collecting localities at the end of Chapter XIII gives some of the best of these. Arnold (1966, p. 13) provides a map and directions to the collecting localities near Grand Ledge. This map is reproduced in Fig. xv–17. Old coal mines and mine dumps in Bay and Saginaw counties used to yield good material, but, with the end of coal mining, most of the old mine areas now are overgrown with vegetation. One old coal mine dump located a few hundred yards south of Expressway U.S. 10 about half way between Bay City and Midland and visible from the road, still yields some specimens (Arnold, 1966, pp. 4–6). Most of the good specimens collected in earlier years now are in the collections of the University of Michigan Museum of Paleontology in Ann Arbor.

The major plant groups especially common in the Early Pennsylvanian Saginaw Formation of Michigan are mentioned in Figure xv–1, and are

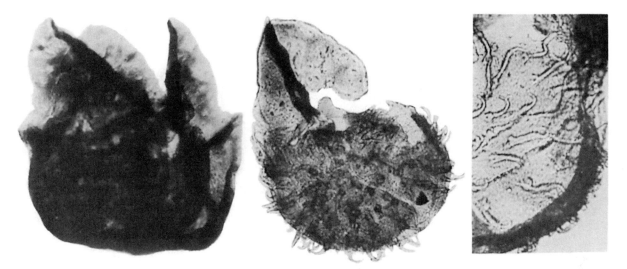

Figure xv–8. Highly magnified spores from a lycopsid, possibly *Lepidodendron*, from Early Mississippian age rocks of the Coldwater Shale and/or Marshall Sandstone near Grindstone City at the tip of the Lower Peninsula "Thumb." Photographed by transmitted light. (All from Chaloner, 1954, courtesy C. A. Arnold, The University of Michigan Museum of Paleontology.)

*Left—Triletes angulatus,* magnified 50 times.

*Center—Triletes subpilosus,* ×50.

*Right—Triletes subpilosus,* ×100.

shown with special emphasis on Figure xv–2. Typical members and fossil remains of each group are illustrated in Figures xv–9 through xv–16. Only the common plant genera are illustrated and described here. Actually, over 100 species have been collected from the Pennsylvanian rocks of the Michigan Coal Basin. Detailed descriptions may be found in the 1949 publication by Arnold.

Descriptions of each main group follow.

LYCOPSIDA (Figs. xv–9, 10), the "Lycopods." These were plants with small, simple, blade-like leaves. The leaves of *Lepidodendron* were arranged in steep spiral fashion up the trunks, whereas those of *Sigillaria* were vertically aligned. Upon falling off, the leaves left cushion-like leaf scars or "scales," likewise arranged up the trunk. Therefore, the many different types of tree-sized lycopods often are called "scale trees" or "seal trees." The scars occur on ridges, alternating with furrows. Reproduction was by means of minute spores contained in pod-like spore cases (sporangia) on fertile leaf stems on or close to the main trunk. The fertile stems often grew together into cone-like structures, forming clusters at the tips of branches as well. Living representatives of the lycopods are relatively small in size and few in number compared with the Pennsylvanian types. *Selaginella* is the most closely similar living genus. The living Clubmoss or Groundpine (*Lycopodium*) is another example. During the Pennsylvanian, some scale trees reached 110 feet in height and 18 inches or more in diameter. The root organs of *Lepidodendron* and *Sigillaria* (common genera in the rocks of Michigan) are indistinguishable, hence are referred to collectively as "*stigmaria.*" Commonly, these "roots" may be seen in their original position of growth in the Pennsylvanian sedimentary deposits. *Stigmaria* often are found embedded in underclays and sands beneath coal seams, suggesting that the trunks and leaves of the trees to which those roots belonged were contributors to the coal deposits above. *Stigmaria* consist of a main "root," really an underground stem, with side rootlets. Upon detachment, the rootlets left pits and scars arranged similarly and somewhat resembling the leaf scars on the trunk above. The lycopods reached the peak of their abundance and diversity in the Pennsylvanian Period.

SPHENOPSIDA (Figs. xv–11, 12), the "scouring rushes" or "horsetail rushes," once a dominant

# Geology of Michigan

Figure xv–9. Restorations of extinct lycopsids from Pennsylvanian age rocks of Michigan.

*Left—Sigillaria.*

*Right—Lepidodendron.*

group of plants, also reached their climax of abundance and diversity in the swamps of the Pennsylvanian Period. They are much less important in the modern flora, being represented today only by the lowly *Equisetum*. The jointed green stems of modern *Equisetum* resemble tiny bamboo shoots and can be pulled apart at the joints or "nodes." Some Pennsylvanian sphenopsids grew to great size compared with their modern descendants, often reaching heights of over 50 feet and diameters of 2 feet. They were common contributors to the Pennsylvanian coal swamps. The jointed stems and trunks were thin-walled and hollow, with closely spaced striations or grooves. Mud or sand fillings of the hollow stems are common fossil forms of this group. Leaves and branches were borne in whorls or circlets at the nodes. Reproduction was by means of minute spores borne in sporangia on special stalks at the ends of stems and branches. A number of genera and species are known from the Pennsylvanian rocks (Saginaw Formation) of Michigan, including many different sizes and leaf types. Un-

Figure xv–10. Lycopsida from the Early Pennsylvanian age Saginaw Formation of Michigan. Illustrations all approximately natural size. Also see Figures xv–9 and v–31. (Fossil photos courtesy of Professor C. A. Arnold, The University of Michigan Museum of Paleontology.)

1, 2, and 3—Portions of trunks of several species of the genus *Sigillaria*. Note leaf attachment marks.

4 through 8—Portions of trunks of several species of the genus *Lepidodendron*. Leaf scars of various shapes.

9 and 10—Stigmaria, the main and lateral "root" organs, or underground stems of large lycopods, often found in the original position of growth in Pennsylvanian sandstones.

11—*Bothrodendron,* a small lycopod fossil.

12—*Lepidostrobus,* the cone-like reproductive body of a lycopod, probably that of *Lepidodendron* in this case.

13—Herbarium sheet with mounted specimens of the modern Ground Pine, *Lycopodium,* which grows only a few inches high. *Selaginella*, a related genus, also grows in Michigan today.

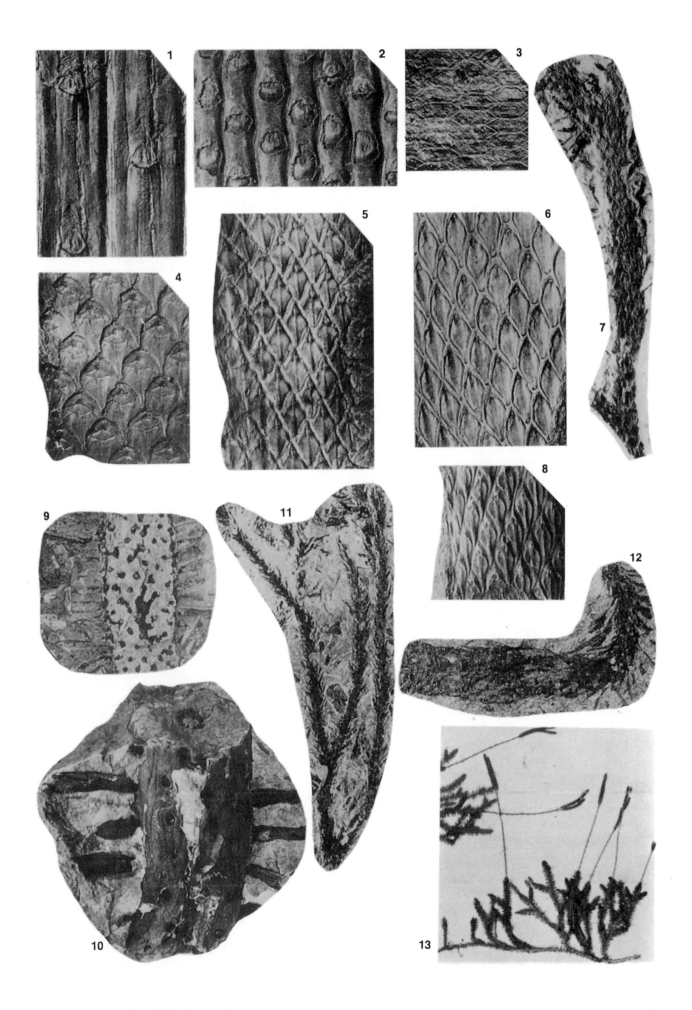

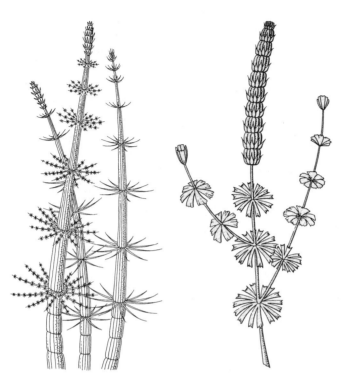

Figure xv–11. Restorations of extinct sphenopsida from Pennsylvanian age rocks of Michigan.

*Left—Calamites,* a tree-sized plant. The small circlets of pointed leaves on small stems are called *Asterophyllites.*

*Right—Sphenophyllum,* a small, low-growing form.

fortunately, the original relationship between separated stems and leaves is often difficult to determine unless the whole plant, stems, leaves and all, is found preserved as a single intact specimen. This latter case is so rare that paleobotanists have had to resort to a peculiar practice in their classification of fossil plants. The whole plant, if known from all its parts (which is extremely rare), is given one scientific name (in Latin). Individual parts of the same plant, if discovered separately, also are assigned names which are used for those parts alone because it is uncertain to what "whole plant" they belonged. This results in a multiplicity of names for the several parts of what may once have been a single plant type. *Calamites* is a good example of this. The whole plant, as well as the distinctive, jointed and vertically striated stem, takes the name just given. The genus *Calamites* included a number of species, some of which were as large as good sized trees. *Annularia* and *Asterophyllites,* 2 other generic names, are given to distinctive types of leaves and twigs, which represent at least 2 different kinds of *Calamites. Asterophyllites* has whorls of pointed leaves at each joint or node of the stem. On the other hand,

Figure xv–12. Sphenopsida from the Early Pennsylvanian age Saginaw Formation of Michigan. All illustrations except number 9 are approximately natural size. (Fossil photos courtesy of Professor C. A. Arnold, The University of Michigan Museum of Paleontology.)

1, 2, and 3—Portions of the trunks of several species of the genus *Calamites.*
4—*Macrostachya,* the organ on *Calamites* which bore the reproductive bodies.
5 and 6—*Asterophyllites.*
7—*Sphenophyllum,* a small, low-growing fossil sphenopsid.
8—*Annularia.*
9—Herbarium sheet with mounted specimens of the modern Horsetail or Scouring rush, *Equisetum,* growing in Michigan today. Inches on ruler indicate size of this plant.

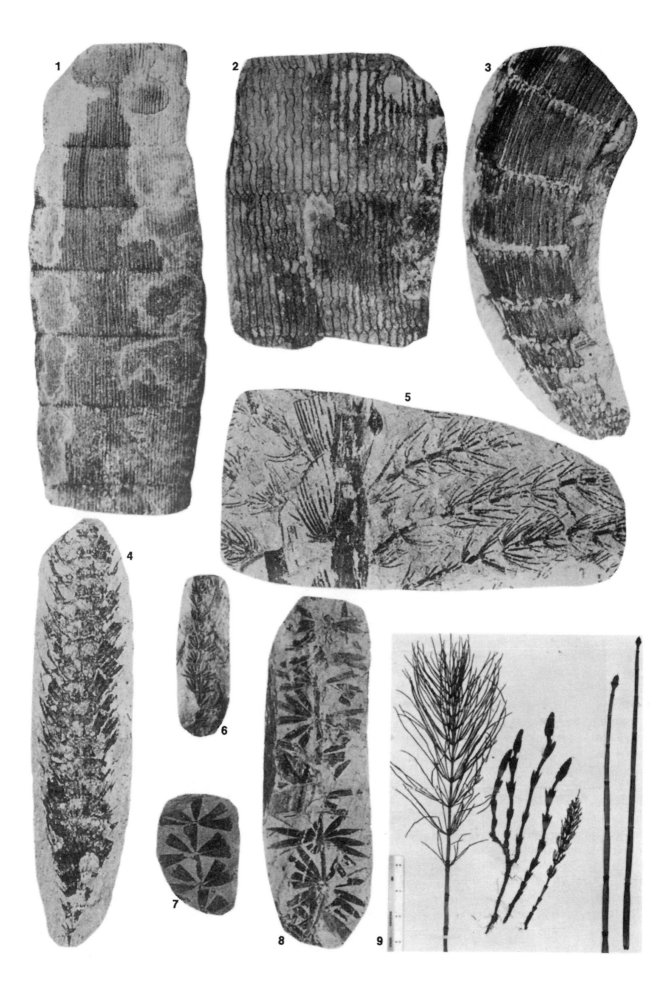

the name *Sphenophyllum* is given to a shrub-like sphenopsid that grew about 2 feet high and was not a *Calamites*. Fragments of *Sphenophyllum* are very common in Michigan, especially at Grand Ledge. The small stems are divided by nodes into segments about a centimeter long. At each node grew circlets of 6–9 wedge shaped leaves with notched tips (Arnold, 1966, p. 9).

FILICINEAE. The True Ferns (Fig. xv–14). Although members of the fern group are common in the modern flora of Michigan, they are rare in the Pennsylvanian age rocks of this state. One genus found at Grand Ledge, *Rhacopteris*, may be a true fern, but the genus also has been classified in the "seed ferns" (Pteridospermales), a structurally and evolutionarily more advanced group described below. The key to the problem lies in whether *Rhacopteris* bore reproductive spores on the undersides of the leaves as do true ferns, or had enlarged, seed-like reproductive bodies on the leaf ends as did seed ferns. These questions are not yet fully answered by the fragmentary fossil record. The problem is discussed more fully below.

PTERIDOSPERMALES (of the *Gymnospermae*) (Figs. xv–13, 14). Members of this group are called the "seed ferns." Although they had delicate, lacy, fern-like foliage, they contrasted with true ferns because their reproductive bodies were contained in much enlarged, seed-like structures on the leaf ends. True ferns, on the other hand, bear their minute reproductive spores in small capsules on the undersides of the fronds. Seed fern leaves were compound, having a number of leaflets arranged in two series along opposite sides of a central axis. The individual leaflets, even though much compressed in the fossil state, have a thick, "fleshy" appearance, somewhat like the leaves of the familiar modern rubber plant. Each leaflet shows distinct, finely divided, closely spaced veins running parallel to one another on either side of a midrib. The stems were simple and woody. Seed ferns ranged in size from low, sprawling shrubs to trees. The whole group is now extinct. Seed ferns were evolutionarily advanced over true ferns in their mode of reproduction. Spores of ferns and the other types of primitive plants mentioned earlier (Lycopsida and Sphenopsida) are simple cells which are freed from the parent plant in

Figure xv–13. Restoration of a typical, extinct pteridosperm ("seed-fern") of the genus *Neuropteris* from Pennsylvanian age rocks of Michigan.

enormous numbers at a very early stage of development. Thus they stand little individual chance for survival, the parent plants depending on the force of numbers for reproduction. Seed ferns, on the other hand, appear to have provided their reproductive bodies with a protective covering. The reproductive bodies themselves may have undergone fertilization and may even have grown to some degree of maturity within the "seed" before being cast loose. Some paleobotanists believe that seed ferns were ancestral to the later and more advanced true-seed-bearing plants, but this suggestion is denied by others. Actually, the exact character and mode of function of the reproductive bodies of seed ferns are incompletely understood at present. The seed fern genus *Neuropteris*, including several distinct species, is common in the Pennsylvanian Saginaw Formation of Michigan. *Neuropteris* is the name for a "whole plant" which is known from associated parts. *Sphenopteris* and *Alethopteris* are names applied to two distinct types of seed fern leaves, both also common in this state.

CORDAITALES (of the *Gymnospermae*) (Figs. xv–15, 16). This once dominant group, common in the Pennsylvanian rocks of Michigan and elsewhere, is now extinct, its last appearance in the

**Fossil Plants**

435

Figure xv–14. Extinct, fossil pteridosperms ("seed-ferns") and other fern-like foliage from the Early Pennsylvanian age Saginaw Formation near Grand Ledge, Michigan. All photos natural size. (Courtesy C. A. Arnold, The University of Michigan Museum of Paleontology.) (Also see Fig. v–31.)

*Left—Rhacopteris michiganensis,* a plant of uncertain relationships with fern-like foliage but neither a true fern nor a seed-fern.

*Center—Neuropteris tenuifolia.*

*Right—Neuropteris.*

geologic record being in rocks of Triassic age. However, many paleobotanists believe this was the group of plants from which the modern conifers evolved. Some students even go so far as to classify members of this group as primitive conifers. The woody trunks and stems were quite like those of modern conifers in structure, except for the presence in Cordaitales of a large and distinct central pith. The pith, as much as an inch in diameter, evidently decayed more rapidly than the outer portion of the trunk after the plant died. As a consequence the resulting cavity commonly became filled with sediment after the trunk fell to the floor of a Pennsylvanian swamp and became buried. Sediment fillings of the pith cavity are common in Pennsylvanian rocks. Tree-sized plants were abundant. The long (up to 3 feet) strap-like leaves with closely spaced, parallel veins, resembled the leaf of a cat-tail. Because of their length and fragile nature whole *Cordaites* leaves are rare in the fossil state but fragments are common. The leaves were borne in spirals on small branchlets. Reproduction was by means of seeds which were heart-shaped in outline, often winged for better dispersal through the air, and borne on short stems separate from but in amongst the leaves. *Cordaites* is a common genus in the Pennsylvanian rocks of the state. Sections of the trunks and portions of the long leaves are most common, the whole plant never being preserved completely intact. *Cordaites* is a "whole plant" name. The name *Cordaianthus* is given to the seed-bearing inflorescence produced by forms of *Cordaites,* and *Cardiocarpon* is the name applied to the winged seeds of the same forms of plants.

The foregoing descriptions of Pennsylvanian

Figure xv–15. Restoration of the extinct cordaitalan tree named *Cordaites*, common in Pennsylvanian age rocks of Michigan.

age fossil plant remains are based upon macroscopic materials, those large enough to be clearly visible to the naked eye. However, microscopic plant spores also have been found in the Pennsylvanian rocks of the state. Arnold (1950) described and illustrated these minute reproductive bodies in great detail. To obtain the specimens, he softened samples of coal from the Pennsylvanian deposits in nitric acid, replaced the acid with a potassium hydroxide solution, and then passed the resulting mud through a fine sieve (25 openings to the inch). The spores were picked out of the residue left on the sieve with the aid of a binocular microscope. The specimens then were treated to make them semitransparent. Thus, they could be photographed through the microscope by means of transmitted light, appearing as shown in Figures xv–18 and 19.

The classification of spores (and pollen) is made difficult by the fact that it often is impossible to be certain what plant was the parent. Thus, distinctive types of spores may be given unique scientific names even though their relationships in the formal classification of the Plant Kingdom are uncertain. The fact that fossil spores are so rarely found in direct association with the parent plant means that the classification of spores in large part must be "artificial" rather than "natural." A natural classification is desirable because it reflects ancestor-descendant and intragroup relationships, as well as being a useful tool for identification. An artificial classification is only a key to recognition, although this alone is very useful. If different types can be told apart, this provides a means for organizing the fossil record to provide stratigraphic and correlation data. There are, however, some cases in which the affinities of spores to parent plant *are* known, at least in a general way. In such cases it may be established that a spore genus was produced by members of a certain major plant group, such as the Lycopsida (see above), even though it cannot be told exactly what genus or species of lycopsid was the parent plant. In other cases, spores of a certain type have been found in direct association with a fossilized portion of a parent plant. Then the relationship is more certainly established, although it may turn out that the same or a closely similar type of spore also was produced by other related plant genera and species. For purposes of the study of ancient environments (paleoecology), however, it may be very significant simply to know that the spore flora of certain rocks originated from one or more members of well known major plant groups. Thus, one could reach paleoecologically significant conclusions such as, for example, "the flora of neighboring lands about the marine basin was dominated by lycopods." This, in turn, would provide a general picture of the nature of land vegetation in some part of geologic time, even though one could not specify whether the lycopod forests consisted of *Lepidodendron, Sigillaria*, or other lycopods.

Returning then to the Pennsylvanian spores of Michigan, Arnold (1950) identified 15 species of

# Fossil Plants

spores from the rocks of the Michigan Basin. All 15 occurred in rocks of the Saginaw Formation (Chapter V, Fig. v-2) which is Early Pennsylvanian in age, although 2 of the 15 also were found in the overlying and younger Eaton Sandstone of the Grand River rock group which now is thought to be Late Pennsylvanian in age. In the Saginaw Formation, the species fall into the following genera: *Triletes* (9 species), *Cystosporites* (2 species), *Lagenicula* (3 species), and *Sporites* (1 species). Both species from the Eaton Sandstone are classified in the genus *Triletes*. Representative spores of each genus are illustrated in Figures xv-18 and 19. According to Arnold (1950, p. 73), *Triletes ramosus*, a common species in the Saginaw Formation, is one of the chief constituents of the spore coal (cannel coal) found near Williamston, not far from Lansing, and described by Berquist (1939). Arnold (1950, pp. 74-88) also concluded that several species of *Triletes* (*ramosus, triangulatus, auritus, mammillarius*, and *Fermii*), 2 species of *Lagenicula* (*rugosa* and *saccata*), and *Cystosporites giganteus*, all were produced by plants of the lycopod group. Included in that group are trees such as *Lepidodendron*. Evidently, the forest cover of Michigan

Figure xv 16. Cordaitales. Fossils from Early Pennsylvanian age Saginaw Formation near Grand Ledge, Michigan. (Photos courtesy C. A. Arnold, The University of Michigan Museum of Paleontology.)

*Top*—*Cordaites*; fragments of leaves. Actual length of rock slab is about 10 inches.

*Bottom*—*Cardiocarpon*, the seed of *Cordaites*. Note winged edge. Actual size about one-half to three-quarters of an inch.

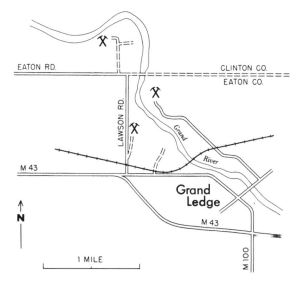

Figure xv-17. Map of the Grand Ledge area showing locations of some of the Michigan quarries where Pennsylvanian age fossil plants occur. Permission to enter must be obtained from the Grand Ledge Clay Product Company. (From C. A. Arnold, in *The Michigan Botanist*, Vol. 5, 1966.)

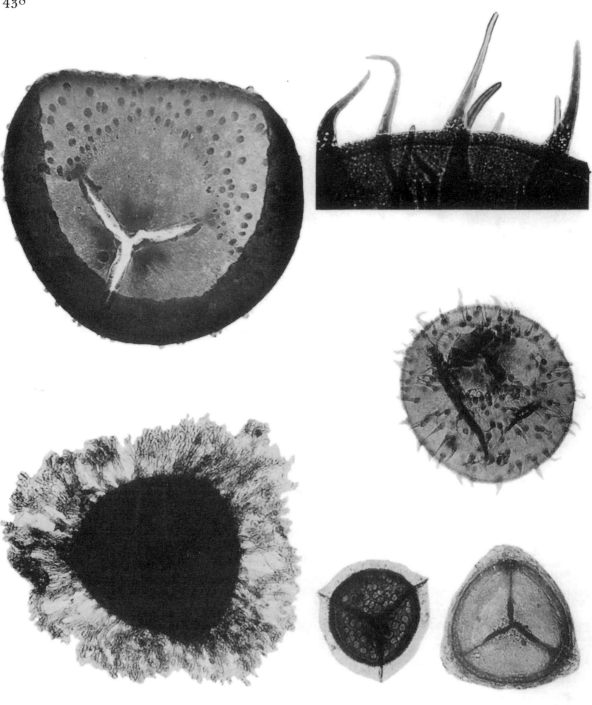

Figure xv-18. Some of the megaspores from Early Pennsylvanian age rocks of the Saginaw Formation. All about ×50, except upper right which is ×200. *Top left—Triletes mamillarius*. Grand Ledge, Michigan. *Bottom left— Triletes ramosus*. From a spore coal at Williamston. *Right top and right center—Lagenicula horrida*. From coal bed at Grand Ledge. *Bottom right—Triletes auritus*. Coal bed; Grand Ledge. *Bottom center—Triletes triangulatus*. From old coal mine near Woodville. (Courtesy C. A. Arnold, The University of Michigan Museum of Paleontology.)

in Pennsylvanian time was rich in lepidodendraceous trees. This type of forest flora was a continuation of the type that was dominant during the Mississippian Period in Michigan and adjoining areas, as revealed by the studies of Chaloner (1954) which were reviewed in the preceding section of this chapter.

This completes the list of plants common in the Pennsylvanian rocks of the state. It is important to notice that the major plant groups described above are not true Flowering Plants (the group called Angiospermae). Yet Flowering Plants today dominate the land and swamp vegetation of large portions of the state, and for that matter of the world. If we could look back directly through time at the landscape of the Pennsylvanian Period that scene would present a very alien aspect. We would miss the deciduous trees such as the oaks, maples, and other broad-leaved types. No flowering weeds or wild flowers would be seen, nor would any of the many forms of grasses so important in nature and to our economy and way of life today. Even the evergreens appear to have been lacking. Some more advanced types of plants may have been in existence on the high hills and mountains well away from the Pennsylvanian swamps, but, if so, the fossil record does not yet prove this. Most of the common and familiar types of modern plants were yet to come in that distant time.

The foregoing sketches of Pennsylvanian plant life were necessarily brief and the reader may wish to expand his knowledge of the Pennsylvanian coal-forming floras of Michigan. If so, he should consult the publications by Professor Arnold listed in the Bibliography.

## Mesozoic Floras

Until recently, no rocks of the Mesozoic Era (Triassic, Jurassic, and Cretaceous periods) had been recognized in Michigan (Chapter VI on the "Lost Interval"). It had long been known from well records that a formation called the "red beds" occurred directly below glacial deposits over an area of several thousand square miles in the central part of the Lower Peninsula (Chapter VI). Unfortunately, there are at present no known surface outcrops of those deposits. Because they lie directly above the Pennsylvanian rocks, in a stratigraphic position that might have been occupied by Permian deposits if they ever existed here, and because Permian rocks elsewhere in North America and the world often are red, it long ago was suggested that the Michigan "red beds" might be of Permian (Latest Paleozoic) age. For a long time there was no fossil evidence to either confirm or deny this suggestion. Recently (1964), however, Professor Aureal T. Cross of the Geology Department at Michigan State University reported some very interesting palynological discoveries he and his students had made during the course of their studies of fossil spores and pollen from the "red beds." Spores and pollen are minute reproductive bodies of plants. Often they survive in the fossil record long after other plant materials have decayed or otherwise been destroyed. Moreover, they are small enough to be brought up whole and intact in the cuttings produced by drilling, and can be separated by certain laboratory techniques from the sedimentary rock materials that contain them. Spore and pollen forms often are sufficiently distinct to allow generic or even specific identification. Palynologists have assembled an impressive record of such fossils from rocks of many different ages throughout the world. Many unique and identifiable types are known to be restricted to rather short spans of geologic time and thus may serve as "time index fossils" for the determination of the geologic ages of the rocks that contain them. Cross' palynological studies of the Michigan "red beds" were only recently begun and are still in progress so the final report is yet to come. However, he has already shown that the "red beds" are not Permian (Late Paleozoic) in age but were deposited during the Mesozoic Era. His studies are not yet conclusive but now seem to indicate that a Late Jurassic age is most probable, especially because so far no pollen from Flowering Plants (Angiosperms) has been found. As was explained earlier in this chapter, the Angiosperms are the dominant, most highly evolved and most adaptable plants of the modern world, but they were the last major plant group to appear in the geologic past. So far the earliest positively identified and accurately dated fossil Angiosperms are Early Cretaceous in age. Thus, the absence of Angiosperm pollen in the "red beds" is a form of negative evidence pointing toward a pre-Cretaceous age for those sedi-

Figure xv-19. *Discinites delectus* from Pennsylvanian age rocks of Michigan. Cone-like, spore-bearing body in upper left is about natural size; magnified directly below. Other views are variously magnified masses of spores and single spores extracted from the sporangium. Common association with fern-like foliage of *Rhacopteris* suggests relationship, but the proper classification of the foliage is uncertain. (Courtesy C. A. Arnold, The University of Michigan Museum of Paleontology.)

ments. The palynological materials that do occur in the "red beds" point more convincingly toward a Late Jurassic age, particularly since many of the genera found are common and typical in the Late Jurassic Morrison Formation of the Western Interior of North America. This recent discovery is all the more interesting because the Jurassic and Cretaceous periods both were times of dominance and maximum abundance of dinosaurs, the first appearance of birds, and of many forms of very primitive early mammals (Chapter VI). The famous dinosaur specimens from the Dinosaur National Monument in Utah and from Como Bluff in southeastern Wyoming were collected from the land laid stream deposits of the Morrison Formation. Giant mollusks of the ammonite group and many other interesting invertebrate and vertebrate animals swarmed in the warm seas of the Jurassic and Cretaceous, too. Perhaps some day a highway cut, gravel pit excavation, or quarrying operation will expose some of the "red beds" of Michigan and offer us a chance to look directly for fossils larger than the tiny spores and pollen that Cross has found.

The tally of plants represented by spores and pollen is still incomplete and the identifications tentative, so the list that follows can only be a "progress report," but it will be sufficient to show that the Mesozoic flora at the time of deposition of the Michigan "red beds" was quite varied. Fern spores have definitely been identified, as have spores of cycads (Fig. xv-1—Classification of Plants). Several varieties of conifers (Phylum Gymnospermae, Class Coniferales) occur in the fossil flora. These include *Tsuga* (the mountain hemlock), and a relative of the modern monkey puzzle tree (an auracarian conifer). A common pollen type is that called *Classopollis*. You will recall from our earlier discussion, in the section on Pennsylvanian plants, that paleobotanists often have difficulty relating the separate parts of a single plant when these occur as isolated fragments, and that sometimes a plant organ may be so similar in several different types of plants in a single group that the single organ alone cannot be positively identified as having belonged to one or another of the members of that group. *Classopollis* pollen is an example of the second type of problem. This pollen is known to have been produced by at least 3 different genera of conifers, *Brachyphyllum, Pagiophyllum,* and *Cheirolepis.*

Thus, any one or more of those 3 genera might be represented in the Michigan "red beds." Illustrations of typical spores and pollen discussed above will be found in Figure xv-20.

## Post-Pleistocene Floras of Michigan

The Cenozoic Era includes over 60 million years of geologic time from the end of the Cretaceous to the present. It is divided into two unequal periods, the long Teritiary Period, followed by the relatively brief Quaternary Period. So far as is now known, the Tertiary Period, including the Paleocene, Eocene, Oligocene, Miocene, and Pliocene epochs, left no rock record here in Michigan (Chapter VI). The Quaternary Period customarily is divided into the Pleistocene Epoch (approximately synonymous with the "Ice Age") beginning about 2 million years ago, and the brief Recent Epoch which began after the retreat of the last continental ice sheet. The geologic record indicates that at least 4 major and many minor advances and retreats of continental ice sheets occurred in North America during the Pleistocene (Chapters VII and VIII). There is little doubt that Michigan was overridden by each ice advance, then bared again when the ice retreated during what are called "interglacial" times. Presumably, too, plants and animals were forced out of this area by each advance, some becoming extinct or at least never returning to this area. Some animal and plant groups, on the other hand, were able to reestablish themselves in the Great Lakes Region after the ice retreated. Unfortunately, each successive major ice advance seems to have destroyed the deposits of the preceding stage here in the heavily glaciated Michigan area so that, with few possible exceptions, all the surficial deposits remaining here were laid down either by the withdrawing ice of the last (Wisconsin) glacial stage or in Post-Wisconsin time. Hence, the first fossil record of plants after that of the Mesozoic "red beds" (see above) in Michigan is one of Late Wisconsin and Post-Wisconsin time, beginning in southern Michigan less than 15,000 years ago. The causes and manner of retreat of the irregular or "lobate" front of the Wisconsin ice sheet were described in Chapters VII and VIII, but it should be recalled that the retreating ice lingered longer where it was thicker in the lower land areas which

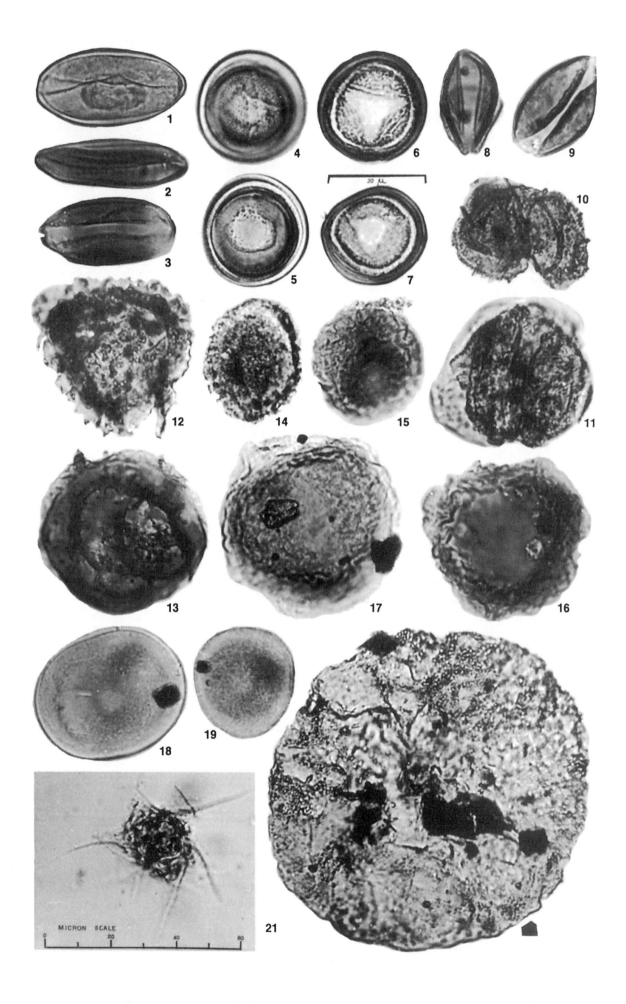

Figure xv–20. Some typical palynomorphs of the Late Jurassic "Red Beds" of the central part of the Michigan Basin. Bar scales on photographs are in microns (1 micron equals 1/1000 millimeter); magnification of objects about ×625. Photographs and identifications are by Professor Aureal T. Cross, Department of Geology, Michigan State University, and Timothy A. Cross. Numbers 12 and 13 are fern spores; number 21 is a dinoflagellate (protozoan with a cellulose body cover); all others are gymnosperm pollen; no angiosperm (flowering plant) pollen yet found.

1. *Cycadopites* sp.; 2. *Cycadopites* sp.; 3. *Cycadopites* sp.;
4. through 7. *Classopollis classoides* Pflug, the most common pollen grain and very typical of Late Jurassic time;
8. and 9. *Ginkgoretectina* sp.; 10. *Platysaccus?* sp.;
11. *Podocarpidites* sp.; 12. *Hamulatisporis* sp.;
13. *Staplinisporites* sp.; 14. *Tsugaepollenites* sp.;
15. *Tsugaepollenites mesozoicus?* Couper; 16. *Tsugaepollenites* sp.;
17. *Callialasporites dampieri* (Balme) Sukh Dev;
18. *Spheripollenites* sp.; 19. *Exesipollenites* sp.;
20. *Araucariacites* sp.; 21. *Baltisphaeridium* sp. (a dinoflagellate).

now are the Great Lakes basins. At the same time the ice front melted back more rapidly, and at any particular time was located farther to the north, in the higher land areas of the central part of the Lower Peninsula because the ice had been thinner there originally. These facts should be kept in mind because it followed from them that when vegetation reoccupied the barren lands left behind by the retreating ice, the pioneer floras became established in central Michigan much earlier and farther north than in the areas occupied by the lingering ice lobes extending southward in the Michigan and Huron-Erie basins on either side.

The Wisconsin ice sheet did not withdraw steadily, but fluctuated back and forth several times. In general, each successive minor readvance was less extensive than its predecessor. These minor fluctuations (such as the Cary, Port Huron, and Valders) affected the vegetation locally, in some cases even overriding forests as at Two Creeks in Wisconsin (Chapter VIII). Moreover, the waters of the Glacial Great Lakes rose and fell several times for a variety of reasons discussed in Chapter VIII. Climates changed, too, favoring certain plants over others. And finally it should be remembered that time is required for a newly exposed land surface to develop soil, to become clothed with vegetation, and for that vegetation to reach its full development. All these factors operated as the Wisconsin ice withdrew from the state.

With this introduction, let us now examine the Late Wisconsin and postglacial history of vegetation in the Great Lakes region.

Some of this history is recorded in deposits of plant material such as peat, often interbedded with other types of sediment. Peat deposits are particularly thick and extensive in the many bogs and swamps within the glacial terrain of the state. Acidic conditions and oxygen depletion retard plant decay in those situations, whereas plants that die on uplands soon are destroyed, leaving little to record their former presence but the organic rich upper zones of some soils. Even in the swampy lowlands decay tends to break down dead plants until only fragments remain. Occasionally, a large log or branch may survive, to serve later as a clue for identification of its plant type, or as a source of plant carbon for use in radiocarbon dating of the enclosing sediments.

By far the best records of the floral history of Michigan in postglacial time come from palynological studies of fossil pollen and spores. These minute plant reproductive bodies often survive decay and are identifiable long after the leaves and stems of the plants that produced them have decayed to an unrecognizable state. Carefully collected samples of sediment from successive superimposed levels often can be analyzed for their pollen and spore content. Certain types of pollen are particularly susceptible to long distance transport by wind. Other types may be especially abundant because of the unusually great amount of pollen produced by certain parent plants. Palynologists can make fairly reliable corrections for such sources of error and then, by identifying and counting the numbers of grains of each type of pollen or spore, reach an estimate of the relative abundance of the various types of plants that lived in the region at the time represented by the level from which the sample was collected. Changes in relative abundance of pollen types, and the appearance or disappearance of certain types from

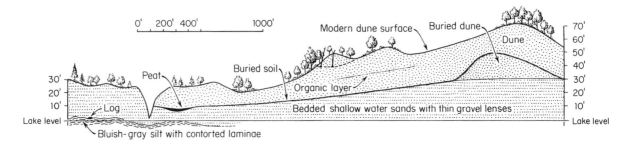

Figure xv-21. The Pilgrim Haven Site south of South Haven, Michigan. (From Zumberge and Potzger, 1956.)

lower to higher levels at a particular site, reflect changes in the regional flora through time. This type of investigation is relatively new, and much work remains to be done before the historical picture even approaches completion, but a number of sites have provided sufficient palynological evidence now to allow the outlines of the Late Pleistocene and Post-Pleistocene floral history of Michigan to be sketched here. Only a few of the most significant studies can be reviewed, but they will serve to indicate the major changes that occurred. The results will be summarized very briefly, but additional detail on such matters as sampling technique, preparation and study methods, location of sites and horizons can be found in the specific publications referred to in the bibliography.

The "Pilgrim Haven" site, about 3.5 miles south of the town of South Haven, along the Lake Michigan shoreline has produced some of the most significant palynological information here in Michigan. Studies were conducted there by Zumberge and Potzger (1956). A stratified sequence of sediments of varied origin crops out in the wave cut shore line of the lake, on the properties of the Larchmoor estate and the Pilgrim Haven church camp. Figure xv-21 shows the geology and stratigraphy of the site. The lowermost deposits are bluish gray silts. The upper layers of the blue silt crop out just above the modern water level, but lower layers extend to an undetermined depth below. A buried log, found resting amid other plant debris on the very top of these fine-grained sediments, produced a radiocarbon date of approximately 11,000 years B.P. (Before Present). That is a date for the time of death of the tree itself, but presumably the log was deposited on the silt a relatively short time after the tree died so that date also serves to indicate in a general way the age of the deposits immediately above and below. Samples of sediment for palynological analysis were collected at 1-inch intervals up the sedimentary section above the silts. These samples yielded a continuous pollen "profile" through a vertical distance of about 107 inches up to the base of an overlying dune sand. Carbon rich organic material from a buried peat zone, some distance above the log, provided not only an exceptionally good pollen sample, but was dated by the radiocarbon method at several levels. The base of the peat gave an age of about 8000 years B.P., the middle section an age of about 5000 years B.P., and top was dated at about 4000 years B.P. Hence the peat zone itself represents about a 4000-year span of time. The total pollen profile, from the level of the log on top of the blue silts to the top of the peat, represents a span of time from about 11,000 B.P. to about 4000 B.P. The pollen profile was "truncated" on both top and bottom, that is to say there were no sediments containing pollen above and below those age levels available for analysis. However, a similar pollen analysis based on samples collected from a thick section of peat in a bog near Hartford, about 15 miles inland east of Pilgrim Haven, produced a similar and partially correlative pollen profile, also studied by Zumberge and Potzger (1956). The Hartford Bog pollen profile extends both backward and forward in time beyond the limits represented at the Pilgrim Haven site. Figures xv-22, 23 presents the results of these studies graphically.

We have altered and modified the 1956 analysis and conclusions of Zumberge and Potzger

# Fossil Plants

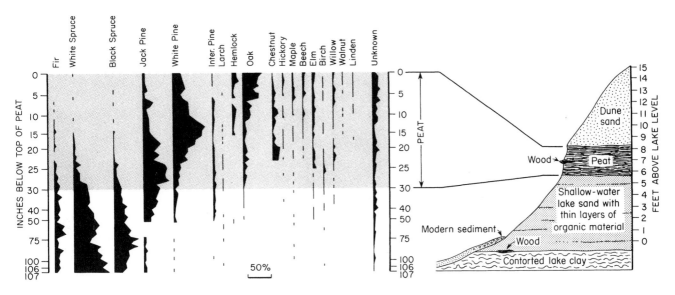

Figure xv-22. Pollen profile from Late Pleistocene and postglacial deposits at Pilgrim Haven site south of South Haven, Michigan. (From Zumberge and Potzger, 1956.)

somewhat to take some more recent data into account. The principal historical events represented at the 2 sites seem to be as follows. The pollen record begins with a period about 13,000 years ago, represented in the lower 9 feet of the Hartford Bog profile, when the Cary glacial ice had just melted back from that area. It was then that meltwater from the retreating ice and pollen from vegetation first began to collect in the depression. At that time the "pioneer" flora consisted of a relatively simple association dominated by white and black spruce, but with a small percentage of fir.

The blue silt now exposed at and just below modern lake level at Pilgrim Haven was deposited in Glacial Lake Chicago sometime before about 12,000 years B.P., when the waters of that lake stood at elevations between 640 and 605 (during the Glenwood, Calumet, and Toleston stages). Lake level then dropped to the low water Kirkfield stage, associated with the Two Creeks interval (at least 11,850 years B.P.), and lake waters withdrew to the west, permitting the accumulation of the 11,000 year old log and associated organic material on top of the silts. Spruce and fir still were the dominant trees. Then the Valders ice moved back into the Great Lakes region from the north, and the water once more rose to cover the South Haven site. During this high water stage, the time of Algonquin (605 feet elevation), several feet of beach sands with lenses of fine gravel were deposited in relatively shallow water. The few thin layers of organic rich material found interbedded with the beach sands and gravels at Pilgrim Haven probably were deposited in a shallow bay protected from strong wave action by a spit or barrier beach that separated it from the open lake which lay to the west. Spruce and fir were declining and jack pine was increasing in abundance. As the ice which held in Lake Algonquin retreated to the north, uncovering once more the Trent Lowland, and eventually the still lower North Bay–Mattawa lowland to the Ottawa River, the lake level began to drop and the shore line receded further and further west of the South Haven area, until neither wave nor wind action affected the region. Vegetation took hold on the now stable site and peat began to accumulate in a low sag, while a soil developed on the drier bordering terrain. The interval from 8000 to 4000 years B.P. was, thus, a time when the lake shore was far enough to the west that little or no deposition of inorganic materials occurred at the site. By 8000 years B.P. spruce and fir had disappeared

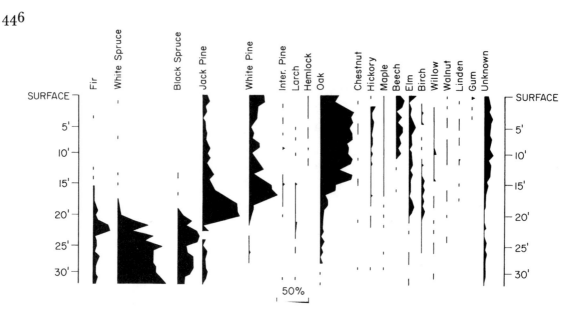

Figure xv-23. Pollen profile from Late Pleistocene and postglacial deposits at Hartford Bog (from Zumberge and Potzger, 1956).

from the area and jack pine had become the dominant tree. But the vegetation continued to change. Jack pine gave way to a mixed pine (including white pine) assemblage and, by 4000 years B.P., a mixed oak-pine assemblage had come to dominate the Pilgrim Haven area. According to recent correlations and dates from Hough (1963) the lowest lake stage, Chippewa (dated at approximately 9500 years B.P.), preceded the beginning of peat accumulation; and the 30-inch thick peat layer and the soil horizon actually represent the long interval of intermediate, but steadily rising lake levels which resulted from the continual uplift of the North Bay outlet, culminating in the Nipissing Great Lakes about 4000 years B.P. As the lake shore approached the South Haven area windblown sand covered the bog, bringing to an end both the accumulation of peat and the formation of the correlable soil, and "truncating" the Pilgrim Haven pollen record. Presumably, the Nipissing shore lay just to the west of the site, although later erosion has removed all trace of the beach itself from the area. Continued wind action, dune formation, and shore erosion have marked the area since Nipissing time.

To continue the plant history we must return to the Hartford Bog area. By 3500 years B.P. the climate of the Great Lakes Region had advanced well into what is known as the "xerothermic" (hypsithermal, "climatic optimum") stage, a time of milder, somewhat warmer and drier conditions than exist here today. This was a time when a rich oak-hickory-pine forest assemblage with many other associated broad-leaved plants flourished. However, from about 2500 B.P. to the present, while the Great Lakes waters continued to drop from the Nipissing levels toward those of the present, the climate seems to have "deteriorated" from the optimum. This last change, as reflected in the pollen profile at Hartford Bog, caused a return from the rich oak-hickory-pine-broad-leaved forest association back to an impoverished oak-pine association. This was the assemblage which greeted the pioneer settlers of the southwestern Lower Peninsula of Michigan upon their arrival.

The foregoing history of the Michigan Late Pleistocene and Post-Pleistocene flora is stated in terms of the "crown cover" of vegetation, that is, the trees forming the forest top. Actually, many other types of pollen and spores occur at the two sites, revealing the character of the whole flora as well. The details are shown in Figure xv-23.

Palynological studies east of the Hartford Bog–Pilgrim Haven area provide fragments of a similar history of forest development and change during late-glacial and postglacial times in the south central and southeastern parts of the Lower Peninsula. In 1956 Ray Coville, of Climax, Michigan, found 3 large vertebrae in a marl pit on the

Coville Farm in Kalamazoo County. Professors W. S. Benninghoff of the Department of Botany and C. W. Hibbard of the Museum of Paleontology and Department of Geology of The University of Michigan studied these bones and reported their findings in the *Papers of the Michigan Academy of Science, Arts and Letters* (1961). Hibbard identified the bones as being the first three vertebrae behind the skull of the now extinct woodland musk ox, *Symbos cavifrons* (Chapter XIV). The third vertebra was destroyed in the process of performing a radiocarbon analysis and provided a date of 13,000 (± 600) years B.P. This date was determined by The University of Michigan Memorial-Phoenix Project Laboratory and was reported first by Hibbard and Hinds (1960, pp. 103-4). The vertebrae, when found, were filled with marl which contained pollen and spores. The nature of the specimens and their occurrence in the field indicated that the marl, with its contained pollen, entered the bones shortly after the musk ox died, hence the radiocarbon date for the bone was also a close date for the pollen as well. Benninghoff was able to identify a wide variety of pollen and spore types, indicating that the flora of the Kalamazoo region and adjoining areas, at the time the musk ox lived there, included the following trees: spruce (86.9 percent), balsam fir, larch, birch, blue beach-hop hornbeam, butternut, hickory, oak, elm, linden (or basswood), and maple. Nontree pollen (or spores) represented the following: broad-leaved cattail, pond weed, grasses, sedges, willow, alder, pigweed, pinks, ragweed, white sage, goldenrod, peat moss, ground pine *(Lycopodium)*, and various ferns. The most striking fact was the dominance of spruce-fir pollen, just as in the lower levels at the Pilgrim Haven and Hartford Bog sites. Benninghoff and Hibbard reached several conclusions as follow. A closed stand (not an open park or glade) of spruce-fir forest dominated the area 13,000 years ago, as indicated by the low percentages of nontree pollen. The hardwood pollen (oak, walnut, hickory, and other broad-leaved deciduous trees), also a minor percentage, was blown in from some distance and does not represent the early vegetation actually living at the site. They also noted that a dominance of spruce is typical in the basal (early, oldest) levels of pollen profiles from many other bogs and lakes in the Lower Peninsula, including Hidden Lake in Livingston County, Powell Bog in Clare County, and Third Sister Lake on the western edge of Ann Arbor in Washtenaw County. Thus, it is apparent that, as at Pilgrim Haven and in Hartford Bog in southwestern Michigan, the vegetation that first occupied the south-central and southeastern parts of the Lower Peninsula, just after the glacial ice of the Wisconsin Ice Sheet had finally withdrawn, was markedly different from that of today. Of course one would have suspected from the presence of the musk ox, even though it was a woodland form rather than the modern tundra type, that environmental conditions were much cooler in Michigan 13,000 years ago than now. Benninghoff and Hibbard (1956, p. 158) concluded that, "the Woodland Musk Ox lived in an extensive spruce forest that contained small proportions of balsam, fir, larch and possibly birch. Local openings in the forest probably occurred on gravelly flood plains and sand plains, where there would have been stands of grasses, with ragweed, sage, goldenrod, other herbs, and willows. Ponds and lakes were occupied by pondweeds and probably by other aquatic plants, and their shores were bordered, in at least some places, by cattails and sedges. The late-glacial landscape of Kalamazoo County had the aspect of the boreal (northern) coniferous forest that now exists north of Lake Superior, but the flora did not necessarily include all of the species now associated with the boreal forest. It is also possible that the late-glacial spruce forest in southern Michigan contained floral elements that today belong to communities well south of the boreal forest, especially if those plants were aggressive pioneer species that could have advanced rapidly across the freshly deglaciated areas."

One final example will illustrate the manner in which studies of pollen profiles, fossil animal remains, and the geology of a site contribute jointly to an historical reconstruction of a part of Michigan's past after the glacial ice receded. Kapp and Kneller (1962) studied an assemblage of plant remains including pollen, and fossil mollusks buried in a stratified sequence of gravel, sand, silt and clay in a low terrace of the Saline River near Milan in southeastern Michigan. A log from a red oak tree, buried about 1 foot above the base of the 10-foot sequence of sediments, provided material for a radiocarbon date. Kapp and Kneller (1962, p. 144) summarized their findings as follows:

### Figure XV–24

#### LATE-GLACIAL AND POSTGLACIAL CLIMATIC CHANGE AND GENERALIZED FOREST HISTORY IN SOUTHERN MICHIGAN

(See Fig. XIV–14 for fossil vertebrate and Early Man records)

| *Approximate Number of Years* (B.P.) | *Dominant Trees in Crown Forest* | *Climate* | *References* |
|---|---|---|---|
| 2500 B.P. to present | Oak—pine | Rapid deterioration from climatic optimum to cooler, more moist conditions | |
| 3500 B.P. | | Climatic optimum (Xerothermic) | Zumberge and Potzger, 1956 |
| 4000 B.P. | Rich broad-leaved forest of oak-hemlock (little pine) in southwestern Michigan | Xerothermic climate | |
| 4080 B.P. | Walnut, sycamore, and butternut, plus oak, basswood, elm, beech and ash in southeastern Michigan | Entering Xerothermic (Hypsithermal) Interval. Warm, dry climate | Kapp and Kneller, 1962 |
| 5000 B.P. | Increase of oak, chestnut and other hardwoods; reduced pine; in southwestern Michigan | Continued warming trend | Zumberge and Potzger, 1956 |
| 6000 B.P. | Pine maximum. Mixed stands including white pine. Spruce and fir nearly gone | Warming trend | |
| 8000 B.P. | Jackpine dominant in southwestern Michigan | | |
| | Pine continues to increase, spruce and fir to decrease in abundance | Moderating postglacial climate | |
| 10,700 B.P. | Spruce and pine dominant and about equally abundant in south-central Michigan | | Oltz and Kapp, 1963 |
| 11,100 B.P. | Pine increasing, spruce still dominant but decreasing in south-central and southwestern Michigan | | Semken, Miller, and Stevens, 1964; Zumberge and Potzger, 1956 |
| 13,000 B.P. or more to about 11,000 B.P. | White and black spruce and fir forest dominant throughout southern Michigan | Climate cold to cool, and relatively moist compared to that of both the Xerothermic Optimum and present periods shown above | Stoutamire and Benninghoff, 1964; Zumberge and Potzger, 1956 |
| 13,200 B.P. | Spruce dominant in southern Michigan | | Benninghoff and Hibbard, 1961 |

We have described an assemblage of plant and animal remains which was exposed upon excavation of a terrace of the Saline River at Milan, Michigan. The mollusks, mostly of fingernail clams, *Sphaerium* spp. and *Pisidium* spp., and some woodland snails, suggest a small stream habitat with adjacent oak-hickory forests. The most obvious plant remains are logs and branches of five kinds of deciduous trees. Wood from the largest log, red oak, has been dated at 4080 ± 200 years B.P. (University of Michigan Memorial-Phoenix Project, M-1149). Although some fruits and seeds were recovered, the most complete evidence for the vegetation and climate of the period is revealed by pollen analysis of the silt enclosing the dated log. Oak, basswood, walnut, elm, sycamore, butternut, beech, and ash are predominant members of the pollen flora. High pollen frequencies of sycamore, butternut, and walnut suggest that the climate may have been warmer than at present in southeastern Michigan. The radiocarbon date agrees well with the period of maximum development of vegetation as a response to the climatic trends of the hypsithermal (xerothermic) interval which has previously been dated in pollen profiles. It is likely that burial of the assemblage occurred because of aggradation of the stream after a rise in the level of ancestral Lake Erie in Nipissing times.

The hypsithermal period referred to by Kapp and Kneller in the quotation above is the same as the period which Zumberge and Potzger, discussing the Pilgrim Haven and Hartford Bog sites, referred to as the "climatic optimum" or xerothermic interval. Note that the radiocarbon dates for the top of the buried peat zone at Pilgrim Haven and for the log in the base of the Saline River terrace deposit agree closely and that the pollen spectra at both sites for that time are closely similar. At both sites the climatic optimum for vegetation occurred about 4000 years B.P.

Not all the details of climatic and glacial fluctuations have been reconstructed, and there is no doubt that future studies will flesh out the bare historical bones of the story considerably, but the evidence now available seems sufficient to support the summary of events, shown in Figure xv–24, in the history of vegetation and climate following the last retreat of ice from the Lower Peninsula of Michigan. Similar events would have followed in about the same sequence but at slightly later dates farther north in the state.

## Summary of the Fossil Record of Plants in Michigan

Figure xv–2 summarizes the geological history of plants in graphic form and shows those portions of that history for which there is a fossil record in Michigan. Although the known Michigan record by no means encompasses the whole of plant history, some especially interesting stages are represented here. The calcareous, coaly, graphitic, and asphaltic deposits of Precambrian algae in the Upper Peninsula are among the most ancient records of life on earth. The "fucoid" markings on Early and Middle Paleozoic marine rocks tell of vast mats of seaweed that floated in the warm, shallow inland seas of that distant age. The scale trees, scouring rushes, true ferns, seed-ferns, and *Cordaites* recall vast swamplands and sluggish streams along the fluctuating margins of widespread Pennsylvanian seas during the great coal-forming age called the "Carboniferous." The tiny but nearly indestructible spores and pollen in the late Jurassic "red beds" of the state are the sole known fossil record of life to survive here from the Age of Dinosaurs. And, finally, pollen from the early spruce-fir forest records the scene in Michigan just after the last retreat of the vast continental ice sheets of the Pleistocene, before the climate in the Great Lakes Region warmed and dried to the "climatic optimum" and then deteriorated to the condition we find in Michigan today.

# Bibliography

AGAR, D. V., 1963, Principles of paleoecology. McGraw-Hill Book Company, Inc., New York, 371 p.

ALESSI, A. J., 1935, Hunting agates around Lake Superior. Rocks and Minerals (magazine), v. 11, p. 139.

ALLEN, J. A., 1913, Ontogenetic and other variations in muskoxen with a systemic review of the muskox group, recent and extinct. American Museum of Natural History Memoirs, v. 1, pp. 101–226.

ALLING, H. L., and L. I. BRIGGS, 1961, Stratigraphy of Upper Silurian Cayugan evaporites. American Association of Petroleum Geologists Bulletin, v. 45, pp. 515–47.

ANDERSON, S. T., 1954, A late-glacial pollen diagram from southern Michigan. Danmarks Geologiske Undersøgelse, Ser. II, v. 80, pp. 140–55.

ARNOLD, C. A., 1947, An introduction to paleobotany. McGraw-Hill Book Company, Inc., New York, 433 p.

———, 1949, Fossil flora of the Michigan Coal Basin. The University of Michigan Museum of Paleontology Contributions, v. 7, pp. 131–269.

———, 1950, Megaspores from the Michigan Coal Basin. The University of Michigan Museum of Paleontology Contributions, v. 8, pp. 59–111.

———, 1954, The Michigan Coal Basin. Michigan Alumnus Quarterly Review, v. 60, pp. 287–96.

———, 1966, Fossil plants in Michigan. The Michigan Botanist, v. 5, pp. 3–13.

BARDACK, D., and R. ZANGERL, 1968, First fossil lamprey: A record from the Pennsylvanian of Illinois. Science, v. 162, pp. 1265–67.

BARGHOORN, E. S., W. G. MEINSCHEIN, and J. W. SCHOPF, 1965, Paleobiology of a Precambrian shale. Science, v. 148, p. 461–72.

BATEMAN, A. M., 1950, Economic mineral deposits. John Wiley and Sons, Inc., New York, 916 p.

BAY, J. W., 1937, Glacial lake levels indicated by terraces of the Huron, Rouge, and Clinton rivers, Michigan. Michigan Academy of Science, Arts, and Letters Papers, v. 22, pp. 411–19.

———, 1938, Glacial history of the streams of southeastern Michigan. Cranbrook Institute of Science Bulletin 12, Bloomfield Hills, Michigan, 66 p.

BAYLEY, R. W., 1958, Geology of the Menominee iron-bearing district, Michigan. Mimeographed abstract for Institute on Lake Superior Geology meeting, Duluth, Minnesota, 5 p.

BENNINGHOFF, W. S., and C. W. HIBBARD, 1961, Fossil pollen associated with a late-glacial woodland musk ox in Michigan. Michigan Academy of Science, Arts, and Letters Papers, v. 46 (Botany), pp. 155–59.

BERQUIST, S. G., 1939, The occurrence of spore coal in the Williamston Basin, Michigan. Journal of Sedimentary Petrology, v. 9, pp. 14–19.

BRETZ, J. H., 1951a, Causes of the glacial lake stages in Saginaw Basin, Michigan. Journal of Geology, v. 59, pp. 244–58.

———, 1951b, The stages of Lake Chicago; their causes and correlations. American Journal of Science, v. 249, pp. 401–29.

———, 1953, Glacial Grand River, Michigan. Michigan Academy of Science, Arts, and Letters Papers, v. 38, pp. 359–82.

———, 1964, Correlation of glacial lake stages in the Huron-Erie and Michigan basins. Journal of Geology, v. 72, pp. 618–27.

BRIGGS, L. I., 1958, Evaporite facies. Journal of Sedimentary Petrology, v. 28, pp. 46–56.

———, 1959, Physical stratigraphy of the lower middle Devonian rocks in the Michigan Basin. Pp. 39–58 in collection of papers compiled by F. D. Shelden for the Michigan Basin Geological Society Annual Geological Excursion Guidebook entitled "Geology of Mackinac Island and Lower and Middle Devonian south of the Straits of Mackinac," published by Michigan Basin Geological Society, 63 p.

BRODERICK, T. M., 1933, Geology, exploration, and mining in the Michigan Copper District. Pp. 29–47 in Guidebook 27, Excursion C-4, 16th International Geological Congress of 1933 (printed by the United States Geological Survey), 101 p.

BROECKER, W. S., and W. R. FARRAND, 1963, Radiocarbon age of the Two Creeks forest bed, Wisconsin. Geological Society of America Bulletin, v. 74, pp. 795–802.

BROWN, A. C., 1965, Mineralogy at the top of the cupriferous zone, White Pine Mine, Ontonagon County, Michigan. Unpublished master's thesis, on file with The Department of Geology and Mineralogy, The University of Michigan, Ann Arbor, 81 p.

BURT, W. H., 1942, A caribou antler from the lower peninsula of Michigan. Journal of Mammalogy, v. 23, p. 214.

———, 1948, The mammals of Michigan. The University of Michigan Press, Ann Arbor, 288 p.

———, 1957, Mammals of the Great Lakes region. The University of Michigan Press, Ann Arbor, 246 p.

———, and R. P. GROSSENHEIDER, 1952, A field guide to the mammals. Houghton Mifflin Company, Boston, 200 p.

BUTLER, B. S., and W. S. BURBANK, 1929, The copper deposits of Michigan. United States Geological Survey Professional Paper No. 144, 238 p.

CARROLL, R. L., 1965, Lungfish burrows from the Michigan Coal Basin. Science, v. 148, pp. 963–64.

CARTER, L., 1958, Rockhounding on Michigan's Northern Peninsula. Lapidary Journal, v. 12, pp. 98 and 100.

CASE, E. C., 1915, On a nearly complete skull of *Symbos cavifrons* Leidy from Michigan. The University of Michigan Museum of Zoology Occasional Papers No. 13, 3 p.

———, 1921, Something about the paleontological collections in the University. Michigan Alumnus, v. 27, pp. 292–300.

———, 1931, Arthrodiran remains from the Devonian of Michigan. The University of Michigan Museum of Paleontology Contributions, v. 3, pp. 163–82.

———, and G. M. STANLEY, 1935, The Bloomfield Hills mastodon. Cranbrook Institute of Science Bulletin, v. 4, pp. 1–8.

———, and others, 1935, Discovery of *Elephas primigenius americanus* in the bed of glacial Lake Mogodore, in Cass County, Michigan. Michigan Academy of Science, Arts, and Letters Papers, v. 20, pp. 449–54.

CHALONER, W. G., 1954, Mississippian megaspores from Michigan and adjacent states. The University of Michigan Museum of Paleontology Contributions, v. 12, pp. 23–35.

CLARK, T. H., and C. W. STEARN, 1968, Geological evolution of North America. The Ronald Press Company, New York, 2d Edition, 570 p.

COHEE, G. V., 1965, Geologic history of the Michigan Basin. Washington Academy of Sciences Journal, v. 55, pp. 211–23.

———, R. N. BURNS, A. BROWN, R. A. BRANT, and D. WRIGHT, 1950, Coal resources of Michigan. United States Geological Survey Circular 77, 56 p.

COLBERT, E. H., 1955, Evolution of the vertebrates. John Wiley and Sons, Inc., New York, 479 p.

CRANE, H. R., 1956, University of Michigan radiocarbon dates I. Science, v. 124, pp. 664–72.

———, 1959, University of Michigan radiocarbon dates IV. American Journal of Science, Radiocarbon Supplement, v. 1, pp. 173–98.

———, and J. B. GRIFFIN, 1958, University of Michigan radiocarbon dates II. Science, v. 127, pp. 1098–1105.

CROSS, AUREAL T., 1964, (Unpublished oral report on Mesozoic spores from the "red beds" of Michigan). Presented at the 1964 meeting of the Geology-Mineralogy Section of the Michigan Academy of Science, Arts, and Letters.

DICE, L. R., 1920, The mammals of Warren Woods, Berrien County, Michigan. The University of Michigan Museum of Zoology Occasional Papers, No. 86, pp. 1–20.

DORR, JR., J. A., 1957, A pleuracanth shark spine from the Early Pennsylvanian, Saginaw Formation, of Michigan. Michigan Academy of Science, Arts, and Letters Papers, v. 42, pp. 99–104.

———, and L. I. BRIGGS, 1953, Historical geology laboratory manual. George Wahr Publishing Company, Ann Arbor, Michigan, 78 p.

———, and E. G. KAUFFMAN, 1963, Rippled toroids from the Napoleon Sandstone Member (Mississippian) of southern Michigan. Journal of Sedimentary Petrology, v. 33, pp. 751–58.

———, and F. MOSER, 1964, Ctenacanth sharks from the mid-Mississippian of Michigan. Michigan Academy of Science, Arts, and Letters Papers, v. 49, pp. 105–13.

DUNBAR, C. O., 1960, Historical geology. John Wiley and Sons, Inc., New York, 2d Edition, 500 p.

———, and J. RODGERS, 1957, Principles of stratigraphy. John Wiley and Sons, Inc., New York, 356 p.

DUSTIN, F., 1932, The gems of Isle Royale, Michigan. Michigan Academy of Science, Arts, and Letters Papers, v. 16, pp. 383–98.

———, 1942, Mineral collecting in unpromising localities. Rocks and Minerals (magazine), v. 17, pp. 86–89.

DUTTON, C. E., 1958, Precambrian geology of parts of Dickinson and Iron counties, Michigan. Michigan Basin Geological Society Guidebook for the annual geological excursion of 1958, published by Michigan Basin Geological Society, 40 p.

EARDLEY, A. J., 1962, Structural geology of North America. Harper and Row, Publishers, New York, 2d Edition, 743 p.

EHLERS, G. M., and R. V. KESLING, 1957, Silurian rocks of the Northern Peninsula of Michigan. Michigan Geological Society, Annual (1957) Excursion Guidebook, lithoprinted by Edwards Brothers, Inc., Ann Arbor, Michigan, 63 p.

ENGEL, A. E. J., B. NAGY, L. A. NAGY, C. G. ENGEL, G. O. W. KREMP, and C. M. DREW, 1968, Alga-like forms in the Onverwacht Series, South Africa: Oldest recognized lifelike forms on Earth. Science, v. 161, pp. 1005–8.

ENSIGN, JR., C. O., 1964, The program of ore dilution control in mining at White Pine Copper Company. Society of Mining Engineers of the American Institute of Mining, Metallurgical and Petroleum Engineers Preprint No. 6414 for paper presented at the Annual Meeting, New York, February, 1965, 29 p.

FARRAND, W. R., 1961, Former shorelines in western and northern Lake Superior. Ph.D. dissertation, The University of Michigan Graduate Library microfilm No. 3827.

———, 1961, Frozen mammoths and modern geology. Science, v. 133, pp. 729–35.

FLINT, R. F., 1956, New radiocarbon dates and Late Pleistocene stratigraphy. American Journal of Science, v. 254, pp. 265–87.

———, 1957, Glacial and Pleistocene geology. John Wiley and Sons, Inc., New York, 553 p.

———, and F. Brandtner, 1961, Climatic changes since the last interglacial. American Journal of Science, v. 259, pp. 321–28.

———, et al., 1945, Glacial map of North America. Geological Society of America, New York.

FOOTE, E. A., 1919, Third volume. Report of the Pioneer Society of Michigan, pp. 402–3.

FRANKFORTER, W. D., 1966, Some recent discoveries of late Pleistocene fossils in western Michigan. Michigan Academy of Science, Arts, and Letters Papers, v. 51, pp. 209–20.

FRYE, J. C., and H. B. WILLMAN, 1960, Classification of the Wisconsinan stage in the Lake Michigan glacial lobe. Illinois State Geological Survey Circular 285, 16 p.

GARRELS, R. M., 1951, A textbook of geology. Harper and Brothers, New York, 511 p.

GATES, D. M., 1939, A deposit of mammal bones under Sleeping Bear Dune. Kansas Academy of Science Transactions, v. 42, pp. 337–38.

GATES, F. C., 1950, The disappearing Sleeping Bear Dune. Ecology, v. 31, pp. 386–92.

GIDLEY, J. W., 1908, Description of two new species of Pleistocene ruminants of the genera *Ovibos* and *Boötherium* with notes on the latter genus. United States National Museum Proceedings, v. 34, pp. 681–83.

GILLULY, J., A. C. WATERS, and A. O. WOODFORD, 1968: Principles of geology. W. H. Freeman and Company, San Francisco, 3d Edition, 687 p.

GILMORE, C. W., 1906, Notes on some recent additions to the exhibition series of vertebrate fossils. United States National Museum Proceedings, v. 30, p. 610.

GRIFFIN, J. B., 1965, Late Quaternary prehistory in the northeastern woodlands. Pp. 655–67 in The Quaternary of the United States, Princeton University Press, Princeton, New Jersey, 922 p.

HAMBLIN, W. K., 1958, The Cambrian sandstones of northern Michigan. Michigan Department of Conservation, Geological Survey Division Publication 51, 141 p.

———, 1961, Paleogeographic evolution of the Lake Superior region from Late Keweenawan to Late Cambrian time. Geological Society of America Bulletin, v. 72, pp. 1–18.

———, 1965, Basement control of Keweenawan and Cambrian sedimentation in Lake Superior region. American Association of Petroleum Geologists Bulletin, v. 49, pp. 950–58.

———, and W. J. HOMER, 1961, Sources of the Keweenawan conglomerates of northern Michigan. Journal of Geology, v. 69, pp. 204–11.

HANDLEY, JR., C. O., 1953, Marine mammals in Michigan Pleistocene beaches. Journal of Mammalogy, v. 34, pp. 252–53.

HATT, R. T., 1924, The land vertebrate communities of western Leelanau County, Michigan, with annotated list of the mammals of the county. Michigan Academy of Science, Arts, and Letters Papers, v. 3, pp. 369–402.

———, 1963, The mastodon of Pontiac. Cranbrook Institute of Science News Letter, v. 36, pp. 62–64.

HAY, O. P., 1914, The Pleistocene mammals of Iowa. Geological Survey of Iowa Annual Report for 1912, v. 23, 662 p.

———, 1915, Contributions to the knowledge of the mammals of the Pleistocene of North America. United States National Museum Proceedings, v. 48, pp. 515–75.

———, 1923, The Pleistocene of North America and its vertebrated animals from the states east of the Mississippi River and from the Canadian provinces east of longitude 95. Carnegie Institute of Washington Publication 322, 499 p.

HEDBERG, H. D., 1964, Geologic aspects of origin of petroleum. American Association of Petroleum Geologists Bulletin, v. 48, pp. 1755–1803.

HIBBARD, C. W., (unpublished), Partial list of specimens of fossil vertebrates from Michigan. In the files of The University of Michigan Museum of Paleontology, Ann Arbor.

———, 1951, Animal life in Michigan during the Ice Age. Michigan Alumnus Quarterly Review, v. 57, pp. 200–208.

———, 1952, Remains of the Barren Ground caribou in Pleistocene deposits of Michigan. Michigan Academy of Science, Arts, and Letters Papers, v. 37, pp. 235–37.

———, and F. J. HINDS, 1960, A radiocarbon date for a woodland musk ox in Michigan. Michigan Academy of Science, Arts, and Letters Papers, v. 45, pp. 103–8.

HIBBARD, E. A., 1958, Occurrence of the extinct moose, *Cervalces*, in the Pleistocene of Michigan. Michigan Academy of Science, Arts, and Letters Papers, v. 43, pp. 33–37.

HINSDALE, W. B., 1925, Primitive man in Michigan. The University of Michigan, University Museum, Michigan Handbook Series, No. 1, 194 p.

HOARE, R. D., 1964, Radiocarbon date on Pleistocene peccary find in Sandusky County, Ohio. Ohio Journal of Science, v. 64, p. 427.

HOTCHKISS, W. O., 1933, Lake Superior Region. In International Geological Congress, XVI Session, United States, 1933, Guidebook 27, Excursion C-4, printed in Washington, 1933, by the United States Geological Survey, 101 p.

HOUGH, J. L., 1958, Geology of the Great Lakes. University of Illinois Press, Urbana, 313 p.

———, 1963, The prehistoric Great Lakes of North America. American Scientist, v. 51, pp. 84–109.

———, 1966, Correlation of glacial lake stages in the Huron-Erie and Michigan basins. Journal of Geology, v. 74, pp. 62–77.

HUBBARD, BELA, 1841, Joint documents of the House of Representatives of Michigan. State of Michigan, Lansing, pp. 1–559.

HUBBS, C. L., 1940, The cranium of a fresh-water sheepshead from post-glacial marl in Cheboygan County, Michigan. Michigan Academy of Science, Arts, and Letters Papers, v. 25, pp. 293–96.

———, and K. F. LAGLER, 1958, Fishes of the Great Lakes Region. Cranbrook Institute of Science Bulletin 26, 274 p. (Revised edition, 1964 University of Michigan Press.)

HUSSEY, R. C., 1930, Items. Science, New Series, v. 72, No. 1875, p. xiv.

———, 1947, Historical geology. McGraw-Hill Book Company, Inc., New York, 2d Edition, 465 p.

———, 1952, The Middle and Upper Ordovician rocks of Michigan. Michigan Department of Conservation, Geological Survey Division, Publication 46, Geological Series 39, 89 p.

IMBRIE, J., and N. D. NEWELL, 1964, Approaches to paleoecology. John Wiley and Sons, Inc., New York, 432 p.

INDEPENDENT PETROLEUM ASSOCIATION OF AMERICA, 1963, The oil producing industry in your state (1963 Edition, 1962 statistics). P.O. Box 1019, Tulsa, Oklahoma, 74101, 110 p.

JAMES, H. L., 1954, Sedimentary facies of iron-formation. Economic Geology, v. 49, pp. 235–93.

———, 1958, Stratigraphy of Pre-Keweenawan rocks in parts of northern Michigan. United States Geological Survey Professional Paper 314-C, 44 p.

———, L. D. CLARK, C. A. LAMEY, and F. J. PETTIJOHN, 1961, Geology of central Dickinson County, Michigan. United States Geological Survey Professional Paper 310, 176 p.

KAPP, R. O., and W. A. KNELLER, 1962, A buried biotic assemblage from an old Saline River terrace at Milan, Michigan. Michigan Academy of Science, Arts, and Letters Papers, v. 47, pp. 135–45.

KELLEY, R. W., 1962, Michigan's sand dunes—a geologic sketch. Michigan Department of Conservation, Geological Survey Division, booklet, 22 p.

———, and H. J. HARDENBURG, 1962, Collecting minerals in Michigan. Michigan Department of Conservation, Geological Survey Division, booklet, 26 p.

KELLY, W. A., 1936, The Pennsylvanian system of Michigan. Michigan Geological Survey Publication 40, Geological Series 34, part 2, pp. 149–226.

KING, P. B., 1959, The evolution of North America. Princeton University Press, Princeton, New Jersey, 189 p.

KNOPF, A., 1949, Time in earth history. Pp. 1–9 in Genetics, Paleontology, and Evolution, edited by

# Bibliography

G. L. Jepsen, E. Mayr, and G. G. Simpson, Princeton University Press, Princeton, New Jersey, 474 p.

KRAUS, E. H., W. F. HUNT, and L. S. RAMSDELL, 1959, Mineralogy. McGraw-Hill Book Company, Inc., New York, 5th Edition, 686 p.

KRUMBEIN, W. C., and L. L. SLOSS, 1963, Stratigraphy and sedimentation. W. H. Freeman and Company, California, 2d Edition, 660 p.

LANDES, K. K., 1959a, The Mackinac breccia. Pp. 19-24 in collection of articles in Michigan Basin Geological Society 1959 Annual Geological Excursion Guidebook entitled Geology of Mackinac Island and Lower and Middle Devonian south of the Straits of Mackinac, compiled by F. D. Shelden, published by Michigan Basin Geological Society, 63 p.

———, 1959b, Petroleum Geology. John Wiley and Sons, Inc., New York, 2d Edition, 443 p.

———, G. M. EHLERS, and G. M. STANLEY, Geology of the Mackinac Straits Region. Michigan Department of Conservation, Geological Survey Division, Publication 44, Geological Series 37, 204 p.

LANE, A. C., 1902, Report of the State Board of the Geological Survey of Michigan for the year 1901. Geological Survey of Michigan, Lansing, Michigan, pp. 252-53.

———, 1905, Mastodon in Bay County. Geological Survey of Michigan, 7th Annual Report, 1905, 553 p.

LANMAN, J. H., 1939, History of Michigan, civil and topographical, in a compendious form; with a view of the surrounding lakes. E. French, New York, 397 p.

LAPHAM, I. A., 1885, On the number of teeth of the *Mastodon giganteus*. Boston Society of Natural History Proceedings, v. 5, p. 133.

LEET, L. D., and S. JUDSON, 1965, Physical geology. Prentice-Hall, Inc., New Jersey, 3d Edition, 406 pp.

LEITH, C. K., R. J. LUND, and A. LEITH, 1935, Pre-Cambrian rocks of the Lake Superior Basin. United States Geological Survey Professional Paper 184, 34 p.

LEVERETT, F., and F. B. TAYLOR, 1915, Pleistocene of Indiana and Michigan and the history of the Great Lakes. United States Geological Survey Monograph 53, 529 p.

LIBBY, W. F., 1955, Radiocarbon dating. University of Chicago Press, Chicago, Illinois, 175 p.

LOWENSTAM, H. A., 1950, Niagaran reefs of the Great Lakes area. Journal of Geology, v. 58, pp. 430-87.

———, 1957, Niagaran reefs of the Great Lakes area. (Same title as 1950.) Geological Society of America Memoir No. 67, v. 2, pp. 215-48.

LUOMA, W. E., 1946, Mineral collecting in northern Michigan. Rocks and Minerals (magazine), v. 21, p. 664.

MACALPINE, A., 1940, A census of mastodon remains in Michigan. Michigan Academy of Science, Arts, and Letters Papers, v. 25, pp. 481-90.

MACCURDY, H. M., 1919, Mastodon remains found in Gratiot County, Michigan. Michigan Academy of Science, Report n. 21, pp. 109-10.

MANDARINO, J. A., 1950, The minerals of the Champion, Michigan area. Rocks and Minerals (magazine), v. 25, pp. 563-65.

MARKERT, R., 1960, Rockhounds paradise, U.S.A. Earth Science (magazine), v. 13, pp. 101-3.

MARTIN, H., 1957, Geological map of Michigan. Michigan Department of Conservation.

———, and M. T. STRAIGHT, 1956, An index of Michigan geology—1823-1955. Michigan Department of Conservation, Geological Survey Division, Publication No. 50, 461 p.

MASON, R. J., 1958, Late Pleistocene geochronology and the Paleo-Indian penetration of the Lower Michigan Peninsula. The University of Michigan Museum of Anthropology Anthropological Papers, No. 11, 48 p.

MATHER, K. F., 1964, The earth beneath us. Random House, New York, 319 p.

MICHIGAN ACADEMY OF SCIENCE, ARTS, AND LETTERS AND MICHIGAN GEOLOGICAL SOCIETY, 1947, Michigan Copper Country. Guidebook for Annual Field Trip of June, 1947, mimeographed, 27 p.

MICHIGAN GEOLOGICAL SURVEY, 1964, Our rock riches. Michigan Department of Conservation, Geological Survey Division Bulletin 1, 109 p.

MIHELCIC, J. F., 1954, Collecting Michigan minerals: Earth Science Digest, v. 7, n. 7, pp. 7-10.

MIKULA, E. J., 1964, Evidence of Pleistocene habitation by woodland caribou in southern Michigan. Journal of Mammalogy, v. 45, pp. 494-95.

MONNETT, V. B., 1948, Mississippian Marshall Formation of Michigan. American Association of Petroleum Geologists Bulletin, v. 32, pp. 629-88.

MOODY, P. A., 1962, Introduction to evolution. Harper and Brothers, New York, 2d Edition, 553 p.

MOORE, R. C. (Editor), 1953—, Treatise on invertebrate paleontology. The Geological Society of America and the University of Kansas Press, a series to consist ultimately of about 25 volumes.

———, 1958, Introduction to historical geology. McGraw-Hill Book Co., Inc., New York, 2d Edition, 656 p.

MORTENSON, F. N., 1953, MMS (Michigan Mineralogical Society) field trip to Michigan's Upper Peninsula. Rocks and Minerals (magazine), v. 28, pp. 591-96.

NEWCOMBE, R. B., 1933, Oil and gas fields of Michigan. Michigan Department of Conservation, Geological Survey Division, Publication 38, Geological Series 32, 293 p.

NIKOL'SKII, G. V., 1954, Special ichthyology. 2d Edition, Moscow, 1954, translated from Russian to English in 1961, and published in Jerusalem in 1961 for the National Science Foundation, Washington, D.C., and the Smithsonian Institution, by the Israel program for scientific translations, 538 p.

OLSON, J. S., 1958a, Lake Michigan dune development; 1. Wind velocity profiles. Journal of Geology, v. 66, pp. 254–63.

———, 1958b, Lake Michigan dune development; 2. Plants as agents and tools in geomorphology. Journal of Geology, v. 66, pp. 345–51.

———, 1958c, Lake Michigan dune development; 3. Lake-level, beach, and dune oscillations. Journal of Geology, v. 66, pp. 473–83.

OLTZ, JR., D. F., and R. O. KAPP, 1963, Plant remains associated with mastodon and mammoth remains in central Michigan. American Midland Naturalist, v. 70, pp. 339–46.

PETTIJOHN, F. J., 1957, Paleocurrents of Lake Superior Precambrian quartzites. Geological Society of America Bulletin, v. 68, pp. 469–80.

PILLERI, G., 1961, Das Gehirn (endocraniolausguss) von *Castoroides ohioensis* (Foster, 1838) (Rodentia, Castoridae) und vergleichend-anatomische beziehungen zum gehirn des rezenten kanadischen Bibers. Acta Anatomica, v. 44, n. 2, pp. 36–82.

POINDEXTER, O. F., H. M. MARTIN, and S. G. BERQUIST, 1952, Rocks and minerals of Michigan. Michigan Department of Conservation, Geological Survey Division, Publication 42 (3d Edition, out of print), 124 p.

POTTS, R., 1959, Michigan mammoth. Nature Magazine, v. 52, pp. 471–72.

POTZGER, J. E., 1946, Phytosociology of the primeval forest in central-northern Wisconsin and upper Michigan, and a brief postglacial history of the lake and forest formation. Ecological Monographs, v. 16, n. 10, pp. 211–50.

POUGH, F. H., 1953, A field guide to rocks and minerals. Houghton-Mifflin Company, Boston, 333 p.

PRATT, W. P., 1954, Some mineral localities in central Dickinson County, Michigan. Rocks and Minerals (magazine), v. 29, pp. 345–50.

PRUITT, JR., W. O., 1954, Additional animal remains from under Sleeping Bear Dune, Leelanau County, Michigan. Michigan Academy of Science, Arts, and Letters Papers, v. 39, pp. 253–56.

ROMER, A. S., 1941, Man and the vertebrates. The University of Chicago Press, Chicago, Illinois, 3d Edition, 405 p.

———, 1966, Vertebrate paleontology. The University of Chicago Press, Chicago, Illinois, 3d Edition, 468 p.

RUSSEL, I. C., and F. B. TAYLOR, 1915, The Ann Arbor Folio. United States Geological Survey, Geological Atlas of the United States, Folio No. 155.

RUTTEN, M. G., 1957, Origin of life on earth, its evolution and actualism. Evolution, v. 11, pp. 56–59.

———, 1962, The geological aspects of the origin of life on earth. Elsevier Monographs, Geo-Sciences Section, Subseries: Geology, Elsevier Publishing Company, Amsterdam-New York, 146 p.

SCHOPF, J. W., E. S. BARGHOORN, M. D. MASER, and R. O. GORDON, 1965, Electron microscopy of fossil bacteria two billion years old. Science, v. 149, pp. 1365–67.

SCOTT, I. D., 1921, Inland lakes of Michigan. Michigan Geological and Biological Survey Publication 30, Geological Series 25, published as part of the Annual Report of the Board of the Geological Survey for 1920, 383 p.

———, and K. W. DOW, 1937, Dunes of the Herring Lake embayment, Michigan. Michigan Academy of Science, Arts, and Letters Papers, v. 22, pp. 437–50.

———, and E. B. STEVENSON, 1931, Shore accretions from the geological standpoint. Michigan State Bar Journal, March, 1931, pp. 215–29.

SCOTT, W. B., 1913, A history of land mammals of the Western Hemisphere. The Macmillan Company, New York, 693 p.

SEMKEN, H. A., B. B. MILLER, and J. B. STEVENS, 1964, Late Wisconsin woodland musk oxen in association with pollen and invertebrates from Michigan. Journal of Paleontology, v. 38, pp. 823–35.

SHARP, R. P., 1960, Glaciers. Condon Lectures, Oregon State System of Higher Education, Eugene, Oregon, 78 p.

SHAW, A. B., 1964, Time in stratigraphy. McGraw-Hill Book Company, Inc., New York, 365 p.

SHERZER, W. H., 1927, A new find of the wooly elephant in Michigan. Science, No. 1695, p. 616.

SHIMER, H. W., and R. R. SHROCK, 1944, Index fossils of North America. John Wiley and Sons, Inc., New York, 837 p.

SHROCK, R. R., 1948, Sequence in layered rocks. McGraw-Hill Book Company, Inc., New York, 507 p.

———, and W. H. TWENHOFEL, 1953, Principles of invertebrate paleontology. McGraw-Hill Book Company, Inc., New York, 2d Edition, 816 p.

SIMPSON, G. G., 1945, The principles of classification and a classification of mammals. American Museum of Natural History Bulletin, v. 85, 350 p.

SINKANKAS, J., 1964, Mineralogy for amateurs. D. Van-Nostrand Co., Inc., Princeton, New Jersey, 585 p.

SKEELS, M. A., 1962, The mastodons and mammoths of Michigan. Michigan Academy of Science, Arts and Letters Papers, v. 47, pp. 101–33.

SNELGROVE, A. K. (Editor), 1957, Geological exploration—15 papers delivered at the Institute on Lake

Superior geology, 1956. Michigan College of Mining and Technology Press, Houghton, Michigan, 109 p.

SORENSON, H. O., and E. T. CARLSON, 1959, Michigan mineral industries—1958. Michigan Department of Conservation, Geological Survey Division, 93 p.

STANLEY, G. M., 1936, Geology of the Cranbrook area. Cranbrook Institute of Science Bulletin, v. 6, 56 p.

———, 1945, Prehistoric Mackinac Island. Michigan Department of Conservation, Geological Survey Division, Publication 43, Geological Series 36, 74 p.

STEVENS, M. (Skeels), 1964, Thoracic armor of a new arthrodire *(Holonema)* from the Devonian of Presque Isle County, Michigan. Michigan Academy of Science, Arts, and Letters Papers, v. 49, pp. 163–75.

STEVENSON, E. B., 1931, The dunes of the Manistique area. Michigan Academy of Science, Arts and Letters Papers, v. 14, pp. 475–85.

STIRTON, R. A., 1959, Time, life and man. John Wiley and Sons, Inc., New York, 558 p.

———, 1965, Cranial morphology of *Castoroides*. In: Mining and Metallurgical Institute of India, Dr. D. N. Wadia Commemorative Volume, pp. 273–89.

STOUTAMIRE, W. P., and W. S. BENNINGHOFF, 1964, Biotic assemblage associated with a mastodon skull from Oakland County, Michigan. Michigan Academy of Science, Arts, and Letters Papers, v. 49, pp. 47–60.

STRONG, E. A., 1872, Notes upon the fossil remains of the Lower Carboniferous limestone exposed at Grand Rapids, Michigan. Kent Science Institute, Grand Rapids, Michigan, Miscellaneous Papers No. 3, 6 p.

SWINEFORD, A. P., 1876, History and review of the copper, iron, silver, slate and other material interests of the south shore of Lake Superior. The Mining Journal, Marquette, Michigan, 280 p.

TURNBULL, W. D., 1958, Notice of a late Wisconsin mastodon. Journal of Geology, v. 66, pp. 96–97.

TYLER, S. A., and E. S. BARGHOORN, 1954, Occurrence of structurally preserved plants in the pre-Cambrian rocks on the Canadian Shield. Science, v. 119, pp. 606–8.

———, E. S. BARGHOORN, and L. P. BARRETT, 1957, Anthracitic coal from Precambrian Upper Huronian black shales of the Iron River District, northern Michigan. Geological Society of America Bulletin, v. 68, pp. 1293–1304.

VANHISE, C. R., and C. K. LEITH, 1911, The geology of the Lake Superior region. United States Geological Survey Monograph 52, 641 p.

WAGNER, G., 1903, Observations on *Platygonus compressus* Leconte. Journal of Geology, v. 11, pp. 777–82.

WARREN, J. C., 1855, Supernumary tooth in *Mastodon giganteus*. American Journal of Science, v. 19, pp. 348–53.

WAYNE, W. J., and J. H. ZUMBERGE, 1965, Pleistocene geology of Indiana and Michigan. Pp. 63–83 in The Quaternary of the United States, Princeton University Press, Princeton, New Jersey, 922 p.

WHITE, W. S., 1960a, The White Pine copper deposit. Economic Geology, v. 55, pp. 402–9.

———, 1960b, The Keweenawan lavas of Lake Superior, an example of flood basalts: American Journal of Science, Bradley, v. 258-A, pp. 367-74.

———, and J. C. WRIGHT, 1954, The White Pine copper deposit, Ontonagon County, Michigan. Economic Geology, v. 49, pp. 675–716.

WILSON, I. T., and J. E. POTZGER, 1943, Pollen study of sediments from Douglas Lake, Cheboygan County, and Middle Fish Lake, Montmorency County, Michigan. Indiana Academy of Science, v. 52, pp. 87–92.

WILSON, R. L., 1967, The Pleistocene vertebrates of Michigan. Michigan Academy of Science, Arts, and Letters, Papers, v. 52, pp. 197–234.

WINCHELL, A., 1861, First biennial report of the state geologist (Michigan). State of Michigan, Lansing, 132 p.

———, 1863, Description of elephantine molars in the museum of the university. Canadian Naturalist, v. 8, pp. 398–400.

———, 1864, Notice of the remains of a mastodon recently discovered in Michigan. American Journal of Science, v. 38, pp. 223–24.

———, 1888, Extinct peccary in Michigan. American Geologist, Personal and scientific news, v. 1, p. 67.

WOOD, N. A., 1914, Two undescribed specimens of *Castoroides ohioensis* Foster from Michigan. Science, new series, v. 39, pp. 758–59.

WRIGHT, JR., H. E., 1964, Aspects of the early postglacial forest succession in the Great Lakes region. Ecology, v. 45, pp. 439–48.

ZUMBERGE, J. H., 1960, Correlation of Wisconsin drifts in Illinois, Indiana, Michigan and Ohio. Geological Society of America Bulletin, v. 71, pp. 1177–88.

———, and J. E. POTZGER, 1955, Pollen profiles, radiocarbon dating, and geologic chronology of the Lake Michigan basin. Science, v. 121, pp. 309–11.

———, and J. E. POTZGER, 1956, Late Wisconsin chronology of the Lake Michigan basin correlated with pollen studies. Geological Society of America Bulletin, v. 67, pp. 271–88.

———, S. H. SPURR, W. N. MELHORN, and others, 1956, The northwestern part of the southern peninsula of Michigan. Pp. 1–36 and L1–L20 in Guidebook for the Friends of the Pleistocene (Midwest Section) field trip on May 11, 12, 13, 1956, privately mimeographed for limited distribution, 56 p.

# Index

Ablation, 144
Abrasion, wind, 198
Absolute time, age, and dates, 5, 13, 18–22, 32, 159, 417, 420
Abyssal environment, 334
Acadian Eugeosyncline, 24
Acanthodian fish; Acanthodii, 358, 366, 396
Acidic conditions, 443
Acidizing, 237
Acorn worm, 308
Actinopterygii, 360
Adirondack Highlands, 25
Adrian, 255, 370, 403
Aerial environment, 334
Aestivate, 361, 368
African Rift Valley, 23
Aftonian Interglacial, 158
Agassiz, Louis, 141
Agate Harbor, 58, 262
Age correlation of rocks, 18, 238, 332
Agglomerated, 64
Agnatha (class), and agnathan fishes, 309, 355, 357–58, 390
Ahmeek, 261–62
Air currents, flow, 219, 224
Alabaster, 111, 113, 126, 258
Alaska, 142
Alberta, 260
Albion, 225, 242, 255, 349
Albion-Scipio "Trend," 237, 240–41, 243
Algae 50, 418–19, 421
Algal deposits, modern and ancient, 36, 47, 50, 108, 333, 344, 416–17, 423, 425–26
Alger County, 258, 345
Algoman Orogeny, 40, 44
Algonac, 190
Algonkian, 31
Allendale, 191
Allendale delta, 172
Allouez, 261
Alma, 184, 257, 371
Alpena, 119, 121, 154, 186, 258, 348, 398, 423
Alpena County, 258, 348, 390–91, 393, 398
Alpena Limestone, 119, 121, 123, 189
Alpena Portland Cement Company Quarry, 348
Alps, 141
Amherstburg, Ontario, 348
Amherstburg Formation, 90, 115, 119
Amherstburg Sea, 90
*Amia,* 360
Ammonoids; Ammonoidea, 301, 324–25, 332, 334–35
Amphibia, origin of, 362
Amphibians; Amphibia, 309, 350, 362
Amphibolite, 36
Amphineura, 298
*Amphioxus,* 308–9
Amygdaloidal, 72, 267
Amygdaloidal basalt, 54, 56, 57
Amygdaloidal lodes, 74
Amygdules, 54–56, 72
Ancient Island, 177, 179, 217

Anderdon Formation, 115, 119
Angiosperms; Angiospermae, 139, 417, 419, 439
Angiosperms, first appearance, 417
Angle of repose, loose sand, 204
Angular unconformity, 91, 97, 102, 128, 279
Anhydrite, 108, 109, 111, 113
Animal classification, categories, 288
Animikie Period, 44
Ann Arbor, 154, 162, 184, 191, 193, 255, 370, 407, 447
Annelid worms; Annelida, 297, 339
*Annularia*, 418, 432–33
Anthozoa, 293; *see also* Corals
Anthracite, 132
Antiarchs; Antiarchii, 358, 366
Anticlinal structures, folds, 6, 8, 50, 77, 231–32
Anticlinal traps, 232, 239, 242
Antrim County, 152
Antrim Shale and Sea, 119, 123–24, 139, 240, 423, 425
Aphanitic rock texture, 266, 269
Appalachian Eugeosyncline, 24
Appalachian Geosyncline, 24, 26, 102, 140
Appalachian Mountain Region, 140
Appalachian Orogeny, 135
Appalachian Piedmont, 38
Aquatic environments, 334, 342–43, 344
Aquatic plants, 195
Aquifer, 182–83
Arachnoidea, 305
Arch, 179, 208, 213
Archaeocyathids; Archaeocyatha, 291, 334
Archean, 31
Archeozoic, 31
Arch Lake, 178
Arch Rock, 179, 187, 213
Arenac County, 161, 257–58, 398
Arkona, Ontario, 348
Arkose, 36
Arrowheads, Indian, 385–88
Artesian water wells, 183–84
Arthrodires; Arthrodira, 358, 365–66
Arthropoda, 302–4, 326–29
Artifacts, early Indian, 385–88
Assemblages of fossils, 335–36, 342, 344
Asteroidea; *see* Starfish
*Asterophyllites*, 131, 418, 432–33
Atlantic, town of, 72, 261
Atomic arrangement in minerals, 244
Au Gres, 257
Au Sable, 94
Au Sable Point, 94
Au Sable River, 191
Austin Mine, 345
Au Train, 94, 175
Au Train Falls, 98, 100, 345
Au Train Formation, 94, 98, 100
Au Train Sandstone, 90
Aves; *see* Birds

Bacteria, 418
Bacterial action, 229
Bacterial decay, 130

Bad River Gabbro, 56, 71
Badwater Greenstone, 46
Baltic, 72
Baltic Lode, 75–76
Baraga, 93
Baraga County, 50, 260
Bark River, town of, 346
Barnacles, 304
Bars, 213
Basalt, 36, 51, 56, 71, 266, 269
Basaltic lava flows, 59
Basalt porphyry, 266
Baselevel, 189–90
Basement rocks, 26, 38, 89, 240, 243
Basic intrusive, 36
Basins, 281
Bass, Largemouth, 401
Bass Island, 144
Bass Island Group, 104, 123
Bathyal environment, 334
Batoidea, 360
Battle Creek, 127, 151, 184, 255, 349
Battlefield Beach, 178, 217
Bay City, 126, 257, 428
Bay County, 132, 257, 403, 428
Bay Furnace, 70, 218
Baymouth bar, 195, 213, 217, 223
Bayport, 257
Bayport Limestone, 125, 144, 367
Bayview, 176, 348
Beach, 115, 117, 119
Beach drift, 208
Beach environment, 334
Beach erosion, 213
Beach lines, 221
Beach ridges, recessional, 214
Beacon, 259–60
Bear, 374, 408
Beaver, Giant; *Casteroides*, 340, 350, 368, 374–75, 377, 406–7
Beaver, modern; *Castor*, 374, 378
Beaver Island, 89
Bedford delta, 125
Bedford Shale, 124
Belemnites, 302
Bellevue, 255
Beneficiating, 63–64, 68
Benthonic environment, 334–35, 344
Benton Harbor, 176, 201
Benzie County, 213, 222–23
Berea Sandstone, 124–25, 240
Bergland, 59
Bering Land Bridge, 380, 388
Berrien County, 255, 369, 389, 407, 412, 414
Berrien Springs, 414
Best Wall Gypsum Company, 126, 367
Betsie Lake, 194, 222
Bichler Quarry, 345
Biddles Point, 179
Big Bay de Noc, 347
Big Hill, 346

**Index**

Big Iron River, 61
Big Sable Point, 195
Bijiki Iron Formation, 48
Biochemical Sedimentary rocks, 275
Biocoenose, 335, 344
Bioherms, 107, 121
Birds, 140, 309, 350, 362, 402–3
Birds, oldest, 140
Birds, Pleistocene, Michigan, 402
Bismarck Township, 187
Bison, 374, 383, 414
Bituminous, 132
Black Lake, 195
Black River, 98, 102
Black River Formation, 239–40, 358
Blastoids; Blastoidea, 126, 306–7, 330–31, 335, 344
Bloomfield Hills, 255
Blowouts, in dunes, 200–201, 204, 219, 222, 225
Blue Ridge Esker, 38, 154, 156
Body temperature control, 363
Bog, quaking, 196
Bois Blanc, 117
Bois Blanc Formation, 115, 123
Bony Falls, 345
Bony Falls Formation, 90
Bony fish; Osteichthyes, 359, 367, 400
*Boötherium*, 376–77
*Bothryolepis*, 366
Bottomland, 189
Boulder clay, 5, 46, 149
Bowfin; *Amia*, 360
Brachiopods; Brachiopoda, 102, 128, 295–96, 318–21, 332, 334–37, 344
Bradyodont sharks; Bradyodonti, 359, 366
Branch County, 125, 255, 349, 412
Branchial arches and clefts, 307
Breakwaters, 213
Breccia, 105, 115, 123, 178–79, 208
Breen Mine, 345
Brick, 125
Brighton, 153–54
Brines, 111
Briquets, 64
British Landing Road, 177
Brockway Mountain Drive, 52, 56, 58
Brown coal, 132
Bruce Peninsula, 104, 164
Bryophyta, 418
Bryozoan colony, 337, 339
Bryozoans; Bryozoa, 295, 316–17, 332, 334, 336, 344
Buckeye Field, 243
Buffalo, New York, 171
Burnt Bluff, 107, 215
Burnt Bluff Group, 105, 107
Burrows, 286, 423
Burt Lake, 172, 194, 214, 414

Cable tool drilling, 236
Cadillac, 184
*Calamites*, 131, 418, 432–33
Calcite Quarry, 119, 121
Calcium chloride, 113
Calcium magnesium chloride, 113
Calhoun County, 125, 132, 237, 243, 255, 349, 398
Calumet, 71, 261
Calumet and Helca lode, 72, 76
Calving, 144
Cambrian fossil invertebrates, 345
Cambrian Period, 91
Canadian Shield, 25, 32, 33, 81, 165
Cape Hurd, 104
Caps, oil and gas, 231
Carbonate banks, 119
Carbonates, 61
Carbon 14 dating; *see* Radiocarbon dates
Caribou, 340, 374–75, 377, 412
Carpoidea, 307
Carpsucker, 401
Cartier, 76
Carver Pond, 177
Cary Glacial Sub-stage, 158, 161
Case Township, 187
Casing, well, 236
Caspian Sea, 109
Cass County, 389
Cast, fossil, 286
Castle Rock, 123, 187, 210
Castle Rock Sandstone, 94
Catfish, 374, 401
Cave, 93, 121, 178–79
Cave collapse, 186
Cave environment, 334
Cedarville, 121
Cement, 105, 119, 121, 125
Cenozoic Era, 136, 238, 369
Centipedes, 305
Central, town and mine, 76
Central Stable Region, 25–26, 81
Cephalochorda, 309
Cephalopods; Cephalopoda, 117, 300, 324–25, 344, 364; *see also* Ammonoids; Nautiloids
*Cervalces; see* Moose, Scott's
Chalcocite, 77
Chalcocite copper ore, 63, 70
Champion, 48, 67
Champion Mine, 48
Chandler Falls, 345
Channel sands, 128
Channels, stream, 233
Chapel Rock, 93–95, 97
Chapel Rock Sandstone, 93, 97–98
Charcoal iron smelters, 70
Charity Island, 398
Charlevoix, 148, 153, 176
Charlevoix County, 148, 152, 163, 391, 393, 398
Chassel, 261
Chattermarks, glacial, 144
Cheboygan County, 123, 151, 172, 403, 414
Chelsea, 151, 156
Chemical elements, common, 245, 254
Chemical sedimentary rocks, 275
Chert, 62

Chicago Sanitary Canal, 224
Chimaeras; Chaemaerae, 359–60
Chimney Rock, 179
China Sea, 26
Chipmunk, 374, 406
Chippewa County, 258, 345–46
Chippewa Point, 346
Chips, oil well, 89, 234
Chitons, 298
Choanichthyes, 361
Chondrichthyan fishes; Chondrichthyes, 359, 366, 398
Chondrostei, 360
Chordates; Chordata, 307–8, 328–29, 350, 390
Cincinnati Arch, 26, 108, 238
Cladoselachan sharks; Cladoselachii, 359, 366
Clams, 127, 299, 300, 322–25, 332, 344
Clare, 203
Clare County, 447
Clarksburg Volcanics, 42, 44
Classification of invertebrate animals, 288–309
Classification of plants, general, 418–19
Clastic sedimentary rocks, 267, 274
Clay, 125, 134
Cleavage, slaty, 49
Cleavage of minerals, 248
Cliff, town of, 262
Cliff Fissure, 76
Cliffs, wave-cut; see Wave-cut cliffs
Climatic change, postglacial, 443, 448
Climatic deterioration, 158
Climatic optimum, 446
Climax, town of, 377, 446
Clinton County, 349, 400
Clinton River, 190–91
Closure, structural, 232, 239
Clovis points, 388
Club-mosses, 418
Coal deposits, mines, seams, 11, 102, 126, 130, 286, 342–43, 417–19, 427–29
Coal-forming regions, modern, 130, 132
Coal swamp floras, 426
Coecilians, 362
Coelacanth, 361
Coelenterata, 291; see also Corals
Coldwater, 255
Coldwater Formation, 240
Coldwater Shale, 125, 126, 134, 336, 366, 429
Coleoidea, 301
Collapse of caves, 123
Collecting locality list, fossil invertebrates, 344–49
Collecting locality list, minerals, 254–62
Colonial coral, 293, 336, 339
Color, mineral, 248
Color, sedimentary rock, 279
Colorado River, 13
Columbia, 349
Columnar basalt, 59
Commensalism, 340
Compaction fold, 233
Compression fold, 233
Conches, 299

Cone of depression, water table, 184
Conglomerate, 36, 267
Conglomerate copper lodes, 72, 74
Conifers; Coniferales, 419
Conifers, first, 139
Continental deposits, 126
Continental drift, 28
Continental environments, 334, 342–43
Continental glacial ice sheets, 29, 165
Continental margins, 25
Continuity, disruption of, 6
Continuity, original, 14
Conularids, 344
Convection Current Hypothesis, 28
Copper culture, Indian, 75, 372
Copper deposits, 70
Copper Harbor, 56, 58, 76, 262, 419
Copper Harbor Conglomerate, 57
Copper implements, Indian, 386–87
Copper lodes, 72
Copper mines, 80
Copper sulphide, 77
Coquina, 267
Corals, 117, 119, 121, 123, 312–15, 334, 339, 341;
    see also Reefs and reef-building corals
Coral-Stromatoporoid association, 338
*Cordaites;* Cordaitales, 419, 434–37
Cordell, 347
Core, well, 29–30, 89, 157, 234
Core barrel and bit, 236
Cornell, 345
Correlation, principles of, 10
Correlation, rock age, 332
Correlation key horizon, 367
Correlation of parts, law of, 287
Crabs, 305
Crack in the Island, 178–79
Cranbrook Institute of Science, 255
Craton, 25
Cratonic, 281
Crayfish, 304
Cretaceous Period, 136, 362
Crevasse fillings, 147
Crinoids; Crinoidea, 117, 126, 306–7, 330–31, 335–37, 339, 344
Cross-bedding, 5, 44, 46, 77, 91, 93, 97, 127, 276, 278
Cross-cutting principle, 17
Crossopterygians, 361
Crust, Earth, 29–30
Crustaceans; Crustacea, 303, 328–29
Crustal depression, 165
Crustal downwarp, Michigan Basin, 238
Crustal rebound, 166
Crystal Falls District, 68
Crystal forms, mineral, 248
Crystal Lake, 193–94
Cuesta, 98, 100, 104
Culture, Early Archaic Indian, 389
Cupriferous zone, 63
Currents, wave and shore, 205
Custer Pond, 177

# Index

463

Cut River Bridge, 215
Cuttlefish, 302
Cyamoidea, 307
Cycads; Cycadales; Cycadeoidales, 419
Cyclic deposits; Cyclothems, 128, 368, 427
Cystoids; Cystoidea, 307, 330-31, 344

Dacite, 269
Dan's Point, 57
Davison, 414
Dead River Basin, 39
Deep River Field, 240
Deer, 374
Deermouse, 407
Defiance Glacial moraine, 5, 149, 160-62
Deflation, wind, 198, 201
Deglaciation, 166
Delaware, town and mine, 262
Delta, 125, 128, 190
Delta County, 345-47, 390, 412
Deltaic, 102
Dense igneous rock texture, 266
Detrital sedimentary rocks, 267, 274
Detroit, 111, 154, 193, 254
Detroit River Group, 8, 111, 113, 117, 123, 240
Deuteric ore deposits, 75
Devil's Icebox; see Quinnesec Mine
Devil's Kitchen, 178-79
Devil's Washtub, 57, 419
Devonian fossil invertebrates, 348
Devonian Period, 113
Devonian seascape dioramas, 364-65
Devonian vertebrate faunas, 365-66
DeWitt, 158
Diabase, 36, 50
Diagenetic, 78
Diaphaneity in minerals, 249
Dickinson County, 39, 46-47, 50, 64, 94, 258-59, 345, 417
Dickinson Group, 42-44
Differential compaction, 232
Dikes, 18, 36, 50, 67, 68
Dinosaurs, 140
Diorite, 269
Dipnoi, 361
Direct shipping ore, 62, 65, 67
Disconformity, 279
Dismal Swamp, Virginia, 130
Divining rod, 185
Dixboro, 154, 162
Dodgeville, 261
Dogfish, freshwater, 360
Dolomite, 36, 113
Dolomitization, 230, 240, 243
Dome, 232
Door Peninsula, 104, 161
Dow Chemical Company, 113
Dowsers; Dowsing, 185
Drainage network, haphazard, 187
Drake Well, 236
Dreikanters, 199

Dresbach, 97
Dresbach Sandstone, 93
Driller's log, 235
Drilling, well, 89, 153, 236
Drilling mud, 236
Drum fish, 403
Drumlin, 147-48, 163
Drumlin field, 152
Drummond Island, 104, 121, 346
Dry holes, 234, 239
Duck, Scaup, 403
Duluth, 172
Duluth Gabbro, 56, 71
Dundee, 121
Dundee Limestone, 11, 119, 121, 239-40, 243, 365
Dune, 115, 117, 198, 205, 213
Dune areas in Michigan, 202
Dune development cycle, 224, 227
Dune growth, 217, 223
Dune migration, 200
Dune sand, 200
Dunes, coastal, 203
Dunes, inland, 202
Dunes, perched, 199, 203-4
Dunes, stabilized, 201, 225
Dynamothermal metamorphism, 44

Eagle, Bald, 403
Eagle Harbor, 53, 56, 58, 226, 262
Eagle River, 262
Early Man in Michigan, 372, 380, 388
Early Man in Michigan, artifacts of, 385-88
Early Man in Michigan, time of entry, 389
Earthquakes, 25, 30
Earthworm, 297
East Branch Arkose, 39, 42-44
East Lansing, 255
Eaton County, 132, 255, 349, 437
Eaton Sandstone, 130
Eau Claire, 97
Echidna, spiny, 363
Echinoderms; Echinoderma; Echinodermata, 126, 305-6, 330-31, 335
Ecologic niches, 333, 344
Eden, 98
Eden Shale, 102
Edrioasteroids; Edrioasteroidea, 307, 330-31, 339
Eggs, large-yolked, shelled, 362
Elasmobranchii, 359
Elberta, 194, 222
Electrical conductivity, 235
Elements, common chemical, 245, 254
Elephants, 411
Eleutherozoa, 305
Elk, 287, 374, 412
Elk Lake, 176
Ellsworth Shale, 119, 124, 134
Emmet County, 172, 258, 348, 393
Empire Mine, 64
End moraine, 147-48, 160-61
Endogenous sedimentary rocks, 267, 273, 275

**Geology of Michigan**

Engadine Dolomite, 90, 104, 108, 144
Environment, major changes of, 341–43
Environmental adjustment, principle of, 11
Environments, ancient, 333, 436
Environments, fossil indices to, 279
Environments of animals, 334, 344
Environments of fossil vertebrates, 363
Environments of sedimentary deposition, 280
Eocrinoidea, 307
Epeirogenic, 281
Epicontinental (epeiric) seas, 26, 81, 134, 229, 365
Epoch, 15
Equigranular, 266
*Equisetum*, 418, 430, 432–33
Era, 15
Erie Basin, 171
Erosion, 9
Erosion surface, 93, 102, 128
Erratics, 148
Escanaba, 345
Escanaba River, 102, 345, 358
Esker, 147, 151, 154, 156
Eugeosyncline, 281
Eurypterids; Eurypterida, 305, 328–29
Eutherian mammals, 363
Evaporation, 181
Evaporation basin, 109, 125
Evaporation cycle, 109
Evaporite sediments, 102, 108, 110–11, 115, 367
Evolution concept, 15, 309, 332, 355
Evolution of vertebrates; *see* Fossil vertebrates
Exogenous sedimentary rocks, 267, 273–74
Exploration, oil and gas, 234
Extrusive; extruded, 18

Faceted, 46
Faceted and striated glacial cobbles, 149, 276
Facies, 11, 108
Facies, reef, 111
Facies, sedimentary, 117
Facies change, 90, 124, 233
Facies fossils, 310, 332
Facies maps, 235
Fairy Arch, 178–79
Farmington, 370, 403, 407
Fault, 9, 67
Fault, Keweenaw, 52
Fault gouge, 233
Fault zones, 240, 243
Fayette, 70, 104–5, 107, 215
Felch, 51, 68, 259
Felsite, 36, 51, 71, 269
Fenton Lake, 370, 374, 400, 403
Fern Creek Formation, 44
Fern Creek Tillite, 46
Ferns, true, 418, 434–35
Ferron Point, 123, 187
Ferron Point Formation, 6
Fetch of wind, 205
Filicineae, 418, 434–35
Findlay arch, 26, 238

Firn, 143
Fish, 309, 350
Fish, jawless, 357
Fish, Paleozoic, 352
Fish, Pleistocene, 402
Fissure lode, 72
Fissures, 56, 74, 233
Five Mile Point, 262
Flanders Pit, 154
Fleas, 305
Fletcher State Park, 186
Flint, 370
Float copper, 74
Flood basalts, 56
Floodplain, 128, 132, 189, 428
Floodplain deposits, 368
Florence, 68
Flotation process, 64
Flow casts, 276
Flowering plants, 139, 419, 439
Flowering plants, first appearance, 417
Fluted Point Hunters, 388–89
Fluted points, Indian, 385–86, 388
Fluviatile environment, 334, 342–43
Flux stone, 105, 119, 121
Flying reptiles, 139
Folds; Folding, 9, 67, 233
Foliated, 268
Foliation, metamorphic, 49
Folsom points, 388
Food pyramid, 350
Fool's gold, 61
Foraminifera, 289–90, 332, 334
Ford Lake, 191
Ford River, 346
Foredune ridges, 195, 203–4, 217, 225, 227
Formation, rock, 90
Fort Holmes, 177, 217
Fort Mackinac, 177, 179, 217
Fort Wayne moraine, 149, 161
Fort Wilkins State Park, 58, 262
Fossil, definition of, 285
Fossil, living, 332
Fossil assemblages, 117, 335–36, 342, 344
Fossiliferous rocks, 337
Fossil invertebrate localities list, 344–49
Fossil invertebrate localities map, 310
Fossil invertebrates, Cambrian, 345
Fossil invertebrates, Devonian, 348
Fossil invertebrates, Mississippian, 126, 348–49
Fossil invertebrates, Ordovician, 345–47
Fossil invertebrates, Pennsylvanian, 131, 349
Fossil invertebrates, Silurian, 347
Fossil plants in Michigan, 366, 416
Fossils, facies, 310, 332
Fossils, finding, 352
Fossils, geological uses, 310
Fossils, guide, 310
Fossils, marine, 334, 344
Fossils, naming, 287
Fossils, preservation of, 285

# Index

Fossils, restoration, 351–52
Fossils, significance, 309
Fossils, study, 286
Fossils, time parallel index, 10, 310, 332
Fossils and ancient environments, 278, 333
Fossil tracks and trails, 286, 423
Fossil vertebrate faunas of Michigan, 363
Fossil vertebrates, 350
Fossil vertebrates, documentation, 354
Fossil vertebrates, geological history, 355
Fossil vertebrates, locality and specimen lists, 389–415
Fossil vertebrates, relative abundance, 350
Fossil vertebrates, value, 354
Fox, 374, 408
Fracture, mineral, 249
Fractured zones, 240, 243
Franconia, 97
Franconia Sandstone, 94
Frankfort, 176, 194, 213, 222
Franklin Eugeosyncline, 24
Freda Sandstone, 61, 71, 77, 79
Freshwater environment, 334, 342–43
Friendship Altar, 178–79
Frogs, 362
Fucoids, 423, 428
Fungi, 418, 421

Gabbro, 36, 50, 56, 269
Garden Island Formation, 113
Garden Peninsula, 70, 104, 107, 215
Garden Village, 347
Gars and Garpike, 360
Gas; *see* Oil and gas
Gas bubble cavities, 57, 72
Gastropoda; *see* Snails
Gaylord, 163, 384
Generic names, 288
Genessee County, 132, 349, 369–70, 374, 383, 401, 409, 412–14
Geologic map of Michigan, bedrock, 86
Geologic structure of Michigan, 238
Geologic time, 13
Geophysical methods, 234
Georgian Bay, 104, 164, 175
Geosynclines, 26, 28, 281
Geosyncline subsidence, 64
Gill arches, clefts, slits, 307, 355, 359
Ginkgos; Ginkgoales, 419
Glacial deposits, 148–49
Glacial erratics, 144
Glacial Great Lakes, major stages of, 164, 167
Glacial grooves, 144, 146
Glacial ice, formation of, 142
Glacial ice deposits, 147
Glacial ice flow, 146–47, 167
Glacial Lake Algoma, 176
Glacial Lake Algonquin, 107, 177, 213–14, 217, 389, 445
Glacial Lake Arkona, 171, 383
Glacial Lake Calumet, 170
Glacial Lake Chicago, 169, 445
Glacial Lake Chippewa, 174, 446

Glacial Lake Grassmere, 172
Glacial Lake Iroquois, 172
Glacial Lake Lundy, 170
Glacial Lake Maumee, 161, 169, 191
Glacial Lake Nipissing, 174–75, 177, 195, 213–14, 217, 222, 225–26, 384, 446
Glacial lake outlets, 168
Glacial Lake Saginaw, 154, 170, 203
Glacial Lake Stanley, 174
Glacial Lake Warren, 171
Glacial Lake Whittlesey, 161, 170, 191, 383
Glacial map of Michigan, 160
Glacial meltwater, 149, 156
Glacial moraines, 151
Glacial striations, 147
Glacial till, 46, 162, 276
Glacial varves, 158
Glaciation, 41, 46
Glaciation, continental, hypothesis, 159
Glaciation, multiple, evidence, 142
Glacier, alpine, 143
Glacier, continental, 144
Glacier, mountain, 143
Glacier, valley, 143
Gladwin County, 161
Glassy texture, 266
Gneiss; Gneissic, 36, 268
Goethite, 61
Gogebic Range, 67–68
Gouge, fault, 233
Grade of coal, 132
Grain size, 273, 275
Grand Canyon, 13, 38
Grand Haven, 225
Grand Hotel, 178–79
Grand Island, 218
Grand Lake, 121
Grand Ledge, 128–29, 132–34, 255, 342, 349, 352, 361, 368, 401, 428, 437–38
Grand Ledge Clay Product Company, 129, 133, 349
Grand Marais, 218, 258
Grand Rapids, 111, 113, 115, 126, 175, 255, 367, 397–98, 400, 412
Grand Rapids Gypsum Company, 115
Grand River, 132, 168, 171, 190, 191
Grand River Group, 130
Grand River outlet, 171
Grand Sable Dunes, 204
Grand Traverse Bay, 148, 195
Grand Traverse County, 258
Grand Valley, 171
Grandville, 255
Granite, 11, 18, 36, 45, 266
Granodiorite, 269
Graphite, 418–19
Graptolites; Graptozoa, 307–9, 328–29, 332, 334–35
Grasses, 139
Gratiot County, 113, 257, 369, 371
Gravity, 30, 182, 189, 230
Gravity survey, 234
Graywacke sediment and rock, 36, 281

Great Lakes, 164
Great Sand Bay, 226
Great Valley, California, 23
Green Bay, 104, 161, 164
Greenland Ice Cap, 5
Greenoaks Pit, 154
Greenstone, 37
Greenwood, 259
Greenwood Falls, 61
Grindstone City, 349, 398, 425–26, 429
Groins, 213, 221
Groos, 345
Gros Cap, 200
Grosse Isle, 184
Ground moraine, 147, 160
Ground water, 182
Ground water extraction, 184
Ground water problems in Michigan, 185
Ground water solution, 179
Ground water solution, ancient, 187
Groveland Mine, 64
Grunerite, 62
Gull Lake, 151
Gull Point, 261
Gunflint Formation, 418
Gusher, 232
Gymnospermae, 418
Gypsum, 109, 111, 113, 125–26, 137, 367
Gypsum mining, 115

Hagfish, 358, 390
Halfmoon Lake, 196
*Halimeda*, 108
Halite, 108, 109, 113
Hamlin Lake, 194
Hancock, 261
Hanks Pond, 178
Hanover, 349, 397–98, 425
Harbor Springs, 176, 214
Hard coal, 132
Harding Sandstone, 358
Hardness, mineral, 247
Hard ores, 67
Harris, 148, 346
Harrisville, 221
Hartford Bog Site, 444–47
Harvey, 42, 93, 423
Haymeadow Creek, 346
Heavy medium separation, 64
Hell, town of, 196
Hematite, 61
Hemichorda, 307, 328–29
Hendricks Dolomite, 105, 107
Hendricks Quarry, 347
Herring Creek, 213
Herring lakes and embayment, 193, 213, 222–23
Hiawatha Graywacke, 46
Hidden Lake, 447
High dunes, 203, 217, 222
Highland Lake, 196
Hillsdale, 255

Hillsdale County, 237, 243, 255, 349, 414
Himalayas, 23
Hinge lines, postglacial uplift, 166, 217
Hinkins Hill, 346
Historical interpretation principles, 2
Holland, 161, 349
Holostei, 360
*Homo sapiens*, 140
Hooked spits, 195, 213, 219, 221
Horizontality principle, 5
Horizontal layers, 5
Horn coral, 293, 339; see also Corals
Horsetail rushes, 418, 429–30, 432
Houghton, 71
Houghton, Douglass, 76
Houghton County, 70, 261
Howell Anticline, 243
Hudson Bay, 32–33, 165
Hudson River Valley, 171–72
Humboldt, 64, 67
Humic acids, 130
Hungry Hollow, Ontario, 348
Huron County, 144, 257, 398, 425
Huronian Period, 44, 65–66
Huron Mountains, 56
Huron Portland Cement Company, 119, 121
Huron River, 190–91, 193
*Hydra*, 292
Hydrafracting, 237
Hydraulic dredging, 154
Hydrocarbons, 228
Hydrologic cycle, 181
Hydrostatic head, 183
Hydrothermal solutions, 74–75, 77
Hyoid arch, 355
Hyomandibular, 355, 359
Hypogene, 74
Hypotheses in geology, 77–79
Hypsithermal, 446

Ice, work of, 147
Ice Age, 141
Icebergs, 144
Ice block pits, 147, 154, 193, 196
Ice channel deposits, 151, 154
Ice contact slopes, 153
Ice lobes, 161, 389, 441
Ice scoured bedrock, 146
Ice sheet, Wisconsin continental, 441–43
Ice tunnel, 147
Ida, 255
Identification of minerals, 250
Igneous intrusion, 9, 68
Igneous Rock Chart, 269–70
Igneous rocks, 2, 265, 269
Igneous rocks, age relationships, 16
Igneous rocks, relative dating, 16
Igneous rock textures, 266
Illinoian Glacial, 158–59
Illinois Basin, 119
Imlay City, 168

Impermeable, 231
Index fossils, 310
Indiana Dunes State Park, Indiana, 217
Indian Early Archaic Culture, 389
Indian River, 172, 214
Indians, 74–76, 350, 363; *see also* Early Man
Indians, early distribution in Michigan, 372
Ingham, 132
Ingham County, 158, 255, 349, 407, 412
Inland lakes, 191, 213
Inland seas, 81
Inner core, Earth, 29
Insects, 305
Interfacial angles, constancy, 248
Interglacials, 158
Interlobate areas, 161
Intermittent streams, 183
Internal nostrils, 361
International Salt Company mines, 113, 254
Intertonguing, 90
Intracratonic basins; *see* Michigan Basin
Intrusions, 17, 50
Invertebrate fossils; *see* Fossil invertebrates
Iodine, 113
Ionia Sandstone, 130
Iosco County, 126, 258, 384, 409
Iowan Glacial Sub-stage, 158
Irish Hills, 154
Iron County, 46
Iron deposits, 65
Iron formation, 46, 61–62, 64, 94
Iron Mountain, town of, 68, 258, 345
Iron ores, Michigan, 61
Iron oxides, 61
Iron ranges, 68
Iron River, 68
Iron River, town of, 418, 425–26
Ironwood, 68
Ironwood Iron Formation, 68
Isabella, 347
Ishpeming, 48, 67, 259
Island arcs, 281
Isle Royale, 51, 56, 70, 72, 76, 260
Isomorphs, mineral, 244
Isopach maps, 235
Isotopes, 19

Jackson, 125, 154, 156, 158, 193, 348–49, 369, 398
Jackson County, 132, 237, 243, 349, 398, 412, 425
Jacobsville Formation, 58
Jacobsville Sandstone, 91, 93–94
*Jamoytius*, 309
Japanese-Philippine archipelago, 26
Jasper, 62
Jasper Hill, 48, 259
Jaspillite, 62
Jawless vertebrate, 355
Jaw mechanics, mastodons and mammoths, 380
Jaws, origin of, 355, 359
Jellyfish, 293
Joints, 179

Jurassic, 134, 137, 238

Kalamazoo, 193, 255, 371
Kalamazoo County, 255, 369, 377, 412, 414–15, 447
Kalamazoo moraine, 160
Kalamazoo River, 190
Kalkaska, 151
Kame, 5, 147, 151, 153–54
Kame terraces, 151
Kangaroo, 363
Kankakee Arch, 26, 119, 238
Kansan Glacial, 158–59
Karaboghaz Gulf, 109
Kara-Kum Desert, 109
Kearsarge, 72, 75–76, 261–62
Keewatin time, 39, 41
Kegomic Quarry, 348
Kelly's Island, 146–47
Kelp, 423
Kelsey Lake, 123
Kent County, 255, 370, 398, 408, 412, 414
Kettle hole lakes, 193
Kettle holes, 151, 154, 196
Keweenawan lavas, 56
Keweenawan Period and Series, 41, 51–52, 70–71, 419
Keweenaw Bay, 52, 91, 93, 418
Keweenaw County, 261
Keweenaw Fault, 52–53, 58, 67
Keweenaw Peninsula, 51–52, 80
Keweenaw Upland, 53
Key horizon, 10
Key to mineral identification, 253–54
Kidney construction, 359
Killins Gravel Pit, 5, 154
Kimeridgian, 136
Kingston Mine, 75–76
Kirkfield low water stage, 172–73
Kitchi Formation, 39
Knife Lake sedimentation, 40
Knife Lake (Timiskaming) Period, 39
Knob and Kettle, 151
Koehler Limestone, 366
Kona Dolomite, 417, 423

Lachine, 121, 154
Lacustrine environment, 334
Lagarde, relations of, 76
Lag gravel, 198–99, 203
Lake environment, 334
Lake Erie, 165
Lake Fanny Hooe, 58
Lake Gogebic, 59
Lake Hamlin, 213
Lake Huron, 165, 208
Lake Michigan, 165, 206–7
Lake Michigan shoreline features, 217, 227
Lake Michigan shoreline history, 224
Lake of the Clouds, 54
Lake Ontario, 165
Lake Orion, 193
Lakes, filling by vegetation, 196

Lake St. Clair, 190
Lake Superior, 52, 157, 165, 208
Lake Superior Syncline, 52, 54, 56–57, 70–71, 165
Lampreys, 355, 358, 390
Lancelets, 309
Land plants, origin, 417
Landslides, 25
L'Anse, 91, 93, 418
Lansing, 128, 255, 349, 428
Lapeer County, 113, 369, 407, 413
Larval stage, 362
Lateral moraine, 148
Lattice, mineral crystal, 244
*Lattimeria*, 361
Laughing Whitefish Falls, 94, 98, 100
Laurentian, 39
Laurentian Orogeny, 41
Laurium, 52
Lava flows, 18, 51, 54–55, 57, 59, 71; *see also* Basalt
Lean ores, 62
Leelanau County, 148, 163, 370, 374, 401, 412
Leer, 186
Lenawee County, 255, 369–70, 383, 403, 407, 409
*Lepidodendron*, 131, 418, 429–31
Levering, 8
Lice, 305
Life, origin of, 417
Lignite, 132
Lima Esker, 156
Lime, 105
Lime Island, 347
Lime Lake, 152
Lime-secreting algae, 50
Limestone, 11, 119, 121, 144
Limonite, 61–62
Limpets, 299
*Lingula*, 128, 332, 343
*Lingula* zone, 129
Linnaeus, Carolus, 287
Lithologic continuity, principle of, 10
Lithologic similarity, principle of, 10
Little Bay De Noc, 175, 346
Little Traverse Bay, 8, 176, 214, 348
Littoral environment, 334, 342–44
Liverworts, 418
Living fossil, 332
Livingston County, 153, 412, 447
Lobe-finned fish, 361
Lobsters, 304
Logs, well record, 89
Long Lake, 123, 187
Longshore current, 207–8, 213
Loretto, 68, 258
Lost Interval, 136
Lover's Leap, 179
Lower Herring Lake, 213
Lucas Formation, 11, 115, 239
Luce County, 258, 347
Ludington, 221
Ludington State Park, 194, 213, 225
Lungfish, 351, 361, 400

Lungfish burrows, 352, 361, 368, 401
Luster of minerals, 249
*Lycopodium*, 418, 429–31
Lycopods; Lycopsids; Lycopsida, 418, 429–31

Machaeridia, 307
Mackinac Breccia, 105, 115, 123, 178, 187, 210, 217
Mackinac City, 338
Mackinac County, 258, 347
Mackinac Island, 115, 166, 175–77, 213, 217, 385, 408–9
Mackinac Straits; *see* Straits of Mackinac
Mackinac Straits Bridge, 105, 123, 177, 187
Macomb County, 102, 241
Macropetalichthyida, 358
Magma, 2, 17, 45, 51, 55, 72
Magnesium, 113
Magnetic separation, 64
Magnetic survey, 234
Magnetism in minerals, 249
Magnetite, 61–62, 202
Malachite, 76
Mammal-like reptiles, 139
Mammals; Mammalia, 139, 140, 309, 362, 403
Mammals of Michigan, Pleistocene, 404–5
Mammoths, 340, 371, 374, 376, 378, 383, 411
Mammoths, Columbian and Imperial, 381
Mammoths, discovery sites, 379
Mammoths, extinction of, 381
Mammoths, frozen, 286
Mammoths, teeth and tusks, 380, 381
Mammoths illustrated, 373
Man; *see* Early Man; *see also* Indians
Mancelona, 151
Manchester, 414
Mandan, 262
Mandibular arch, 355
Manistee County, 111, 113
Manistee Lake, 193
Manistee moraine, 217
Manistique, 70, 104, 107, 121, 217, 258, 347
Manistique Lake, 347
Manitoulin Dolomite, 107
Manitoulin Island, 104
Mankato Glacial Sub-stage, 158
Mantle, Earth interior, 29, 30
Marble, 11
Marine environments, 334, 342–44
Marine fossils, 334, 344
Marker bed, 10
Marl Lake, 414
Marquette, 42, 46, 67, 93, 175, 259, 345, 417, 423
Marquette Bay, 93
Marquette County, 50, 64, 259
Marquette Range, 64, 67–68
Marshall, 349, 397
Marshall Formation, 111, 125–27
Marshall Sandstone, 125, 127, 156, 184, 240, 366, 425, 429
Marsupialia, 363
Marten, 374, 408
Mason, 158

# Index

469

Mason County, 113
Mason Esker, 158
Mass, property of, 59
Mass, town of, 260
Mastodon, discovery sites, 379
Mastodons, 333, 340, 350, 368, 371, 374, 378, 383, 410–11
Mastodons, extinction of, 381
Mastodons, teeth and tusks, 380
Mastodon stomach contents, 381
Maybee, 254
M–Discontinuity; Moho, 29
Meandering streams, 132, 189, 193, 222
Medusa, 291, 348
Member rock units, 90
Menominee County, 345–46
Menominee Range, 67–68
Mesabi Range, 67
Mesnard Quartzite, 42, 44, 46, 93
Mesozoic Era, 136, 238, 369
Mesozoic floras, 439–43
Metadiabase, 37
Metagabbro, 37
Metallic mineral luster, 249
Metamorphic intensity facies, 271–72
Metamorphic rock chart, 269, 271
Metamorphic rocks, 2, 11, 265
Metamorphic rocks, genetic relations of, 271
Metamorphic rock textures, 267–68
Metatheria, 363
Metronite Quarry, 51
Metropolitan, 259
Michigamme, 260
Michigamme Slate Formation, 48, 67, 418–19, 425
Michigan Alkali Company Quarry, 348
Michigan Basin, 24–27, 87, 108, 111, 113, 281
Michigan City, Indiana, 204
Michigan Formation, 111, 113, 115, 240, 367
Michigan Geological Survey, 234
Michigan Intracratonic Basin; see Michigan Basin
Michigan Stray Sand, 240
Mid-bay bars, 217, 223
Midland, 113, 203, 407, 428
Midland County, 111, 113
Migration from source rock, oil and gas, 231, 239
Milan, 347–49, 374
Milford, 255
Millersburg, 187
Millington Township, 370
Millipedes, 305
Milwaukee, 175
Mineral collecting in Michigan, 251
Mineral defined, 244
Mineral identification key, 253–54
Mineral identification kit, 250
Mineral locality list, 254–62
Mineral locality map, Lower Peninsula, 254
Mineral locality map, Upper Peninsula, 258
Mineral physical properties, 246–50
Mineral properties, special, 249–50
Minerals, chemical composition, 256–57
Minerals, common, 245

Minerals, occurrence, 256–57
Minerals of Michigan, list, 263–64
Mineral study methods, 245
Mineral uses, 256–57
Miners, 94
Miner's Castle, 94, 95, 208, 345
Miner's Castle Sandstone, 94, 97–98, 100
Miner's Falls, 98, 100
Minnesota Mine, 76
Miogeosyncline, 24, 281
Missaukee County, 241
Mississippian fossil invertebrates, 348–49
Mississippian Period, 123, 127
Mississippian plants, 425
Mississippian vertebrate faunas, 366
Mississippi delta, 25
Mites, 305
Mohawk, town of, 75
Mohawk Valley, 171–72
Mohs's mineral hardness scale, 247
Mold, 286
Molding sand, 134
Mollusca, 300–302; see also Cephalopods; Chitons; Clams; Scaphopods; Snails
Mona Schist Formation, 39, 43
Monotremata, 363
Monroe County, 104, 254, 396, 398
Moorland, 370, 414
Moose; *Alces*, 287, 374, 412
Moose, Extinct Scott's; *Cervalces*, 340, 374–75, 383, 412–13
Moraine lake, 375
Morrison Creek, 419
Morrison Formation, 362
Moscow, 349
Mosses, 418
Mountain glaciers, 29
Mt. Baldhead, 200
Mt. Bohemia, 71
Mt. McSaupa, 176
Mt. Morris, 383, 409
Mt. Pleasant Field, 237
Mt. Simon, 97
Mouse, 374
Mowhawk Mine, 261
Mud, drilling, 236
Mudcracks, 5, 276, 279
Mud Lake, 348
Mud Lake Quarry, 8
Mullet Lake, 172, 195
Multiple working hypotheses, 79
Munising, 14, 70, 94, 97, 208, 218
Munising Bay, 218
Munising Falls, 100
Munising Formation, 14, 93–94, 98, 100, 102
Munising moraine, 160
Munising Sea, 94, 97
Muskegon, 163
Muskegon County, 111, 377
Muskegon oil field, 243
Muskellunge, 401

Musk oxen, 335, 350–51, 371, 374, 376–77, 383, 414–15, 447
Muskrat, 374–407
Mystery Valley, 123

Napoleon, 127, 349
Napoleon Sandstone, 125, 127, 366
National City, 111, 113, 115, 126, 258
Native copper, 70, 72, 76
Natural gas, 228–29
Natural salines, 113
Nautiloids; Nautiloidea, 300, 324–25, 334–36, 344
Nautilus, Chambered, 300
Nearshore, 115, 117
Nebraskan glacial, 158
Neebish Channel, West, 346
Negative area, 25
Negaunee, 49, 67, 259
Neritic environment, 334
Nervous system, dorsal, 307
Nesting bowls, 238
*Neuropteris*, 131, 419, 434–35
Newago County, 414
Newberry, 346
New Richmond Sandstone, 102
Newts, 362
Niagara Falls, 104, 164
Niagaran Escarpment, 104, 107, 164, 172
Niagaran limestone, 240
Niagaran reefs, 105, 108, 240, 344
Nipissing dunes, 195
Nodulizing, 64
Nonconformities, 279
Nonesuch Shale Formation, 61, 63, 71, 77, 79, 419–21, 427
Nonfoliated, 268
Nonmarine environments, 334, 342–43
North Bay, 175
North Bay–Mattawa outlet, 175
North Channel, 190
Northern Michigan Highlands, 51, 58, 79, 91, 93–94, 97
North Lake, 196
Northville, 161, 184
Northville oil and gas field, 243
Norway, town of, 44, 46, 68, 258
Norway Lake Granite Gneiss, 39, 42
Norwood, 152, 391, 393
Nostril openings, 361
Notochord, 307
Nucleus, continental, 38

Oakland County, 255, 369–70, 389, 403, 407, 412
Obsidian, 266, 269
Oceanic environment, 334–35, 344
Ocqueoc Falls, 117
Octopus, 302
Offlap deposit, 128, 283
Offshore, 117
Offshore bars, 195
Ogemaw County, 89
Ohio Basin, 111

Oil and gas, 102, 125, 134
Oil and gas, Michigan, 237, 241
Oil and gas accumulation, 231
Oil and gas exploration and production, 234–35
Oil and gas migration, 229
Oil and gas origin, 228
Oil and gas seeps, 230, 234
Oil and gas sources, 228
Oil well repressurizing, 236
Oil wells, 231
Ojibway, Ontario, 111
Ojibway Mine, 262
Omer, 257
Onaway, 153, 187, 366, 390
Oneota Dolomite, 102
Onondagan, 117
Ontonagon, 260, 420
Ontonagon Boulder, 72
Ontonagon County, 70, 77, 259, 261
Onychophora, 298
Opaque minerals, 249
Opossum, 353, 363
Orchard Lake, 193
Ordovician fossil invertebrates, 345–47
Ordovician Period, 98
Ore, 63
Original Continuity, principle of, 83, 90, 115
Osceola, 72, 76
Oscoda, 191, 384, 409
Osteichthyes, class, 359, 367, 400
Ostracoderms; *see* Agnatha
Ostracods; Ostracoda, 304, 328–29
Otsego County, 151, 409
Ottawa County, 255, 349
Ouachita Eugeosyncline, 24
Outer Conglomerate, 56, 71
Outwash deposits, 5, 148–49, 154, 162
Outwash plain, 147, 151, 160
Overlap and offlap deposits, 128, 282–83
Owosso, 407, 410
Oxbow bends and lakes, 130, 193
Oxford, 154
Oxides, 61
Oysters, 300
Ozark, 144
Ozark Uplift and Dome, 25, 108

Pacific Coast Ranges, 140
Pacific Eugeosyncline, 24
Painesdale, 261
Paleoecology, 333, 436
Paleogeographic map, Early Cambrian, 94
Paleogeographic map, Late Cambrian, 97
Paleogeographic map, Late Keweenawan, 79
Paleogeographic map, Late Silurian, 110
Paleogeographic map, Mississippian, 125
Paleogeographic maps, Devonian, 118
Paleogeographic maps, Permian to Pleistocene, 138
Paleogeography, 77, 81, 83, 235
Paleolithic hunters, Siberian, 388
Paleoniscoidea, 360

# Index

Paleospondyloidea, 358
Paleozoic, thickness, 87
Paleozoic continental interior, 25
Paleozoic Era, 81
Paleozoic Era summary, 134
Paleozoic geosynclines, 25
Paleozoic rock formations, 83
Palludal environment, 334, 342–43
Palmer, 64
Paracrinoidea, 307
Parma, 349
Partridge Point, 348
Peat, 11, 130, 132–33, 195, 342–43, 417, 443
Peat bog, 196
Peccaries; *Platygonus*, 340, 374, 376, 383, 411–12
Pegmatite dikes, 50
Pelagic environment, 334–35
Pelecypoda; *see* Clams
Pelletizing, 64
Pelmatozoa, 305
Penn-Dixie Cement Company Quarry, 348
Pennsylvanian coal swamps, 130–31
Pennsylvanian fossil invertebrates, 349
Pennsylvanian fossil plants, 131, 418–37
Pennsylvanian fossil vertebrates, 367
Pennsylvanian Period, 126–34
Penokean Highlands, 51
Penokean Orogeny, 40, 49–50, 58, 67, 91
Penokean Range, 51
Pentwater Field, 240
Peridotite, 37
Periods, geologic, 15
Perkins, 346
Permeable, 231
Permian Period, 134
Permineralization, 286
Petoskey, 8, 121, 214, 258, 348
Petoskey Limestone, 8
Petoskey stones, 121
Petroleum; *see* Oil and gas
Pewabic Mine, 72, 76, 345
Phase changes in ice, 146
Phoenix, 262
Phoronida, 295
Photosynthesis, origin of, 421
Phylogenetic classification, 287
Physical properties of minerals, 246–50
Pictured Rocks, iv, 14, 93–94, 98, 208
Pigeon, town of, 367
Pilgrim Haven Site, 444–47
Pillow structure, 39, 43
Pinch-out, 233
Pinckney State Recreation Area, 193, 196
Pitted outwash plain, 148, 151, 153
Placental mammals, 363
Placoderm; Placodermi, 309, 355, 358, 365–66, 390–91, 393, 395
Plankton, 229
Plants, classification of, 418–19
Plants, Devonian, 423
Plants, history of, 416, 420

Plants, Mesozoic, 439–43
Plants, Mississippian, 425
Plants, Pennsylvanian, 131, 418–37
Plants, Pleistocene and postglacial, 441–49
Plants, Precambrian, 416–23, 427
Plastic Earth, 28
Platypus, duck-billed, 363
Pleistocene, subdivisions of, 158
Pleistocene and postglacial floras, 441–49
Pleistocene diorama, 368
Pleistocene Epoch, 141, 158, 369
Pleistocene glacial ice, extent, 143
Pleistocene in Michigan, 159
Pleistocene mammals, 404–5
Pleistocene vertebrate faunas, 371
Pleistocene vertebrates, associated, 370
Pleistocene vertebrates, discovery sites, 382
Pleistocene vertebrates radiocarbon dated, 369, 372
Pleuracanths; Pleuracanthodii, 359
Pogonophora, 309
Pointe Aux Barques, 208, 349
Pointe Aux Chenes Shale, 90, 104–5, 111
Pointe Aux Pins, 178
Point Lookout, 179
Pollen, 439
Pollen, Jurassic, 441–43
Pollen, post-Pleistocene, 369, 371–72, 447–49
Pollen analyses, 389
Pollen profiles, post-Pleistocene, 444–47
Polymorphs, mineral, 244
Polyp, 291, 293
Pontiac, 193
Pontiac's Lookout, 179
Porcupine Mountains and State Park, 51, 54, 56, 61–62, 70, 77
Porifera; *see* Sponges
Porosity, 230, 240
Porphyrins, 421
Porphyritic, 45, 266
Portage Lake Lava series, 55, 59
Porter Field, 243
Port Huron, 257
Port Huron Glacial Sub-stage, 158
Port Huron Moraine, 161
Posen, 121, 123, 186–87, 189, 395
Positive area, 25
Postglacial crustal rebound, 166
Postglacial lake sequence, 168
Postglacial time, 164
Potash, 113
Pottsville time, 128
Pouched mammals, 363
Powell Bog, 447
Prairie du Chien Group, 98
Preadaptation, 361
Precambrian basement, 87, 240, 243
Precambrian eras, 31
Precambrian in Michigan, 38
Precambrian plants, 416–23, 427
Precambrian proportion of time, 41
Precambrian rocks, 36

Precipitation, 181
Preclassical Wisconsin Sub-stage, 158
Preglacial stream valleys, 165
Pre-Paleozoic land surface, 89
Preservation of fossils, 285
Presque Isle, 91, 93
Presque Isle County, 172, 187, 258, 391, 398
Presque Isle River, 61
Presque Isle Township, 123
Pressure surface, 183
Primates, 140
Principles of historical interpretation, 2
Proglacial lake, 147
Proglacial outwash, 151
Proterozoic Era, 31
Protozoa, 289; see also Foraminifera; Radiolaria
Psilopsida, 418
Pteridospermales, 419, 434–35
Pyroclastic, 266

Quartz, 62
Quartzite, 11, 37, 49
Quillback fish, 401
Quinnesec, 68, 102, 258
Quinnesec Mine, 67, 94, 97, 102

Raccoon, 374, 408
Radioactive, 235
Radiocarbon dates, 19-20, 22, 159, 172, 369, 371–72, 374, 377, 389, 413–15, 444–49
Radiocarbon Dating Laboratory, 21
Radiogenic dating, 5, 13, 18–22, 32, 159, 417, 420
Radiolaria, 289, 291
Raindrop impressions, 57, 276, 279
Rainy Lake, 123, 187, 189
Randville, 68
Randville Dolomite, 47, 51, 417
Rank of coal, 132
Ratfish, 359–60
Rayfinned fish, 360
Raynolds Point, 346
Rays (fish), 359–60
Rebound, postglacial crustal, 166
Recessional moraine, 148
Red beds, 136, 137, 441–43
Redhorse fish, 401
Reducing conditions, 119
Reef banks, 111
Reef oil and gas, 243
Reefs, modern, 107
Reefs, Niagaran, 105, 108, 240, 344
Reefs, patch, 107
Reefs and reef-building corals, 102, 105, 108, 230, 233, 240, 293, 332–34, 341
Reese, 154
Refracted wave, 219
Regional metamorphism, 44
Regressing sea, 283
Rejuvenated stream, 191
Relation, Jesuit, 76
Relative age, 5–6, 10

Relative dating, 20
Relative time, 13, 310
Reptiles, Age of, 136, 139, 362, 369
Reptiles, Michigan Pleistocene, 402
Reptiles; Reptilia, 309, 350, 362, 403
Republic, 64, 259
Reservoir rocks, Michigan, 239
Reservoir rocks, oil and gas, 229–31
Resistivity, electrical, 235
Rhyolite, 71, 269
Richmond Group, 98, 102
Riparian rights, 180
Rippled toroid, 127
Ripplemarks, 5, 57–58, 93, 98, 276
Rivers, 187
River terraces, 193
Riverton Iron Formation, 68
Robinson's Falls, 177
Robinson's Folly, 179
Rochester, 154, 191
Rock cycle, 2
Rock identification, 269
Rockland, 260
Rockport, 6, 348, 390–91, 393
Rockport Quarry Limestone, 6, 117
Rocks, definition of, 265
Rock salt, 108, 111
Rock textures, 265
Rock types, major, 2, 265
Rockwood, 254
Rocky Mountain Miogeosyncline, 24, 26
Rocky Mountains, 23, 38, 140
Rogers City, 6, 117, 119, 121, 163, 258, 391, 398
Rogers City Limestone, 6, 119, 239–40, 243, 365
Room and pillar mining, 113
Roscommon County, 241
Rosy Mound Dune, 225
Rouge Park, 191, 193
Rouge River, 193
Rounding of sediment grains, 91, 277
Round Lake, 175
Runoff, rainwater, 181, 198

Saginaw, 133, 154, 257
Saginaw Bay, 113, 154, 171, 240
Saginaw County, 132, 257, 398, 428
Saginaw Formation, 128–29, 132, 361, 368, 427–28
Saginaw Formation, historical interpretation, 342–43
Saginaw Group, 130
St. Anthony's Rock, 187, 211
St. Charles, 133
St. Clair County, 38, 102, 111, 237, 241, 243, 257
St. Clair outlet, 176
St. Clair River, 172, 175
St. Clair River delta, 190
St. Croix River, 172
St. Ignace, 115, 123, 187, 210–11, 215, 258
St. Ignace Dolomite, 90, 104, 111
St. Ignace's Rock, 123
Saint Jacques, 104
St. Joseph County, 369–70, 414

St. Joseph Island, Ontario, 346
St. Lawrence Lowland, 175
"St. Lawrence" Sea, 172
St. Lawrence Valley, 172
St. Louis, 111, 184
St. Mary's River, 70, 347
St. Peter Sandstone, 90, 98, 102
Salamanders, 362
Salina evaporite series, 240
Salina Formation, 113
Salina Group, 90, 104, 108–9
Saline River, 369, 374, 447–49
Salinity, 109
Salt, 102, 111, 113
Sand Barrens, 161
Sandbars, 193
Sand dollar, 305
Sand dune, 200
Sand stabilization, 200, 226
Sand stabilizing vegetation, 219
Sangamon interglacial, 158
Sanilac Arch, 179
Sanilac County, 412
Sarcopterygii, 361
Satin spar, 113
Saturation point, 109
Saugatuck, 200
Sault Point, 345
Sault Ste. Marie, 70, 175, 346
Sawheidle Quarry, 347
Scale Model Theory, 29
Scale trees, 418, 429–31
Scaphopods; Scaphopoda, 298–99
Schist, 11, 37, 270
Schistose, 268
Schoolcraft County, 258
Scofield, 254
Scolecodonts, 298
Scorpions, 305
Scotts, town of, 377, 414
Scott's Cove, 178–79
Scouring action, glacial ice, 43, 46
Scouring rushes, 418, 429–30, 432
Sea cave, 94, 97, 208
Sea cliff, 208
Sea cucumber, 305
Sea lilies; *see* Crinoids
Seal rocks, oil and gas, 231
Seal trees, 429–31
Sea mice, 298
Sea of Japan, 26
Sea squirts, 309
Sea urchin, 305
Sea water, 109
Seaweed, 117, 126, 418, 423, 428
Secondary permeability, 240
Secondary porosity, 240
Secondary recovery of copper, 76
Secondary solution, 240
Sedimentary facies, 282
Sedimentary rock, 2, 265, 272

Sedimentary rock color, 279
Sedimentary rock, historical significance, 277
Sedimentary rock structures, 273, 276
Sedimentary rock textures, 267, 273
Sediment deposition, tectonic environments of, 281
Seed ferns, 131, 419, 434–35
Seeds, true, 418
Seeps, oil and gas, 230
Seiches, 224
Seismograph surveys, 234
Selachii, 360
*Selaginella*, 418, 429–30
Selenite, 113
Seney, 258
Shaffer, 346
Shakopee Dolomite, 102
Shale, 11, 37
Shallow water, 128
Shape of minerals, 248
Shark, freshwater, 368
Sharks, 350, 359–60, 364, 366–67, 397–400
Shiawassee County, 132, 349, 407
Shields, Precambrian, 32
Shoreline features, 208
Shoreline processes, 205
Shoreline tilting, 166
Short Rifle Range, 177
Shrew, 374, 403
Shrimp, 304
Shrinkage cracks, 59
Shrinking Earth, 28
Siamo Slate, 49
Siberia, 142
Sibley Quarry, 348
Siderite, 61–62
Sierra Nevada, 140
*Sigillaria*, 131, 418, 429–31
Silica, Ohio, 90, 348
Silica Formation, 90
Silica mineral, 62
Sill, 37, 50, 68
Silurian fossil invertebrates, 347
Silurian Period, 102, 113
Silurian Reefs, 102
Silver City, 261
Similarity of sequence, 10
Sinkhole, 121, 123, 189, 193
Sinkholes, location of, 186
Sintering, 64
Six Mile Amphibolite, 42
Six Mile lava flows, 43
Skates, fish, 359–60
Skull Cave, 178–79, 213
Slate, 11, 37, 49
Slaty, 268
Sleeping Bear Dune, 176, 199, 203–4, 369–70, 374, 401, 406–8
Sloughs, 130
Slugs, 299
Smelters, 107
Snails, 117, 299, 322–23, 332, 344

**Geology of Michigan**

Snow lines, permanent, 142
Soft coal, 132
Soft ores, 64
Soil environment, 334, 343
Soil zones, 128
Solberg Schist, 42
Solberg sediments, 43
Solitary coral, 336; see also Corals
Solution, underground, 123
Solution fissures, 179
Soo; see Sault Ste. Marie
Sorting, 91, 273, 275
Source rocks, oil and gas, 229, 240
South Channel (St. Clair River), 190
South Haven, 444–47
South Range, 55, 261
Spalding, 345
Species, names, 288
Specific gravity of minerals, 247
Specularite, 67
*Sphenophyllum*, 131, 418, 432–33
Sphenopsida, 418, 429–34
Spiders, 305
Spiraling trough ore beneficiation, 64
Spit, 179, 195, 213, 219
Spit, hooked, 221
Sponges; Porifera, 291–92, 344, 364
Spores, 136, 429
Spores, Jurassic, 441–43
Spores, Pennsylvanian, 436–40
Spores, post-Pleistocene, 447–49
Spring-fed, 183
Spruce forest, 371
Squid, 302
Squirrels, 374, 406
Stabilizing vegetation, dune, 201, 225
Stable interior, 25
Stack; see Wave-cut stacks
Stamp sands, 76
Starfish, 126, 305, 330–31, 364
Stegoselachii, 358
Steno's Law, 248
Stigmaria, 129, 429–31
Stomach contents, mastodon, 381
Stone Age, 76
Stoney Point Quarry, 349
Stonington Peninsula, 102
Stony Creek Park, 154
Storm beach, 206
Straits of Mackinac, 104, 119, 123, 175–76, 187
Stratigraphic columns, 87
Stratigraphic trap, oil and gas, 232–33, 239
Streak of minerals, 248
Stream capacity, 189
Stream channel deposits, 368
Stream competence, 189
Stream erosion, 189
Stream fluctuation, 183
Streams, 187
Striae, glacial; striated; striations, 46–47, 73
*Stromatactis* in reefs, 341, 344

Stromatoporoid; Stromatoporoidea, 119, 292, 311, 333–34, 338, 341
Structural trap, oil and gas, 232–33
Structure contour maps, 235
Sturgeon Point, 221
Sturgeon Quartzite, 46
Sublittoral environment, 334, 344
Subsequent stream, 164
Subsurface Laboratory, 89
Sucker, 401
Sugarloaf Stack, 178–79, 213
Sulphur Island, Ontario, 346
Sunbury Shale, 124
Sunday Quartzite, 44, 46
Sunken Lake, 186
Sun's energy, 182
Sunset Rock, 179
Superposition, Law of, 5, 6, 14, 16, 18, 28, 71, 83, 97
Surface geology, 234
Surface roughness, 217
Susquehanna River, 172
Swamp, 182, 191
Swamp, Dismal, of Virginia and North Carolina, 130
Swamp environment, 334, 342–43
Swamplands, 427
Swamps, Pennsylvanian, 368
Swan Creek Mine, 133
Swell and swale, 151
Sylvania, Ohio, 348
Sylvania Sandstone, 90, 115, 117
Symbiosis, 340
*Symbos*; see Musk oxen
Syncline, 6, 67, 165
Syncline, Lake Superior; see Lake Superior Syncline

Taconic Mountains, 102
Taconite, 62
Tadpole stage, 362
Tahquamenon Falls, 94, 98
Tahquamenon River, 94
Tamarack, town of, 261
Tawas City, 126, 221, 258
Tawas Point, 221
Taylor Iron Mine, 50
Tazewell Glacial Sub-stage, 158, 161
Tectonic environments, 282
Tectonic features of North America, 24
Teeth, mastodon and mammoth, 380–81
Tekonsha moraine, 160
Teleostei, 360
Temperature control, body, 363
Terminal moraine, 148
Terrace; see Wave-cut terraces
Terrestrial environment, 334, 342–43
Tertiary Period, 238
Thallophyta, 418
Thanatocoenose, 335
Third Sister Lake, 447
Thread Lake, 370
Three Oaks, 255
Thumb area of Michigan, 168, 171, 208

Thumb Delta, 124–25
Thunder Bay, 163
Thunder Bay Quarries Company, 348
Ticks, 305
Tidal flat deposits, 128, 130, 368
Tidal swamp, 128, 132
Tight formation, for oil and gas, 231
Tile manufacturing, 125, 133–34
Till, glacial, 5, 47, 148
Tillite, 37, 46–47
Tilting of shorelines, 166
Time, absolute; see Absolute time
Time, geologic, 13
Time, relative, 310
Time, sense of, 13
Time correlation of rocks, 10, 11, 332
Time-parallel fossil zones, 11, 332
Time scale, geologic, 15, end paper
Tinley moraine, 160
Titusville, Pennsylvania, 236
Toads, 362
Toivola, 261
Tolerance, ranges of, fossils, 333, 344
Tooth-shells; see Scaphopods
Torch Lake, 76, 148, 152–53, 176
Tower Hill, 201
Tracks, fossil, 286, 423
Trails of animals, fossil, 286, 423
Transcontinental arches, 25
Transgressive and regressive cycles of seas; see Overlap and offlap deposits
Transparent minerals, 249
Transpiration, 181
Traps, oil and gas, 71, 231, 233
Traverse Bay, 176
Traverse City, 166, 258
Traverse Group, 107–8, 115, 117, 119, 123, 239, 365
Trempealeau Formation, 97–98
Trenary, 345
Trent Lowland, 171, 175
Trenton, city of, 98, 102, 109, 348, 390
Trenton Formation, 102, 239–40
Triassic, 136
Trilobite; Trilobita, 94, 102, 117, 303–4, 326–27, 334–37, 344
Trivial name of animal, 288
Troughs of sedimentation, 26
Trout, 400
Tunicate, 308
Turkey, 403
Turner, 258
Turtles, 403
Tuscola County, 132, 349, 370, 403, 408
Tusks, mastodon and mammoth, 380
Tusk-shells, 299
Two Creeks interstadial, 158, 161, 172, 443

Ubly channel, 171
Unconformities, 44, 51, 91, 93, 98, 102, 113, 125, 233, 279, 284, 367
Unconformity, erosional, 9

Underclay, 128, 429
Underground environment, 334, 343
Uniformitarianism, principle of, 3, 5, 19, 141, 274, 333, 350
Union, 349
Union Lake Skull, 389
Uplift, 9
Upper Herring Lake, 213, 217
Uranium, 19, 20
Urochorda, 308

Valders Glacial Sub-stage (stadial), 158, 161, 163, 165, 172–73, 443, 445
Valley train, 151
Valparaiso moraine, 160
Van Buren County, 412, 414, 444–47
Vanderbilt, 151
Van Riper State Park, 260
Variable porosity trap, 239
Varves, 19, 157
Vascular plants, 418
Vegetation, sand stabilizing, 219
Vegetation filling lakes, 195
Ventifacts, 199
Vents, volcanic, 56
Verne Limestone, 130
Vertebrate faunas, Devonian, 365–66
Vertebrate faunas, Mississippian, 366
Vertebrate faunas, Pennsylvanian, 367
Vertebrate fossils, finding, 352
Vertebrate fossils in Michigan, locality and specimen lists, 389–415
Vertebrates, Pleistocene, associated, 370–71
Vertebrates, Pleistocene, discovery sites, 382
Vertebrates, Pleistocene, radiocarbon dated, 369, 372
Vertebrates, relationships and time distribution, 356
Vertebrates; Vertebrata, 307, 309, 350, 390
Vesicles; Vesicular, 51, 72, 266–67
Vesicular basalt, 51
Volcanic activity, 25, 37, 46, 51
Volcanic archipelago, 24, 26
Volcanic ash, 18
Vole, 374, 407
Vulcan, 68, 258
Vulcan Iron Formation, 68, 94, 102

Wabash moraine, 160
Wakefield, 61
Walpole Island, 190
Walruses, 340, 350, 383–85, 408–9
Walruses, entry and occurrence in Great Lakes, 384
Wamplers Lake State Park, 154
Wapiti; see Elk
Warren Dunes State Park, 176, 201, 207, 217, 226
Washtenaw County, 255, 370, 407, 412, 414, 447–49
Water, doctrines of use, 180
Water, surface, 187
Water conservation, 198
Water flushing of oil and gas, 230
Water in Michigan, work of, 180
Waterloo Recreation Area, 193

**Geology of Michigan**

Water pollution, 198
Water problems, 180
Water rights, 180
Water shortage, 181
Watersmeet moraine, 160
Water table, 182, 195
Water underground, 182
Water use, 180
Water waves, 205
Water witching, 184
Waucedah, 94, 148, 345
Wave action, 206
Wave action on and over reefs, 344
Wave amplitude, 206
Wave-cut cliffs, 208, 214–15, 217–18
Wave-cut stacks, 94, 179, 208, 210, 213, 217
Wave-cut terraces, 94, 208, 213, 217–18
Wave erosion features, 208
Wavelength, 206
Wayne County, 111, 113, 254, 348, 393, 398
Wedge-out, 233
Wells, 27
Wells, oil and gas, 89
Whales, 340, 350, 376, 383, 409
Whales, entry and occurrence in Great Lakes, 384
Whitefish, 401
Whitefish Bay, 345
Whitefish River, 175
White Marble Lime Company Quarry, 347
White Pigeon, town of, 370
White Pine, town of, 261, 420
White Pine Copper Mine, 62–63, 77

White Pine Fault, 77
Whitmore Lake, 153, 193
Whittaker-Gooding Pit, 162
Wildcat wells, 235, 237, 241
Williamston, 437
Wilson, town of, 346
Wind, 198
Wind erosion, 219
Wind-frosted grains, 97, 198
Wind in Michigan, work of, 180
Wind velocity studies, 217
Wisconsinan glacial stage, 158
Wisconsin Highlands, 25, 124
Wisconsin ice sheet, 441–43
Wisconsin Upland, 108
Wolf, 374–75, 408
Wolverine, 151
Woodchuck, 374, 406
Woodville Sandstone, 130
Worm burrows, 295
Wyandotte, 113, 254, 390, 396, 398

Xerothermic, 446

Yarmouth Interglacial, 158
Ypsilanti, 191
Ypsilanti delta, 191

Zero isobases, 166
Zone of aeration, 182
Zone of saturation, 182–83